P9-APJ-899
WITHDRAWN

Handbook of Plastics Test Methods

The first edition of this book was prepared for the Plastics Institute by G. C. Ives, J. A. Mead and M. M. Riley.

The revision for this third edition was undertaken by the following staff from Rapra Technology Ltd:

Roger Brown
David Hands
Steve Hawley
Steve Holding
Ivan James
Ron Norman
Keith Paul
Brian Wain

Handbook of Plastics Test Methods

Third edition

Edited by Roger P. Brown
Technical Manager, Rapra Technology Ltd

Copublished in the United States with
John Wiley & Sons, Inc., New York

In association with
The Plastics & Rubber Institute

Longman Scientific & Technical
Longman Group UK Limited
Longman House, Burnt Mill, Harlow
Essex CM20 2JE, England
and Associated Companies throughout the world

Copublished in the United States with
John Wiley & Sons, Inc., 605 Third Avenue, New York, NY 10158

First published in Great Britain 1971 by Iliffe Books, an imprint of the Butterworth Group
Second edition by George Godwin 1981
Third edition by Longman Scientific & Technical 1988

British Library Cataloguing in Publication Data

Handbook of plastics test methods.—3rd ed.
1. Plastics. Testing – Standards
I. Brown, R.P. (Roger Philip), *1940–*
II. Plastics and Rubber Institute
668.4′197

ISBN 0-582-03015-3

Library of Congress Cataloging in Publication Data

Handbook of plastics test methods/edited by Roger P. Brown. — 3rd ed.
p. cm.
"In association with the Plastics & Rubber Institute."
Includes bibliographies and index.
ISBN 0-470-21134-2 (Wiley)
1. Plastics—Testing. I. Brown, Roger (Roger P.) II. Plastics and Rubber Institute.
TA455.P5H35 1988
520.1′923′0287—dc19 88-22018
CIP

Printed and Bound in Great Britain
at the Bath Press, Avon

Contents

Preface

When *Handbook of Plastics Test Methods* was first published in 1971 for the Plastics Institute (now the Plastics & Rubber Institute) it was quickly accepted as the standard work on the subject and has remained in that position to the present time. In the preface, G. C. Ives asked the question 'Is there any need for a handbook of plastics test methods?' The success of the book leaves no doubt that the answer is 'Yes', and we believe that there will continue to be a need.

In the decade or so prior to 1971 there had been drastic changes and extensive additions to the standard methods for testing plastics. Since 1971 development continued and in particular International Standards have assumed increasing prominence, so that it was inevitable that a revision of the *Handbook* was needed. Testing staff at the Rubber and Plastics Research Association were very pleased to undertake this important task when it was not possible for the original authors to do so, and the second edition appeared in 1981.

Work on standardisation of test methods of national and international level has shown no signs of decreasing and there has been continued development of new procedures and, especially in recent years, of the instrumentation used to make tests and to process the data. Once again, therefore, it has become necessary to update this work and once again staff at Rapra Technology were happy to provide the new text.

The style and scope of the book remains substantially unaltered: since it deals with physical testing, chemical analysis has been excluded completely but polymer characterisation tests, which are perhaps on the borderline of the two disciplines, have been retained. Similarly, such tests as thermal ageing, which are normally thought of as physical although they may induce chemical change, are included. There has been no question of including elastomers as there exists a recent volume on that subject, *Physical Testing of Rubber*, which may be considered complementary to the present work. Apparatus requirements are discussed as an intrinsic part of test methods but attention is also drawn to the *Rapra Guide to Rubber and Plastics Test Equipment*, which gives information on the sources of supply of test equipment for both plastics and rubbers. Both of these works are to be found cited as references. Tests on cellular materials have been excluded on the basis that this distinct class of materials should be treated separately, both rubbers and plastics being considered together.

The bulk of the book covers tests on materials either using specially formed test pieces or test pieces cut from products. A chapter on finished and semi-finished products has been included but inevitably could not be exhaustive because product tests are too specialised for general treatment. The chapter on non-destructive testing is also primarily directed at products. Similarly, tests restricted to specific shapes or forms, or to a single polymer, have generally only been accorded a passing reference. The testing of compounding ingredients and aids to processing-solvents, plasticisers, lubricants, stabilisers, pigments and the like – have not been covered except in so far as their evaluation logically forms part of the testing of a plastics material. Much of the revision for the third edition has been concerned with changes and additions to the nationally and internationally standardised methods but in several chapters considerable re-working has been necessary because of developments in the approach to measuring a property, the equipment used or the importance of a topic. In the case of fire testing it has been appropriate to expand the coverage to a complete chapter.

The aim of the book is to present an up-to-date account of plastics testing procedures; within the limitations of scope discussed above, it is intended to be comprehensive in covering all the tests in common, and sometimes not so commond, use. Inevitably, most of the tests are the standard ones, intended primarily for quality control use and often of a somewhat arbitrary nature. However, where appropriate, the requirements for obtaining meaningful design data are discussed. The subject of plastics testing could have been treated in a number of ways but it is hoped that the one selected will be found logical and convenient to two classes of reader in particular: the person in industry directly involved with testing plastics and the student of plastics technology. In addition, it is hoped that the book will be of value to those indirectly involved with the evaluation of plastics, such as design engineers.

Since the time of the original *Handbook*, considerable advance has been made in aligning material standards with international (ISO) standards and therefore the latter have become increasingly important. Wherever appropriate the international method has been described followed by consideration of national deviations. As regards national standards, the emphasis has been placed on British standards together with American (ASTM) and German (DIN) methods, as these are those most frequently encountered by the authors and certainly among the most widely used and technically advanced. Generally, test methods peculiar to particular commercial companies have not been considered at all. The trend towards the adoption of International Standards is reducing the regrettable differences between the standards of different countries and organisations, and it is perhaps appropriate to make a plea for the adoption of recognised standards without modification where there is really no strong technical reason for change.

Technical literature does not stand still – between writing and publication it is inevitable that some standards will have been revised. To counteract this as far as possible, the likely trends in test methods have been estimated from the draft standards in circulation and the known activities of the various committees. It should be emphasised that the latest edition of a standard should be called up for all contractual arrangements.

Acknowledgements

The concept and firm foundation for this book is due to the authors of the original edition whilst the PRI must take credit for promoting its production from the outset. The authors are also most grateful to Rapra Technology for their permission to carry out the task and their support for the project.

The chapter on Standards and standards organisations has been adapted from *Physical Testing of Rubbers* by kind permission of Elsevier Applied Science Publishers Ltd.

We are most grateful for permission to use illustrations taken from other works or supplied by companies in industry and these are acknowledged individually. Figures from British Standards, as cited, are reproduced by kind permission of the British Standards Institution, 2 Park Street, London W1A 2BS from whom complete copies of the relevant standards can be obtained. Those illustrating ASTM test methods are copyright American Society for Testing and Materials, 1916 Race Street, Philadelphia, PA 19103 and are reprinted or adapted with permission.

Thanks are due to colleagues at Rapra and friends in industry for helpful advice and criticism. Finally, we thank the ladies who typed the text which constituted a considerable addition to their normal work.

PUBLISHER'S NOTE

Chapter One

Introduction

1.1 PHILOSOPHY

Why test plastics? For that matter why test anything? Why can we not rely on experience and good workmanship? It cannot be denied that testing costs money and, as will be seen throughout what follows, practically every test has its limitations and can only be applied with discretion to the everyday conditions of service.

In the case of an established material and application, there would ideally be no reason to continue testing were it not for the unfortunate fact that all men and machines are fallible and liable to vary in performance for a variety of reasons. Hence there is a need to test routinely to detect unacceptable deviations as a quality control measure. Plastics are certainly no exception in this matter and being complex materials require particularly careful control to ensure a consistent product. There is currently a growing trend towards greater demands for quality assurance and increasing consumer protection legislation which is resulting in more testing rather than less.

For a new material, a new application or a new product it is clearly prudent if not absolutely essential to prove the performance before unleashing the product on an unsuspecting customer. In fact he will put up very considerable sales resistance if you do not have this evidence and hence there is good reason for the very considerable amount of testing which is carried out to prove fitness for purpose.

The philosophy can be taken further in that at the design stage physical property data is necessary to correlate with the calculated stresses, the expected environment and so on. Without such data one would be reduced to inspired guesses with its uncertainty of possible failure or gross over-design with its accompanying wastage.

In the case of plastics these needs are particularly great because of the rapid change within the industry. The plastics in use today are very often not precisely the same as those available 10 years ago, even if the polymer is basically the same, and there are continuing refinements in processing. Also, plastics are being used in more and more new applications, and frequently more critical applications, than hitherto. Thus, in many circumstances there is not much experience upon which to rely and this makes it very difficult to promote the use of plastics in, for instance, structural applications where a guaranteed 50-year performance may be wanted.

Even after all the design, proof of fitness for purpose and quality control, failures and disputes have been known to happen and there arises a fourth need for testing which is to carry out failure analysis to find the reason for the problem. Hopefully, such testing feeds back to aid improved design or quality control procedures.

If it is accepted that we must test then why write a whole book on the unfortunate necessity? Well, to start with we have at least four reasons for testing and the requirements will not be the same in each case.[1] For example, whereas a quality control test is preferably simple, rapid and inexpensive but the result need not have fundamental meaning, for design purposes extreme speed is of relatively minor importance but the results must be relevant to a variety of configurations and conditions which implies more fundamental understanding and multi-point rather than single-point data. The first essential for a test to predict performance is that it relates to service and for failure analysis that it discriminates well. The emphasis on different requirements will result in a different approach to designing and selecting the test procedure in each case and will often involve different apparatus. Consequently there is rather more to discuss than one procedure for each basic property.

Plastics are not just like any other material, they have their own, often complicated, characteristics and behaviour which necessitates their own carefully formulated test procedures and it is not satisfactory to transfer methods and philosophies taken from other technologies if meaningful results are to be obtained. In this context, it must be remembered that by plastics we mean a whole family of materials which range from soft thermoplastics to very rigid thermosets and which includes materials with various types of reinforcement. Very often the same procedure cannot sensibly be applied to them all.

Throughout this book, when dealing with practically every test method described, it will be necessary to mention problems specific to that technique and peculiar limitations of the data so obtained. However, there are a number of topics in addition to the basic reasons for testing which are pertinent to testing in general and as it is desirable and convenient to introduce them at the beginning they will be considered in the following sections of the introduction.

1.2 PREPARATION OF TEST PIECES

The statement that, for example, a plastic material has a tensile strength of 10 units is unfortunately not an absolute fact. The uncertainty starts at the raw material stage because, as any plastics technologist knows, the strength of a moulding and indeed other properties, can be altered simply by changing the cycle under which it is made. It follows therefore that if test results on any two materials are to be comparable they must be obtained on mouldings produced under the same temperature, cycle time, etc. This is particularly important when materials are being compared for cost or performance. Ideally results are obtained on mouldings produced under conditions comparable to those to be used in production and certainly results on test pieces produced under ideal but not commercially viable conditions would be misleading. Unfortunately the tester very often does not have this information pertaining to the history of the material when it is presented to him in moulded form and must then be rather cautious as to conclusions he draws from his results.

A somewhat similar uncertainty arises from the actual source of the material tested. This is generally a small quantity which has been taken from a much larger whole. Again the tester frequently does not know whether his sample is representative of the bulk material.

Both these problems will be considered further in later sections but it is worth emphasising at the start that a test result on a plastic material cannot be taken at face value unless there is knowledge of how it was sampled and how the test moulding was produced. The fabrication details should, if known, be quoted with the result.

There are further complications when the actual test pieces used have been produced from the moulding by cutting, sawing, routing, etc. These operations may change the physical properties through excess heating or driving off of moisture which could result in unrepresentative test data. Many properties are influenced by the condition of the surface of the test piece. Notches and scratches act as stress raisers and lower mechanical properties, sometimes seriously. Quite obviously the surface condition is crucial when measuring optical properties. Methods for preparation of test pieces are considered in Ch. 3.

By no means all plastics articles have identical properties along their principal axis, i.e. there are some which are anisotropic. It is easy to envisage this in a fabric reinforced laminate because woven fabrics are usually themselves rather stronger in the warp direction than in the weft, unless a special weave is incorporated. However, materials that are homogeneous with respect to composition can also be anisotropic. An injection moulded bar with the gate at one end is likely to show pronounced orientation of the polymer molecules in the length direction unless special precautions are taken, and the tensile strength, for instance, will be higher along the bar than across it. It is therefore important to state the direction of testing when relevant or, preferably, to examine properties in two, or in some cases three, orthogonal directions.

The story of the problems or pitfalls of which the tester and the user of test results must be wary continues with the conditioning of the test pieces and the atmosphere in which the testing is carried out. Most plastics are affected by quite small changes in temperature and it is essential that comparisons are only made between results obtained at substantially the same temperature. Many materials and properties are affected by moisture content and it is then necessary both to condition and test under known relative humidity conditions. A quite short time is needed for a test piece to reach equilibrium temperature with its surroundings but very considerable periods may be needed to reach moisture equilibrium.

Even if carefully conditioned, the age of a material may be important. There are the fairly obvious hazards of degradation by light and the changes which even moderate temperatures will bring about. Some materials crystallise very slowly at room temperature and since polymer properties often differ significantly in the crystalline and amorphous forms, it is necessary to test at equilibrium or when the rate of crystallisation is so slow as to be without significance. Similarly, the process of plasticiser–polymer gelation may proceed slowly, even at room temperature, after the processing cycle, so that plasticised PVC for instance, must be examined only after some specified minimum time. Storage, conditioning and test atmospheres are considered in Ch. 4.

1.3 SIZE AND SHAPE OF TEST PIECES

It might be thought that, since most properties are reduced to units of length, area or volume to yield the basic data for the material, the precise size of test pieces cannot be of importance – it should all come out in the calculation. The statement of results as per unit thickness, for instance, implies that the property is proportional to thickness but this in fact is likely to be very misleading because the properties of a given material moulded in thin sections and those moulded in very thick sections may be quite different. In a thermosetting material the degree of cure is likely to be less in the thicker secion, while with both thermosets and thermoplastics, the extent of locked-in strain is greater in the thicker test piece because of the slower cooling of the centre with respect to the skin. Clearly the measured properties may vary between thick and thin mouldings as a result of effects other than the ratio of the cross-sectional areas.

There are a host of other reasons why size or shape will influence the result and whilst it is not appropriate to try to enumerate them all, a few examples serve as a warning. It is not difficult to envisage that the two dumb-bell shapes shown in Fig. 1.1 would give different results because (b) has undesirable 'stress raisers' at the sharp 'shoulders' which would result in a lower breaking load. This may be an extreme shape but quite a variety of tensile test pieces are in use and they do not all yield identical results. Apart from shape they vary in actual size and may be produced from different thicknesses of material.

Water absorption can be largely a surface phenomenon and it might be thought that logically the property would be calculated per unit surface area. In fact very often it is started as a percentage w/w! Thus, rigorous control of test piece size is necessary so that the surface area is standard and the fictitious percentages may be compared. The mechanical property of flexural strength is calculated from the force required to break a bar at its mid-point, the breadth and thickness of the bar and the distance between the outer supports (Fig. 1.2):

$$\text{bending strength} = \frac{3FS}{2bt^2} \qquad [1.1]$$

Why, then, worry about the precise values of S, b and t as long as they are measured accurately? In fact the classical bending formula above only holds in the case of ideal three-point bending for certain ratios of S/t. Thus, whatever the absolute merits of the property, for comparison of data the test pieces must be essentially identical.

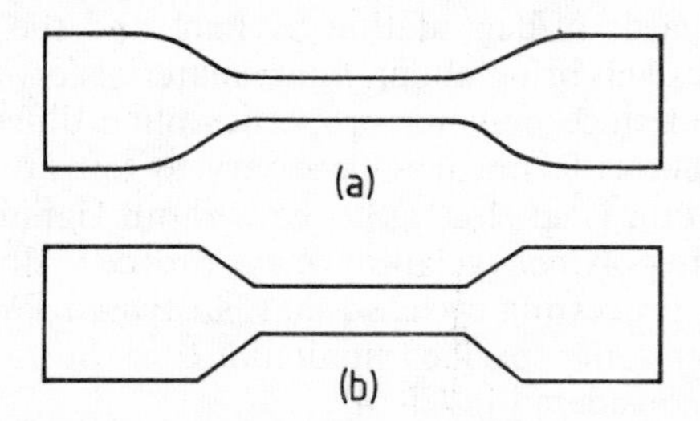

Fig. 1.1 Possible shapes for tensile test pieces

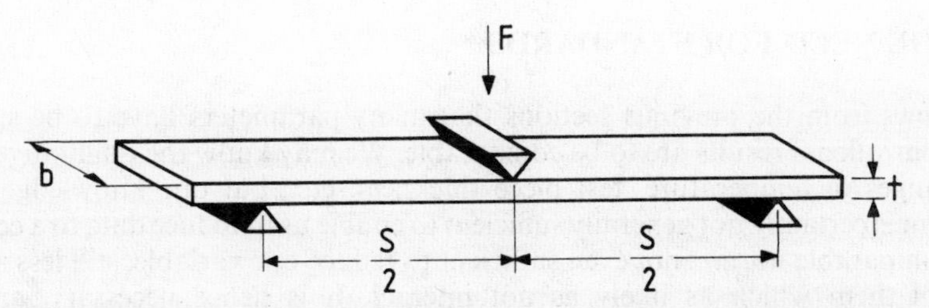

Fig. 1.2 Flexural strength measurement – general arrangement

Electrical strength is calculated from the breakdown voltage of a flat test piece and its thickness, often in volts per millimetre. This implies that the breakdown voltage is proportional to thickness but Fig. 1.3 shows this is not the case. It is necessary to know the slope of the voltage/thickness curve, i.e. obtain multi-point data, for the results to be relevant to products of different thickness. For comparison purposes the thickness must be carefully controlled and even then the comparison is only valid at that thickness unless the materials exhibit similarly shaped curves.

These are just typical examples of the influence of test piece shape and size on measured properties; there are many more (and also many more outside influences which affect the properties considered above), but these serve to illustrate the extreme care that is necessary in checking all details concerning the execution of any given test and in comparing and interpreting results.

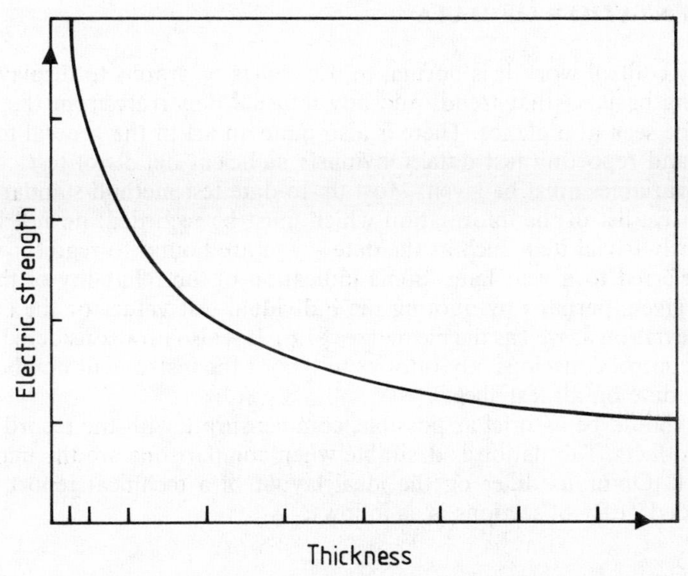

Fig. 1.3 Electric strength as a function of thickness

1.4 THE NEED FOR STANDARDS

It follows from the previous sections that many parameters have to be specified and controlled if results are to be comparable. We may know the qualitative effects of changes of temperature, test piece thickness, etc., but our knowledge at the moment is certainly not generally sufficient to enable us to reduce data to a common and comparable norm – not even sufficient to reduce one variable, still less a whole host of them which as likely as not interact. It is hence necessary, and also convenient, for all workers to measure each property in exactly the same way and the prime mechanism to achieve this is to use standards, and more particularly test method standards.

A test method standard aims to specify all the apparatus parameters, the test piece details, the steps in the procedure and the presentation of the results which are important and hence allow any two laboratories to achieve comparable data. In this way supplier and user can be guaranteed to speak the same language, thus facilitating trade by giving some definite meaning to sales contracts. This is not only a national matter, for if agreement on methods can be achieved continentally or better, universally, then international trade also benefits. Many test method standards have been formulated both nationally and internationally. They by no means all completely achieve their objective in that alternative test pieces and test speeds are included as a necessary compromise to obtain general consensus, but nevertheless they form a very sound foundation on which to base our testing. All the effort in producing them and the common sense which they represent is wasted if we do not use them whenever possible. Standards, specifications and the organisations which produce them are dealt with in detail in Ch. 2 together with the units used to express results.

1.5 PRESENTATION OF DATA

For quality control work it is normal to use charts or graphs to display data on a continuing basis so that trends and any unusual departure from the expected levels can be seen at a glance. There is also quite an art in the general manner of recording and reporting test data. Obviously sufficient details of test conditions and other variables must be given. Most up-to-date test method standards give a comprehensive list of the information which must be recorded: do not leave out an apparently trivial item such as the date – you are bound to regret it when the data are referred to a year later. Some indication of the reliability of the results should be given, perhaps by quoting the individual test values or the calculated standard deviation as well as the means (see §1.6). It is also now considered essential in quality control conscious laboratories to record the instrument number and its calibration date on all test sheets.

Reports should be as brief as possible, commensurate with the recording of all the relevant facts. Tabulation is desirable when comparisons are the main object of the work. Opinions differ on the ideal layout of a technical report, but one recommended order of sections is as follows:

Summary
(Contents)

1. Introduction
2. Materials/products examined
3. Test method details
4. Results
5. Conclusions
6. Recommendations
(Appendices)

The summary is for the busy managing director who has not the necessary time to digest all the report; it is not identical with the conclusions because the summary is a précis of the report as a whole. It should not, therefore, contain any data or opinions which are not in the body of the report. The introduction essentially gives the reasons why the work was done and the various happenings which led up to it. It is always important to record precisely what was examined, not just in terms of a type description, but also batch numbers and other relevant data. Sections 3 and 4 are self-explanatory and so also is Section 5, a statement of the conclusions to be drawn from the work in the light of the data of the report and any other information available from experience. A section giving recommendations, for example for further investigations, if often desirable. The managing director (or anyone else for that matter) will not be pleased if the general standard of English is poor!

If the work being reported contains, for example, a mass of numerical data, it is advisable to collect this together into an appropriate number of appendices suitably cross-referred in the text, so that the main body of the report keeps within reasonable proportions; Section 4 can then be restricted to summary tables of mean values, for instance. Graphical representation of results should always be considered as a concise way of reporting data and one which will often assist the drawing of the correct conclusions. Graphs are usually best placed with the appendices.

In certain cases, a section of bibliography or literature cited should be included, between Section 6 and any appendices. In large reports, a contents list immediately after the summary is very useful.

While it is not desirable to dispense with individual style, numbers should be presented so that clarity, elegance and uniformity take precedence. As a simple example, a quarter of a metre should be written 0·25 m and not ·25 m. For guidance on the presentation of numerical values attention is drawn to BS 1957.[2] The units should be stated correctly. Care should be taken that symbols, signs and abbreviations are unambiguous and conform to accepted practice. ISO 1000[3] and BS 5555[4] give information on SI units; symbols and quantities are covered by the ISO 31 series[5] and the BS 5775 series.[6] A glossary of terms and common abbreviations used in the plastics industry is contained in BS 1755[7] and abbreviations for polymers and compounding ingredients in BS 3502[8] and BS 4589.[9]

1.6 LIMITATIONS OF TEST RESULTS

When we have taken all these precautions – prepared the test pieces correctly, followed the prescribed method to the letter, reported all the facts clearly without mixing our units – what is the value of the data obtained? Essentially the figures for strength, resistance, etc., derived from our measurements relate only to conditions which simulate precisely those under which we performed our tests, and strictly only to the particular sample tested. There are two aspects to the value, or rather the significance, of our results. First, there is significance in the statistical sense (that dreaded subject which must inevitably rear its head): the result is useless unless we know its significance in terms of the extent to which the sample is representative of the material and what reliance can be placed on the result, taking account of experimental error and material variation. The essential subject of statistics will be approached gently in the next section. The second aspect of significance is the relevance of the result in terms of material or product performance.

It has already been pointed out, and will be further amplified throughout the book, that the property value obtained may vary according to the test method used. Many of our test methods use quite arbitrary conditions and procedures. The data obtained from this routine type of test, whilst admirable for quality control and perhaps as an indication of service performance if interpreted carefully, will rarely give the designer the values upon which to base his calculations. The more nearly a test approaches the real conditions of service the more relevant or significant it is likely to be in terms of predicting service performance. The more fundamental a result is in terms of it being independent of test piece shape or conditions the more relevant it will be for design purposes.

Frequently, pressures of economics and time prevent test procedures being very close to service conditions. The largest gap between most published property data and performance behaviour is in the time scale: just how long will a given component withstand a certain stress without failing? Trouble-free service over a period of years may be essential, yet tests have lasted but a few seconds or minutes. Accelerated tests can often provide very useful guides and frequently are the only solution to such demands. However, it must always be remembered that to produce this very acceleration, some test variable or variables have had to be intensified, for example the temperature raised, the nature of the environment changed or the frequency of stressing increased. These necessary changes may in themselves induce effects which would never occur at the usual ambient temperature, etc., and thus misleading data may result.

What with the limitations due to the effect of test piece size and shape, the time-scale of the test, the influence of accelerating effects and the nearness or otherwise of test conditions to those of service it is quite clear that one cannot emphasise too forcefully that all measured properties should be most critically assessed to establish their true relevance and applicability.

1.7 THE CONCEPT OF STATISTICS

1.7.1 Introduction

No responsible person even remotely connected with science or technology should be without a working knowledge of statistical principles; anyone ignorant of the basic ideas of this subject runs the risk of undertaking his work inefficiently and of being unable to draw the correct conclusion from the results he obtains. An example in plastics testing illustrates the second point quite simply. Suppose that two laminated sheet materials A and B of equal thickness are to be compared for strength. The edgewise Izod values are determined in the usual way, in quintuplicate, with the following results:

	A	B
	1·65 J	1·09 J
	1·37 J	1·48 J
	1·41 J	1·38 J
	1·26 J	1·26 J
	1·60 J	1·39 J
Mean	1·46 J	1·32 J

On the basis of the two mean values, A may be selected as the material with the better impact strength; however, reflection indicates that the picture may not be so clear-cut. Impact strength tests on laminated sheet, by this method, invariably show a scatter of results (the spread of individual values) and this is certainly true of the figures above. It is instructive to compare the ranges of the sets of values:

A: 1·26 – 1·65 J
B: 1·09 – 1·48 J

There is seen to be a region 1·26 – 1·48 J, within which a test figure taken at random might be placed in either set of figures; three out of five of the individual values for A fall in this region and four out of five for B! The essential difference, therefore, lies in two high figures, 1·60 and 1·65, for A and one low figure, 1·09, for B. Are these the all important values, pointing out the relative merits of the two materials, or are they the 'odd men out', being outside the common region as a result of some chance happening – say, test pieces cut from a fabric reinforced laminate in such a way that they contain one more or one less thread than the average? We can tell from these figures whether there is likely to be a real or significant difference between the two sets of data only by statistically analysing them to find out whether there is a sound reason for thinking that all ten figures have been drawn from one set of data (i.e. there is no difference between the materials), rather than from two different sets (i.e. A is more impact-resistant than B). The greater part of the study of statistics is to learn to assess the magnitude of chance error in relation to the particular effect under examination.

It is not appropriate to include here a treatise on statistics but it is worth while to outline some of the basic tools and concepts and make reference to relevant textbooks and standards. A fairly substantial account of statistics in terms of polymer testing is given in a complementary volume, *Physical Testing of Rubbers.*[10] General statistical textbooks are not, of course, written with particular reference to plastics testing. Two volumes edited by Davies[11,12] are good examples of the many comprehensive books available. Extremely good accounts of basic statistical techniques can be found in standards, notably BS 2846: Parts 1–7[13] (ISO equivalents, 2602, 2854, 3207, 3301, 3494, DIS 5479...).

1.7.2 Variability

All measurements are subject to variability. We need to know the sources of variability and make a reliable estimate of its magnitude. From this information we may then judge the reliability of our results and hence their significance.

The sources of variability are numerous. In testing a sheet of plastic for tensile strength it is necessary to take into account the intrinsic variability of the material, arising from the fact that it is not perfectly homogeneous, and the variability due to the testing procedure, including test piece preparation, machine accuracy and operator error. This sheet is only a sample taken from the total population of sheets which could be made from a batch of material. Hence, if more than one sheet is tested we introduce variability due to variations in moulding. Different batches may show variations because of changes in the ingredients or the mixing process. Different testing machines, different operators and different laboratories may all introduce further variability.

The tester is always trying to reduce the variability of his results by carefully standardising preparation and testing procedures, frequent calibration of the apparatus and training of operators. Distinction should be made between: (a) the random variation due, for example, to real material variations, and (b) bias or systematic error due, for example, to a machine being wrongly calibrated.

It is necessary to distinguish between accuracy and precision. Accuracy is the closeness of the result to the true result (should you be so lucky as to know the true result) whilst precision is the closeness of agreement between results from repeat measurements. To keep variability to a minimum we want our test method to be as reproducible as possible, i.e. we want it to have good precision. However, it is not much good having high precision if the test has a large bias and hence poor accuracy. So we want both and indeed they are related in that poor precision (reproducibility) will contribute to lowering the accuracy.

For many properties the distribution of results will be normal or Gaussian so that, with a sufficient number of results, a plot of frequency (of the measured value) against the value itself will follow the form shown in Fig. 1.4. The arithmetic mean $\bar{x}$ of a set of n values is defined by

$$\bar{x} = \frac{1}{n} \sum_{i=1} x_i \qquad [1.2]$$

i.e. the sum of all the values divided by the number of readings.

Other measures of central tendency are the median, which is the middle value when the results are arranged in ascending or descending order, and the mode, which is the most frequently occurring value. The mean, median and mode are coincident for a normal distribution but not for a skew distribution.

The most useful measure of dispersion or variability is the standard deviation, s, which is calculated from

$$s = \sqrt{\left[\sum_{i=1} (x_i - \bar{x})^2 \middle/ (n-1) \right]} \qquad [1.3]$$

In Fig. 1.4 it can be seen that approximately 68 per cent of all values lie within the range $\pm s$ of the mean, 95 per cent within $\pm 2s$ of the mean and 99·7 per cent within $\pm 3s$. It is often more convenient to use the coefficient of variation, v, which

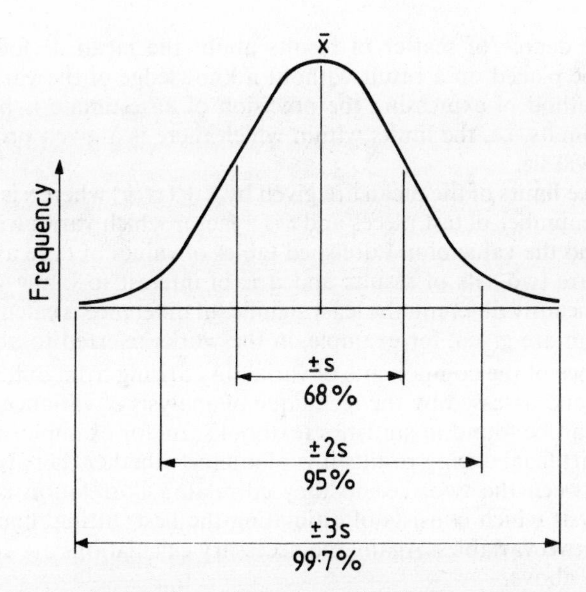

Fig. 1.4 Normal frequency distribution curve

is the standard deviation expressed as a percentage of the mean, i.e.

$$v = \frac{s}{\bar{x}} \times 100 \text{ per cent} \qquad [1.4]$$

The variance is the standard deviation squared.

Unexpectedly high or low results sometimes occur, however well the test is organised. Such results should not be rejected unless some physical reason for their occurrence, for example a flaw in the test piece, can be determined. Even a flawed test piece would not be rejected if it was typical of the product or material.

1.7.3 Significance and Discrimination

Significance in the statistical sense is concerned with whether observed differences in results are real or can reasonably be attributed to chance.

In the example of impact strength values given in §1.7.1 a difference in mean results for two sheets was recorded. The results would be said to be 'significantly different' if it could be shown that the odds were heavily against the difference being due to chance, i.e. there was a high probability that the difference was real. It is usual to make statistical tests at the 5 per cent or 1 per cent level (i.e. less than 1 in 20 or 1 in 100 probability of the difference being due to chance). Significance is dependent on the magnitude of the difference and also on the amount of variability, so that for a test showing a wide scatter of results a relatively large difference between means would be needed to be proved significant. In the same way the significance or the reliability which can be attributed to a mean result depends on the degree of scatter of results about the mean. It follows that no reliability can be placed on a result without a knowledge of the variability.

The usual method of expressing the precision of an estimate is by calculating the confidence limits, i.e. the limits within which there is a given probability that the true value will lie.

The confidence limits of the mean are given by $\bar{x} \pm (ts/n)$ where s is the standard deviation, n the number of test pieces and t is a factor which varies with the degree of confidence and the value of n. Published tables of values of t are available.[10–13]

When there are two sets of results and it is of interest to know whether their means are significantly different, the least significant difference is calculated. Details of the calculation are given, for example, in the works referred to above.

The significance of the components of variability arising from difference sources can be isolated and assessed by the technique of analysis of variance. Accounts of this technique can be found in statistics textbooks. In, for example, a comparison of natural and artificial ageing results it is of interest whether there is a significant relationship between the two: i.e. do they correlate? Correlation is assessed by regression analysis which consists of estimating the best 'fitting' line through the plotted pairs of two variables. Again, the necessary calculations are covered in the references given above.

To discriminate between two materials in respect of any property it is necessary to prove a significant difference between two sets of results. We can regard the discriminating power of a test as a measure of the ease with which it will show up real differences between the materials. The better the reproducibility of the test the smaller the difference between results which can be proved significant and hence the better the discriminating power. Perhaps more obviously, discriminating power also depends on the magnitude of the difference between results obtained on the two materials. For example, different abrasion machines often show marked differences in the relative performance of materials. If the reproducibility with two machines was similar the test with the greater discriminating power would be the one that gave the greater relative difference between the materials.

1.7.4 Design of Experiments

It should by now be apparent that it is virtually essential to have some grasp of the basic principles of statistics in order to make sensible judgements about one's own or other people's test results. However, no amount of clever analysis will compensate for poor experiment design and planning. The usefulness of a series of tests is increased and the difficulty of analysing the results eased if the experiment is planned carefully and properly in the first place.

Much of experiment planning is common sense. The object is to eliminate as many unwanted sources of variability as possible while not missing any of the important factors which should be studied. A simple example of common sense is to test aged and unaged samples on the same occasion to avoid variability due to different operators on different days.

It is not surprising that statistical principles play as great a part in experiment design as they do in the analysis of results. The test must be repeated often enough to give the required significance of results. The differences in materials or methods that are under investigation must not be masked by changes in uncontrolled variables. The experiment plan must be such that statistical analysis can be sensibly carried out on the results, i.e. the experiment and the method of analysis are planned together.

Each experiment must be designed to suit the particular circumstances and requirements but there is a considerable number of 'standard' approaches which can be used. Once again the reader is referred to works giving a fuller coverage of the subject.[10,12,14,15]

1.8 QUALITY CONTROL

1.8.1 General Considerations

The manufacturer needs some form of quality control:

1. In order to produce a consistent product, satisfy his customers and meet contractual obligations;
2. In order to monitor the manufacturing process and to help to identify, and thus rectify, faults at the earliest possible stage.

Quality control or quality assurance is concerned with maintaining the quality of products to set standards. This embraces design, the control of incoming materials, the control of the manufacturing processes and the inspection and testing of the final product. Hence, the quality engineer is concerned with testing, specifications, statistical control schemes, inspection and the monitoring of production methods. Quality control is therefore a discipline in which testing plays but a part and hence in this section discussion will be restricted to making reference to publications, particularly standards, which are relevant and to certain aspects which are pertinent to testing.

A comprehensive tome on quality control is by Juran *et al.*[16] whilst a rather shorter book giving a practical approach to the subject is that by Caplen.[17] There are many others of varying scope and with the current increase in awareness of the importance of quality control, many more can be expected. Quality assurance is certainly about people, ultimately it being human fallibility that limits the control of quality and Drury and Fox[18] have considered the subject from this angle. A considerable number of standards have been established in this field and a collection of a number of British standards is given in BS Handbook 22.[19] For other standards related to quality control the ISO and national catalogues should be consulted. The level of quality control applied in the plastics industry has been investigated by Wain. His reports cover the plastics moulding industry,[20] the polymer and compound supply industry[21] and the views of the customers of the polymer industry.[22] Despite the number of meetings and conferences on quality in the polymer industry in recent times there appears to be a lack of up-to-date comprehensive texts on quality control in our industry although there is an *ASTM Special Publication.*[23]

The selection of the actual tests to employ for quality control must be a compromise between scientific interest and economic necessity. Most of the standardised test methods are principally intended for quality control use and probably, in terms of quantity, the majority of testing is carried out for quality assurance purposes. If the control can be exercised adequately with a non-destructive test or tests, only the actual cost of the latter is relevant for no product is actually destroyed. This, unfortunately, is rarely the case. It is, however, often possible to reduce to a reasonable minimum the number of different types of test carried out regularly, even though full specification tests must be carried out at less frequent intervals.

1.8.2 Sampling

In general, it is necessary to take samples from production, and perhaps from deliveries of incoming materials, and test these on the basis that they are representative of the total quantity. For routine quality control, samples must be taken repetitively in time and there is need for a long-term sampling plan and a continuing method for assessing the results.

The whole point of efficient sampling is to select small quantities which are truly representative of the much larger whole. The majority of standard test pieces are quite small, weighing about 50 g or less; usually five is the maximum number used in any one property measurement. Therefore the combined mass of test pieces is very small in relation, say, to the contents of a 50 kg sack and far smaller in relation to the production batch from which the selected sack was taken. Similarly, small test pieces from a large sheet represent a very small fraction of the total area, and the sheet is only one taken from perhaps hundreds produced.

When sampling from a large number of items for quality control purposes it is usual to use statistical tables to decide the number to be taken which will provide any given level of probability of out-of-specification items being present. Information on such statistical sampling schemes will be found in the texts on quality control referenced earlier. Selecting the items for test should be at random and a book of random numbers (a set of tables designed to pick numbers at random without the risk of unconscious bias) can be useful. Care should also be taken that the sampling procedure is not biased, for example by sampling at set times which might coincide with a shift change or other external influence.

Having selected the number of items for test, care must also be taken in the way in which the test pieces are obtained. When powders are sampled, devices must be used to take representative portions from the sack, drum or other container, bearing in mind that coarse particles tend to separate out. When cutting from sheet, film, rod, tube, etc., the test pieces should generally be randomised over the area or length, although for some purposes it may be desirable to take the opposite approach and plan the position of test pieces, for example if edge effects were being investigated. One particular, and important, case is where a sheet may not be isotropic when the direction of the test pieces relative to the axes of the sheet is important.

The number of test pieces or repeat tests to take per unit item selected is usually decided by the test method standard being used. The current methods are not consistent although there is a trend towards three or five as the preferred number. The latter has much to recommend it for most methods being just about large enough to make reasonable statistical assessments of variability. Testing larger numbers will not yield a proportional increase in precision but may be desirable for very variable properties such as fatigue life. Using one test piece only is rarely satisfactory but this is all that is specified in some cases, probably influenced by financial considerations. An odd number of tests is advantageous if the median is to be extracted. In a continuous quality control scheme the number of test pieces used at each point is perhaps less important than the frequency of sampling and it may be preferable to use fewer test pieces at each sampling time and sample more often.

1.8.3 Quality Control of the Test Laboratory

Testing laboratories are as much in need of quality control as is the production unit. The test procedures must be controlled, the operators trained and the equipment calibrated. Good test method standards go a long way to ensure that the test procedures are carried out reproducibly but, particularly in the areas of test piece preparation and conditioning, there is often need to supplement the standards with in-house standard procedures which should be written down, not passed on verbally. Operators should be examined for competence in performing each test and a formal record kept of their training and examination.

It is self-evident that apparatus should be correctly calibrated. This is generally appreciated for such items as the force scales of tensile machines and often the manufacturer offers a calibration service. There are also British and ISO standards[24,25] dealing specifically with the requirements and accuracy of such machines. It is the little things that get overlooked: the calibration of dial gauges and thermometers, the check that pointers have not been bent or that dirt is not causing undue friction. Many of these seem trivial but if errors from such sources were not present there would not be the poor interlaboratory agreement so often witnessed in practice. Errors seem to arise from the most unlikely sources and a good defence is a suspicious mind and a passion for detail. A good insurance is a formal maintenance and calibration scheme for all apparatus, including written records of all checks and measurements and clear labelling of the apparatus as to its state of calibration. BS 5781[26] gives a specification for measurement and calibration systems.

In fact this formal procedure should be taken further in that the laboratory should have its quality manual which details all the procedures which must be followed to ensure reliable results. This will include formal training of staff, properly documented test procedures and the control of samples, test sheets and reports as well as calibration of the apparatus. The best way of achieving this is to be subjected to the disciplines of a recognised accreditation scheme. Major purchasers have applied their own accreditation systems for many years but it is only relatively recently that national schemes have reached prominence.

The British National Laboratory Accreditation Scheme (NATLAS) applies comprehensive and rigorous procedures. These are given in deceptively short form in BS 6460[27,28] but in practice considerable effort and discipline is needed to reach the required standards. ISO Guide 25[29] is the international equivalent of BS 6460 and a number of countries now have national schemes which are essentially based on it. The International Laboratory Accreditation Conference (ILAC) has emerged as the major forum for discussions of laboratory accreditation. They have produced or are producing reports on many aspects of accreditation including proficiency testing, quality manuals and calibration intervals. A bibliography on laboratory accreditation has been produced by Bryan *et al.*[30]

Even when working with strict calibration and control, it is only by taking part in comparison exercises with other laboratories that proof of how you measure up is obtained. It is important to appreciate that interlaboratory testing programmes may be organised for various purposes as the details of how they are formulated will be different. An accreditation body may run trials in line with ISO Guide 43[31] to demonstrate the proficiency of laboratories whereas a standards committee may be interested in aiding the development of a new test method.

A very important need for interlaboratory testing is to produce precision statements for standard test methods. It has now been accepted by ISO TC61, the international committee for plastics, that ideally all test methods should include precision statements to advise users of the level of variability which can be expected. This practice has been followed by ASTM for several years and now ISO faces the enormous task of generating the necessary data. Essentially, a precision statement provides, from the results of a controlled interlaboratory trial, measures of repeatability, which refers to the precision within one laboratory, and reproducibility which refers to the precision between different laboratories. There is an international standard ISO 5725[32] (British equivalent BS 5497[33]) which sets out procedures for generating the precision statements.

International trials have, in the past, shown up quite alarming differences between reputable laboratories and it is quite clear that it is in any laboratory's interest periodically to compare their results with others, preferably before a disagreement results in dispute. Interlaboratory trials produce very valuable information on the variability of tests and maybe also on the parameters which influence variability. Unfortunately, the results of trials are rarely published but when precision statements are produced internationally there will at least be a guide to all testers of the reasonable variations to expect.

REFERENCES

1. Brown, R. P. (1984) *Polymer Testing*, **4**, 2.
2. BS 1957 (1953) *Presentation of Numerical Values.*
3. ISO 1000 (1981) *SI Units and Recommendations for the Use of their Multiples and of Certain other Units.*
4. BS 5555 (1981) *SI Units and Recommendations for the Use of their Multiples and of Certain other Units.*
5. ISO 31 (in several parts) *Quantities and Units.*
6. BS 5775 (in several parts) *Specification for Quantities, Units and Symbols.*
7. BS 1755 (in two parts) *Glossary of Terms used in the Plastics Industry.*
8. BS 3502 (in two parts) *Schedule of Common Names and Abbreviations for Plastics and Rubbers.*
9. BS 4589 (1970) *Abbreviations for Rubber and Plastics Compounding Materials.*
10. Brown, R. P. (1987) *Physical Testing of Rubbers*, Elsevier Applied Science, Ch. 3.
11. Davies, O. L. and Goldsmith, P. L. (1976) *Statistical Methods in Research and Production*, Longman.
12. Davies, O. L. (1978) *The Design and Analysis of Industrial Experiments*, Longman.
13. BS 2846: Parts 1–7, *Guide to the Statistical Interpretation of Data.*

14. Cochran, W. G. and Cox, G. M. (1956) *Experimental Designs*, Wiley.
15. Cox, D. R. (1958) *Planning of Experiments*, Wiley.
16. Juran, J. M., Gruna, F. M. and Bingham, R. S. (1975) *Quality Control Handbook*, McGraw-Hill.
17. Caplen, R. H. (1978) *A Practical Approach to Quality Control*, Business Books.
18. Drury, C. G. and Fox, J. G. (1975) *Human Reliability in Quality Control*, Taylor and Francis.
19. B.S. Handbook 22, *Quality Control.*
20. Wain, B. J. (1980) *RAPRA Members Report*, no. 56.
21. Thorn, A. D. and Wain, B. J. (1982) *RAPRA Members Report*, no. 80.
22. Wain, B. J. (1980) *RAPRA Members Report*, no. 58.
23. Green, F. T., Miller, R. W. and Turner, V. L. (Eds.) (1985) *ASTM Special Publication*, no. 846.
24. BS 5214 : Part 1 (1975) *Testing Machines for Rubbers and Plastics, Constant Rate of Traverse Machines.*
25. ISO 5893, 1985, *Rubber and Plastics Test Equipment – Tensile, Flexural and Compression Types (Constant Rate of Traverse).*
26. BS 5781 (1979) *Measurement and Calibration Systems.*
27. BS 6460 : Part 1 (1983) *Specification for the General Requirements for the Technical Competence of Testing Laboratories.*
28. BS 6460 : Part 2. To be published.
29. ISO Guide 25 (1982) *General Requirements for the Technical Competence of Testing Laboratories.*
30. Bryan, J., Drake, L., Hall, W. and Thomas, O. (1982) *Bibliography on Laboratory Accreditation.* National Bureau of Standards.
31. ISO/IEC Guide 43 (1984) *Development and Operation of Laboratory Proficiency Testing.*
32. ISO 5725 (1981) *Determination of Repeatability and Reproducibility by Interlaboratory Tests.*
33. BS 5497 : Part 1 (1979) *Guide for the Determination of Repeatability and Reproducibility for a Standard Test Method.*

Chapter Two

Standards and Standards Organisations

2.1 STANDARDS – TEST METHODS AND SPECIFICATIONS

2.1.1 Introduction

It does not take much imagination to realise that if there were no standardised test methods trade would be severely impaired, progress would be stunted and chaos would ensue. Fortunately, technologists appear to have a strong sense of order and from the early days of the industry have supported the development of standard procedures and the use of these in product specifications; not that the effort put in should make us complacent, as there is still plenty of room for confusion for the unwary. Because of increased demand for product reliability and fears of liability legislation, standards have become even more important in recent years. The British government, amongst others, has expressed its commitment to standards and there has been much discussion of the role of standards in industrial strategy.[1]

To avoid misunderstanding over terminology it is as well to note that the British Standards Institution calls all its documents standards and the word 'specification' is reserved for those standards which specify minimum requirements for materials or products. Other types of standard are Methods of Test, Glossaries of Terms and Codes of Practice. It follows that a specification may refer to several methods of test and that a commercially written specification can refer to nationally standardised test methods.

In terms of trade it is ultimately specifications which are important, with test methods acting as building blocks. For this reason it has made chronological sense for the test method standards to be developed first and indeed this has generally been the case in practice. Now that test methods are well developed at the national and international level it can be argued that rather more of the effort available should be put into specifications, especially in the current economic climate where less money than before is available. Certainly there is no case for the development of standard test methods which are of academic interest only and unlikely to be generally used. This does not mean that such methods should not be developed. Tests are needed for a number of purposes and not all justify the standardisation process. In a discussion of the requirements for physical testing of polymers[2] the different needs for test methods and the particular role of standards has been considered. There is no doubt that further improvements in plastics test procedures are needed and progress in recent times has not been as rapid as many would like.

However, it will be apparent throughout this book that considerable activity in test method standardisation is still taking place.

2.1.2 Test Methods

In this book we are concerned with methods of test and only indirectly with specifications. Leaving aside for the moment the various sources of standard test methods, one can recognise different styles or types of published methods. This is not a matter of accident but rather one of progression; the most obvious yardstick being the number of options left open to the user. In the simplest case a particular apparatus is specified, one set of mandatory test conditions given and no choice allowed as to the parameters to be reported; this is the form in which the specification writer needs a test method. Unfortunately for those who want a quiet life many national and international test methods have become rather more complex. This is partially a result of compromise but more importantly because the measurements being described are not intrinsically simple and the method will be required for a number of different purposes and probably for many different end products. The specification user must therefore select the particular conditions which best suit his individual purposes. In practice he frequently fails to do this either because he omitted to read the standard carefully enough or because his understanding of it was somewhat limited. As more advanced concepts are being introduced into test method standards so there is an increase in the practice of including explanatory notes. I fear that these do not always achieve their desired aim.

We can conveniently distinguish three different circumstances in which a standard method is used: (a) purely for quality control; (b) as a performance requirement; and (c) for development purposes. In the first case the prime consideration is that precisely the same procedure is always used and also that this procedure is relatively simple and rapid. The test conditions may be completely arbitrary but one set of conditions and one set only is required. If the test is intended, apart from a quality control function, to be a measure of the performance of the product then test conditions will be chosen which have some relevance to the product end use. For development work it is highly probable that a series of conditions will be wanted in the hope that data of use in designing future products will be realised. Committees preparing standard test methods have all these possibilities in mind and the penalty for the user of the standard is that he must understand the subject sufficiently well to make an intelligent selection of conditions to suit his particular purpose. The following example may not stand up to too close an inspection but serves I think to illustrate the point. If a test for resistance to liquids is considered, one would expect a quality control procedure to involve one liquid at one temperature for a relatively short time. The liquid might be a standard fuel and the test involve 24 h exposure with weight change being measured. If the testing was intended to have a performance function then the liquid met in service would be used, for example commercial petrol, and testing continued long enough for equilibrium absorption to be reached. Apart from volume change, other relevant physical properties would be measured before and after exposure. For development purposes testing would be further extended to cover a number of fluids each tested at several temperatures. An international or national standard would attempt to cater for these and other possibilities and would hence include a choice of measuring procedure, test temperature, duration of exposure, properties to be monitored and test liquids. Preferred test parameters might be indicated for use when there were no outside factors influencing the choice.

This is not the place to discuss in any detail what should or should not be included in standard test methods or how they should be written. The quality and the style of those in current existence varies very considerably but it is possible to detect certain general trends in recent years. Standards have become more involved as more factors which cause variability are identified and control of these is specified. At the same time some apparatus has been specified in a more general way, stating what its performance must be without restricting its design or construction to any particular form. This can only be done when all the important parameters have been identified. Standards also become more complicated as the underlying principles of the property being measured become better understood and as more meaningful results are demanded by product designers. It is very much a matter of opinion as to whether we at present have taken this progress too far or not far enough.

2.2 ORGANISATIONS PRODUCING STANDARDS

Generally, the sources of standards can be placed into three groups:

International organisations
National organisations
Individual companies

Despite the argument that in terms of trade it is commercial specifications which are most important, it appears proper and logical to discuss these groups in descending order of scope, i.e. from the international downwards. In practice of course a new test method usually proceeds in the opposite direction from humble beginnings in particular laboratories via national recognition to international status, by that time having become much modified.

2.2.1 International Standards

The ultimate state of unity would be for all countries to be using the same standards. This would obviously be of great value in smoothing the course of international trade and make it easier for technologists to exchange technical information. It is also a very ambitious concept that the countries of the world can compromise on their national procedures and overcome the very great difficulties of language in a field where language is the most important tool of trade.

The International Standards Organisation
In most fields, including plastics, the principal body attempting to achieve the ideal of international agreement is the International Standards Organisation (ISO) which is hence our most important organisation in the standards field. The ISO in its present form is not a very old organisation, being formed in 1946 after previous attempts at setting up this sort of body had met with little success.

The work of ISO is administered by a permanent central secretariat which has headquarters at 2, rue de Varembe, 1221 Geneva 20 and has as members more than seventy national standards bodies, one body per country. Apart from central committees concerned with planning, certification, etc., the technical work of ISO is carried out by technical committees each relating to a particular area of industry. The secretariat of each technical committee is held by a member country and each member may join any committee either as a participating (P) member or observing (O) member. The P members have voting rights at committee meetings. The choice of P or O member depends on the country's interest and the finance available.

The committee for plastics is TC61 with the United States of America holding the secretariat; rubber is covered by TC45. TC61 normally meets once per year, member countries acting as hosts more or less in rotation. The delegates to the technical committee are decided by the national standards body.

Most ISO technical committees operate with an infrastructure of subcommittees, working groups and task groups and TC61 is no exception. There are at present ten subcommittees in TC61 as listed below, the secretariat of each being taken by a member country.

SC1 Terminology
SC2 Mechanical Properties
SC4 Burning Behaviour
SC5 Physical/Chemical Properties
SC6 Ageing, Chemical and Environmental Resistance
SC9 Thermoplastic Materials
SC10 Cellular Materials
SC11 Products
SC12 Thermosetting Materials
SC13 Composites and Reinforcement Fibres

Within each subcommittee there is a number of working groups, each with a leader, to take care of the details of one aspect of the subcommittee's subject. For example in SC2 there is a working group for impact tests, another for tensile properties and so on. Task groups, appointed by a working group for a specific task are in fact now little used in TC61. It can be seen that the subcommittees of most interest to physical testing are TCs 2, 4, 5 and 6 but other subcommittees have of course an interest in tests to be included in specifications, particularly specialised product-oriented methods.

The order of progress towards an International standard is that after consideration at working group and subcommittee level a document is proposed to the plenary session of the technical committee for circulation to members for postal vote as a draft proposal (DP). If approved, balloting and commenting take place and the votes and comments are considered by the working group at the next meeting of the technical committee. If agreement is reached the revised document is again circulated this time to all ISO member countries as a draft international standard (DIS). The comments and votes on this are considered at a subsequent meeting and if approved at the plenary session the final revision is sent to the central secretariat for approval by the ISO council and publication as an international standard. It is immediately obvious that this process is slow. Although certain short cuts are possible, it is not unusual for a document to pass through a second draft proposal or second draft standard stage if agreement proves difficult.

ISO standards were first published in 1972; before that time ISO recommendation was the title used. It is not obligatory for the ISO standards to be incorporated into a national system but obviously the whole aim is a little defeated if this is not done. The BSI takes a very positive attitude in this direction following the dictum: 'Do it once, do it right, do it internationally.' Wherever possible British Standards at least agree with ISO standards technically and the aim is to reproduce them verbatim. To become an ISO standard requires that a document receives the approval of 75 per cent of the members casting a vote. This fairly high percentage helps to ensure that unsatisfactory documents do not get through. Normally, if the UK gives a positive vote at ISO that standard is then adopted as regards technical content into the BSI system.

It is difficult enough to reach agreement on a standard procedure within one country and the problems internationally are considerably greater, not being helped by language difficulties – the official languages of ISO are English, French and Russian. Therefore the slow pace of production of an ISO standard is hardly surprising. Differences between national documents and ISO methods are often not a matter of disagreement but simply that the two time scales are different and things have got out of step. However, there are encouraging signs that progress towards more complete rationalisation is being achieved. There are at present more than 210 ISO standards published in the plastics field. These are listed in the *ISO Catalogue* and additions during the year are publicised in the British Standards publication *BSI News*. In the UK ISO standards can be obtained from the BSI and in other countries through the national standards body. Standards for plastics can also be purchased grouped into two volumes as *ISO Handbook 21*.

Other International Standards

In the electrical field the International Electrotechnical Commission (IEC) performs the same function as ISO. The work of this body is of interest where plastics are used in electrical insulation, etc. As regards electrical test methods for plastics, both ISO and BSI adopt the basic procedures and principles standardised by IEC.

Britain has particular interest in the more limited scope of European standardisation. The European Committee for Standardisation (CEN) was founded in 1961 and comprises the national standards bodies of EEC and EFTA countries. CENELEC is the equivalent body in the electrical field. To many people the concept of European standards is an unnecessary complication, it being argued that there is no need for any activity in between ISO and the national bodies. However, the work of CEN is likely to assume importance where documents are drawn up at the specific request of the EEC in their programme for the removal of technical trade barriers. Such standards will be used in EEC directives which, if ratified, are binding on the community members. CEN standards are voluntary, being adopted only by approving countries. There are also fundamental differences in the voting and approval systems in CEN and EEC and it is apparent that those concerned with standardisation will have to watch the European developments carefully if unsatisfactory standards and legislation are not to be forced upon us.

There are many other international organisations concerned with standards and a short guide to these is given in BSO Part 3, which makes clear the confusing abbreviations of titles in use.

2.2.2 National Standards

Although generally each country has one principal standards organisation which provides the official membership of ISO, other organisations can be issuing standards at national level. It is usual to include government departments in this category. It is not practical, and indeed not necessary, to consider here the national standards bodies of all countries but a list of ISO members is given in Appendix B on page 431. The operations of the British Standards Institution (BSI) will be described in some detail, which, apart from being of particular interest to those trading in Britain, serve to illustrate how the process of generating standards as a national level can be undertaken. It must not be assumed that other countries operate in even roughly the same format. It should be noted, however, that the BSI is one of the longest established and most highly rated of national standards bodies.

The British Standards Institution
The BSI was formed in 1901 and has now developed to the point where it covers an astonishing range of subjects from virtually all branches of industry. Apart from its main function of producing standards it also operates a quality assurance division which operates BSI's certification and assessment schemes and a comprehensive test house.

BSI receives government support but raises the majority of its income from membership fees, the sale of standards and fees from certification and testing services.

Membership is open to virtually anyone, various categories of organisations being defined for the purpose of computing membership fees. It is in fact rather difficult to keep up to date on standards matters without being in membership. Details of new standards, amendments and articles on standards matters generally are published monthly in *BSI News* which is circulated to members. There are also annual editions of the *BSI Catalogue* which lists all the British Standards available and is issued to members and may be purchased by non-members. It is appropriate to mention here BSO *A Standard for Standards*[3–5] which gives in considerable detail an account of BSI structure, procedure and editorial practice.

Structure
The permanent staff of BSI, administrative, technical and editorial, operate under a Director General, with the overall responsibility for policy being vested in an Executive Board. Matters of policy relating to convenient groups of industries (e.g. building, engineering) are dealt with by Divisional Councils and within each of these councils is a number of Industry Standards Committees. Of particular interest to us is the Plastics Industry Standards Committee PLM/– which operates within the umbrella of the Multitechnics Council M/–.

We now have an introduction to the numbering system used for BSI committees. The PL refers to plastics and the M for multitechnics. PLM/– has responsibility for authorising the initiation of a standards project and for final approval of drafts before publication. The preparation of standards is carried out by technical committees and subcomittees reporting to the industry committee. Hence we have for example PLM/7 styrene plastics and PLM/9 plastics pipes and fittings. The BSI supplies the secretariat for technical committees and the members are nominated by industry, government departments and research associations. The industry representatives are usually nominated by trade associations and not the individual company, although individual experts are sometimes co-opted.

Testing of plastics is the concern of PLM/17 which currently has five active panels, each dealing with a specific area of testing. Panels rather than subcommittees are used purely for administrative convenience. These *ad hoc* panels, despite being almost off the end of the numbering system, are very important and PLM/17 would get very bogged down without them. PLM/36 is concerned with the presentation of plastics design data and hence formulates procedures suitable for the generation of such data. However, it has been inactive for several years, largely because of a cut back in funding.

Other committees which should be mentioned in the context of testing are the PLM/RUM series which are joint committees between the plastics and rubber industries. PLM/RUM/10 deals with methods of test for cellular materials, PLM/RUM/6 with accuracy of test machines and PLM/RUM/9 with electrical tests. Unfortunately systems are never as simple as we would like; specialised tests may be considered in product committees and not all products containing plastics are covered in PLM committees. For example, coated fabrics, including testing are the subject of RUM/13. Lastly, the subcommittee PLM/–/1, reporting to the industry committee, exists to co-ordinate international work and hence is the prime link with ISO affairs.

Preparation of British Standards

Consideration of an initial draft is carried out in a technical committee, subcommittee or panel as appropriate. This initial draft may have come from one of a number of sources, for example, being based on work carried out by one of the bodies represented on the committee. The draft works upwards to the main committee, being presented there as a private circulation. When agreement is reached the document is circulated as a draft British Standard (DC) to industry for comment. These drafts have no official standing and nowadays are always prepared after careful consideration of the position in ISO. The comments are considered by the technical committee and the amended document passed to the industry committee for approval. The BSI editorial department then vets the document for consistency of layout, etc., before it goes for printing. This procedure is generally efficient and quicker than the ISO procedure but it must be admitted that unfortunate delays do occur, often as a result of too great a pressure of work on the BSI resources.

The technical committee also has the responsibility of keeping its published standards up to date, which for a major revision would mean a complete re-issue and would be dealt with in the same manner as a new standard. Relatively minor changes are dealt with by amendment slips the publication of which, as for new standards and revisions, is announced in *BSI News*.

Each standard is given a separate number although there may be more than one part to a standard, each issued separately. The year of publication is added so that different editions of the same standards can be recognised. The *BSI*

Catalogue gives a complete list of standards in numerical order, but unfortunately not grouped into subjects. For a complete picture of the up-to-date situation it is necessary to consult the catalogue and the subsequent issues of *BSI News*, remembering that the catalogue might be 18 months out of date.

Other British National Standards
Standards or specifications issued by individual companies are not considered to be of national status, however large or multinational the concern might be. Specifications issued by local authorities and nationalised industries would be in the same bracket. Rapra Technology Limited (The Rubber and Plastics Research Association), The British Plastics Federation and the Plastics and Rubber Institute do not issue standards.

Government departments, although contributing a great deal to the work of BSI, produce a large amount of their own standardisation. The reasons for this are really similar to those which apply to individual companies – they are unable to wait for the BS system or they have specific requirements unique to themselves. This latter reason applies particularly to the armed forces. However, a memorandum of understanding between the government and BSI signed in 1982 states that the government will seek to use British Standards rather than to develop its own. Unfortunately the various standards issued by the government departments are rather confusing to outsiders, as indeed are the departments themselves, which appear to undergo frequent change of titles or scope.

USA Standards
American standards, particularly those of ASTM, are widely used in many parts of the world and indeed many companies adopt wholesale methods from ASTM under their own name. The national standards system in the USA differs in many respects from the British, in particular the organisation which publishes the standards of most interest, the American Society for Testing and Materials (ASTM), is not the official national standards body having ISO membership. That function is fulfilled by the American National Standards Institute (ANSI).

American National Standards Institute (ANSI). ANSI is the premier USA standardisation body and is the counterpart of BSI in being the official ISO representative. It was until a few years ago known as the American Standards Association. ANSI does not itself write standards but approves as American standards those produced by ASTM and other similar organisations. Not all ASTM standards are approved in this way and approval may take place years after the introduction of the standard by ASTM.

American Society for Testing and Materials (ASTM). Apart from work at ISO, it is the ASTM that most people in the polymer industry think of as representing American standards. ASTM has a membership drawn from similar sources to those of BSI and operates through more than 140 technical committees which in turn have a subcommittee structure. D20 is the committee for plastics.

ASTM standards can be obtained individually but are more usually seen as ASTM books, each being a collection of standards covering a particular subject or related group of subjects. The books are revised annually and although some standards remain unchanged for years there is always a significant amount of new or revised matter. Hence it is advisable to use only the current edition. Although this is rather expensive it is easier for the user than keeping British Standards up to date by studying *BSI News*. There are currently some sixty-six volumes of ASTM standards, those concerning plastics in particular being volumes 8.01, 8.02, 8.03 and 8.04. The ASTM is active in the technical field apart from purely producing standards. It organises conferences and publishes numerous books and reports as well as the journal *Standardisation News*.

German Standards
The reason for including a mention of German standards as opposed to those of France, Russia or any other country in an English language book is that it has been the author's experience that DIN standards are those most commonly met with in Britain after BS and ASTM.

The principal German standards organisation Deutsche Institut für Normung was founded in 1917. It operates rather similarly to the BSI in that it has subscription-paying members from industry, trade associations, etc., and is independent of individual pressures. It has standards committees and some 3800 technical committees and study groups, membership of which is honorary with representatives from all interested areas.

The DIN series of standards, of which there are many thousands, are catalogued into subject groups numerically. For example the 6770 and 6790 groups cover plastics testing. The standards are published separately and are all listed in the *DIN Catalogue*. Many DIN standards are also published in English and Spanish and a few in French. There is a separate yearly book which lists those standards which have been translated.

The use of DIN standards is generally voluntary unless they are incorporated into legislation. The DIN mark on a product signifies that it complies with the relevant DIN standard and manufacturers can be forbidden to use the mark if they do not stick to the rules. The mark is hence similar in concept to the BSI Kitemark although the policing system appears to be less rigorous. The DIN Testing and Supervision Mark appears to be more equivalent to the Kitemark.

Other National Standards
It is not possible to discuss all of the national standards bodies but it is necessary to be able to identify the source of a standard from the abbreviations like BS and DIN which are used. Taking the whole world and including government standards the total becomes enormous and very confusing. The letters used by ISO members are given in Appendix A.

The standards mostly used by any laboratory will depend on whom they are trading with. Information can be gained from the national bodies listed in Appendix A.

2.2.3 Company Standards

There must be literally millions of company standards in existence. Although they have relatively little significance in a national or international sense, they are the basis of many commercial contracts and hence are perhaps the most important standards of all. Among their number are some of the best examples of standardisation and also some of the worst; sadly the worst appear very frequently.

Using a commercial standard is like using any standard, the user must be careful that he has the latest edition and that he has read it very carefully and missed none of the detail. A common fault in commercial standards is that rather a lot of detail is missing, for example there might be insufficient information in a test method to be sure that you are carrying it out correctly. All one can do is to talk to the originator of the standard.

It would save a great deal of pain and confusion if those writing commercial specifications would wherever possible use published standard test methods, preferably those of ISO. Special tests will often be needed but there is no point in inventing one's own procedure for a straightforward test which has been well standardised. Perhaps a lot of the trouble is that in some cases those writing specifications are not well versed in standardisation outside of their own organisation and that many engineers have a poor understanding of plastics and their properties.

2.3 UNITS

In the second edition of this book it was noted that it should be unnecessary to state that SI units will be used – that will be assumed to be the case. However, the imperial system lingers on in various isolated outposts. The universal adoption of SI units virtually eliminates the need to include a section on units because there is no question of conversions or explanations of obscure systems. Nevertheless it is appropriate to make reference to relevant information.

The basic reference[6] is to ISO 1000 which details all the units, multiples and submultiples to be used. BS 5555[7] is identical. PD 5686[8] gives advice on the use of SI units given in ISO 1000, explanatory information for British readers and details of EEC directives on units of measurement, whilst BS 3763[9] gives basic features of the SI as promulgated by the Bureau Internationale des Poids et Mesures.

Many organisations have produced their own version of how SI units should be presented but these were generally printed when metric units were relatively new in British industry and would no longer seem to be warranted. However, certain special considerations will apply in any particular industry and both ISO committees TC45 and TC61 have actively considered the subject. Their conclusions on units which are normal to their materials and products have been included in their own procedural documents.

Where there is a need to convert to or from SI units reference can be made to the conversion factors found in BS 350,[10] or PD 6203.[11] The detailed conversion tables of BS 350: Part 2 have been withdrawn as obsolete!

REFERENCES

1. *Standards, Quality and International Competitiveness* (1982) Department of Trade.
2. Brown, R. P. (1984) *Polymer Testing*, **4**, 2.

3. BSO (1981) *A Standard for Standards*: Part 1: *General Principles of Standardisation.*
4. BSO: Part 2 (1981) *BSI and its Committee Procedures.*
5. BSO: Part 3 (1981) *Drafting and Presentation of British Standards.*
6. ISO 1000 (1981) *SI Units and Recommendations for the Use of their Multiples and of Certain other Units.*
7. BS 5555 (1981) *SI Units and Recommendations for the Use of their Multiples and of Certain other Units.*
8. PD 5685 (1978) *The Use of SI Units.*
9. BS 3763 (1976) *The International System of Units.*
10. BS 350: Part 1 (1974) (1983) *Basis of Tables.*
11. PD 6203 1967 (1982) *Additional Tables for SI Conversion.*

Chapter Three

Preparation of Test Pieces

3.1 INTRODUCTION

In Ch. 6 it is mentioned that the degree of crystallinity can be estimated from density measurements, the density of the amorphous and the crystalline forms generally being significantly different. Since the degree of crystallinity in a test piece depends on the manner in which it has been prepared – temperature, pressure, profile of the mould cavity, rate of cooling, etc. – it therefore follows that the magnitude of this apparently most simple and straightforward of properties, the mass per unit volume, varies according to the history of the test piece under examination. This comment applies to practically every property measured on the massive polymer. Reference is made in other chapters to the influences of test piece shape and pre-conditioning and to these can be added such considerations as mechanical surface finish and, for certain properties, cleanliness of surface.

While for the purposes of assessing merit in a specific end use, the condition of material under examination should simulate realism, for the purposes of a standard test the conditions should approach, as it were, idealism – that is, history, shape, condition, etc., should be carefully defined. Much effort is wasted on producing standard test data, even within the limited confines of the value of such data, because of insufficient attention to the preparation of the test pieces. The subject as a whole has been discussed by Cohen.[1]

By and large the examination of plastics 'raw materials' is outside the scope of this book but, in passing, it is worth commenting on the necessity of, for instance, pre-drying hygroscopic powders or flake before making up solutions for viscosity measurement. Similarly, moulding granules may need pre-drying, e.g. of urea formaldehyde or nylon, if satisfactory test mouldings are to be obtained.

Many tests are normally performed on specially moulded test pieces; in fact all the tests usually described as being on moulding materials are really on test pieces moulded from the latter, with the exception of the few which may be broadly described as checking the 'mouldability' of the material – particle size, bulk factor, volatile content, cup flow and melt behaviour generally (see Chs. 5, 7 and 16).

3.2 MIXING AND MOULDING

3.2.1 General Considerations

Nearly all of the common plastics moulding materials (both thermosets and thermoplastics) arriving at the physical test laboratory are in the form of granules or powder which may be directly moulded to test pieces or test sheets from which test pieces can be machined. As such, very little mixing is necessary prior to moulding but for one or two thermoplastic materials, notably PVC and ABS, it is essential to pre-mill the granules to sheet form prior to compression moulding test sheets if consistent and meaningful results are to be obtained. This is recognised in ISO 293 (1974)[2] *Compression Moulding Test Specimens of Thermoplastics Materials*, Appendix B of BS 2571 (1963)[3] *Flexible PVC Compounds*, Method 901A of BS 2782 (1983)[4] (which is identical to ISO 293), ASTM D3463[5] *Compression Molding Test Specimens of Rigid Acrylonitrile–Butadiene–Styrene (ABS) Plastics* and ASTM D3010 (1981)[6] *Preparing Compression Molding Test Specimens of Rigid Poly(vinyl chloride) Compounds*.

In all of these a two roll mill is specified but the ISO and BS 2782 test methods leave the temperature conditions of milling open, the test sample to be mixed at a 'suitable' temperature. The ASTM methods are much more precise, not only specifying the temperature (182 °C ± 2 °C for rigid PVC and from 165 °C to 170 °C for ABS) but also the roll clearance and rolling bank size together with an indication of the time on the mill. Appendix B of BS 2571 also gives an indication of roll temperature and milling time.

Harrison *et al.*[7] have explored the particular problems associated with producing test pieces from flexible PVC compounds for physical testing. They found that the biggest single influence on tensile strength was the mill temperature, an increase of which led to higher tensile strength. Press temperature had little effect on physical properties. It was also noted by these workers that as the filler level increased or the plasticiser content decreased the effect of both mill and press temperature decreased.

In some cases where liquid resins are used to make test sheets, pre-mixing of the liquid resin system is necessary prior to casting, e.g. preparing cast poly(methyl methacrylate) sheet, or using a polyester resin to prepare a low-pressure laminated test sheet. Such techniques are not often encountered and usually refer the user to the manufacturer of the resin for instructions. Such is the case in ISO 1268 (1974)[8] *Preparation of Glass Fibre Reinforced, Resin Bonded, Low Pressure Laminated Plastics or Panels for Test Purposes* (Methods 920A to 920C of BS 2782 1977 are identical to ISO 1268).

When considering the moulding of test pieces or test sheets the most important decision is to settle which method of moulding is appropriate, for the orientation induced in an injection-moulded test piece may yield quite different test results from those from a relatively non-orientated compression moulded or extruded test piece. Hayes[9] illustrates this well with some bending strength measurements on $5 \times \frac{1}{2} \times \frac{1}{4}$ in (127 × 12·5 × 7 mm) bars of toughened polystyrene, compression moulded in one case and injection moulded from one end in the other (Table 3.1). In the same paper, the effect of degree of orientation is demonstrated by measuring the same property on bars produced at different moulding temperatures and pressures, the greatest degree of orientation resulting from the lowest temperature and highest pressure (Table 3.2).

Similar studies have been made *inter alia* by Budesheim and Knappe,[10] Horsley *et al.*,[11] Bryant and Hulse,[12] Bossu *et al.*,[13] Hogberg,[14] Morgan and Vale,[15] Williams and Mighton,[16] Dasch,[17,18] Koda,[19] and Malac.[20] Hogberg, for instance,

Table 3.1 Comparison of bending strengths of compression and injection moulded toughened polystyrene

	Bending strength (lbf/in²)	(MPa)	Deflection at break (in)	(mm)
Compression moulded	6570	45·30	0·71	18·0
Injection moulded	8170	56·33	>1·00	>25·4

(After Hayes.[9])

Table 3.2 Breaking strengths at different moulding temperatures and pressures

Temperature (°C)	Pressure (lbf/in²)	(MPa)	Breaking strength (lbf/in²)	(MPa)
180	10 000	69·0	8 170	56·33
200	8 000	55·0	7 430	51·23
220	5 000	34·0	7 460	51·45
240	4 500	31·0	7 010	48·33

(After Hayes.[9])

examined impact strength measured on bars in which, in one case, flow had occurred in the direction of the bar and, in the other, the direction of flow was perpendicular thereto. Four styrene–acrylonitrile copolymers were examined and the ratios of the values were between 2½ and 3 to 1, in favour of the bars where flow had been along their length. Amongst the effects studied by Williams and Mighton was influence of moulding conditions on the stability of mouldings. Koda's study, on polycarbonates, embraced abrasion resistance, hardness, density and heat shrinkage measurements.

The subject, in relation to injection moulding, has been comprehensively reviewed by Whisson.[21] Wintergerst[22] endeavoured to anneal test pieces of high impact polystyrene back to a condition of 'basic strength' from the injection-moulded condition.

On the other hand, moulding conditions may have a more macroscopic effect on the material; a filled or reinforced composition can yield rather different data depending on whether the filler is subject to fracture and whether the process is rigorous enough to cause it. Work reported by Spiwak[23] well demonstrates these points. A range of thermosetting materials was examined, using compression-moulded test pieces and two sets prepared by transfer moulding, the size of one gate (B) being twice that of the other (A). The figures shown in Table 3.3 have been selected from Spiwak's work. A somewhat similar exercise has been described by Elmer and Harrington using phenolic materials in an interlaboratory study.[24]

Panov[25] investigated the effect of test piece size on the tensile strengths of a range of plastics materials and found that the strengths decreased with increasing cross-sectional area of the test piece – over 20 per cent decrease was observed for

Table 3.3 Impact and bending strengths of test pieces made of different materials and by different moulding techniques

Material	Impact strength: Compression moulded (ft lbf/in)	Compression moulded (J/mm)	Transfer moulded (gate size A) (ft lbf/in)	Transfer moulded (gate size A) (J/mm)	Transfer moulded (gate size $B = 2 \times A$) (ft lbf/in)	Transfer moulded (gate size $B = 2 \times A$) (J/mm)
PF with short cotton fibre	0·44	0·023	0·45	0·024	0·44	0·023
PF with long asbestos fibre	2·31	0·123	1·23	0·066	1·17	0·062
DAP with short acrylic fibre	0·64	0·034	0·66	0·035	0·66	0·035
DAP with long glass fibre	3·68	0·196	0·68	0·036	0·88	0·047
	Bending strength					
	(lbf/in)	(MPa)	(lbf/in)	(MPa)	(lbf/in^2)	(MPa)
PF with short cotton fibre	9 200	63·5	9 930	68·5	9 210	63·5
PF with long asbestos fibre	12 100	83·5	6 400	44·0	7 280	50·0
DAP with short acrylic fibre	10 160	70·0	11 590	80·0	9 940	68·5
DAP with long glass fibre	14 800	102·0	10 220	20·5	8 210	56·5

(After Spiwak,[23] courtesy Society of Plastics Engineers Inc.)

polystyrene and unplasticised PVC by increasing the area from 25 mm^2 to 400 mm^2.

Parts of a book edited by Ogorkiewicz[26] deal with the effect of processing variables on the properties of thermoplastics.

Crawford and co-workers[27] made a comprehensive study of the effect of injection-moulding parameters and their influence on mechanical properties. They describe the design of a mould to produce a set of injection mouldings whose properties can be related generally to product performance.

3.2.2 Standard Procedures

Until 1970 it was traditional to include moulding conditions in the specification for the material under test, certainly in British Standards, although ISO and ASTM had for some years a small number of published documents dealing exclusively with the moulding of test pieces. During the past 10 years a number of international and national standards on the moulding of test pieces have been published; material specifications, as they are revised, tend to refer to these methods rather than include them in appendices, as has been the practice hitherto.

ISO Standards
There are six ISO standards relating to the moulding of test pieces. ISO 293[2] lays down the general principles to be followed when compression moulding test pieces of thermoplastics materials. The exact conditions are not given because they are considered to be part of the specification for the material under test, or to be agreed between purchaser and supplier.

ISO 294[28] deals with injection moulding of test pieces of thermoplastics and, like ISO 293, only lays down general principles to be followed. Although a warning is given under 'Scope' that many factors can influence the characteristics of injected-moulded test pieces, no details are given and one is referred to the material specification for exact moulding conditions.

ISO 295(1974)[29] is intended to serve as a basis for ensuring comparability of results between test houses when moulding thermosetting materials. It includes three annexes:

A Preparation of test specimens from phenolic moulding materials
B Preparation of test specimens from aminoplastic moulding materials
C Preparation of test specimens from polyester and epoxy resin moulding materials

These annexes give recommended moulding conditions for the various materials.

ISO 2557 is somewhat complementary to ISO 293 and ISO 294 and recognises the need to be able to produce test pieces which have defined levels of shrinkage. This is intended to achieve greater consistency in test results with those materials which show significant differences in mechanical properties with level of internal strain in the test pieces. It lays down procedures for preparing test pieces with a defined level of shrinkage as measured by finding the longitudinal shrinkage of test pieces by a standardised (annealing) process. Moulding conditions are adjusted to achieve the required level of shrinkage as specified by the material specification or as agreed between the parties concerned.

Part 1 of the Standard[30] describes the preparation of test specimens in the form of parallelepipedic bars by both compression and injection moulding and Part 2[31] covers test specimens in the form of plates by injection moulding.

ISO 1268[8] *Preparation of Glass Fibre Reinforced, Resin Bonded, Low Pressure Laminated Plates or Panels for Test Purposes* specifies the procedure for the preparation of test sheets by bonding glass cloth or mats with low-pressure thermosetting resins. Three methods are given: Method A for glass reinforcement impregnated with liquid resin moulded in a press, Method B for moulding pre-pregs and Method C for hand lay-up. The basic design of suitable moulds is given and detailed instructions are included for the hand lay-up method.

ISO 3167[32] describes how a Type 1 tensile test specimen of ISO R527 (see Ch. 8), can, with only simple machining, be made suitable for a variety of other tests. Emphasis is placed on strict control of moulding conditions and reference is made to ISO 2557 (see above) and to ISO 2818 (see below). The use of such multipurpose test pieces is attractive for a busy test laboratory where one set of moulded test pieces can be produced under identical conditions followed by machining to yield test pieces suitable for a wide range of physical tests. ISO 3167 proposes that such test pieces may be used for tensile, flexural, compressive, impact, thermal, environmental stress cracking and certain flammability properties. properties.

British Standards
At the present time British Standards on moulding test pieces closely follow the ISO practice. Thus:

BS 2782, Method 901A[4] is the same as ISO 293
BS 2782, Method 902A[33] is the same as ISO 295
BS 2782, Method 910A[34] is the same as ISO 294
BS 2782, Methods 920A, 920B and 920C[35] are the same as those given in ISO 1268
BS 2782 Methods 940A[36] and 940B[37] are the same as ISO 2557 Parts 1 and 2, respectively.

In addition, some British Standards for material include procedures for the preparation of test pieces, notably Appendix B of BS 2571[3] *Flexible PVC Compounds*, BS 771[38] *Phenolic Moulding Materials*, BS 3126 (1987)[39] *Method of Specifying Toughened Polystyrene (SB) Moulding and Materials* and BS 1493 (1987)[40] *General Purpose Polystyrene Moulding Materials.*

ASTM Standards
As mentioned earlier, the ASTM standards for the moulding of test pieces are more comprehensive and detailed than the ISO standards. There are no fewer than eleven standards which are related to the moulding of test pieces.

ASTM D647 (1981)[41] on design of moulds for test pieces, gives details of compression and injection tools including fully dimensioned drawings. It is referred to in the other ASTM methods for moulding test pieces.

ASTM D958 (1981)[42] covers two optional procedures for finding the temperature condition of moulds either by a thermometer or by a pyrometer.

ASTM D956–51 (1981)[43] on compression moulding of AMC test pieces is roughly equivalent to ISO 295 but more precise.

ASTM D796 (1981)[44] on compression moulding of PMC test pieces is, again, roughly equivalent to ISO 295 but more precise.

ASTM D2292 (1985)[45] (styrene–butadiene moulding and extrusion materials) uses a simple picture-frame flash mould similar to that given in ISO 293.

ASTM D3027–72 (1981)[46] (alkyl moulding compounds) calls for use of the positive moulds described in D647 and closely specifies moulding conditions.

ASTM D3463[5] (rigid ABS plastics) gives precise details of moulding test pieces and has already been mentioned above in connection with pre-milling.

Similarly, ASTM D3010[6] (test sample plaques of rigid PVC compounds) closely specifies the procedure to be followed for moulding test pieces of rigid PVC, including milling prior to moulding to ensure homogeneous test pieces.

ASTM D1897[47] is similar to ISO 294 but is much more precise and allows the use of screw injection machines. It contains more details of moulding cycles than ISO 294 but these have been based on interlaboratory tests using polystyrene and polycarbonate only.

The preparation of test pieces of thermosetting materials by injection moulding is covered in ASTM D3419.[48] This standard specifies the type of machine and mould to be used and gives useful guidance on moulding temperatures. Similarly, ASTM D1896[49] lays down the procedure for moulding test pieces by transfer moulding and specifies radiofrequency pre-heating. Again, recommended moulding temperatures and cycles are given.

German Standards
DIN 53451[50] deals with the compression moulding of test specimens of thermosetting plastics and is broadly similar to ISO 295.

Like the traditional British Standards, German standards tend to describe the preparation of test specimens in the test methods and materials specifications.

3.3 CUTTING FROM SHEET

When the sample to be tested has already been moulded to some 'massive form', for example calendered sheet, laminate, casting or large commercial moulding, test pieces of the appropriate form may have to be cut from the sample.

In many ways this is a more complex problem than moulding specimens directly, for the machining of test specimens is a difficult operation to standardise, especially in terms of the quality of the finishing. No simple recourse to using a stainless steel mould with mirror finish is possible here; yet many properties are profoundly influenced by the presence of surface flaws such as nicks and scratches.

The point is well illustrated by reference to the apparently simple operation of cutting out specimens from soft thin film with a die. Patterson[51] gives the results of a correlation exercise in tensile testing of 1.5 mil polyethylene, 1 mil polyester and 4 mil polycarbonate (1 mil = 0·001 in or 0·025 mm) undertaken by four laboratories using four different techniques: die cut, manual razor cut, hand-driven rotary and shear cutter.

Yield strength data showed the least scatter – 10 per cent or less covering all four techniques and laboratories. However, much greater divergencies were found with ultimate tensile strength (30 per cent or more) and elongation at break (up to approximately 75 per cent), with the laboratories being relatively consistent and the techniques accounting for the majority if not all the variation. Die-cut specimens were clearly the worst in this exercise, though this should not be taken as a total indictment of the technique; indeed, although in this work simple rectangular specimens were examined, many more complex shapes could not be conveniently prepared by any other technique. More importantly, the lesson to be learned is of the necessity to ensure that cutting dies are in good sharp condition and 'clean' cuts are made, for example by a clicking press.

At one time standards did not specify or even recommend methods of cutting and machining test pieces from sheet and other products, but since 1970 this omission has been substantially rectified. For example, ISO 1184[52] BS 2782 Methods 326A to 326C[53] and ASTM D882,[54] draw attention to the need for care in preparing test pieces for the tensile testing of plastic films and the ISO and BS standards suggest suitable methods of cutting the test pieces. Similarly, BS 2782 Methods 320A to 320F,[55] for tensile properties, have sections on the preparation of test pieces. The current revision of ISO R527[56] (DIS 527) *Tensile Properties* by ISO TC61 also has a section giving guidance on preparation of test pieces.

It is often necessary to cut rectangular blanks from sheet samples which are subsequently machined to test pieces of specified form and size (§3.4). Such test piece blanks are cut using hand saws, circular saws and diamond-impregnated discs. With care and careful hand finishing by file and abrasive paper it is possible to produce finished test pieces from such sawn blanks which are acceptable for many tests, but some form of machining using a properly designed machine, speeds up the preparation of test pieces and generally leads to more consistent test results.

3.4 MACHINING FROM SHEET AND FINISHED PRODUCTS

For flexible, semi-flexible and some of the 'softer' rigid plastics sheet and finished products (up to 1·5 mm thick), test pieces can be cut by using a sharp and properly maintained die cutter and clicking press. This method cannot be used satisfactorily for the more rigid materials and recourse must then be made to machining. Sometimes, too, the product when tested is too thick and then one or both surfaces have to be machined to reduce the thickness to the required level.

Both milling and grinding have been found suitable and standard machine tools may be used provided they can be operated at a high enough speed and be fitted with suitable cutters. Tungsten carbide or diamond tipped milling cutters give the longest life but high-grade steel cutters may be used with many materials and will give an acceptable service life. Tool rotation speeds vary according to the material machined and are best found by trial and error but generally lie within the range from 8000 to 30 000 r.p.m.

The technique of grinding (or buffing) plastics is not so widely practised as for rubber where it is quite commonly used for reducing the thickness of rubber test pieces. When grinding plastics it is important to use the right grade of abrasive wheel; open grit wheels give the best results but in the absence of specific information the advice of the abrasive wheel manufacturer should be sought both as to the correct abrasion grade and the optimum speed of rotation of the wheel.

Standard machine tools and surface grinders can be used for making rectangular bar test pieces; the production of curved test pieces (e.g. dumb-bells) require some form of copying milling machine or router. Many variations have been used but all essentially consist of a motor-driven cutting tool against which the blank is machined, guided by a pin following a template of the required shape.

When machining some plastics, a coolant is necessary if a satisfactory surface finish is to be obtained. A jet of compressed air is suitable in many cases, but if a liquid coolant is used care must be taken to ensure that it does not affect the plastics material being machined.

There is much room for experiment in the machining of plastics before precise details of machining conditions can be laid down for any given material, but the need for standardised procedures is recognised by the publication of ISO 2818[57] *Preparation of Test Specimens by Machining.* This gives machining conditions which have been found suitable for thermoplastics and thermosets where the thickness does not exceed 10 mm and also includes diagrams for a manually guided copy milling machine and a tubular abrasive machine tool. ISO 2818 does not offer any recommendation about reducing the thickness of a test piece blank or sheet if it is greater than 10 mm, and it is considered that this is a weakness of the standard. BS 2782 Method 930A[58] is identical to ISO 2818.

Apparatus and materials for the preparation of test pieces have been discussed by Brown,[59] who also mentions the techniques available for machining the notches in impact test pieces – an important subject because in some plastics small variations in notch geometry can cause large variation in impact test results. Shaping or milling is recommended and hand-operated notch cutters are now available commercially.

Finally, we return to the subject of reducing the thickness of sheet where this is necessary to produce test pieces of the specified dimensions. The specification for the material under test, or the test method should be consulted before cutting because some tests (e.g. BS 2782 Method 320E[60] for tensile properties) requires that one face of the manufactured sheet should be left intact, while others (e.g. BS 2782 Method 359[61] for Charpy impact resistance) specify that both surfaces shall be machined uniformly to give the required thickness.

REFERENCES

1. Cohen, L. A. (1967) in *Testing of Polymers*, vol. 3 (Eds. J. V. Schmidtz and W. E. Brown), Interscience, p. 15.
2. ISO 293 (1974) *Compression Moulding Test Specimens of Thermoplastic Materials.*
3. BS 2571 (1963) *Flexible PVC Compounds.*
4. BS 2782 Method 901A (1983) *Compression Moulding Test Specimens of Thermoplastic Materials.*
5. ASTM D3463 (1981) *Compression Moulding Test Specimens of Rigid Acrylonitrile–Butadiene–Styrene (ABS) Plastics.*
6. ASTM D3010 (1981) *Preparing Compression-Molded Test Sample Plaques of Rigid Poly(vinyl chloride) Compounds.*
7. Harrison, W. H., Johnston, W. R. and Wadey, B. L. (1981) *Journal of Vinyl Technology*, **3** (3), 170.
8. ISO 1268 (1974) *Preparation of Glass Fibre Reinforced Resin Bonded, Low Pressure Laminated Plastics or Panels for Test Purposes.*
9. Hayes, R. (1954) *Transactions of the Plastics Institute, London*, **27** (49), 219.
10. Budesheim, H. and Knappe, W. (1959) *Kunststoffe*, **49** (6), 257.
11. Horsley, R. A., Lee, D. J. A. and Wright, P. B. (1959) *Physical Properties of Polymers*, SCI Monograph no. 5, Society of Chemical Industry, 63.
12. Bryant, K. C. and Hulse, G. (Eds.) (1955) *Plastics Progress Papers and Discussions at the British Plastics Convention, 1955*, Illiffe.
13. Bossu, B., Chatain, M., Dubois, P. and Rougeaux, J. (1959) *International Symposium on Plastics Testing Standards*, STP no. 247, ASTM, p. 67.
14. Hogburg, H. (1959) ibid., p. 95.
15. Morgan, D. E. and Vale, C. P. (1959) *Physical Properties of Polymers*, SCI Monograph no. 5, Society of Chemical Industry, 169.
16. Williams, J. L. and Mighton, J. W. (1953) *Symposium on Plastics Testing – Present and Future*, STP no. 132, ASTM 32.
17. Dasch, J. (1967) *Kunststoffe*, **57** (2), 117.
18. Dasch, J. (1968) *Kunststoffe*, **58** (11), 769.
19. Koda, H. (1968) *Journal of Applied Polymer Science*, **12** (10), 2257.
20. Malac, J. (1969) *Journal of Applied Polymer Science*, **13** (8), 1767.
21. Whisson, R. R. (1967) *RAPRA Technical Review*, 42.

22. Wintergerst, S. (1967) *Kunststoffe*, **57** (3), 188.
23. Spiwak, L. (1963) *Society of Plastics Engineers Journal*, **19** (6), 557.
24. Elmer, C. and Harrington, E. C., Jr. (1960) *ASTM Bulletin*, 249, October, 35.
25. Panov, P. (1967) *Plaste und Kautschuk*, **14** (7), 491.
26. Orgorkiewicz, R. M. (Ed.) (1969) *Thermoplastics: Effect of Processing*, Iliffe.
27. Crawford, R., Klewpatinond, V. and Benham, P. P. (1978) *Plastics and Rubber Processing*, December, 133.
28. ISO 294 (1975) *Injection Moulding Test Specimens of Thermoplastic Materials.*
29. ISO 295 (1974) *Compression Moulding Test Specimens of Thermosetting Materials.*
30. ISO 2557/1 (1976) *Amorphous Thermoplastic Moulding Materials – Preparation of Test Specimens with a Defined Level of Shrinkage – Parallelepipedic Bars (Injection Moulding and Compression Moulding).*
31. ISO 2557/2 (1979) *Amorphous Thermoplastic Moulding Materials – Preparation of Test Specimens with a Defined Level of Shrinkage – Rectangular Plates (Injection Moulding).*
32. ISO 3167 (1983) *Preparation and use of Multipurpose Test Specimens.*
33. BS 2782 Method 902A (1983) *Compression Moulding Test Specimens of Thermosetting Materials.*
34. BS 2782 Method 910A (1983) *Injection Moulding Test Specimens of Thermoplastic Materials.*
35. BS 2782 Methods 920A to 920C (1983) *Preparation of Glass Fibre Reinforced, Resin Bonded, Low Pressure Laminated Plates or Panels for Test Purposes.*
36. BS 2782 Method 920A (1981) *Preparation of Test Specimens of Amorphous Thermoplastic Moulding Material with a Defined Level of Shrinkage in the Form of Parallelepipedic Bars by Compression Moulding and Injection Moulding.*
37. BS 2782 Method 940B (1981) *Preparation of Test Specimens of Amorphous Thermoplastic Moulding Material with a Defined Level of Shrinkage in the Form of Rectangular Plates by Injection Moulding.*
38. BS 771 (1980) *Phenolic Moulding Materials.*
39. BS 3126 (1959) *Toughened Polystyrene Moulding Materials.*
40. BS 1493 (1967) *General Purpose Polystyrene Moulding Materials.*
41. ASTM D647 (1981) *Design of Molds for Test Specimens of Plastics Molding Materials.*
42. ASTM D958 (1981) *Determining Temperature of Standard ASTM Test Molds for Test Specimens of Plastics.*
43. ASTM D956–51 (1981) *Compression Molding of Test Specimens of Amine Molding Compounds.*
44. ASTM D796 (1981) *Compression Molding of Test Specimens of Phenolic Moulding Compounds.*
45. ASTM D2292 (1985) *Compression Molding of Test Specimens of Styrene–Butadiene Molding and Extrusion Materials.*
46. ASTM D3027–72 (1981) *Compression Molding Test Specimens of Alkyd Molding Compounds.*
47. ASTM D1897 (1981) *Injection Molding Test Specimens of Thermoplastic Molding and Extrusion Materials.*
48. ASTM D3419 (1981) *In-Line Screw Injection Molding Test Specimens from Thermosetting Compounds.*
49. ASTM D1896 (1981) *Transfer Molding Test Specimens of Thermosetting Compounds.*
50. DIN 53451 (1972) *Plastics – Preparation of Compression Moulded Test Specimens of Thermosetting Moulding Materials.*
51. Patterson, C. D., Jr (1964) *Materials, Research and Standards*, **4** (4), 159.
52. ISO 1184 (1983) *Plastics – Determination of Tensile Properties of Films.*
53. BS 2782 Methods 326A to 326C (1983) *Determination of Tensile Strength and Elongation of Plastics Films.*
54. ASTM D882 (1983) *Tensile Properties of Thin Plastic Sheeting.*

55. BS 2782 Methods 320A to 320F (1986) *Tensile Strength Elongation and Elastic Modulus.*
56. ISO R527 (1966) *Plastics – Determination of Tensile Properties.*
57. ISO 2818 (1980) *Plastics – Preparation of Test Specimens by Machining.*
58. BS 2782 Method 930A (1983) *Preparation of Test Specimens by Machining.*
59. Brown, R. P. (Ed.) (1979) *Guide to Rubber and Plastics Test Equipment*, RAPRA.
60. BS 2782 Method 320E (1983) *Tensile Properties of Plastics.*
61. BS 2782 Method 359 (1984) *Determination of Charpy Impact Resistance of Rigid Plastics and Ebonite (Charpy Impact Flexural Test).*

Chapter Four

Conditioning and Test Atmospheres

4.1 INTRODUCTION

Plastics materials, generally, are affected by ambient conditions and, in the normal test laboratory, these variables may be limited to varying temperature and humidity; hazards of irradiation, be they ultraviolet from sunlight, gamma rays from nuclear power of heat from rocket propulsion, are more logically treated under the general heading, 'Heat Ageing and Environmental Resistance' (see Chs. 16 and 17).

Sensitivity to the effects of temperature is primarily a consequence of the fact that most present-day plastics materials are based on polymers built upon a backbone chain consisting solely or predominantly of covalently bonded carbon atoms. Crosslinking, secondary intermolecular forces and crystallinity can influence reaction to temperature quite significantly, but by comparison with most other 'materials of construction' (in the broadest sense of the phrase) plastics are *temperature-sensitive.* A number of them contain hydroxyl groups, e.g. cellulose acetate, polyvinyl acetate and polyvinyl alcohol, while others contain amino groups, such as the nylons and casein, and these groups have a distinct affinity for water molecules, such as are present in air having any finite moisture content or humidity; the water acts as a plasticiser and brings about a change of properties. One of the most spectacular cases is the increase in impact strength of about twentyfold which may be achieved by allowing dry Nylon 6 to absorb moisture to saturation. Even if the polymer itself is not subject to the influences of temperature or humidity, the other materials present in the product may well be – particularly to humidity. Wood flour or chopped rag filler and cotton, paper or glass fibre reinforcement are well known examples.

It will be obvious, therefore, that to obtain results of any meaning, from the point of view of defining the physical condition of the material when tested and providing data comparable with others obtained similarly, test pieces should be subjected to standard pre-conditioning to bring them into an equilibrium state with a specified atmosphere. The need for pre-conditioning (hereafter 'conditioning') clearly depends on the nature of the materials under test; in many cases control of humidity is unimportant because the materials are to all intents and purposes completely hydrophobic.

4.2 STORAGE

The properties of vulcanised rubber change most rapidly immediately after vulcanisation and later, assuming that no accelerating influences are present, the changes become so slow as to be negligible over a period of, say, a few weeks. Hence it is desirable that a minimum period is allowed between vulcanisation and testing. Consequently most standard test methods for rubbers require that no test shall be made less than 16 h after vulcanisation and any sample preparation, and that the maximum time between vulcanisation and testing shall be 28 days (90 days in the case of tests on products).

Similar considerations do not apply to most plastics materials which, by and large, do not show significant changes in properties with storage after moulding; however, a notable exception to this applies in the case of plasticised PVC. Walter[1] found that the elastic properties of PVC gels changed over a period of 2 years storage at ordinary temperatures, which change could not be related to plasticiser loss. Heap and Norman[2] investigated the changes in BS softness after moulding plasticised PVC and concluded that the mechanism was not due to loss of plasticiser nor to laminar inhomogeneity. They suggested a hypothesis of droplet separation but the precise mechanism is still not fully understood. This phenomenon is recognised in the case of BS softness measurements (see Ch. 8) on flexible PVC by specifying a 7-day conditioning period between moulding the test piece and carrying out the test.

Common sense and good housekeeping dictate that samples of moulding powder should be stored in cool and dry conditions prior to moulding and the same also applies to samples of semi-finished and finished products. With the exception mentioned above, test methods for plastics do not usually specify permissible maximum and minimum storage times prior to testing. However, some product specifications offer recommendations on this subject. For example, BS 771 (1980)[3] for phenolic moulding materials recommends that tests should be commenced within 24 h of preparing test pieces and BS 1763 (1975),[4] for calendered PVC sheeting, recommends that the sample should be wound onto a tube of adequate diameter and requires that this sample should be stored so that it is out of contact with other plastics materials. In any case it is advisable to take note of the storage time and conditions of storage for further reference in the event of anomalous test results being obtained.

4.3 CONDITIONING

4.3.1 Definitions

In the past there have been wide variations in conditioning procedures because national standards naturally tend to simulate average ambient weather conditions of the country of origin. This is not only to facilitate the actual conditioning and testing operations but also to produce data relevant to the likely environment of use of the plastics material. In general, too, there has been a tendency for each specification for each different material to dictate its own, often different, conditions. Internationally, standardisation has been even more difficult, 'ambient' in North Canada is rather different from 'ambient' in the Middle East. Nevertheless agreement has been reached for a limited number of alternative standard conditions to be applied internationally.

Before discussing the various standard conditions it is worth defining the terms used. The preferred unit of temperature is the Celsius degree (°C) although the SI system specifies kelvin (K). The water content of the air is invariably specified as relative humidity (r.h.) and this is defined in BS 1339 (1965)[5] as 'the ratio of actual vapour pressure to the saturation vapour pressure over a plane liquid water surface at the same (dry bulb) temperature, expressed as a percentage'.

Further definitions are given in BS 1339 (1965)[5] and ASTM E41 (1986).[6]

4.3.2 Standard Procedures

ISO Standards

The ISO standard on conditioning most referred to is ISO 291 (1977).[7] This specifies four atmospheres as follows:

23 °C 50 per cent r.h. Pressure 86 to 106 kPa (recommended atmosphere)
27 °C 65 per cent r.h. Pressure 86 to 106 kPa (for tropical countries)

and, where humidity has no influence on the properties being examined, either of the above temperatures used with uncontrolled relative humidity.

Two levels of tolerance are allowed in the standard. 'Normal' requires ±2 °C on temperature and ±5 per cent on relative humidity, while 'close tolerance' requires ±1 °C on temperature and ±2 per cent on relative humidity. Periods of conditioning are as stated in the appropriate international standard, but where this is not so ISO 291 specifies a minimum conditioning period of 88 h for the atmospheres with controlled relative humidity and a minimum of 4 h for the others. The Annexe to ISO 291 points out that for certain materials these conditioning periods do not allow equilibrium to be reached and suggests that, in such cases, the test piece either be dried at high temperature or be conditioned at 23 °C, 50 per cent r.h. until equilibrium is reached. To achieve this it is recommended to condition to constant mass (within 0·1 per cent) for two determinations made at an interval of d^2 weeks where d is the thickness of the test piece (in mm). Both of these recommendations have disadvantages; drying out completely at high temperature often results in mechanical properties quite different from those obtained at moisture equilibrium, while conditioning to equilibrium at 23 °C 50 per cent r.h. can be time-consuming, extending to weeks or even to months.

These problems have been recognised in respect of polyamides 66, 610 and 6 by the publication of ISO 1110 (1975)[8] which offers a procedure for accelerated conditioning of these materials by heating test pieces between 7 h and 7 days at from 95 °C to 100 °C in aqueous potassium acetate solutions of specified concentrations depending on polymer type and thickness of test piece. The disadvantages of the method are: (a) that one has to know the type of polyamide being tested before the method can be applied with confidence; (b) there is no guarantee that mechanical properties obtained will be the same as those obtained by conditioning at 23 °C 50 per cent r.h. for a long period; and (c) electrical tests on accelerated conditioned test pieces may not be valid because of residual salt in the test pieces. The UK disapproved of this conditioning procedure and consequently it does not appear as a method in BS 2782.

ISO 291 makes reference to ISO 554 (1976)[9] which offers the choice of six standard atmospheres (Table 4.1). It must be noted that ISO 554 makes no

Table 4.1 ISO 554[9] *Standard Atmospheres for Conditioning*
Each condition has a permitted choice of two tolerances:
(a) Ordinary: ±2 °C for temperature and ±5 per cent r.h.
(b) Reduced: ±1 °C for temperature and ±2 per cent r.h.

Designation	**Temperature** (°C)	**Relative humidity** (%)	**Pressure** (kPa)	**Remarks**
23/50	23	50		Recommended atmosphere
27/65	27	65	between 86	For tropical countries
20/65	20	65	and 106	Used in certain fields of application

allowance for the use of uncontrolled humidity, yet ISO 291 does so; since ISO 554 is the general standard for conditioning for all ISO standards there would appear to be some inconsistency which needs attention by ISO.

British Standards
Until comparatively recently there has been no one standard conditioning procedure for plastics; individual material specifications had to be consulted for details of the levels of temperature and humidity to be used in each case. Even in the 1970 edition of BS 2782 each test method carried its own instructions for conditioning. This state of affairs has now been rectified by the publication of Part 0 of BS 2782 (1982)[10] which includes standard atmospheres for conditioning and testing identical to ISO 291.

ASTM Standards
ASTM E171[11] specifies the same conditioning atmospheres as ISO 554 and also gives a list of conditioning procedures for various materials.

The ASTM standard specific for the conditioning of plastics is ASTM D618 (1981)[12] which goes into the subject very thoroughly. The following definitions are given:

Standard Laboratory Temperature. A temperature of 23 °C with a normal tolerance ±2 °C and a closer tolerance ±1 °C if required.

Standard Laboratory Atmosphere. An atmosphere having a temperature of 23 °C (tolerances as above) and a relative humidity of 50 per cent, the latter with a normal tolerance of ±5 per cent and a closer tolerance of ±2 per cent when required.

Room Temperature. A temperature in the range 20°–30 °C.

The standard lays down a useful shorthand for identifying the type of conditioning used. A first number indicates the number of hours of conditioning, a second the temperature and a third the relative humidity; the last mentioned may be replaced in certain circumstances by words indicating special treatments. Thus:

(1) 96/23/50	96 h at 23 °C and 50 per cent r.h.
(2) 48/50/water	48 h at 50 °C in water
(3) 48/50 + 96/23/50	48 h at 50 °C followed by 96 h at 23 °C and 50 per cent r.h.

For the actual test condition, the conditioning data as above are followed by a colon, then a capital T, a number indicating the test temperature, and after it another number indicating the relative humidity if controlled, e.g.

96/23/50: T–35/90	Conditions as (1) above, and test at 35 °C and 90 per cent r.h.

ASTM D618 lays down minimum conditioning times according to the thickness of test specimens and for the most commonly used atmosphere (t/23/50), t is 40 h for specimens 7 mm or under, and 88 h for specimens above 7 mm. This is known as Procedure A, but there are five other alternatives to meet the needs of rapidity or to simulate special atmospheres or use environments. Corresponding instructions for test conditions are given.

As in British Standard practice, there is no ASTM specification for accelerated conditioning of polyamides. ASTM D789 (1981)[13] for nylon moulding materials requires the test for flexural modulus of elasticity (the only mechanical property specified) to be determined on test pieces moulded and immediately stored in watertight containers prior to the test which is carried out at 23 °C $\pm$ 1 °C and 50 $\pm$ 2 per cent r.h.

German Standards

The most important DIN standard on conditioning, to which most German test methods for plastics refer, is DIN 50014 (1985).[14] This specifies a conditioning atmosphere of 23 °C $\pm$ 2 °C, 50 per cent r.h. at a pressure of from 800 to 1060 mbar.

DIN 50005 (1975)[15] lists a selection of climates for conditioning and testing of plastics and other electrical insulating materials.

DIN 53714 (1978)[16] offers a procedure for the accelerated conditioning of polyamide test pieces. This procedure is more precise than that of ISO 1110 in that it requires the test pieces to be brought to a specified water content by treatment in water-saturated air in a closed vessel for a few days at a temperature not exceeding 40 °C, followed by a further 4 weeks in saturated air sealed in polythene bags and, finally, by a further 4 weeks in a standard atmosphere of 23 °C and 50 per cent r.h.

4.4 TESTING CONDITIONS

The object of conditioning is to bring the test piece as nearly as possible into equilibrium with a standard atmosphere so as to eliminate one source of variation in test results either within one test using replicate test pieces or between test laboratories. It is therefore desirable that the test is carried out in a test atmosphere identical to that used for conditioning. ISO 291 (1977)[7] states that this should be the case unless otherwise specified but goes on to say that in all cases the test shall be made immediately after the removal of test specimens from the conditioning enclosure. The sister standard for rubber, ISO 471 (1983)[17] allows for testing in a less rigorous atmosphere than that used for conditioning where such procedure does not affect the result. This recognises the fact that it is more difficult, and costly, to maintain a test room at constant temperature and (particularly) humidity than a small enclosure. No such concession is made in ISO 291 but it is common practice in the plastics testing laboratory only to control temperature, carrying out the test immediately after removal from the conditioning cabinet. This is sound practice for most plastics materials but it is not satisfactory to follow this procedure if the test piece has been conditioned at elevated or subnormal temperatures unless the test piece is very bulky and the test can be made extremely rapidly.

British Standard[10] and ASTM[12] practices follow that of ISO.

The majority of testing is carried out in one of the normal standard atmospheres but ISO 3205 (1976)[18] and ASTM D618 (1981)[12] list peferred subnormal and elevated temperatures. The British Standard[10] practice follows that of ISO 3205. It will be noted from Table 4.2 that ISO 3205 and ASTM D618 do not entirely agree either in individual temperatures or in tolerances. These standards do not specify precisely the conditioning periods to be used with the subnormal and elevated temperatures. ASTM D618 merely states that test pieces should be maintained at the test temperature for a time not less than that required to ensure thermal equilibrium and not more than 5 h prior to testing, but no advice is given on how the time to reach equilibrium can be estimated. Tables of approximate times required to reach equilibrium in both air and liquid media have been given[19] for a wide range of temperatures and various test piece geometries, and these are produced in full as an Appendix at the end of this chapter.

For further information on non-ambient temperature testing the reader is referred to §16.9.

4.5 APPARATUS FOR CONDITIONING

It is not feasible, in a small section of this book, to describe the construction of conditioning chambers and design of temperature- and humidity-controlled test rooms; this is a science in itself which, particularly in the case of the test room or laboratory, is best left to the expert. However, considering the lack of any reference to apparatus in standards such as ISO 291, it is worth discussing briefly certain aspects of the subject.

4.5.1 Air-Conditioned Rooms

It is difficult to run a plastics physical testing laboratory without air conditioning of the room in terms of temperature. Nevertheless large numbers of laboratories do not have this facility, presumably on the grounds of cost. However, if the room is reasonably well isolated by doors, does not have excessive window area and is of moderate size, the cost of installing self-contained heating and cooling units is

Table 4.2 Preferred subnormal and elevated temperatures according to ISO 3205[18] and ASTM D618[12]

ISO 3205 Preferred test temperature[a] (°C)		ASTM D618 Preferred test temperature (°C)	
−269	Tolerance	−70	
−196	depends on	−55	
−161	equipment	−40	±2
−70		−25	
−55		0	
−40	±3	35	±1
−25		50	
−10		70	
0		90	
5		105	
25		120	±2
40		130	
55	±2	155	
70		180	
85		200	
100		225	
105		250	±3
125		275	
150	±3	300	
175		325	±4
200		350	±5
225		400	±6
250		450	±8
275	±5	500	±10
300		600	±12
350			
400			
450			
500	±10		
600			
700			
800			
900	±15		
1000			

[a] ISO 3205 also has an Annex which lists temperatures used in particular fields, culled from documents prepared by various ISO Technical Committees. Thirty-two temperatures are quoted in the Annex, ranging from 165 °C to 950 °C.

surprisingly low. Complete air conditioning of both temperature and humidity is inevitably much more expensive and is not necessary unless the majority of physical testing is to be carried out on moisture-sensitive materials such as cellulose acetate and polyamides. Even then, use of suitable enclosures (§4.5.2) can be quite adequate and give satisfactory results.

As mentioned earlier it is necessary to consult an expert when considering the installation of air conditioning but it should be noted that there is a British Standard covering the design of controlled atmosphere laboratories.[20]

4.5.2 Enclosures

There has been a tendency to treat conditioning chambers in the 'do-it-yourself' manner – a suitable box, a dish of salt solution, a circulation fan, a low wattage heater with a controller, a thermometer and some form of dial hygrometer all slung together. This is not to decry salt solutions as a means of controlling humidity; they can provide a cheap and accurate method if used properly and particularly if the salt solution has a large surface area. The often overlooked factors are in the design of the chamber: lack of thermal insulation, doors which when opened immediately cause an almost complete interchange of atmosphere with that of the surrounding room, shelves and corners which prevent adequate air circulation, fans which do not move the air sufficiently, heaters which radiate to test pieces, and hygrometers which do not indicate relative humidity correctly.

Two types of humidity cabinet are in common use, salt-tray cabinets and moisture-injection types, which are covered by BS 3718 (1984)[21] and BS 3898 (1984),[22] respectively. The salt-tray cabinet is by far the simpler of the two types, being essentially a temperature-controlled enclosure in which the humidity is controlled by the use of saturated salt solutions. Despite their simplicity, much care is required in their design and operation if accurate conditions are to be realised and much useful detail is to be found in BS 3718.

ISO R483 (1966),[23] which refers specifically to plastics, is similar to BS 3718 but additionally describes the use of glycerol solutions to replace saturated salt solutions. ASTM E104 (1985)[24] covers the same ground as the other two standards and also describes the use of aqueous sulphuric acid solutions as well as glycerol and salt solutions. It is, however, of less use than BS 2718 and ISO R483 because the 'usual' plastics conditioning temperature of 23 °C is not included. At the condition of most interest, i.e. 23 °C, 50 per cent r.h., sodium dichromate is just within the tolerance of ± 5 per cent r.h. but no salt is listed which will give 50 per cent r.h. ISO R483 claims that a glycerol solution will achieve this tolerance if its refractive index is maintained at $1{\cdot}444 \pm 0{\cdot}002$.

The injection type of humidity cabinet uses a humidity sensing device to control the injection of moisture into the cabinet from a reservoir. The humidity sensing device is often a wet-and-dry bulb hygrometer with mercury contact thermometers. Humidity levels are more readily changed with this type of equipment and some have the facility to cycle both humidity and temperature, so extending the range of tests which can be carried out. BS 4864 (1986)[25] describes enclosures with a greater operating range than those of BS 3898 but this type of apparatus is more pertinent to accelerated ageing than to conditioning as defined in this chapter.

4.5.3 Thermometers

The ordinary mercury-in-glass thermometer as covered by BS 593 (1981)[26] is in such common use that it is taken for granted. In practice, variability of test results obtained at a set temperature can often be traced to the misuse of thermometers, which should be calibrated frequently, inspected for separation of mercury and immersed to the correct depth. The worst errors occur with low-temperature thermometers (not covered by BS 593) and particular care needs to be taken when conditioning or testing at sub-zero temperatures. Other standards covering thermometers are ISO 653 (1980)[27] and ISO 654 (1980)[28] and the British equivalent BS 5074 (1985)[29] which incorporates these two ISO standards. BS 1704 (1985)[30] for general purpose thermometers is the same as ISO 1770 (1981).[31]

Apart from the ordinary liquid-in-glass thermometer, there are many other types of temperature-measuring instrument and these require just as much care in use. Useful descriptions of the various types may be found in *The Instrument Manual*[32] and a guide to the selection and use of thermometers may be found in BS 1041.[33–36]

4.5.4 Hygrometers

The most useful standard hygrometer is the wet-and-dry bulb type which should be used in conditions where air is circulating around the instrument at a velocity not less than 3 m/s. Hygrometric tables for use with wet-and-dry bulb hygrometers are published in BS 4833 (1986)[37] which also contains a bibliography.

In recent years a large number of different types of hygrometer have become available commercially and many of these are described in *The Instrument Manual.*[32] Simple 'mechanical' types of hygrometer, such as those which depend on the change of dimensions of hair or paper, can be useful; they are small in size and relatively inexpensive but they require frequent calibration and attention if their accuracy is to be maintained.

4.5.5 Apparatus for Elevated and Subnormal Temperatures

Generally, conditioning at elevated or subnormal temperatures indicates that the test will be carried out at that same temperature and hence in the same enclosure as is used for conditioning. Under these circumstances the conditioning enclosure usually forms part of the testing apparatus and is therefore likely to take one of many forms, depending on the test in question. Comments on the types of enclosure available are given by Brown[38] and the requirements for particular tests will be discussed in the relevant sections of this book.

4.6 MECHANICAL CONDITIONING

Brown[39] has drawn attention to the need for mechanical conditioning when performing certain dynamic mechanical tests on rubbers. This is because it has been shown that the stress–strain curves of vulcanised rubbers containing filler are permanently damaged when they are deformed.

There does not appear to be a need for the mechanical conditioning of plastics, judging by the lack of any published standards on this subject. However, in any mechanical test involving repeated stressing of the test piece, the test results should be observed carefully to see whether mechanical conditioning is necessary to achieve meaningful results from the material under test.

APPENDIX A: THERMAL EQUILIBRIUM IN CONDITIONING

If a product is used at non-ambient temperatures, it is logical that it should be tested under the same service conditions. However, elevated or subnormal temperature testing inevitably requires more complex, and hence more expensive apparatus and normally takes a considerably longer time to carry out. In this respect the conditioning period is a crucial consideration and is often the limiting factor to efficient testing. This is of course of particular importance in quality control testing where both time and money may be of paramount importance.

If the conditioning time used is too short the effect of temperature change on the material will be underrated, and furthermore the variability of the test results is liable to increase. Alternatively, if excessive periods of conditioning are allowed, not only is the test made more expensive but it may become more difficult to distinguish time-dependent effects on the material. Generally, technologists have used a rule-of-thumb method to arrive at conditioning times, frequently biased by financial considerations. Their difficulty has been the almost complete lack of available information and the complication of working with a multitude of different test piece geometries. Relatively few standards are available in which conditioning times are specified and the derivations of the figures quoted are never given. For example, BS 903 : Part A2 *Tensile Stress Strain Properties of Rubber* states: 'If the test is to be carried out at a temperature other than 20 °C the test piece shall be conditioned at the test temperature obtained immediately prior to testing for a period sufficient to reach substantial temperature equilibrium.'

The data required are the times for the centre of the test piece to come within the tolerance allowed at the temperature in question, calculated for a variety of materials, each tested in liquid and gaseous media over a range of temperature. Furthermore the figures will be required for all the test piece geometries in normal use. The absence of these figures or, in some cases, substitution of inspired guesses at the true values, has been the result of both lack of basic thermal data and the difficulty of deriving the required information. Thermal property data are certainly still rather sparse but considerable advance has been made in the technique of calculating temperature distributions during heating or cooling.

In this Appendix the methods described by Hands[40] have been extended and applied to standard test pieces for rubbers and plastics subjected to step temperature changes in conditioning chambers and baths (Tables 4.A1 to 4.A3).

Test Pieces and Conditions

To make individual calculations for every test piece in current use would clearly be impractical. Fortunately, nearly all test pieces used for rubber and plastics fall into three basic geometries. These are cylinders, characterised by diameter and length, flat sheets, characterised by thickness, and flat strips. A test piece may be considered to approximate to an infinite sheet when both its length and width are greater than four times its thickness, or to an infinite strip when its length only is greater than four times its thickness. The principal form of test piece which does not obviously fall into one of these three geometries is the dumb-bell used in tensile testing. However, for our purpose dumb-bells may be considered as flat strips. For most purposes it is the central parallel part of a dumb-bell which is of importance and this width may be taken as the strip width. If the tab ends of the dumb-bell are to be considered as well then the overall width is used.

Generally, when a measurement at a non-ambient temperature is to be made, the test piece, at room temperature, is introduced into a cabinet pre-set at the test temperature. It must be expected that the introduction of a test piece will alter the temperature of the cabinet to some extent but in theory at least the test piece is subjected to a step change of temperature. If time is required for the cabinet to regain its set temperature then this must be considered additional to the times given in the tables. The equilibrium times given have been calculated for step changes from 20 °C, and normal variations of room temperature around 20 °C may be neglected. The range from −50 °C to +250 °C was chosen to cover the test temperatures most commonly specified.

Tolerances allowed on temperatures of test are commonly ±1 °C or ±2 °C. Consequently the equilibrium times calculated are those to reach within 1 °C of the set temperature, which should be adequate for all normal circumstances.

The majority of temperature-controlled cabinets used for testing use either air or a liquid as the heat transfer medium, although occasionally test pieces may be in contact with metal platens. Separate figures are quoted in the tables for both air and liquid media. Although a variety of liquids are used, their heat transfer properties have been considered to be very similar and are simply quoted as oil in the tables.

Although it is essential that the test piece is given sufficient time to reach equilibrium, in practice the conditioning time is not critical to the nearest minute and consequently all times in the tables have been rounded up to the next highest multiple of five minutes. For intermediate temperature changes the time for the next highest temperature should be used.

Cylinders (Table 4.A1, p. 53–4)
Cylindrical test pieces are frequently used in compression stress–strain and compression set tests and also for some resilience and abrasion tests. In the table cylinders have been characterised by diameter and length, specific sizes having been chosen to correspond with currently used standard test pieces. For example the 28·7 mm × 12·7 mm cylinder is that specified in BS 903 Part A4.

Flat Sheets (Table 4.A2, p. 55–6)
Flat sheets refer to test pieces where both lengths and width are large compared with thickness. This covers, for example, specimens for falling weight impact strength measurements and sheets used for hardness and indentation tests. Because the thickness used is often not closely specified, results are given for a range of thicknesses up to 25 mm; for any intermediate thickness the time given for the next thickest sheet in the tables should be used.

Flat Strips (Table 4.A3, p. 57–62)
Table 4.A3 is the most important table because the majority of specimens for mechanical tests are of this form. There are, for example, the flexural tests and many impact tests and, the largest group of all, the tensile tests. Dumb-bells, although shaped, are essentially of strip form and may be treated for most purposes as strips of the same width as the central parallel portion. For example, the

dumb-bell specified in BS 2782 Method 320B would be taken as a 10 mm wide strip. Also, ring test pieces may be considered as strips with dimensions equal to the cross-sections of the ring. Results are given for a range of widths up to 25 mm. Above this the strip could be treated as a flat sheet. At each width a range of thicknesses are tabulated, which includes some specific to a particular test. Again, for any intermediate dimensions the time for the next largest strip should be used. It should also be noted that the terms 'width' and 'thickness' are completely interchangeable.

Assumed Thermophysical Properties

	Crystalline plastic	*Amorphous plastic*	*Rubber*
Thermal conductivity			
cal/cm s °C	8×10^{-4}	4×10^{-4}	5×10^{-4}
(J/cm s °C)	($33{\cdot}48 \times 10^{-4}$)	($16{\cdot}74 \times 10^{-4}$)	($20{\cdot}92 \times 10^{-4}$)
Thermal diffusivity			
cm^2/s	$1{\cdot}2 \times 10^{-3}$	$0{\cdot}9 \times 10^{-3}$	$1{\cdot}0 \times 10^{-3}$

Assumed values for surface heat transfer coefficient

in air: 5×10^{-4} cal/cm^2 s °C ($20{\cdot}92 \times 10^{-4}$ J/cm^2 s °C)

in oil: $1{\cdot}8 \times 10^{-2}$ cal/cm^2 s °C ($7{\cdot}53 \times 10^{-2}$ J/cm^2 s °C)

Table 4.A1 Cylinders

Diameter (mm)	Height (mm)	Temperature (°C)	Time (min) to 1 °C off equilibrium: Crystalline plastic		Amorphous plastic		Rubber	
			Air	Oil	Air	Oil	Air	Oil
64	38	−50	135	60	130	80	130	75
		0	100	45	95	65	95	60
		50	115	50	105	70	105	65
		100	140	60	130	85	130	80
		150	155	65	145	90	145	85
		200	165	70	155	95	155	90
		250	170	75	160	100	160	90
40	30	−50	85	30	75	40	75	35
		0	60	25	55	35	55	30
		50	70	25	60	35	60	30
		100	85	30	75	45	75	35
		150	95	35	85	45	85	40
		200	100	35	90	50	90	45
		250	105	40	90	50	95	45
37	10·2	−50	40	10	35	10	35	10
		0	30	10	25	10	25	10
		50	35	10	25	10	30	10
		100	40	10	35	10	35	10
		150	45	10	35	10	40	10
		200	50	10	40	15	40	10
		250	50	10	40	15	45	15
32	16·5	−50	50	15	45	20	45	15
		0	40	10	30	15	35	15
		50	45	15	35	15	35	15
		100	55	15	45	20	45	20
		150	60	15	50	20	50	20
		200	65	20	50	20	55	20
		250	65	20	55	25	55	20
29	25	−50	60	20	50	25	50	20
		0	45	15	40	20	40	15
		50	50	15	40	20	45	20
		100	60	20	50	25	55	25
		150	70	20	55	25	60	25
		200	70	25	60	30	65	25
		250	75	25	65	30	65	25
28·7	12·7	−50	40	10	35	15	35	10
		0	30	10	25	10	25	10
		50	35	10	30	10	30	10
		100	45	10	35	15	35	15
		150	50	10	40	15	40	15
		200	50	15	40	15	45	15
		250	55	15	40	15	45	15

Table 4.A1 Cylinders (*cont.*)

Diameter (mm)	Height (mm)	Temperature (°C)	Time (min) to 1 °C off equilibrium Crystalline plastic Air	Oil	Amorphous plastic Air	Oil	Rubber Air	Oil
25	20	−50	50	15	40	20	40	15
		0	35	10	30	15	30	15
		50	40	10	35	15	35	15
		100	50	15	40	20	45	15
		150	55	15	45	20	45	20
		200	50	15	50	20	50	20
		250	60	15	50	20	50	20
25	8	−50	30	5	25	10	25	5
		0	20	5	20	5	20	5
		50	25	5	20	5	20	5
		100	30	5	25	10	25	5
		150	35	5	25	10	30	10
		200	35	10	30	10	30	10
		250	35	10	30	10	30	10
25	6·3	−50	25	5	20	5	20	5
		0	20	5	15	5	15	5
		50	20	5	15	5	20	5
		100	25	5	20	5	20	5
		150	30	5	20	5	25	5
		200	30	5	25	5	25	5
		250	30	5	25	5	25	5
13	12·6	−50	25	5	20	5	20	5
		0	20	5	15	5	15	5
		50	20	5	15	5	20	5
		100	30	5	20	10	20	5
		150	30	5	25	10	25	10
		200	30	5	25	10	25	10
		250	35	5	25	10	25	10
13	6·3	−50	20	5	15	5	15	5
		0	15	5	10	5	10	5
		50	15	5	15	5	15	5
		100	20	5	15	5	15	5
		150	20	5	15	5	20	5
		200	25	5	20	5	20	5
		250	25	5	20	5	20	5
9·5	9·5	−50	5	5	15	5	15	5
		0	5	5	10	5	10	5
		50	5	5	10	5	15	5
		100	5	5	15	5	15	5
		150	5	5	15	5	20	5
		200	5	5	15	5	20	5
		250	5	5	20	5	20	5

Table 4.A2 Flat sheets

Thickness (mm)	Temperature (°C)	Time (min) to 1 °C off equilibrium: Crystalline plastic		Amorphous plastic		Rubber	
		Air	Oil	Air	Oil	Air	Oil
25	−50	115	80	145	100	135	90
	0	80	65	105	85	95	75
	50	90	70	120	90	110	80
	100	115	80	150	100	140	90
	150	130	85	165	105	155	95
	200	135	85	180	110	160	100
	250	140	90	185	115	170	105
15	−50	60	30	80	40	70	35
	0	40	25	55	30	50	30
	50	45	30	65	35	60	30
	100	60	30	80	40	75	35
	150	65	35	90	40	80	40
	200	70	35	95	40	85	40
	250	75	35	100	45	90	40
10	−50	35	15	50	20	45	15
	0	25	15	35	15	30	15
	50	30	15	40	15	35	15
	100	40	15	50	20	45	20
	150	40	15	55	20	50	20
	200	45	15	60	20	55	20
	250	45	20	60	20	55	20
8	−50	30	10	40	15	35	10
	0	20	10	30	10	25	10
	50	25	10	30	10	30	10
	100	30	10	40	15	35	10
	150	35	10	45	15	40	10
	200	35	10	45	15	40	15
	250	35	15	50	15	45	15
5	−50	20	5	25	5	20	5
	0	15	5	20	5	15	5
	50	15	5	20	5	20	5
	100	20	5	25	5	20	5
	150	20	5	25	5	25	5
	200	20	5	30	5	25	5
	250	20	5	30	10	25	5
3	−50	10	5	15	5	15	5
	0	10	5	10	5	10	5
	50	10	5	15	5	10	5
	100	10	5	15	5	15	5
	150	15	5	15	5	15	5
	200	15	5	20	5	15	5
	250	15	5	20	5	15	5

Table 4.A2 Flat sheets (*cont.*)

		Time (min) to 1 °C off equilibrium					
		Crystalline plastic		Amorphous plastic		Rubber	
Thickness (mm)	Temperature (°C)	Air	Oil	Air	Oil	Air	Oil
2	−50	10	5	10	5	10	5
	0	5	5	10	5	10	5
	50	5	5	10	5	10	5
	100	10	5	10	5	10	5
	150	10	5	10	5	10	5
	200	10	5	15	5	10	5
	250	10	5	15	5	10	5
1	−50	5	5	5	5	5	5
	0	5	5	5	5	5	5
	50	5	5	5	5	5	5
	100	5	5	5	5	5	5
	150	5	5	5	5	5	5
	200	5	5	5	5	5	5
	250	5	5	10	5	5	5
0·2	−50	5	5	5	5	5	5
	0	5	5	5	5	5	5
	50	5	5	5	5	5	5
	100	5	5	5	5	5	5
	150	5	5	5	5	5	5
	200	5	5	5	5	5	5
	250	5	5	5	5	5	5

Table 4.A3 Flat strips

Width (mm)	Thickness (mm)	Temperature (°C)	Time (min) to 1 °C off equilibrium: Crystalline plastic Air	Crystalline plastic Oil	Amorphous plastic Air	Amorphous plastic Oil	Rubber Air	Rubber Oil
25·4	12·7	−50	50	10	40	15	45	15
		0	35	10	30	10	30	10
		50	40	10	35	15	35	10
		100	55	10	40	15	45	15
		150	60	15	45	15	50	15
		200	60	15	50	15	50	15
		250	65	15	50	20	55	15
	10	−50	45	10	35	10	35	10
		0	30	5	25	10	25	10
		50	35	10	30	10	30	10
		100	45	10	35	10	35	10
		150	50	10	40	10	40	10
		200	50	10	40	15	40	10
		250	55	10	40	15	45	10
	9·5	−50	40	10	30	10	35	10
		0	30	5	25	10	25	10
		50	35	10	25	10	30	10
		100	40	10	35	10	35	10
		150	45	10	35	10	40	10
		200	50	10	40	10	40	10
		250	50	10	40	10	40	10
	6·5	−50	30	5	25	5	25	5
		0	20	5	15	5	20	5
		50	25	5	20	5	20	5
		100	30	5	25	5	25	5
		150	35	5	25	5	30	5
		200	35	5	25	5	30	5
		250	40	5	30	10	30	5
	5·0	−50	25	5	20	5	20	5
		0	20	5	15	5	15	5
		50	20	5	15	5	15	5
		100	25	5	20	5	20	5
		150	30	5	20	5	20	5
		200	30	5	20	5	25	5
		250	30	5	25	5	25	5
	3·0	−50	15	5	10	5	15	5
		0	10	5	10	5	10	5
		50	15	5	10	5	10	5
		100	15	5	10	5	15	5
		150	20	5	15	5	15	5
		200	20	5	15	5	15	5
		250	20	5	15	5	15	5

Table 4.A3 Flat strips (*cont.*)

Width (mm)	Thickness (mm)	Temperature (°C)	Time (min) to 1 °C off equilibrium: Crystalline plastic Air	Crystalline plastic Oil	Amorphous plastic Air	Amorphous plastic Oil	Rubber Air	Rubber Oil
25·4	2·0	−50	10	5	10	5	10	5
		0	10	5	5	5	10	5
		50	10	5	10	5	10	5
		100	10	5	10	5	10	5
		150	15	5	10	5	10	5
		200	15	5	10	5	10	5
		250	15	5	10	5	10	5
	1·0	−50	5	5	5	5	5	5
		0	5	5	5	5	5	5
		50	5	5	5	5	5	5
		100	5	5	5	5	5	5
		150	10	5	5	5	5	5
		200	10	5	5	5	5	5
		250	10	5	5	5	5	5
15·0	15·0	−50	45	10	35	15	35	10
		0	35	10	25	10	30	10
		50	35	10	30	10	30	10
		100	45	10	35	15	40	10
		150	50	10	40	15	40	15
		200	55	15	40	15	45	15
		250	55	15	45	15	45	15
12·7	12·7	−50	35	10	30	10	30	10
		0	25	5	20	10	25	10
		50	30	10	25	10	25	10
		100	40	10	30	10	30	10
		150	40	10	35	10	35	10
		200	45	10	35	10	35	10
		250	45	10	35	10	40	10
	10·0	−50	35	5	25	10	25	10
		0	25	5	20	5	20	5
		50	25	5	20	5	20	5
		100	35	5	25	10	30	10
		150	35	10	30	10	30	10
		200	40	10	30	10	30	10
		250	40	10	30	10	35	10
	9·5	−50	30	5	25	10	25	10
		0	25	5	20	5	20	5
		50	25	5	20	5	20	5
		100	35	5	25	10	25	10
		150	35	10	30	10	30	10
		200	40	10	30	10	30	10
		250	40	10	30	10	35	10

continued

Table 4.A3 Flat strips (*cont.*)

			Time (min) to 1 °C off equilibrium					
			Crystalline plastic		Amorphous plastic		Rubber	
Width (mm)	Thickness (mm)	Temperature (°C)	Air	Oil	Air	Oil	Air	Oil
12·7	6·5	−50	25	5	20	5	20	5
		0	20	5	15	5	15	5
		50	20	5	15	5	15	5
		100	25	5	20	5	20	5
		150	30	5	20	5	25	5
		200	30	5	25	5	25	5
		250	30	5	25	5	25	5
	5·0	−50	20	5	15	5	15	5
		0	15	5	10	5	15	5
		50	15	5	15	5	15	5
		100	20	5	15	5	20	5
		150	25	5	20	5	20	5
		200	25	5	20	5	20	5
		250	25	5	20	5	20	5
	3·2	−50	15	5	10	5	15	5
		0	10	5	10	5	10	5
		50	15	5	10	5	10	5
		100	15	5	10	5	15	5
		150	15	5	15	5	15	5
		200	20	5	15	5	15	5
		250	20	5	15	5	15	5
	3·0	−50	15	5	10	5	10	5
		0	10	5	10	5	10	5
		50	10	5	10	5	10	5
		100	15	5	10	5	10	5
		150	15	5	10	5	15	5
		200	15	5	15	5	15	5
		250	20	5	15	5	15	5
	2·0	−50	10	5	10	5	10	5
		0	10	5	5	5	5	5
		50	10	5	5	5	10	5
		100	10	5	10	5	10	5
		150	10	5	10	5	10	5
		200	15	5	10	5	10	5
		250	15	5	10	5	10	5
	1·0	−50	5	5	5	5	5	5
		0	5	5	5	5	5	5
		50	5	5	5	5	5	5
		100	5	5	5	5	5	5
		150	10	5	5	5	5	5
		200	10	5	5	5	5	5
		250	10	5	5	5	5	5

Table 4.A3 Flat strips (*cont.*)

Width (mm)	Thickness (mm)	Temperature (°C)	Time (min) to 1 °C off equilibrium Crystalline plastic Air	Oil	Amorphous plastic Air	Oil	Rubber Air	Oil
6·35	12·7	−50	25	5	20	5	20	5
		0	20	5	15	5	15	5
		50	20	5	15	5	15	5
		100	25	5	20	5	20	5
		150	30	5	20	5	25	5
		200	30	5	20	5	25	5
		250	30	5	25	5	25	5
	10·0	−50	25	5	15	5	20	5
		0	15	5	15	5	15	5
		50	20	5	15	5	15	5
		100	25	5	20	5	20	5
		150	25	5	20	5	20	5
		200	25	5	20	5	20	5
		250	30	5	20	5	25	5
	6·5	−50	20	5	15	5	15	5
		0	15	5	10	5	10	5
		50	15	5	10	5	15	5
		100	20	5	15	5	15	5
		150	20	5	15	5	15	5
		200	25	5	15	5	20	5
		250	25	5	20	5	20	5
	5·0	−50	15	5	15	5	15	5
		0	15	5	10	5	10	5
		50	15	5	10	5	10	5
		100	15	5	15	5	15	5
		150	20	5	15	5	15	5
		200	20	5	15	5	15	5
		250	20	5	15	5	15	5
	3·0	−50	15	5	10	5	10	5
		0	10	5	10	5	10	5
		50	10	5	10	5	10	5
		100	15	5	10	5	10	5
		150	15	5	10	5	10	5
		200	15	5	10	5	10	5
		250	15	5	10	5	10	5
	2·0	−50	10	5	10	5	10	5
		0	10	5	5	5	5	5
		50	10	5	5	5	5	5
		100	10	5	10	5	10	5
		150	10	5	10	5	10	5
		200	10	5	10	5	10	5
		250	10	5	10	5	10	5

continued

Table 4.A3 Flat strips (*cont.*)

			Time (min) to 1 °C off equilibrium					
			Crystalline plastic		Amorphous plastic		Rubber	
Width (mm)	Thickness (mm)	Temperature (°C)	Air	Oil	Air	Oil	Air	Oil
6·35	1·52	−50	10	5	5	5	5	5
		0	5	5	5	5	5	5
		50	5	5	5	5	5	5
		100	10	5	5	5	5	5
		150	10	5	5	5	10	5
		200	10	5	10	5	10	5
		250	10	5	10	5	10	5
	1·0	−50	5	5	5	5	5	5
		0	5	5	5	5	5	5
		50	5	5	5	5	5	5
		100	5	5	5	5	5	5
		150	5	5	5	5	5	5
		200	10	5	5	5	5	5
		250	10	5	5	5	5	5
4·0	12·7	−50	20	5	15	5	15	5
		0	15	5	10	5	10	5
		50	15	5	10	5	10	5
		100	20	5	15	5	15	5
		150	20	5	15	5	15	5
		200	20	5	15	5	15	5
		250	20	5	15	5	20	5
	10·0	−50	15	5	15	5	15	5
		0	15	5	10	5	10	5
		50	15	5	10	5	10	5
		100	15	5	15	5	15	5
		150	20	5	15	5	15	5
		200	20	5	15	5	15	5
		250	20	5	15	5	15	5
	6·5	−50	15	5	10	5	10	5
		0	10	5	10	5	10	5
		50	10	5	10	5	10	5
		100	15	5	10	5	10	5
		150	15	5	15	5	15	5
		200	20	5	15	5	15	5
		250	20	5	15	5	15	5
	5·0	−50	15	5	10	5	10	5
		0	10	5	10	5	10	5
		50	10	5	10	5	10	5
		100	15	5	10	5	10	5
		150	15	5	10	5	10	5
		200	15	5	10	5	15	5
		250	15	5	10	5	15	5

Table 4.A3 Flat strips (*cont.*)

Width (mm)	Thickness (mm)	Temperature (°C)	Time (min) to 1 °C off equilibrium: Crystalline plastic Air	Crystalline plastic Oil	Amorphous plastic Air	Amorphous plastic Oil	Rubber Air	Rubber Oil
4·0	3·0	−50	10	5	10	5	10	5
		0	10	5	5	5	5	5
		50	10	5	5	5	10	5
		100	10	5	10	5	10	5
		150	10	5	10	5	10	5
		200	15	5	10	5	10	5
		250	15	5	10	5	10	5
	2·0	−50	10	5	5	5	5	5
		0	5	5	5	5	5	5
		50	10	5	5	5	5	5
		100	10	5	5	5	10	5
		150	10	5	10	5	10	5
		200	10	5	10	5	10	5
		250	10	5	10	5	10	5
	1·0	−50	5	5	5	5	5	5
		0	5	5	5	5	5	5
		50	5	5	5	5	5	5
		100	5	5	5	5	5	5
		150	5	5	5	5	5	5
		200	5	5	5	5	5	5
		250	5	5	5	5	5	5

REFERENCES

1. Walter, A. T. (1954) *Journal of Polymer Science*, **13**, 217.
2. Heap, R. D. and Norman, R. N. (1966) *RAPRA Research Report*, no. 149.
3. BS 771 (1980) *Phenolic Moulding Materials.*
4. BS 1763 (1975) *Flexible, Unsupported, Thin PVC Sheeting.*
5. BS 1339 (1981) *Definitions, Formulae and Constants Relating to the Humidity of the Air.*
6. ASTM E41 (1986) *Standard Definition of Terms Relating to Conditioning.*
7. ISO 291 (1977) *Plastics – Standard Atmospheres for Conditioning and Testing.*
8. ISO 1110 (1975) *Polyamide 66, 610 and 6 – Accelerated Conditioning of Test Specimens.*
9. ISO 554 (1976) *Standard Atmospheres for Conditioning and/or Testing – Specifications.*
10. BS 2782: Part 0 (1982) *Methods of Testing Plastics – Introduction.*
11. ASTM E171 (1982) *Standard Atmospheres for Conditioning and Testing Materials.*
12. ASTM D618 (1981) *Conditioning Plastics and Electrical Insulating Materials for Testing.*
13. ASTM D789 (1981) *Nylon Injection Molding and Extrusion Materials.*
14. DIN 50014 (1985) *Climates and their Technical Applications; Standard Atmospheres.*
15. DIN 50005 (1975) *Testing of Plastics and other Electrical Insulating Materials; Selection of Climates for Preconditioning, Conditioning and Testing of Specimens.*

16. DIN 53714 (1978) *Testing of Plastics; Accelerated Conditioning of Test Specimens of Polyamides.*
17. ISO 471 (1983) *Standard Atmospheres for the Conditioning and Testing of Rubber Test Pieces.*
18. ISO 3205 (1976) *Preferred Test Temperatures.*
19. Brown, R. P. and Hands, D. (1973) *RAPRA Members Journal*, April.
20. BS 4194 (1984) *Design Requirements and Testing of Controlled Atmosphere Laboratories.*
21. BS 3718 (1984) *Laboratory Humidity Ovens (Non-Injection Type).*
22. BS 3898 (1984) *Laboratory Humidity Ovens (Injection Type).*
23. ISO R483 (1966) *Plastics – Methods for Maintaining Constant Relative Humidity in Small Enclosures by Means of Aqueous Solutions.*
24. ASTM E104 (1985) *Maintaining Constant Relative Humidity by Means of Aqueous Solutions.*
25. BS 4864 (1986) *Design and Testing Enclosures for Environmental Testing.*
26. BS 593 (1981) *Laboratory Thermometers.*
27. ISO 653 (1980) *Long Solid Stem Thermometers for Precision Use.*
28. ISO 654 (1980) *Short Solid Stem Thermometers for Precision Use.*
29. BS 5704 (1985) *Short and Long Solid Stem Thermometers for Precision Use.*
30. BS 1704 (1985) *Specification for Solid Stem General Purpose Thermometers.*
31. ISO 1770 (1981) *Specifications for Solid Stem General Purpose Thermometers.*
32. Miller, T. J. (Ed.) (1975) *The Instrument Manual*, United Trade Press.
33. BS 1041 Part 2.1 (1985) *Guide to the Selection and use of Liquid in Glass Thermometers.*
34. BS 1041 : Part 3 (1969) *Industrial Resistance Thermometry.*
35. BS 1041 : Part 4 (1966) *Thermocouples.*
36. BS 1041 : Part 5 (1972) *Radiation Pyrometers.*
37. BS 4833 (1986) *Schedules of Hygrometric Tables for Use in the Testing and Operation of Environmental Enclosures.*
38. Brown, R. P. (1973) *Rapra Guide to Rubber and Plastics Test Equipment*, RAPRA.
39. Brown, R. P. (1986) *The Physical Testing of Rubbers*, Elsevier Applied Science, p. 70.
40. Hands, D. (1971) *RAPRA Technical Review*, no. 60, July.

Chapter Five

Processibility Tests

5.1 INTRODUCTION

The majority of the properties of plastics discussed in this book relate to finished articles. While these are important to users, there are a number of tests which are applied to plastics compositions in the unmoulded or pre-cast state. Certain of the properties of such 'raw materials' are of paramount importance to the processor and fabricator if he is to predict his machinery requirements rather than pursue an entirely *ad hoc* approach. Again, there are tests designed for application to polymers prior to their formulation into moulding compositions. This chapter considers tests which can be grouped under the general heading of 'processibility' and which range from the gel time of a polyester resin to a study of the rheological (flow) behaviour of a polymer in the molten state.

5.2 VISCOELASTIC FLOW BEHAVIOUR

Study of the flow behaviour of molten polymer is essential to the moulder or fabricator if he is to establish the feasibility of any intended process involving melting. If the test is ideal it will forecast a variety of properties such as thermal stability, ease of processing (is the machine powerful enough?) and quality of mouldings. The manufacturer of plastics compositions will likewise be interested in the subject in order to give his customers adequate technical service, for he can exercise a considerable degree of control over processibility by altering parameters such as average molecular weight and molecular weight distribution of polymer and by incorporating various types and quantities of plasticisers, lubricants and stabilisers. Efficient and reliable laboratory-scale test procedures for processibility are essential if experimental work is to be kept within reasonable bounds during the early stages of development of a new polymer or moulding compound. Ultimately, however, when a 'short list' has been prepared, there is still no real substitute for trying out the most promising compositions on a factory scale, using the actual processing equipment to be employed in manufacture.

Molten polymers are viscoelastic materials and hence the study of their flow behaviour is not without complication. In an ideal or Newtonian liquid, to which water at normal temperatures approaches very closely, there is a linear relationship between shear rate (rate of deformation) $\mathrm{d}u/\mathrm{d}r$ and shearing stress τ, such that

$$\tau = \eta \frac{du}{dr} \qquad [5.1]$$

where η is a constant known as the 'coefficient of viscosity', or simply the 'viscosity'; this constant, as is well known, is markedly dependent on temperature. Application of the concept to flow of a Newtonian liquid through a capillary of radius r and length l leads to the familiar Poiseuille formula:

$$\frac{Pr}{2l} = \eta \frac{4Q}{\pi r^3} \qquad [5.2]$$

where P is the pressure on the liquid and Q the volume rate of flow. It is unfortunate that this formula does not apply to a great number of liquid systems, including polymer melts, all of which are non-ideal. Some liquids change their flow behaviour with time, for example one may 'thicken up' after stirring, another 'thin down'. Even ignoring these complications and considering only time-stable phenomena, there are still a number of possibilities, some of which are illustrated in Fig. 5.1.

Many polymer melts follow the 'pseudoplastic' pattern and hence the relationship between shear rate and shear stress is not linear. It follows that, since modern processing techniques for thermoplastics are associated with shear rates between about 100 s^{-1} and 500 000 s^{-1}, there is really no substitute for a comprehensive study of shear stress against shear rate over a range of the latter embracing all likely processing techniques throughout a suitable temperature range. The majority of standard flow tests employ only one (low) shear rate, either fixed by the machine used or dictated in part by the flow behaviour of the material under examination. Clearly, such a one-point observation cannot yield much information for general processibility and, although the test may be useful for quality control measures, it can be misleading even in this context since the precise shapes of all pseudoplastic curves are not identical (Fig. 5.2).

Detailed consideration of the flow behaviour of molten polymers is a science in itself: Bernhardt,[1] and Pearson[2] for instance, study the subject in relation to

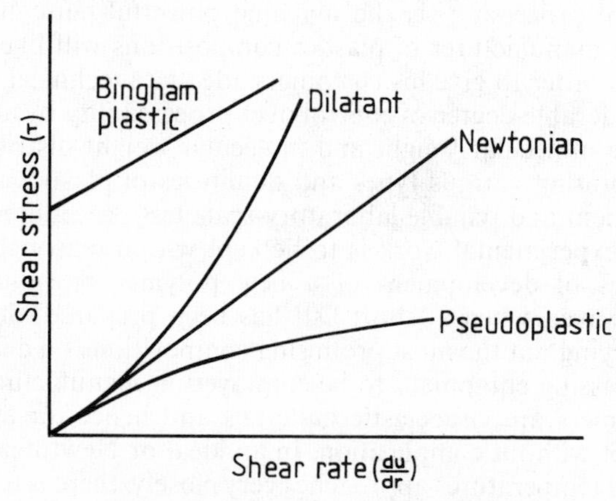

Fig. 5.1 Flow relationships for various types of liquid

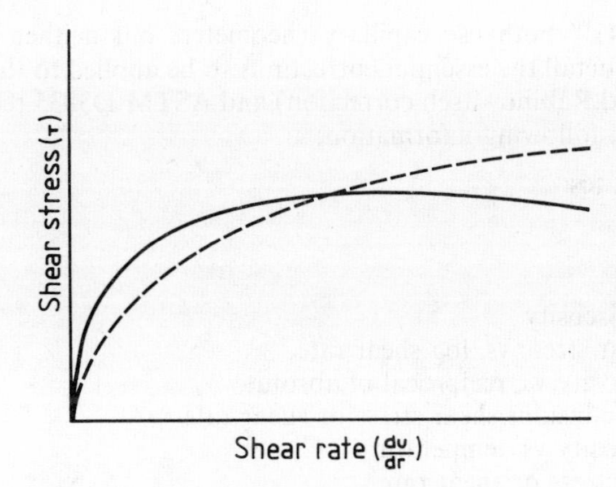

Fig. 5.2 Different pseudoplastic curves

common processing techniques, while Cogswell[3] provides a comprehensive study of modern concepts of the subject as a whole.

So far we have only considered materials whose flow characteristics, if not time-dependent, are such that no change of state is involved. Thermoplastics fit this description, except when heating to fluidise has been carried on for so long that significant decomposition has taken place. When, however, we turn to thermosets, where a fluid mass sets to the solid state while still hot and during processing, it can well be imagined that study of such an unstable system is very complicated.

During the past 10 years there have been increasing studies of the rheological properties of polymer melts and their relationship to processibility. For example, Brydson[4] has outlined the relationship for a number of the common processing techniques for plastics; Paul[5] has suggested that measurement using a variable torque rheometer, capillary or orifice rheometer and oscillating disc rheometer can characterise thermosetting materials for injection moulding and perhaps, for other processes; and Wissbrun[6] has outlined the utility and limitations in the application of rheology to polymer processing with particular reference to development of suitable grades of acetal resin for extrusion blow moulding. More recently Moos[7] has reported on the significance of rheological properties of resins and their measurement for plastics processing practice. This worker has concluded that much more precise rheological information of plastics will be required if modern, computer-controlled, processing machinery is to realise its full potential. Furthermore, the big strides now being taken into the field of computer-aided mould design also needs reliable rheological data on polymers if it is to be successful. Takahashi *et al.*[8] have described a specially designed capillary rheometer which can achieve shear rates up to at least $10^7\,s^{-1}$ and have used it to examine the rheological properties of HDPE, PS and SAN. From the data produced generalised flow curves may be prepared and it is claimed that these adequately explain the flow behaviour of polymers exposed to high shear rates.

With this need to generate more precise rheological data, two standard tests using capillary rheometers have been published. ASTM D3835 (1983)[9] and

DIN 54811 (1984)[10] both use capillary rheometers but neither closely specify dimensions. Both detail the essential corrections to be applied to the data (Bagley end-correction and Rabinowitsch correction) and ASTM D3835 requires that the report contain the following information:

Temperature of test
Shear stress
Shear rate
Viscosity
Intrinsic melt viscosity
Plot of log shear stress vs. log shear rate
Plot of log viscosity vs. reciprocal of absolute temperature at constant shear stress or shear rate
Plot of log viscosity vs. temperature (°C) at constant shear stress or shear rate
Melt viscosity stability
Die swell ratio

Neither ISO nor BS have seen fit to publish similar standards on capillary rheometry although it was proposed at one time to include a recommended procedure for determining rheological properties as a part of BS 4618.[11]

Capillary rheometers instrumented to enable the above parameters to be determined are expensive and require considerable expertise to operate and to interpret the results. It is suspected that these are the main reasons why they have not become more generally used in test laboratories and for quality control. The simpler flow tests employing single-point determinations have enjoyed a much greater popularity and are described in the next section.

5.3 FLOW TESTS FOR THERMOPLASTIC MATERIALS

5.3.1 Melt Flow Rate (MFR)

The determination of melt flow rate (commonly called melt flow index, MFI, is one of the most widely used tests, though it suffers from the disadvantages discussed above. (In the ISO, BS and ASTM tests for MFR values in the range from 0·3 to 20, the corresponding shear rates are from 1 to 50 s^{-1}.) The test exists in a number of variants according to polymer type and molecular weight. Although widely used, the test has limitations; while it is of particular value as a quality control test on thermoplastics having relatively low melt viscosities, it does not necessarily indicate uniformity of other properties of polymers of the same type but made by other manufacturers, nor is MFR a fundamental polymer property.

The determination of melt flow rate of thermoplastics is covered by ISO 1133 (1981),[12] BS 2782 Method 720A (1979),[13] ASTM (D 1238 (1985)[14] and DIN 53735 (1983).[15] These all use the same apparatus and differ only in the number of conditions (temperatures, loadings and reference times) which are permitted.

The principle employed is that of determining the rate of flow of molten polymer through a closely defined extrusion plastometer: Fig. 5.3 shows a cross-section of the important parts.

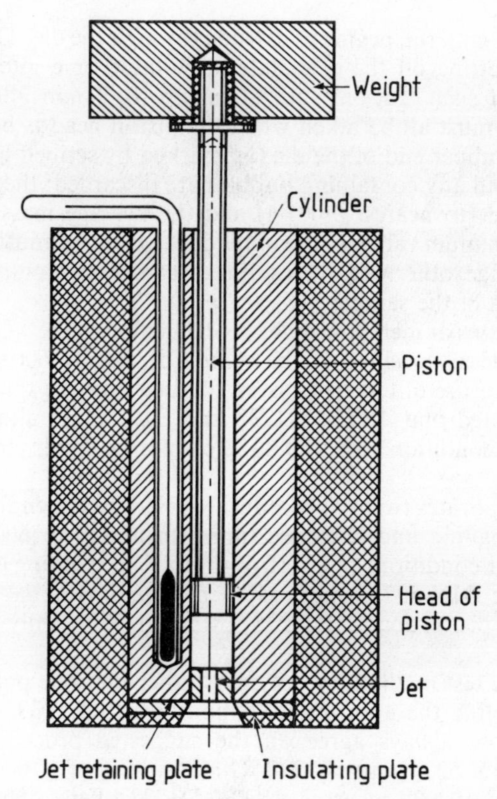

Fig. 5.3 Apparatus for determination of melt flow index: BS 2782

The cylinder is of hardened steel and is fitted with heaters, lagged and controlled for operation at the required temperature (between 125 °C and 300 °C; all $\pm 0{\cdot}5$ °C); the dimensions of the cylinder, particularly its diameter, are closely specified. The piston is of steel and the diameter of its head is $0{\cdot}075 \pm 0{\cdot}015$ mm less than that of the internal diameter of the cylinder (which is 9·5 mm). The die (or 'jet') has an internal diameter of 2·095 mm $\pm$ 0·005 mm or 1·180 mm $\pm$ 0·005 mm (depending on procedure used) and is made of hardened steel or tungsten carbide. Note the close specification for the die diameter (see Poiseuille formula above: although, as explained, it does not apply to polymer melts, often a power law of the type $\tau = K[\mathrm{d}u/\mathrm{d}r]^n$ does so). All surfaces of the apparatus which come into contact with the molten polymer are highly polished. Weights are provided so that the piston can be loaded to give a total weight of between 325 g and 21 600 g, according to the procedure used.

The first requisite is to ensure that the cylinder, piston and jet are scrupulously clean by treatment with hot solvents and wiping with lint-free cloths (abrasives must not be used). The cylinder is maintained at the test temperature for 15 min and then charged with between 4 and 8 g of sample (within a period not exceeding 1 min) and the unloaded piston reinserted. From 4 to 6 min later the weight is added and molten material begins to extrude through the die. The rate of extrusion is measured by cutting off the extrudate at suitable time intervals (given in the standard). Several such 'cut-offs' are taken up to 30 min after insertion of the sample and these must all be taken when the piston head is between 50 mm and 20 mm above the upper end of the die (as marked by scribed lines on the piston). The first cut-off and any containing bubbles are discarded; the remainder, at least three, are weighed (to nearest 0·001 g) and the average mass is calculated. The maximum and minimum values of the individual weighings must be within specified values of the average, otherwise the result is discarded and a further test performed on a fresh portion of the sample.

The melt flow rate (or melt flow index) is calculated as the weight (in g) extruded in a reference time as specified in the standard (usually 600 s). In the report of results the temperature of test and piston loading must be quoted.

It should be noted that ASTM D1238 and DIN 53735 allow only one size of die and therefore condition 1 of ISO 1133 cannot be employed in these standard test methods.

ISO 1133 incorporates twenty procedural variations depending on temperature, die size, piston loading and reference times (from 120 to 600 s). ASTM D1238 allows twenty-four conditions of test (from A to X) depending on temperature and piston load, while DIN 53735 suggests twenty conditions based on variations of temperature and piston loads. BS 2782 Method 720A only recommends eight conditions of test.

When using the test method for polymers other than the polyolefins, caution is necessary in selecting the appropriate conditions. ISO 1133, ASTM D1238 and DIN 53735 do not always agree on the suggested procedures. For example, ISO 1133 and DIN 53735 specify 220 °C and a load of 10 000 g for determining the melt flow rate of ABS whereas ASTM D1238 offers a choice of 200 °C at a loading of 5000 g or 230 °C at 3800 g for the same material.

ASTM D3364 (1983)[16] provides an extension of the ASTM D1238 test for the measurement of flow rates of poly(vinyl chloride) and rheologically unstable thermoplastics. The apparatus differs only in respect of the die which has an 120° angle of entry and is about three times as long as for D1238. In addition it is recommended that the temperature control be maintained at ±0·1 °C. A cylinder temperature of 175 °C and piston loading of 20 000 g is specified but may be changed to 5000 g for PVC compounds having high flow rates (in excess of 10 g/10 min). ASTM D3364 is claimed as being useful for quality control tests on PVC compounds having a wide range of melt viscosities. Measurements are made at shear rates close to 1 s^{-1}. However, for rigid PVC compound ASTM D1238 is recommended using a temperature of 190 °C and a piston load of 21 600 g.

The determination of melt flow rate has become commonplace in most test laboratories and is particularly used for characterising polyolefins. This has resulted in the development of apparatus which automatically measures the flow rate based on volume displacement from the cylinder through the die. ASTM D1238 allows for this as an alternative method (procedure B) is given which permits the use of such equipment. Extensions of the use of the melt flow apparatus have also been proposed to indicate rheological and other properties of thermoplastics. For example Shenoy and Saini[17] have demonstrated that by plotting modified apparent viscosity function (X MFI) against the modified apparent shear rate function

(X/MFI) on a log–log scale a rheogram of the polymer can be estimated. Such rheograms can be useful to the plastics processor if no other rheological data is available.

Another technique (but not covered in any standard) is to determine the MFR at two different loadings at the same temperature. The ratio of these two values can give an indication of the molecular weight distribution of the material, which, in itself, is a useful pointer to its processibility.

5.3.2 Rossi–Peakes Test

For many years this test was practised for determining the flow characteristics of thermoplastic moulding material and was usually used (in the UK) for cellulose acetate. The test never found favour with ISO and although the test appeared in the earlier editions of BS 2782 it has now been withdrawn. The essential form of the Rossi–Peakes apparatus is shown in Fig. 5.4.

To carry out the test, the apparatus is assembled and the temperature adjusted according to the material under examination. A pellet of the test material, of size to match the charge chamber, is introduced therein and the pressure applied immediately. The length of flow after 2 min is determined and then the material is removed and the operation repeated with a fresh pellet. Two pellets are tested at each of three temperatures, at all of which the flow length is between 13 and 38 mm and where at least one measurement is above 25 mm and one below. The results are plotted and the flow temperature is read off where the length of flow is 25·4 mm.

ASTM have preserved this test in ASTM D569 (1982)[18] which offers two procedures, one as above and the other where the degree of flow is measured at a specified temperature. In this latter method the degree of flow is measured as the length of extrudate (to the nearest 0·25 mm) at the temperature and pressure at which the test is carried out. The pressure is fixed at 3·45, 6·89 or 10·35 MPa and the temperature is chosen at some multiple of 5 °C (e.g. 130 °C, 135 °C, 140 °C, 145 °C, etc.) so that the flow lies within 12·7 to 38·1 mm. The wording of this procedure in the standard leaves much to be desired; it does not actually say, for example, that the extrudate has to be measured, yet the very definition of the test requires that this be done.

The main disadvantages of the Rossi–Peakes test are that preparation of the test pellets are difficult and time-consuming and the results do not relate to any standard rheological property.

5.3.3 Other Flow Tests

Although the melt flow index and Rossi–Peakes tests are the only ones generally standardised by ISO, BSI or ASTM for thermoplastics there are many others, most of which suffer from the same disadvantages as those described above, namely that they use operating conditions, especially shear rates, far removed from practice. As far back as 1952 Hayes[19] described two extrusion plastometers and Gray *et al.*[20] in 1968 developed an apparatus where change in apparent viscosity with temperature could be followed continuously and, it was claimed, the minimum

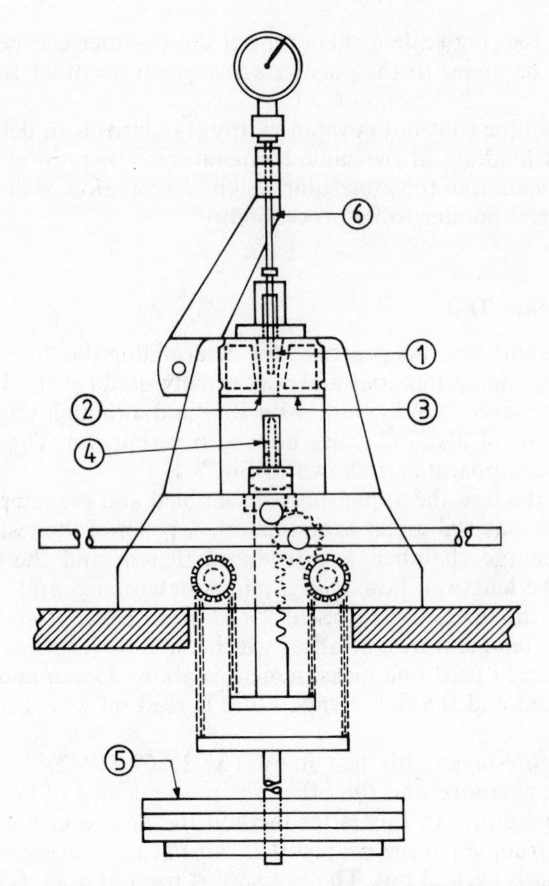

Fig. 5.4 Rossi–Peakes flow tester. The orifice (1) is highly polished and runs in a split cone which is clamped in the heater block (3). A thermometer well is inserted into a hole drilled into one half of the split cone. The charge chamber (2) is situated below the orifice and a heated ram (4) is so positioned that it forces material in the chamber up into the orifice. (5) is a system for applying a pressure of 10·3 MPa to the ram. The rate of flow is measured by (6), consisting of a follower rod and indicator exerting a specified pressure

and maximum processing temperatures could be predicted. Again, the apparatus suffered from determination at low shear rates.

A useful *ad hoc* method of comparative testing is a spiral disc mould where a fair length of flow channel can be provided by a conveniently sized mould fitted to an injection machine. Campbell and Griffiths[21] describe one where the effective length is 77 in (1·956 m), in a mould of overall dimensions 12 in × 9 in (305 × 230 mm). The flow length of the material into the mould, which is vented to atmosphere at its inner end, is determined and related to the test conditions of cycle time, temperatures of mould and barrel, pressure, etc.

During the past 25 years the torque rheometer has found wide use for studying the behaviour of plastics materials. The apparatus consists of a small mixing head, similar in many respects to a miniature internally heated mixer of the Banbury type, driven by a motor via a torque-measuring device. Various sizes of mixing head are available, most being fitted with a feed chute and loaded ram. The mixing chamber is fitted with a thermocouple and means for recording melt temperature and torque against time as the test proceeds are incorporated into the apparatus. The mixing head is usually heated by oil circulation and the temperature is controlled to within 1 °C. Provision is made to vary the rotor speed, modern instruments having a speed range between 10 and 200 r.p.m.

In use, the weighed charge of plastics material, in powder or granule form, is introduced into the heated chamber with the motor running, the load is applied to the feed chute ram and the chart recorder(s) are started. In this way material change with time under dynamic conditions can be determined and this provides useful data, for example, as to what happens during the compounding of a plastics material.

The torque rheometer is available commercially; both the Brabender and Hampden/Rapra instruments are well known. Some torque rheometers can be fitted with a small instrumented extruder in place of the mixing head, thus adding versatility to the machine.

McCabe[22,23] has described rheological measurements with the Brabender instrument, Schramm[24] the measurement of fusion rate of rigid PVC dry blend and Matthan[25] has evaluated the variables of the Brabender Plastograph. Paul[26,27] has described the Hampden/Rapra machine and has shown how it may be used with PVC. Allen and Williams[28] have used a Brabender instrument to predict polymer processing properties and use of their technique can give relative melt viscosity stability and moulding temperatures of a range of thermoplastics.

A further use for the torque rheometer has been proposed by Leskovyansky[29] to determine plate-out from PVC formulations. This is done by mixing the material for a specified period as a fraction of the determined degradation time of the compound under test and then inspecting the interior of the mixing head for signs of sticking which indicates that the compound will plate-out during processing.

Despite the amount of work that has gone into investigating the torque rheometer it has not found favour with the standardisation authorities and, so far, there is no standard test method based on the instrument for thermoplastic materials. ASTM have specified the instrument as a means of determining the flow and cure properties of thermosetting resins (see §5.4).

Over the last 15 years much work has been described with specially designed pieces of apparatus aimed at bridging the gap between laboratory tests and factory operations. Various workers have attached measuring instruments to actual extruders, injection-moulding machines and rheometers, and facilities for studying the effects of a wide range of rates of shear have been described.[30] Often small-scale laboratory extruders, etc., are instrumented and these are available commercially. One such apparatus which has found use in the laboratory is the extrusion capillary rheometer. This is basically a miniature ram extruder fitted with a capillary die and instrumented so that melt temperature and pressure behind the die can be monitored. Speed of extrusion can be varied so that the properties of a plastics melt can be determined under a wide range of shear rates. Dies with different lengths and angles of entry are available to permit study of the effect of die design on output and quality of extrudate.

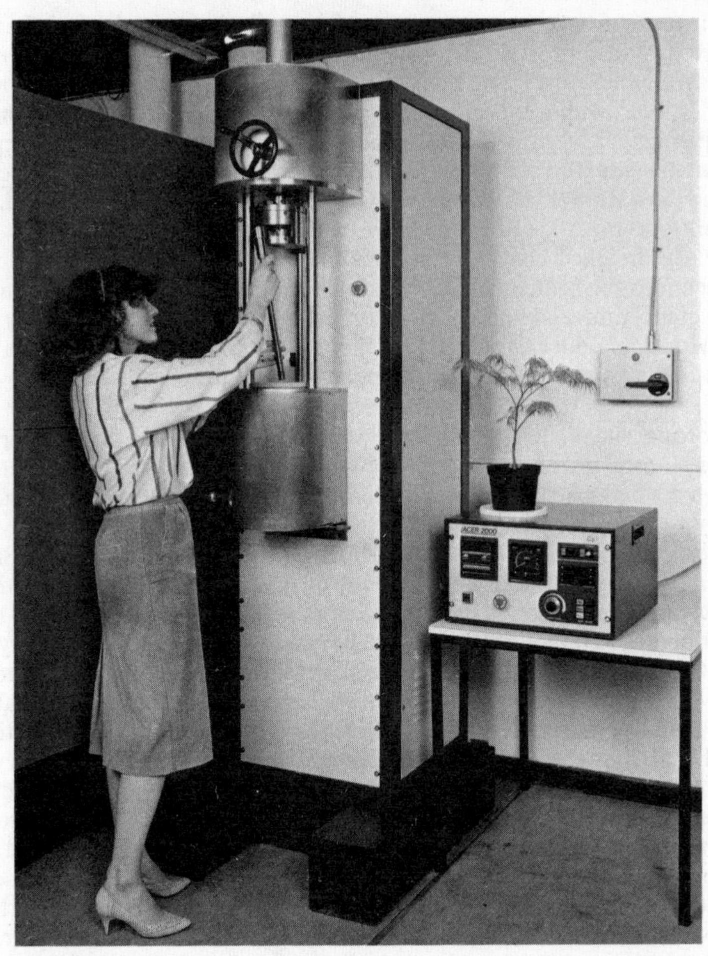

Fig. 5.5 Carter–Baker extrusion capillary rheometer, (courtesy Carter–Baker Enterprises)

Several extrusion capillary rheometers available commercially are capable of operating at shear rates up to $25\,000\,s^{-1}$.[31] One commercially available type of extrusion rheometer is shown in Fig. 5.5.

Two properties of polymers which are of considerable interest to processors are 'die swell' and 'melt fracture'. Die swell is simply defined as the degree of swelling of a polymer melt as it issues from a die. It is important for the design of extrusion dies and as a quality control parameter for polymers used in critical extrusion applications (for example, extrusion below moulding of bottles). Melt fracture is manifested by a curious 'sharkskin' effect seen on the surface of extrudates and clearly must be avoided if extruded products of acceptable quality are to be produced. There is no standardised test method for either of these properties but it is possible to use an extrusion rheometer to investigate both melt fracture and die swell. Melt fracture may be examined by finding the temperature and rate of extrusion at which it just commences, while die swell can be determined by comparison of the diameter of extrudate with the die diameter. At best, such methods serve as useful quality control tests, and great care is needed in interpreting the results to predict performance on full-scale processing equipment. Some control laboratories use the ordinary MFR apparatus (BS 2782 Method 720A) to determine the die swell at the same time as the normal MFR test is carried out.

5.4 FLOW TESTS FOR THERMOSETTING MATERIALS

The tests mentioned in §5.3 have all been designed, basically, for thermoplastics materials, although mention has been made in §5.2 that some workers have used, for example, capillary rheometers to assess the processibility of thermosetting materials. The Rossi–Peakes test (§5.3.2) has been used for thermosets and Von Meysenburge[32] has described a modified form of this test. A number of tests designed to determine the flow properties of thermosetting materials have now been standardised.

5.4.1 Cup Flow Test

A test designed specifically for thermosetting moulding materials is the so-called 'cup flow' determination. Method 720B of BS 2782 (1986)[33] describes the UK practice and limits applicable to phenolic and alkyd moulding materials, excluding 'fast-curing' ones. The cup flow mould is shown in Fig. 5.6. Needless to say, in a flow test such as this the surfaces of the inner male and female members must be smooth and highly polished. The general form and dimensions (including steam coring) are very carefully specified.

The temperature of the mould is laid down as 163 °C ± 1 °C and the thrust on the mould 100 ± 5 kN. The closing speed of the mould when empty must be 130 mm in from 4 to 5 s. The mass of moulding material to use is found by trial and error so that a flash of between 2 and 2·5 g is obtained in the test. The mould is opened just sufficiently to charge the weighted-out sample of material, at from 10 °C to 40 °C, with the aid of a scoop and the mould is closed again immediately; the time between dropping the powder into the mould and the registration of full pressure must be between 5 and 10 s. The time of flow is measured in seconds as that between the first registration of pressure on the hydraulic gauge of the press and the instant the flash is seen to cease moving. This is the 'cup flow' of the material.

The technique described in ASTM D731 (1984)[34] to determine moulding index, follows much the same pattern although the mould cavity differs in design (Fig. 5.7).

An elaborate method is laid down for obtaining the appropriate weight of material to use. For materials of impact strength under 27 J/m (by ASTM D256 Method A (Ch. 8) a cup with flash of thickness from 0·15 to 0·20 mm is moulded, the flash removed and the cup weighed; this weight, multiplied by 1·1, is the charge to use in the test, except for materials of impact strength 27 J/m and above when flash of thickness from 0·51 to 0·66 mm is produced and the cup weight multiplied by 1·05. The preferred temperature is 150 °C, 155 °C or 165 °C (all ±1 °C),

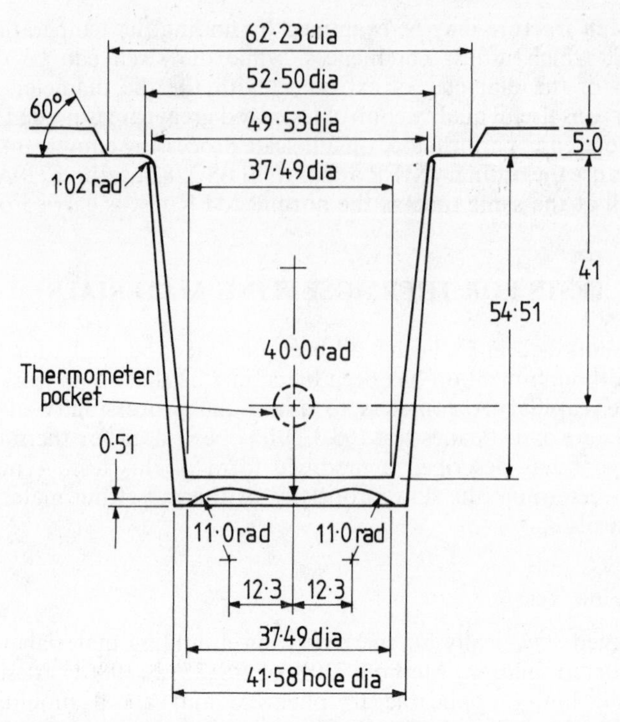

Fig. 5.6 Detail of cup flow mould: BS 2782 (dimensions in millimetres)

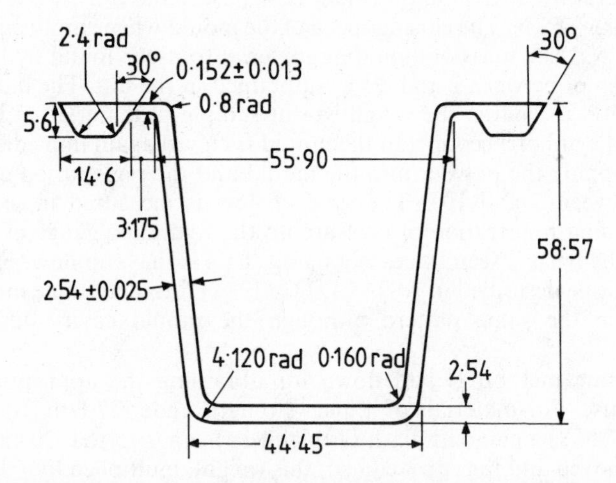

Fig. 5.7 Detail of cup flow mould: ASTM D731 (dimensions in millimetres). All surfaces are highly polished to No. 2 microfinish (SPI–SPE standard for mould finish). Rockwell C–58 steel. Tolerances on dimensions are $\pm 0{\cdot}025$ mm except as noted

according to the material under test. A load is first determined which produces a cup of the required flash thickness and then the next lowest load is selected, from a table of eight specified alternatives, for test purposes. The time of flow in seconds is recorded from the instant the hydraulic gauge registers an applied load of 454 kg to when the prescribed flash thickness has been reached.

The result is expressed as the 'moulding index', the minimum force required to produce the necessary flash, with a subscript of the closing time in seconds, e.g. 6800_{18} kgf for a 6800 kg minimum load and 18 s flow time.

DIN 53465 (1963)[35] also uses a cup flow type of mould and expresses the result as 'closing time' of thermosetting moulding materials.

5.4.2 Spiral Flow Test

A test designed for low pressure thermosetting moulding compounds (moulding pressures less than 6·9 MPa) is described in ASTM D3123 (1983).[36] This utilises a transfer moulding press fitted with a specified spiral flow mould and the test is conducted at a mould temperature of 150 °C ± 3 °C and transfer pressure of 6·9 MPa. After cure the moulding is removed and its length read off to the nearest 6·35 mm which is reported as the spiral flow of the material under test.

5.4.3 Disc Flow Test

Method 105D of BS 2782 (1970)[37] describes a method for determining the flow of synthetic resin impregnated glass fabric. At least four pieces of the sample, each 100 mm square, carefully superimposed to form a rectangular pack, comprise the test piece. This is weighed (to nearest 0·01 g) and placed between stainless steel plates, not less than 125 mm square and approximately 1·6 mm thick, that have been treated with an appropriate type of release agent. The assembly is then placed between the platens of a heated press and pressure is applied within 10 s and for not less than 15 min. Temperatures and pressures are according to the following schedules:

Resin	*Temperature* (°C ± 5 °C)	*Pressure* (MPa ± 10%)
Polyester	120	1·4
Epoxy	160	1·4
Silicone	175	1·4 or 6·9

The laminate is then removed from the press and:

1. If the flash can be cleanly removed from it, this is done and the laminate is weighed (W_2).
2. If the flash cannot be removed cleanly, a piece 50 mm ± 1 mm square is cut from the central area of the laminate and weighed (W_3).

The 'flow' of the test piece is then calculated as follows: if W_1 is the original weight of the test piece then, if (1) above has been followed,

$$\text{flow} = 100\frac{(W_1 - W_2)}{W_1} \text{ per cent} \qquad [5.3a]$$

or, if (2) above has been followed,

$$\text{flow} = 100\frac{(W_1 - 4W_3)}{W_1} \text{ per cent} \qquad [5.3b]$$

5.4.4 Other Tests

ASTM have adopted the torque rheometer for determining the flow and cure properties of thermosetting plastics in ASTM D3795 (1983).[38] This lists the temperatures of the Brabender instrument to be used for the various classes of material together with the head size (30 ml) and rotor speed (40 r.p.m.). The test results are reported for the following parameters:

Charge weight and specific gravity of the test material
Maximum loading torque
Time to reach minimum torque
Minimum torque
Final torque peak
Time from loading the mixer to final torque peak

The standard claims that by studying the temperature and torque characteristics in this way it is possible to predict the behaviour of the material during processing but no precise instructions are given as to how to achieve this. It does, however, state that the results are of considerable use in checking the batch-to-batch uniformity of the moulding compound.

5.5 VISCOSITY – STANDARD TEST METHODS

Mention has already been made (§5.1) of the difficulties of determining meaningful melt viscosities of polymers in relation to their processing characteristics. The molecular weight and molecular weight distribution of a polymer have a marked effect on the processibility of polymers; determination of molecular weight by solution viscosity methods is common and is dealt with in Ch. 6. Ths viscosity of resins, plastisols and organosols (dispersion of PVC resin in plasticiser), which is of interest to the processor handling these materials, is commonly determined by the use of cone and plate, double cone, coaxial cylinder and rotating spindle viscometers. ISO 3219 (1977)[39] specifies a method for determining the viscosity of polymers in the liquid state with a rotational viscometer in which the theoretical shear rate employed in the test is defined. This standard shows how to calculate the theoretical shear rate from the dimensions and rotational frequency but cautions the user that the equations only give results near the true shear rate of Newtonian liquids if the instrument characteristics agree with the standard. BS 2782 Method 730B (1978)[40] is identical to ISO 3219.

It is worth mentioning at this point that the SI unit of viscosity, is the Pascal second (Pa s). The Pascal second is equivalent to the Newton second per square metre ($N\,s/m^2$) which is equal to 10^3 cP (centipoise).

The Brookfield instrument is a well known type of rotational spindle viscometer and is the subject of several standard test methods for polymers in the liquid state.

ISO 2555 (1974)[41] states the method of test for the determination of Brookfield RV viscosity of resins in the liquid state or as an emulsion or dispersion. It specifies the use of the Brookfield RV Model viscometer (RVF, RVF–100 or RVT), chosen according to the product to be tested and the desired experimental accuracy. By selection of the appropriate instrument type it is possible to measure viscosities in the range from 0·02 to 8000 Pa s. This standard has two appendices, Annex A and Annex B. Annex A sets out the general application of the method to resins in the liquid state and Annex B gives the principle, description and characteristics of Brookfield viscometers. This standard only details the method of using the viscometer and requires reference to material standards for details of spindle size, speed of operation and temperature of test. As such it does little more than duplicate the manufacturer's instruction manual which comes with the instrument. During its passage through ISO it met with some opposition, including that of the UK, and consequently does not appear as a method in BS 2782.

ASTM D2393 (1986)[42] specifies the use of the Brookfield viscometer for finding the viscosity of epoxy resins and related components and ASTM D1824 (1983)[43] uses the same apparatus for finding the apparent viscosity of plastisols and organosols at low shear rates.

ASTM D1823 (1982)[44] offers a standard method of test for determining the apparent viscosity of plastisols and organosols at high shear rates by the Burrell–Severs A120 extrusion rheometer but does not describe the apparatus, which is gas-operated. Again, the method appears to duplicate the instruction manual provided with the instrument and does not give any indication of the shear rates obtainable.

5.6 GELATION TIMES OF THERMOSETTING MATERIALS – STANDARD TEST METHODS

The time required for a thermosetting liquid resin to thicken after addition of catalysts and accelerators is clearly of importance to users of these products since, *inter alia*, the gelation (usually shortened to gel) time governs the size of batch of activated resin a processor can mix to satisfy his production line without waste caused by the usage time exceeding the pot life. Most of the standardised methods depend upon stirring the resin until resistance is felt, the gel time being quoted in minutes between the start of the test and the point at which a specified resistance to agitation is obtained.

ISO 2535 (1974)[45] describes the procedure for unsaturated polyester resins at 25 °C using a reference initiator and accelerator. The end point is found by a rotating (from 1 to 2 r.p.m.) glass rod 6 mm in diameter. Suspension of the rod is by a torsion drive so that when the viscosity of the resin reaches 50 Pa s an indication is given and the time can be taken. The result is expressed as 'gel time at 25 °C'. It should be noted that although the UK approved ISO 2535 this method has not been published as a British Standard test method.

ASTM D2471 (1979)[46] describes a standard test method for gel time and peak exothermic temperature of reacting thermosetting resins. This uses wooden probes

2·4 mm in diameter manipulated by hand every 15 s during the test. At the same time a thermocouple is inserted in the resin to indicate the temperature change throughout the test. The method permits the use of mechanical gel time meters if the sample is large enough, but it is claimed that these do not give results consistent with hand stirring. With the hand probe method the end-point is recorded when the reacting resin no longer adheres to the end of a clean probe. Unless otherwise specified, the temperature of test is 23·0 °C ± 1·0 °C.

BS 2782 includes five tests for the determination of gel time. Method 835A (1980)[47] is for the determination of the gelation time of phenolic resins. No set temperature is specified but the user is referred to the standard for the material in question. It is implied that the temperature to be used is between 130 °C and 150 °C based on the methods 111A and 111B of the earlier (1970) edition of BS 2782. The method comprises stirring the heated resin (0·2 g) with a beaded glass rod until a rubbery condition is reached. The time is recorded from the commencement of the test until this rubbery condition is attained.

Method 835B (1980)[48] is similar and is specifically for polyester resins. Here the temperature of test is selected from the ranges 81 °C to 82·5 °C and 116 °C to 118 °C and the amount of resin used is a 25 mm depth in a test tube of specified size.

Method 835C (1980)[49] uses a gel timer which relies on a vertically reciprocating disc and plunger rod coupled to a trip device which stops a timer when the force on the disc is sufficient to match the apparent mass of the partially immersed plunger as indicated by a lag of at least 0·8 mm during a downward stroke. The method is suitable for both epoxy and polyester resins.

Method 835D (1980)[50] is for determining the gel time of thermosetting resins using a hot plate. Here the resin (1 g) is heated on a thermostatically controlled hot plate (to within ±0·5 °C). The resin is stirred by a palette knife until the rubbery state is reached which is deemed to be end of the test. Again the gel time is defined as the time from the initial application of the resin to the hot plate until the rubbery condition of the resin is reached. The standard allows the use of both steam or electrically heated hot plates. The temperature of test is as specified in the relevant materials standard.

Method 835E (1980)[51] is for determining the gelation time of resin impregnated materials used for laminates. Ten pieces of the test material (100 × 25 mm) are piled between caul plates and this assembly placed in a pre-heated press (160°C ± 5°C) until the exuding resin gels are tested by prodding with a piece of thin wire. The time taken from introduction of the material into the press until gelation is taken as the gelation time.

Before concluding this section, mention is made of a method of test for viscosity of highly viscous resins and pastes. ASTM D2730 (1980)[52] is a standard test method for sag flow of highly viscous resins. It is offered for quality control and acceptance tests for highly viscous materials and pastes. The principle of the test is to fill horizontal channels of various depths (from 1·6 mm to 19·1 mm) with the test material, then, after turning the channels to the vertical plane, measuring (in mm) the amount of movement of the material out of the channel after 30 min. The result is expressed as the length of sag flow for each depth of channel.

5.7 OTHER TESTS

Three further properties of interest to the processor of plastics materials are melting point, particle size, and apparent density and bulk factor.

Melting point is of particular interest when processing those plastics and resins which have a fairly sharp change of state, e.g. certain of the polyamides (nylons) and novolac resins. A number of methods have been standardised and these are dealt with in Ch. 16.

The particle size of powders, granules or pellets may be of profound importance for it influences the packing density of a moulding powder, and hence the charge in a fully positive mould, and the gelation characteristics of paste-making PVC polymer, to take two common examples.

Test methods for determining particle size are dealt with in Ch. 7, which also deals with the measurement of apparent density and bulk factor. All of these properties are of importance to the machine designer as well as the processor because they have to be taken into account when designing, for example, screws for injection machines and extruders. Furthermore, having designed a machine to operate with optimum output and quality with a material of certain bulk factor and particle size, it is then in the processor's interest to use routine tests for these properties to control the quality of different batches of moulding powder.

There are undoubtedly other properties which are of interest to the processor but, except for shrinkage which is dealt with in Ch. 16 and those described above, it is considered that they are so restricted in application as to be outside the scope of this book.

REFERENCES

1. Bernhardt, E. C. (1959) *Processing of Thermoplastic Materials*, Reinhold.
2. Pearson, G. H. (1982) *Plastics Polymer Science and Technology* (Ed. M. D. Baijal), Wiley.
3. Cogswell, F. N. (1981) *Polymer Melt Rheology: A Guide for Industrial Practice*, Godwin/Plastics and Rubber Institute.
4. Brysdon, J. A. (1981) *Flow Properties of Polymer Melts*, 2nd edn, Godwin/Plastics and Rubber Institute.
5. Paul, K. T. (1975) *Joint Conference of British Society of Rheology and Rubber Institute, Loughborough, 1975*, p. 49.
6. Wissbrun, K. F. (1977) *Polymer News*, **4** (2), 55.
7. Moos, K. H. (1985) *Kunststoffe*, **75** (1), 3.
8. Takahashi, H., Matsuoka, T. and Kurauchi, T. (1985) *Journal of Applied Polymer Science*, **30** (12), 4669.
9. ASTM D3835 (1983) *Rheological Properties of Thermoplastics with a Capillary Rheometer*.
10. DIN 54811 (1984) *Testing of Plastics – Determining the Fluidity of Plastics Melts by the Capillary Rheometer*.
11. BS 4618 *Recommendations for the Presentation of Plastics Design Data*.
12. ISO 1133 (1981) *Plastics – Determination of the Melt Flow Rate of Thermoplastics*.

13. BS 2782 Method 720A (1979) *Determination of Melt Flow Rate of Thermoplastics.*
14. ASTM D1238 (1985) *Flow Rates of Thermoplastics by Extrusion Plastometer.*
15. DIN 53735 (1983) *Determination of Melt Flow Index of Thermoplastics.*
16. ASTM D3364 (1983) *Flow Rates for Poly(vinyl chloride) and Rheologically Unstable Thermoplastics.*
17. Shenoy, A. and Saini, D. R. (1985) *British Polymer Journal*, **17** (3), 314.
18. ASTM D569 (1982) *Measuring the Flow Properties of Thermoplastic Molding Materials.*
19. Hayes, R. (1952) *Chemical Industry*, **44**, 1069.
20. Gray, T. F., Jr, Holford, T. G. and Coombs, R. L. (1968) *Society of Plastics Engineers Journal*, **24** (9), 35.
21. Campbell, G. and Griffiths, L. I. (1956) *Plastics Progress 1955*, Iliffe, p. 259.
22. McCabe, C. C. (1960) *Transactions of the Society of Rheology*, **IV**, 335.
23. McCabe, C. C. (1960) *Chemistry in Canada*, **12** (10), 4.
24. Schramm, G. (1965) *International Plastics Engineering*, **5** (12), 420.
25. Matthan, J. (1968) *RAPRA Research Report*, no. 165.
26. Paul, K. T. (1972) *RAPRA Bulletin*, February.
27. Paul, K. T. (1973) *RAPRA Members Journal*, November.
28. Allen, E. O. and Williams, R. F. (1971) *Proceedings of SPE Annual Technical Conference 1971*, p. 587.
29. Leskovyansky, P. J. (1984) *Journal of Vinyl Technology*, **6** (2), 82.
30. Ballam, R. L. and Brown, J. J. *Capillary Rheometer*, Instron Engineering Company, Canton, Mass.
31. Rapra (1979) *Rapra Guide to Rubber and Plastics Test Equipment*, p. 18.
32. Von Mysenburge, C. M. (1959) *International Symposium of Plastics Testing Standards*, STP no. 247, ASTM.
33. BS 2782 Method 720B (1986) *Cup Flow of Phenolic and Alkyd Moulding Materials.*
34. ASTM D731 (1984) *Molding Index of Thermosetting Molding Powder.*
35. DIN 53465 (1963) *Determination of the Closing Time of Thermosetting Moulding Materials: Cup Flow Method.*
36. ASTM D3123 (1983) *Spiral Flow of Low Pressure Thermosetting Molding Compounds.*
37. BS 2782 (1970) Method 105D (Reprinted 1986) *Flow of Synthetic Resin Impregnated Glass Fabric.*
38. ASTM D3795 (1983) *Thermal Flow and Cure Properties of Thermosetting Plastics by Torque Rheometer.*
39. ISO 3219 (1977) *Determination of Viscosity of Polymers in Liquid, Emulsified or Dispersed State with a Rotational Viscometer Working at a Defined Shear Rate.*
40. BS 2782 Method 730B (1978) *Determination of the Viscosity of Polymers in the Liquid, Emulsified or Dispersed State using a Rotational Viscometer Working at a Defined Shear Rate.*
41. ISO 2555 (1974) *Resins in the Liquid State or as Emulsions or dispersions. Determination of Brookfield RV Viscosity.*
42. ASTM 2393 (1986) *Viscosity of Epoxy Resins and Related Components.*
43. ASTM D1824 (1983) *Apparent Viscosity of Plastisols and Organosols at High Shear Rates by Extrusion Viscometer.*
44. ASTM D1823 (1962) *Apparent Viscosity of Plastisols and Organosols at High Shear Rates by Extrusion Viscometer.*
45. ISO 2535 (1974) *Unsaturated Polyester Resins – Measurement of Gel Time at 25 °C.*
46. ASTM D2471 (1979) *Gel Time and Peak Exothermic Temperature of reacting Resins.*
47. BS 2782 Method 835A (1980) *Determination of Gelation Time of Phenolic Resins.*
48. BS 2782 Method 835B (1980) *Determination of Gelation Time of Polyester Resins (Manual Method).*
49. BS 2782 Method 835C (1980) *Determination of Gelation Time of Polyester and Epoxide Resins using a Gel Timer.*

50. BS 2782 Method 835D (1980) *Determination of Gelation Time of Thermosetting Resins using a Hot Plate.*
51. BS 2782 Method 835E (1980) *Determination of Gelation Time of Resin Impregnated Materials used for Laminates.*
52. ASTM D2730 (1980) *Sag Flow of Highly Viscous Resins.*

Chapter Six

Polymer Characterisation

6.1 INTRODUCTION

The most profound influences on the properties of a polymer are exerted by the chemical type and structure of the polymer molecule itself. However, this handbook is concerned with physical testing and accordingly it will be assumed that the chemical constitution is known or can be determined by chemical analysis. After these, properties are influenced, to varying degrees, by the morphology and the molecular weight distribution of a polymer.

It is not within the scope of this book to cover the techniques required for a detailed study of the morphology of a polymer, but the degree of crystallinity has a direct relevance to many physical properties covered later. Also the measurement of molecular weight is considered relevant and these aspects of polymer characterisation will be covered in this chapter.

6.2 CRYSTALLINITY

The crystallinity of a polymer is likely to have an influence on a range of parameters, many of which are referred to elsewhere in this handbook:

Density	(Ch. 7)
Rigidity	(Ch. 8)
Short-term tensile strength	(Ch. 8)
Shock (impact) resistance	(Ch. 8)
Softening point	(Ch. 16)
Solvent resistance	(Ch. 17)
Permeability to gases	(Ch. 19)

The degree of crystallinity is normally determined by X-ray diffraction or nuclear magnetic resonance (NMR). These techniques are described fully elsewhere[1-4] and are beyond the scope of this book.

Infrared spectroscopy techniques are applicable to certain polymers where the absorptions of crystalline and amorphous regions differ significantly. This is not a direct measurement and the relationship of absorption, or the ratio of absorptions at two frequencies, with crystallinity, has to be established by calibration. There is also a danger of incorrectly designating frequencies to crystallinity rather than other phenomena.[5]

The most simple method for determining the degree of crystallinity is by density measurement. Again, this technique is dependent upon accurate calibration. It is not difficult to envisage that the more ordered (crystalline) the packing of polymer molecules the greater will be the density of the mass polymer and usually the relationship between density and percentage crystallinity is linear. Nevertheless there are polymers with anomalous behaviour in this respect.

The measurement of density is described in Ch. 7: any method accurate to 0·0001 g/ml and appropriate to the test specimen in question should be suitable.

6.3 MOLECULAR WEIGHT

6.3.1 General Considerations

The molecular weight of a polymer is likely to influence a number of parameters considered elsewhere in this handbook:

Melt viscosity and processibility	(Ch. 5)
Shock (impact) resistance	(Ch. 8)
Loadbearing properties (long term)	(Ch. 11)

It must be emphasised that these are broad generalisations and are not exhaustive. The comments only apply to the normal commercial range of polymer size and not to low degrees of polymerisation where the influence of this parameter is very significant.

Most polymer species will have a molecular weight distribution, rather than a single molecular weight. For a simple distribution it is often possible to relate physical properties to an average molecular weight. However, it is generally better to consider the molecular distribution as a whole and with complex distributions this would be essential.

The molecular weight can be quantified by a variety of averages, number average: $\bar{M}_n$, viscosity average; $\bar{M}_v$, weight average; $\bar{M}_w$ and the z average; $\bar{M}_z$. Each of these progressively give more emphasis to the higher molecular weight species present and they can be defined mathematically:

$$\bar{M}_n = \frac{\sum_i N_i . M_i}{\sum_i N_i} \qquad [6.1]$$

$$\bar{M}_v = \left(\frac{\sum_i N_i (M_i^{\alpha+1})}{\sum_i N_i M_i} \right)^{1/\alpha} \qquad [6.2]$$

$$\bar{M}_w = \frac{\sum_i N_i (M_i^2)}{\sum_i N_i M_i} \qquad [6.3]$$

$$\bar{M}_z = \frac{\sum_i N_i(M_i^3)}{\sum_i N_i(M_i^2)} \qquad [6.4]$$

where α is derived from the Mark–Houwink formula (see §6.3.4), and N_i is the number of molecules of molecular weight M_i.

Within this chapter, consideration will be given to both the primary techniques for molecular weight and the secondary techniques requiring calibration. None of these techniques are straightforward and when considering polymer molecular weight data an appropriate amount of caution is required. The difficulty of performing polymer molecular weight measurements has led to judgements being frequently based on single (occasionally duplicate) measurements without any of the statistical appreciation suggested in Ch. 1.

6.3.2 Primary Techniques for Molecular Weight

The primary methods for measuring the molecular weight of polymers include end-group analysis, ultracentrifuge, light scattering and the colligative properties techniques of measuring changes in osmotic pressure, vapour pressure, freezing point and boiling point. Of these, the most frequently encountered are osmotic and vapour pressure measurements to yield $\bar{M}_n$[6,7] and light scattering to yield $\bar{M}_w$.[8] These techniques have been most used because of their wide range of molecular weight applicability and their relative simplicity, although it should be stressed that the accurate measurement of polymer molecular weight by any technique is extremely difficult.

The introduction of low-angle laser light scattering has significantly simplified the determination of $\bar{M}_w$. Not only has the technique simplified the measurement, but the degree of precision has been greatly enhanced and the agreement between different laboratories is much improved.[9] Low-angle laser light scattering is probably the most accurate and reproducible technique currently available for measuring polymer molecular weights.

In theory, the small cell size permits the use of low-angle laser light scattering as an on-line detector for gel permeation chromatography (see §6.3.3) and elevates the latter technique to that with a primary status. However, the variety of potential errors and artifacts with the combined technique and the ease with which the data can be misinterpreted, suggests that while valuable data can be obtained, combined gel permeation chromatography/low angle laser light scattering should not be relied upon as a primary technique.

A considerable degree of expertise is required successfully to operate the instruments used for the primary measurement of polymer molecular weights and as such, it is not appropriate to describe these techniques in detail here.

6.3.3 Gel Permeation Chromatography

Gel permeation chromatography (GPC), also known as size exclusion chromatography (SEC), is almost certainly the most used and the single most useful technique for measuring polymer molecular weights. This fact is possibly surprising when one realises that the level of reproducibility, even within a single laboratory is poor. However, when used in a purely comparative manner, valuable data on possibly subtle differences in the molecular weight distribution of samples can often be observed.

GPC can be a relatively simple, secondary technique for measuring polymer molecular weights. However, some applications of the technique are extremely difficult and it is not appropriate to attempt to describe the detail of this technique within this handbook. The practice of the technique has been described elsewhere[10-12] and this section will be limited to guidance on the applicability of the techniques and some cautionary comments on the interpretation of the results.

The main elements of a typical GPC system are shown schematically in Fig. 6.1. It can be seen that most of these elements are identical to those required for high performance liquid chromatography (HPLC). For simple GPC applications, the equipment is often used in a multifunctional manner. The main difference between GPC and HPLC is in the mechanism of the separation and hence in the type of column used.

GPC can be applied to characterising virtually any soluble polymer type and some suitable solvents and operating temperatures are listed for the main polymer types in Table 6.1.

The practice of GPC with routine solvents such as tetrahydrofuran (THF) and toluene should not be difficult. In contrast, GPC at high temperature (e.g. for polyolefins) or with polar solvents (e.g. dimethylformamide) can be extremely difficult, requiring considerable expertise in GPC and, probably, dedicated GPC systems.

Important information can be revealed by using GPC; the technique yields data on the total molecular weight distribution of a polymer and this can be vital when the extremes of the distribution could be the controlling influence in the physical properties or the processibility of the sample. However, GPC is best restricted to comparing molecular weight distributions and to obtaining a general guide to the molecular weight distribution of a sample.

Modern GPC systems usually have dedicated computers to assist in interpreting the results and all too often the numerical results are accepted without question. Replication will often show surprising variations and significance levels established for one polymer sample can be very different from another sample of different molecular weight or with a broader/narrower molecular weight distribution.

Also, it must always be remembered that GPC is a secondary technique and the validity of the calibration and the data handling should always be questioned. Frequently, GPC systems are calibrated using narrow molecular weight distribution polystyrenes and the results are either expressed as the 'polystyrene equivalent' or a mathematical procedure is applied to allow for the difference in chemical type. The Universal Calibration Procedure[13,14] is the best known of these corrections, unfortunately; in applying this procedure, Mark–Houwink parameters (see §6.3.4) of questionable origin are often employed.

Standard Test Methods

Standard test methods for GPC are only covered by the following ASTM methods:

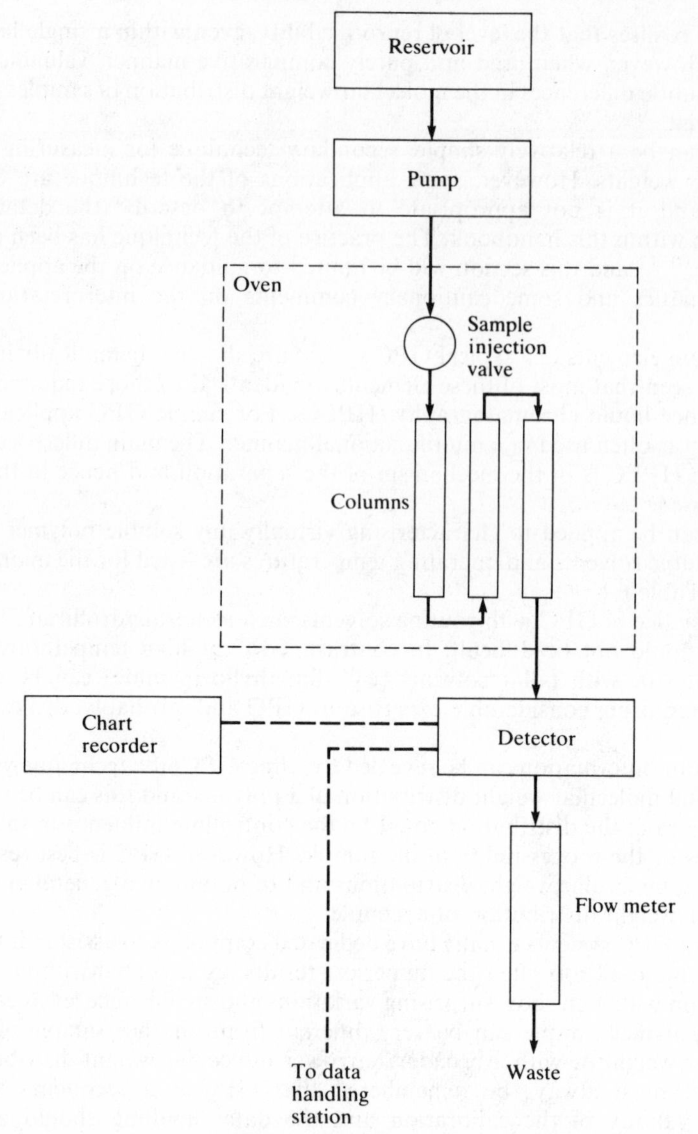

Fig. 6.1 Schematic diagram of GPC system

ASTM D3016–78, *Practice for Use of Liquid Exclusion Chromatography Terms and Relationships*

ASTM D3536–76, *Molecular Weight Averages and Molecular Weight Distribution of Polystyrene by Liquid Exclusion Chromatography (Gel Permeation Chromatography – GPC)*

ASTM D3593–80, *Molecular Weight Averages and Molecular Weight Distribution of Certain Polymers by Liquid Size-Exclusion Chromatography (Gel Permeation Chromatography – GPC) Using Universal Calibration*

Table 6.1 Solvents/temperatures suitable for the GPC of common polymer types

Polymer type	Solvent[a]	Temperature (°C)
ABS	DMF	80
Asphalt	THF or ODCB	Room or 140
Cellulose Acetate	THF	Room
Elastomers	Toluene	Room or 80
EPDM	ODCB	140
EVA	THF or ODCB	Room or 140
Phenolic Resins	THF	Room
Polyacrylates	THF or DMF	Room or 80
Polyacrylonitrile	DMF	80
Polyamides	*m*-Cresol	120
Polybutadiene	Toluene	80
Polycarbonate	THF	Room
Polyether sulphone	DMF	80
Polyethylene	ODCB	140
Polyethylene glycol	DMF	80
Polyethylene oxide	DMF	80
Polyethylene terephthalate	*m*-Cresol	120
Polymethylmethacrylate	THF	Room
Polypropylene	ODCB	140
Polystyrene	THF	Room
Polyurethanes	DMF	80
Polyvinylacetate	THF	Room
Polyvinylchloride	THF	Room
Polyvinylidene difluoride	DMF	80
Siloxanes	Toluene	Room or 80
SBR	Toluene	Room or 80
Waxes	ODCB	140

[a] THF, tetrahydrofuran; ODCB, 1,2-dichlorobenzene, trichlorobenzene also frequently used; DMF, dimethylformamide, usually with 100 ppm lithium bromide orthochlorophenol and fluorinated alcohols are usually alternative solvents to metacresol.

Unfortunately the latter two methods refer to a level of technology which has generally been replaced by more high performance GPC systems. Also these methods only relate to GPC using tetrahydrofuran as the solvent.

6.3.4 Solution Viscosity

Earlier editions of this handbook have had to contend with the multitude of terms employed in solution viscometry. However, the ISO or IUPAC terminology now appears to be generally accepted and only these terms will be referred to in this edition. Even within the IUPAC system, there are a large variety of ways of expressing the solution viscosity of a sample; these are listed and defined in Table 6.2.

Table 6.2 Definition of viscosity terms

Term	Definition	
Viscosity/density ratio or Kinematic viscosity	η/ρ	where η = dynamic or absolute viscosity and ρ = density
Viscosity ratio	η/η_0	where η = viscosity of the polymer solution and η_0 = viscosity of the pure solvent
Viscosity relative increment	$(\eta - \eta_0)/\eta$	
Viscosity number	$(\eta - \eta_0)/\eta_0 . c$	where c = polymer concentration (g/ml)
Logarithmic viscosity number	$\ln(\eta/\eta_0)/c$	
Limiting viscosity number	$[\eta]$	viscosity number at zero concentration

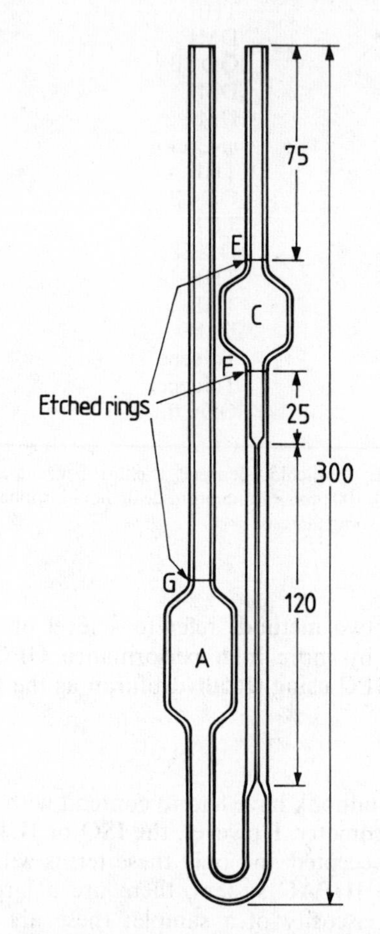

Fig. 6.2 U-tube viscometer (dimensions in millimetres)

A considerable range of viscometer types and sizes are available and the main types are described in detail in BS 188 and ISO 3105. The choice of type of viscometer is usually a matter of convenience, however, an appropriate size must be selected to give acceptable efflux times (i.e. the period of time for a liquid meniscus to travel between two etched marks under the action of gravity). The above standard methods include tables which relate liquid viscosities and suitable viscometer sizes. Unfortunately, within the UK, the viscometer manufacturers tend to follow the coding used within the equivalent ASTM D445–83. Similar general guidance is given in DIN 51550 (1972). Diagrams of the three viscometer types covered in BS 188 and ISO 3105 are shown in Figs. 6.2–6.4; the etched timing marks being E and F in all three cases.

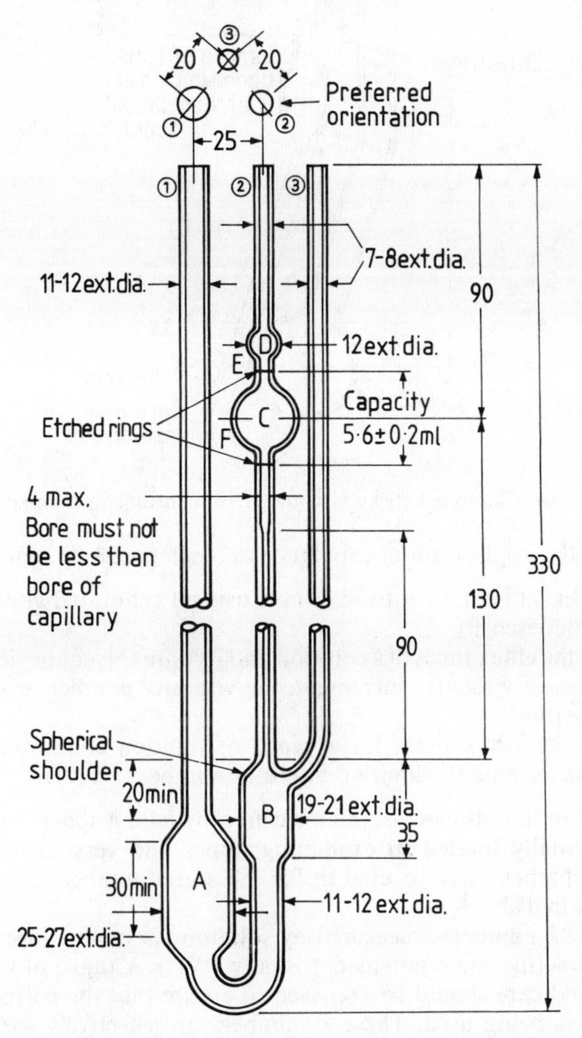

Fig. 6.3 Suspended level viscometer (dimensions in millimetres)

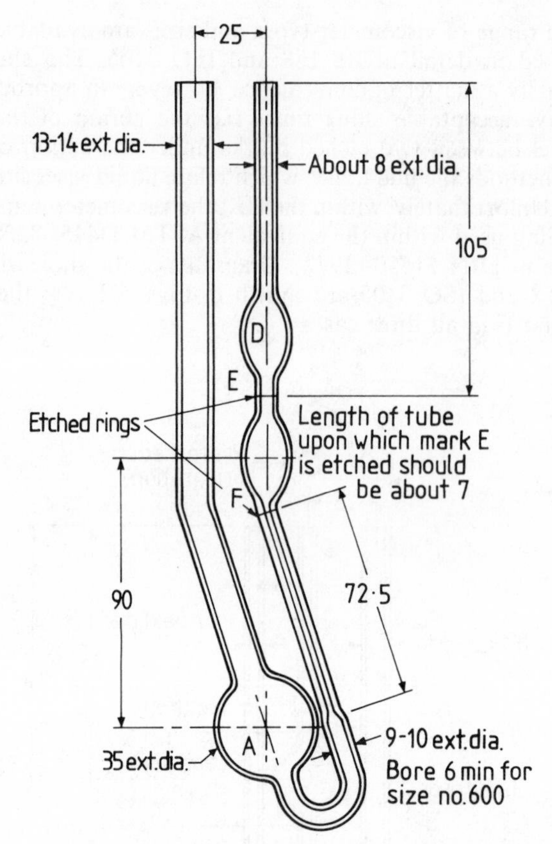

Fig. 6.4 Cannon–Fenske viscometer (dimensions in millimetres)

In practice, the application of capillary viscometers can be reduced to:

1. Measuring the efflux time for the sample using a calibrated viscometer to yield the kinematic viscosity.
2. Comparing the efflux times of a solution and its pure solvent to yield the viscosity ratio, the viscosity relative increment, the viscosity number or the logarithmic viscosity number.
3. Comparing the efflux times for a range of solution concentrations with the solvent, to determine the limiting viscosity number (LVN).

The viscosities of liquids can also be measured by falling sphere viscometry. This technique is usually limited to examining dopes and very viscous liquids; the method is comprehensively covered in BS 188 and a number of specific standard methods given in Table 6.3.

Confusingly, '*k* numbers', measured by solution viscometry, are still frequently employed to describe some polymers (usually PVC). A range of *k* numbers have been defined and care should be exercised to ensure that the correct definition of the *k* number is being used. These *k* numbers are effectively variations on the definition of viscosity number, but using different solution concentrations.

Table 6.3 Standardised procedures for determining solution viscosity

Polymer type	Solvent	Temperature (°C)	Viscosity result	Source
ISO				
PVC	Cyclohexanone	25	Viscosity number (VN)	ISO 174 (1974)
Polyamides	Formic acid or *m*-cresol	25	VN	ISO 307 (1984)
Cellulose acetate	Dichloromethane and methanol	25	VN	ISO 1157 (1975)
Polyethylene and polypropylene	Decahydronaph-thalene	135	VN and LVN	ISO 1191 (1975)
Polycarbonate	Dichloromethane	25	VN	ISO 1628/4 (1986)
Poly(alkylene-terephthalate)	*o*-Chlorophenol	25	VN	ISO 1628/5 (1986)
Methylmeth-acrylate polymers and copolymers	Chloroform	25	VN and LVN	ISO 1233 (1975)
BSI				
PVC	Cyclohexane	25	VN	BS 2782 404A (1970)
Polystyrene	Toluene	25	Kinematic	BS 2782 404B (1970)
Polyamides	Formic acid	25	VN	BS 2782 404C (1970)
Polyamides	*m*-Cresol	25	VN	BS 2782 404D (1970)
Cellulose acetate	Dichloromethane and methanol	25	Viscosity ratio	BS 2880 (1957)
ASTM				
Cellulose nitrate	Ethanol, toluene ethylacetate	25	Kinematic[a] viscosity	ASTM D301–72 (1983)
Polyvinyidene chloride	*o*-Dichloro-benzene	120	Kinematic[a] viscosity	ASTM D729–83
Polyamides	Formic acid or *m*-cresol	25	Viscosity ratio	ASTM D789–81
CAP/CAB[b]	—	25	LVN or kinematic viscosity	ASTM 817–72 (1983)
Cellulose acetate	—	25	Kinematic[a] viscosity	ASTM 871–72 (1983)

Table 6.3 Standardised procedures for determining solution viscosity (*cont.*)

Polymer type	Solvent	Temperature (°C)	Viscosity result	Source
Ethyl cellulose	Ethanol toluene	25	Kinematic viscosity	ASTM 914–72 (1983)
PVC[c]	Cyclohexane	30	log VN	ASTM D1243–79 (1984)
Methyl cellulose	Water or alkali	20	Kinematic viscosity	ASTM D1347–72 (1983)
Polyethylene	Decahydronaph-thalene	130	Various	ASTM 1601–78
DIN				
Polystyrene	Benzene	25	*k* no.	DIN 7441 (1977)
Polycarbonate	Methylene chloride	25	*k* no.	DIN 7744 (1980)
PMMA	Chloroform	25	VN	DIN 7745 (1980)
PVC	Cyclohexanone	25	*k* no.	DIN 53729 (1983)
Polyamide	*m*-Cresol	25	VN	DIN 53727 (1980)
Cellulose acetate	Methylene chloride methanol	25	VN	DIN 53728: Part 1 (1970)
Poly(ethylene terephthalate)	*o*-Chlorophenol	25	VN	DIN 53728: Part 3 (1985)
Polyethylene and polypropylene	Decahydronaph-thylene	135	VN or LVN	DIN 53728: Part 4 (1975)

[a] A falling sphere method, see also ASTM D1343–69 (1985) *Viscosity of Cellulose Derivatives by Ball-Drop Method.*
[b] Cellulose acetate propionate or cellulose acetate butyrate.
[c] See also ASTM D3591–77 (1985) *Determining Limiting Viscosity Number of Poly(vinyl chloride) (PVC) in Formulated Compounds.*
[d] The *k* number is defined by the expression: $\log(\eta/\eta_0) = [75\,k/(1 + 1{\cdot}5\,k{\cdot}c)] + k{\cdot}c$, where c = polymer concentration.

From a measured LVN it is often possible to calculate a viscosity average molecular weight, $\bar{M}_v$ (defined in §6.3.3). A number of relationships have been derived to calculate $\bar{M}_v$ from LVN, but the best known is the Mark–Houwink equation:

$$[\eta] = KM^{\alpha} \qquad [6.5]$$

The values of K and α for a large number of polymer/solvent/temperature combinations have been published and these constants are also the constants used in the Universal Calibration Procedure for GPC (see §6.3.3). As with GPC, it is important to question the applicability of the values of K and α to the samples being considered: are the Mark–Houwink parameters, derived for an anionic polymerisation produced narrow molecular weight distribution polystyrene, applicable to a free radical polymerisation produced broad distribution polystyrene?

Capillary solution viscometry is a particularly useful technique for characterising polymers, especially in that it is simple, does not require any expensive equipment and is a highly reproducible measurement. However, to achieve this high level of reproducibility (and agreement between laboratories) extreme cleanliness and adherance to the appropriate standard method is required (e.g. temperature control to $\pm 0{\cdot}05\,°C$ is frequently required).

Standard Test Methods

General guidance on the viscometers available and on the practice of solution viscometry is given in the following standard test methods:

ISO. ISO 3104 (1976) *Petroleum Products: Transparent and Opaque Liquids – Determination of Kinematic Viscosity and Calculation of Dynamic Viscosity.* ISO 3105 (1976) *Glass Capillary Kinematic Viscometers – Specification and Operating Instructions.* ISO 1628/1 (1984) *Guidelines for the Standardization of Methods for Determination of Viscosity Number and Limiting Viscosity Number in Dilute Solution* – Part 1: *General Conditions.*

BSI. BS 188 (1977) *Methods for the Determination of the Viscosity of Liquids.* BS 2000: Part 71 (1982) *Kinematic Viscosity of Transparent and Opaque Liquids and Calculation of Dynamic Viscosity.* BS 2782: Part 7 Method 730A *Determination of Reduced Viscosity (Viscosity Number) and Intrinsic Viscosity of Plastics in Dilute Solution.*

ASTM. ASTM D445–83 *Kinematic Viscosity of Transparent and Opaque Liquids (and Calculation of Dynamic Viscosity).* ASTM D446–85a *Specification for Operating Instructions for Glass Capillary Kinematic Viscometers.* ASTM D2857–70 (1977) *Dilute Solution Viscosity of Polymers.*

DIN. DIN 51550 (1978) *Viscometry; Determination of Viscosity; General Principles.* DIN 51562: Part 1 (1983) *Viscometry; Determination of Kinematic Viscosity Using the Standard Design Ubbelohde Viscometer.* DIN 51562: Part 3 (1985) *Viscometry; Measurement of Kinematic Viscosity by means of the Ubbelohde Viscometer; Viscosity Relative Increment at Short Flow Times.* DIN 53012 (1981) *Viscometry; Capillary Viscometry of Newtonian Liquids; Sources of Error and Corrections.*

Table 6.3 summarises some of the internationally and nationally standardised methods for measuring the viscosities of specified polymers in dilute solution.

REFERENCES

1. Kavesh, S. and Schultz, J. M. (1969) *Polymer Engineering and Science*, **9**, 452.
2. Alexander, L. E. (1969) *X-ray Diffraction Methods in Polymer Science*, Wiley Interscience, Ch. 3.

3. Marx, C. L. and Cooper, S. L. (1973) *Makromoleculare Chemie*, **168**, 339.
4. Kamel, J. and Charlesby, A. (1981) *Journal of Polymer Science*, Polymer Physics Edition, **19**, 803.
5. Maddams, W. F. (1982) in *Analysis of Polymer Systems* (Eds L. S. Bark and N. S. Allen), Applied Science, Ch. 3.
6. Ulrich, R. D. (1975) in *Techniques and Methods of Polymer Evaluation*, vol. 4, *Polymer Molecular Weights* (Ed. P. E. Slade), Dekker, Ch. 2.
7. Glover, C. A. (1975) ibid., Ch. 4.
8. Huglin, M. B. (Ed.) (1972) *Light Scattering from Polymer Solutions*, Academic Press.
9. Dumelow, T., Holding, S. R. and Maisey, L. J. (1983) *Polymer*, **24**, 307.
10. Holding, S. R. (1984) *Endeavour*, **8**, 17.
11. Dawkins, J. V. and Yeadon, G. (1978) in *Developments in Polymer Characterisation* (Ed. J. V. Dawkins), Vol. 1, Applied Science, Ch. 3.
12. Cooper, A. R. (1982) in *Analysis of Polymer Systems* (Eds. L. S. Bark and N. S. Allen), Applied Science, Ch. 8.
13. Benoit, H., Grubisic, Z., Rempp, P., Dekker, D. and Zilliox, J. G. (1966) *Journal of Chemical Physics*, **63**, 507.
14. Dawkins, J. V. (1977) *European Polymer Journal*, **13**, 837.

Chapter Seven

Density and Dimensions

7.1 DENSITY

7.1.1 Definitions

Density	Mass per unit volume (at defined temperature) Unit: kg/m^3
Relative density (formerly known as 'specific gravity' – a term no longer used in ISO and BSI standards)	Mass (of substance) compared with the mass of an equal volume of a specific (reference) substance – most often water. In this instance the temperature of both the substance and the reference substance must be stated, e.g. relative density 23/23 °C. Being a ratio, the property is dimensionless.

7.1.2 Standard Test Methods

The determination of the density of solid plastics, excluding cellular plastics, is detailed in ISO 1183 (1987),[1] which gives four methods according to the nature of the plastics material. The British equivalents are contained in BS 2782, Methods 620A to 620E (1980),[2] which closely follow the present ISO format. American standards are to be found in ASTM D792 (1966)[3] and D1505 (1985),[4] and the German standard in DIN 53479 (1976).[5] The four basic methods are described below.

Displacement Method

This is Method A of ISO 1183. It is suitable for film, sheet, rod, tube and moulded articles, the density being determined at one of the standard laboratory temperatures of 20 °C, 23 °C and 27 °C.

A specimen, with smooth surfaces free from crevices and dust and preferably of mass approximately 5 g, is weighed in air (W_1) and then in freshly boiled distilled water (W_2) at the requisite temperature, after sufficient time has been allowed for the specimen to reach this temperature.

If the density of the material under examination is less than that of water, a sinker is attached and the mass of this in water (W_3) and of the sinker plus the specimen also in water (W_4) are taken. Fine wire is used for the suspension of the specimen (and sinker) – the upthrust on part of the wire is sufficiently small to be ignored. All air bubbles must be removed, for example by use of a minute quantity of detergent,

$$\text{density} = \frac{W_1}{W_1 - W_2} \times \rho \qquad [7.1a]$$

When a sinker is used, this becomes:

$$\text{density} = \frac{W_1}{W_1 + W_3 + W_4} \times \rho \qquad [7.1b]$$

where ρ is the density of the water, usually taken to be $1000\,kg/m^3$ but, for more accurate work, set to the value indicated in the table in the ISO standard.

Provision is made for using a liquid other than water if a polymer–water interaction occurs, in which case ρ is given by the density of that liquid at the test temperature.

The same basic procedure is followed in BS 2782 Method 620A and ASTM D792 Methods A1 to A3, with only slight changes in emphasis rather than principle.

Because density is frequently measured as a quality control parameter, relative density balances have been developed which enable the measurement to be made much more quickly. These devices are based on the displacement principle.

Pyknometer Method

This is Method B of ISO 1183. It is suitable for plastics in a pre-formed condition, such as powder, pellets, flake, etc. A pyknometer (Fig. 7.1) is accurately weighed (W_1), a suitable quantity of the material (1–5 g) is added and the whole is re-weighed (W_2). The material is covered with freshly boiled distilled water and the pyknometer is then placed in a vacuum desiccator to remove all the air. The pyknometer is filled with water and placed in a constant temperature bath at 23 °C ± 0·1 °C (or other standard laboratory temperature). When the temperature has been achieved the capillary is filled or emptied as necessary to the appropriate level and the pyknometer is re-weighed (W_3). The pyknometer is emptied, filled with freshly

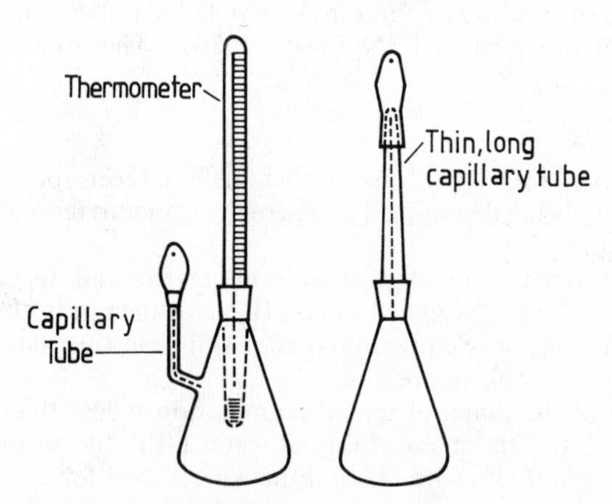

Fig. 7.1 Typical pyknometers

boiled distilled water and re-weighed. The mass (W_4) is recorded:

$$\text{relative density } 23/23\,^\circ\text{C} = \frac{(W_2 - W_1) \times d}{(W_3 - W_4) + (W_2 - W_1)} \qquad [7.2]$$

For water d is exactly 1, but if another displacement liquid is used, for example when the material is not inert to water, d is the relative density of this liquid.

Density $\rho^{23\,^\circ C}$ (kg/m^3) = relative density 23/23 °C × 997·6 where 997·6 is the density of the water (in kg/m^3) at 23 °C.

The equivalent British Standard is BS 2782, Method 620B. This procedure is also contained in ASTM D792, Method B.

The pyknometer method is used for the determination of the density of liquid resins, as detailed in ISO 1675 (1975)[6] and BS 2782, 620E. BS 733: Part 1 (1983)[7] gives details of pyknometer size, construction and design and is identical to ISO 3507.

Sink–Float Method

This is Method C of ISO 1183. It is suitable for plastics in forms similar to those required for Method A and for pellets.

A series of liquids is prepared having a range of densities covering the anticipated range of interest and kept at 23 °C ± 0·1 °C (or other standard laboratory temperature). A small quantity of wetting agent is permitted if required. A test piece is placed in each liquid and the densities of the densest liquid in which the test piece sinks and the least dense liquid in which it floats are determined as quickly as possible by a standard method. The density of the plastic is, then, between these two densities.

Greater precision can be obtained by preparing a further set of liquids whose densities cover the range determined as above and the whole procedure repeated until the required precision is attained.

BS 2782 Method 620C uses the same procedure.

Density Gradient Method

This is Method D of ISO 1183. It is suitable for plastics in the same forms as Methods A and C.

Two solutions are made (A and B), of which A has a density lower than the lowest density of interest by about 20 per cent and B a density higher than the highest of interest by about 30 per cent. One litre of A is placed in vessel 1 and one litre of B in vessel 2, the taps being closed (Fig. 7.2).

The magnetic stirrer in vessel 1 is started (to give thorough mixing but not bubble formation) and the connecting tap is opened, when probably some of solution B will flow into solution A. When hydrostatic equilibrium has been established the right-hand tap is opened so that, in conjunction with the capillary bore, the column takes about 2 h to fill. While this happens, solution B flows into, and is mixed with, solution A in vessel 1 which therefore increases in density; hydrostatic equilibrium between the two vessels must be maintained throughout and, for this, a wide-bore connecting tube is necessary with slow filling. Finally the filling tube, with the upper end closed, is removed carefully and slowly (see also Payne and Stephenson[8]).

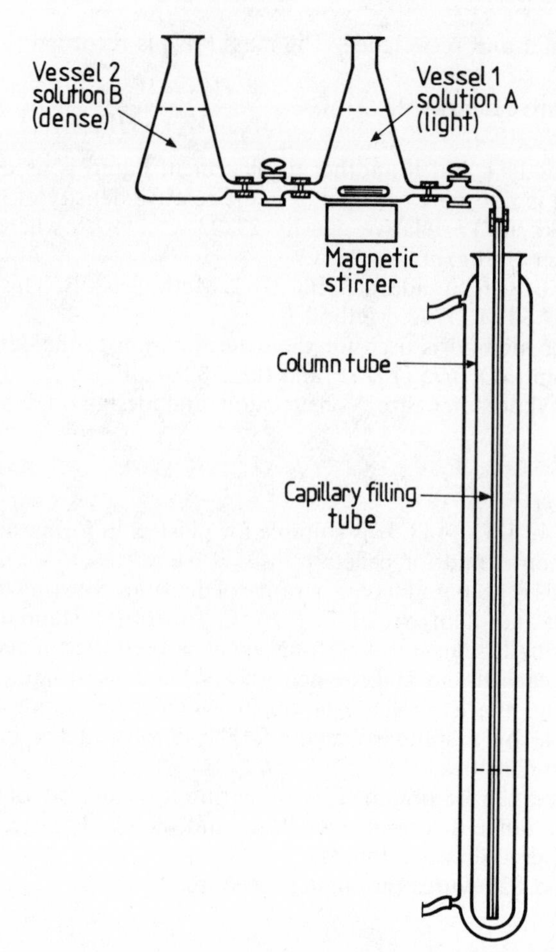

Fig. 7.2 Method for filling density column

In use, careful control of temperature ($\pm 0{\cdot}1\,^{\circ}C$) is of course essential. Total density gradients of more than 0·2 g/ml or less than 0·02 g/ml are not recommended. Needless to say, the mixed liquids must be without effect on the material under test, be transparent, of low volatility and low viscosity. The tube is calibrated with spherical glass floats, not greater than 5 mm in diameter, the densities of which may, for instance, be obtained by a method such as that described in the previous section.

These floats are placed gently in the column and after at least 24 h the height of each above a reference level is determined by a cathetometer. The measured heights are plotted against density and a smooth curve, without discontinuities, should result. With careful use, a column should last several months.

Test specimens should be of easily identifiable shape and size the latter being such that the centres of their volumes can be estimated to within 1 mm. At least two are placed gently in the column and their heights are recorded after not less than 10 min. The densities are read off the calibration graph and, in the case of polyethylene film for instance, should be within 0·0005 g/ml. Care must be taken to avoid air bubbles, dirt and interference between specimens and floats.

The column is cleared as and when necessary by a wire gauze basket, of the widest practicable mesh, drawn upwards through the column at a uniform rate not exceeding 30 mm/min after which the basket is returned to the bottom of the column which is then recalibrated.

Table 7.1 lists typical liquid systems that are used, but is not meant to be exhaustive.

BS 2782 Method 620D (1980),[2] which makes reference to BS 3715 (1964),[9] and ASTM D1505 (1985)[4] also employ this technique.

7.1.3 Special Methods

Apparent Powder Density

Two standard methods are available for determining the apparent density of moulding powders.

The first is a test developed for moulding material that can be poured from a funnel. This is detailed in ISO 60 (1977).[10] A funnel of the form shown in Fig. 7.3 is used. It is mounted vertically with its lower orifice 20–30 mm above the top of a measuring cylinder (capacity 100 ml; internal diameter 40–50 mm) and coaxial with it. With the lower orifice closed, 110–120 ml of well mixed powder is poured into the funnel and then the powder is allowed to flow into the measuring cylinder, assisted if necessary by being loosened with a rod. When the cylinder is full, a straight-edged blade is drawn across the top of the cylinder to remove excess and then the contents are weighed. The mean of two determinations is taken and expressed in g/ml.

Table 7.1 Liquid systems for density gradient columns

System	Density range (g/ml)
Methanol/benzyl alcohol	0·80/0·92
Isopropanol/water	0·79–1·00
Isopropanol/diethylene glycol	0·79–1·11
Ethanol/carbon tetrachloride	0·79/1·59
Ethanol/water	0·79–1·00
Toluene/carbon tetrachloride	0·87–1·59
Water/sodium bromide	1·00–1·41
Water/calcium nitrate	1·00–1·60
Zinc chloride–ethanol/water	0·80–1·70
Carbon tetrachloride/1,3-dibromopropane	1·60–1·99
1,3-Dibromopropane/ethylene bromide	1·99–2·18
Ethylene bromide/bromoform	2·18–2·89
Carbon tetrachloride/bromoform	1·60–2·89

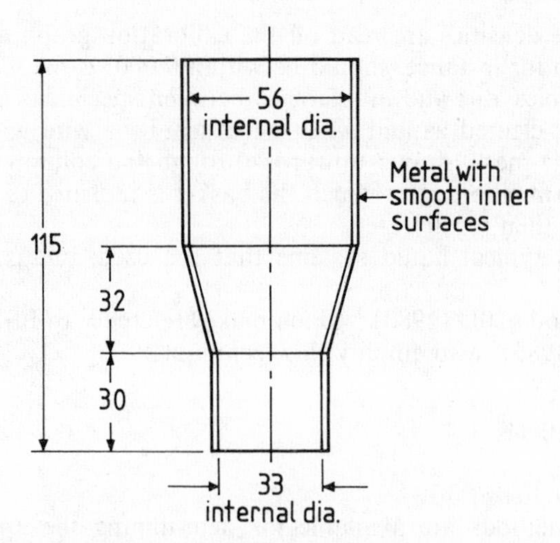

Fig. 7.3 Standard funnel for apparent powder density (dimensions in millimetres)

BS 2782, Method 621A (1978)[11] and one of the methods in DIN 53466 (1984)[12] are identical to this method.

ASTM D1895 (1969)[13] offers two similar methods, but both with different sized funnels from the above and some other, minor, variations. One of these funnels (of Method A) is used to measure pourability by timing the rate of flow of the powder out of the funnel.

The second method is for materials that cannot be poured from a funnel and is given in ISO 61 (1976).[14] A cylinder is used which is 1000 ml in capacity and of internal diameter 90 ± 2 mm. Into this fits a plunger of slightly smaller diameter and with a total mass of 2300 g. A cylinder closed at its lower end and weighted with lead shot may be used.

Sixty grams of the powder is dropped, little by little, into the cylinder so that it is evenly distributed and has a level surface. The plunger is then lowere onto the powder and allowed to remain three 1 min before the height of the powder is measured with the plunger still in position. From the height of the powder, the diameter of the cylinder and the weight of the powder, the apparent density is calculated.

BS 2782, Method 621B (1978)[11] and one of the methods in DIN 53466 (1984)[12] are identical to this method, while Method C of ASTM D1895 (1969)[13] is essentially the same.

ISO 1068 (1975)[15] for PVC resins is a cylinder method for determining the compacted bulk density; a shaking machine is used to tamp down the material under a piston. BS 2782, Method 621D[16] is identical to this.

Bulk Factor

The bulk factor of a moulding is defined as the ratio of the volume of a given mass of moulding material to its volume in the moulded form. It is thus the ratio of the density of the moulded material to its apparent density before moulding.

ISO 171 (1980),[17] BS 2782, Method 621C (1983)[18] and the method of ASTM D1895 all require determination of apparent powder density by the corresponding techniques (as above) and moulded density by the appropriate method (§7.1.2). The German practice is similar and is to be found in the remaining method of DIN 53466 (1984).[12]

7.2 DIMENSIONS

7.2.1 Standard Test Methods

Fabricated Dimensions

There are within ISO three standards covering the measurement of film and sheet dimensions: ISO 4591 (1979)[19] dealing with gravimetric thickness, ISO 4592 (1979)[20] dealing with length and width measurement and ISO 4593 (1979)[21] dealing with thickness measurement by a mechanical scanning technique. These correspond exactly to BS 2782 (1982) Methods 631A,[22] 632A[23] and 630A,[24] respectively.

In ISO 4591 a square of material of side 100 mm is cut from the film or sheeting and its mass determined to 0·1 mg. The density is also measured to ISO 1183[1] and then the gravimetric thickness, t_s, can be found from the expression

$$t_s = \frac{100 \times m}{\rho}$$

When the mass, m, is in grams and the density ρ in grams per cubic centimetre (g/cm^3) then t_s is in microns (micrometres, μm).

In ISO 4592 a procedure is given for measuring the length of a roll to 0·1 m and the width of 0·1 mm if the width is less than 100 mm (but greater than 5 mm) or to 1 mm if the width is greater than 100 mm.

In ISO 4593 three levels of precision are defined for the measuring device which may be mechanical, optical or electronic. For films of less than 100 μm the measurement must be made to 1 μm, above this and up to 250 μm measurement to 2 μm is required and above 250 μm, 3 μm is acceptable.

In the UK film thickness is often referred to as its ‘gauge’; a 100 gauge film has a thickness of 25·4 μm.

For rigid plastics such specialist methods are not required and the normal micrometer or dial gauge is most frequently employed, usually graduated in 0·02 mm (0·001 in). Micrometers are accurately standardised in BS 870 (1950)[25] their use is described in ASTM D374 (1979),[26] Methods A and B. Likewise, dial gauges (or comparator gauges) are dealt with in BS 907 (1965)[27] and Method C of ASTM D374. Appropriate German standards are DIN 863 (1977)[28] and DIN 878 (1979).[29]

The measurement of soft materials which readily yield, particularly highly plasticised PVC and cellular plastics, presents difficulties in defining what the thickness (or width) really is. A figure under no compressive force is difficult to obtain, for optical means are hardly suitable for routine rapid working. The use of micrometers without ratchets clearly involves a largely subjective judgement of how hard the spindle is screwed down. Ratchet micrometers are little better, slipping at some arbitrary force. Many product specifications require the use of dial gauges, specifying the force on the plunger (which includes any spring loading) and the dimensions of the bearing faces ('anvils'), and there is much variation from one document to another.

Particle Size

The particle size of powders, granules or pellets may be of profound importance for it influences, for instance, the packing density of a moulding powder and hence the charge in a fully positive mould, and the gelation characteristics of a polymer used in paste form (such as PVC). The simplest method, and the most widely used on a routine basis, is that of sieving. This technique is not normally recommended for particle sizes much below about 75 μm diameter (though ASTM D1921 (1963)[30] suggests 38 μm as minimum) where it is necessary to use sedimentation, optical or other methods.[31–35]

A word of warning is necessary in the use of test sieves, which are usually numbered to indicate directly or indirectly the number of apertures per unit dimension. Sieve screens may be made of woven silk or fine wire mesh and these two types differ quite considerably in their resistance to distortion under pressure. Wire mesh is most often specified, but again it is necessary to appreciate that a simple statement of the number of holes per unit dimension does not suffice since the diameter of the wire affects that all-important factor, the aperture size. Nor is specification of the latter alone adequate, since the diameter of the wire used to achieve that size controls the actual shape of the aperture, as the weave of the mesh must of necessity yield a non-planar hole. Obviously a coarse wire gives a more distorted (and larger) hole than one obtained from a fine wire, though in the main plane of the mesh both holes may be identical. For this reason it is absolutely essential to tie sieve analysis – as particle sizing by test sieves is termed – to the particular type of sieve used, for example as specified in BS 410 (1976)[36] which follows the recommendations of ISO Technical Committee TC24 and, in particular, ISO 565 (1972).[37]

There is a variety of ways of using test sieves, well illustrated by reference to the relevant ASTM standards, D1705 (1982)[38] and D1921 (1936).[30]

DIN 53477 (1965)[39] gives a method for sieve analysis of granular thermosetting compression moulding materials while ISO 4576 (1978)[40] describes a method of determining gross particle content of aqueous dispersions by sieve analysis and ISO 4610 (1977)[41] uses an air-jet sieve for vinyl chloride resins.

For normal dry sieving an appropriate range of sieves (usually 20 cm diameter) – appropriate, that is, in total span of aperture sizes covered and in the intervals of sizes between them – are 'nested' together, decreasing in aperture size from top to bottom; a tray or pan is fitted under the smallest mesh sieve.

A convenient quantity (say, 100 g) of sample is weighed out and transferred to the top sieve and the cover is placed on the top sieve. Although hand tapping of the assembled nest of sieves may be used it is both tedious and inefficient and therefore a mechanical vibrator is recommended (ASTM 1921 specifies one with a rotary motion and tapping at 150 taps per minute). After 10 min or some appropriate period to attain equilibrium, the sieves and tray are separated and the contents weighed to the nearest 0·1 g either by direct weighing in the sieves (having already weighed the empty sieves) or by transferring the contents to weighing bottles, etc.; in this latter case, extreme care must be taken to remove all particles from the walls and mesh of the sieve with a soft brush.

Normally the total recovered contents amount to 98 per cent or more of the original sample weight, in which case any lost is added to the figure for material passing through the finest sieve used, i.e. onto the tray.

An obvious variant of this method is to simplify it by employing just one, or a very limited number of sieves.

With vacuum dry sieving, smaller sieves, normally of from 7 to 8 cm diameter, are used and, after being dried in a desiccator, are weighed individually before the test is started. They are then inserted in a conical adaptor of appropriate diameter to take the nested sieves and fitted with a lead at its smaller diameter to a vertical standpipe attached to a vacuum system. The suction of the latter is adjusted so that the finest sieve is not appreciably strained and then the vibrator is started. An appropriate quantity of sample is accurately weighed (1 g for very fine powders of 5–10 g for coarser ones) and carefully transferred to the top sieve; tapping of the sieve and light brushing of the sample assists. When no more powder is transferred through the top sieve, suction and vibration are ceased, the sieve is removed and any powder adhering underneath is brushed on to the next sieve below, when the operation is repeated and so on. The sieves with retained powders are weighed, whence the weights of the powder are obtained by difference.

A method is described in ASTM D1705 (1982)[38] for powdered polymers and copolymers of vinyl chloride. This uses the wet sieving technique, which eliminates static charges and troubles associated therewith such as agglomeration (it also reduces the tendency to 'flying' of powder and hence loss of sample).

Sieves are used as described above and are weighed prior to testing. An appropriate quantity of polymer is weighed out (25–100 g) into a beaker and approximately 300 ml of 0·5 per cent wetting agent solution (anionic sulphate or sulphonate type) is added with stirring. This mixture is poured into the top sieve and the powder is washed through this and the lower ones by more solution and then water. The sieves are dried in an oven and weighed.

In ASTM D1457 (1983)[42] for polytetrafluoroethylene materials the apparatus is illustrated in support of the description of the test method.

7.2.2 Non-Standard Methods

In addition to the standardised methods for particle size analysis given above, a number of non-standardised methods are used[43] such as direct microscopic measurement – now a much more powerful technique than formerly with the development of increasingly sophisticated image analysis systems – sedimentation, stream scanning or multiple scattering. More indirect methods include surface area measurement and hydrodynamic chromatography.

Specialist methods for continuously measuring film or sheet thickness such as beta-ray absorption, capacitive and magnetic methods have been developed[44,45] while for measuring wall thickness several methods are described in the literature including ultrasonic and infrared techniques.[46–54] A recent paper describes the development of X-ray fluorescence techniques for measuring coating thickness.[55]

7.2.3 Surface Roughness

It is not often that the surface roughness of a plastics material is required to be known so it is not surprising to find that no standard test methods exist.

For transparent plastics the surface roughness is of great importance since the light transmission and reflection characteristics are considerably dependent on it. In this case, however, the direct measurement of transmission, haze and gloss is carried out by methods of the type described in Ch. 14, rather than the measurement of roughness *per se*.

If measurements are attempted, then those devloped for use with metals may be appropriate, although it must be borne in mind that the greater deformability of most plastics compared with metals may render the method unsuitable unless it can be modified to avoid deformation. The measurement of surface roughness is covered by ISO 468 (1982),[56] and also by the British equivalent, BS 1134 (1972),[57] which is in two parts, the first dealing with instrumentation and the second forming a general explanation. The British Standard is, therefore, a useful introduction to the subject.

A recent paper[58] describes the use of a projector which can be attached to a microscope for making measurements of surface topography and texture.

7.2.4 Extensometry

The measurement of extension – or other forms of deformation – is an essential part of several tests, especially short-term mechanical tests of the type to be described in the next chapter. The method of measurement is, however, considerably dependent on the geometry of the test piece and the absolute deformation to be measured, and so the subject is best discussed in the context of the specific tests applied. Suffice it to say here that the precision required must be specified in the individual test method and is unlikely to be the same as that required for test piece dimensions.

7.2.5 Dimensional Stability

A number of standard test methods exist for the measurement of the dimensional stability of plastics but as these are intimately linked to the effect of temperature on plastics a description of them is deferred until Ch. 16.

7.2.6 Dispersion

The dispersion of pigments in plastics mouldings can be of profound importance in determining the suitability of the moulding for the job intended. In recognition of this, BS 2782 Methods 823A and B (1978)[59] describe two methods for assessing the uniformity of dispersion of carbon black in polyethylene, although the methods could be modified to be of more general application. The methods consist of examining a thin section of sample under a microscope using $\times 100$ linear magnification, the sections being obtained either by melting and pressing on a hot stage, as in Method A, or by microtomy, as in Method B. A visual assessment of the uniformity of dispersion is made based on the reference photographs presented in the standard.

It is appropriate to mention the valuable role the microscope can play in throwing light on the cause(s) of failure of a particular component. Examination of the failed component under a microscope may reveal the presence of weld or spider lines, poor pigment dispersion, incomplete blending of two polymers, insufficient rubber phase in a 'high impact' grade of polymer, etc. This use of the microscope as a powerful diagnostic tool is, sadly, often overlooked.

REFERENCES

1. ISO 1183 (1987) *Methods for Determining the Density and Relative Density of Plastics Excluding Cellular Plastics.*
2. BS 2782 Methods 620A to E (1980) *Determination of Density of Solid Plastics Excluding Cellular Plastics.*
3. ASTM D792 (1966) *Specific Gravity and Density of Plastics by Displacement.*
4. ASTM D1505 (1985) *Density of Plastics by the Density Gradient Technique.*
5. DIN 53479 (1976) *Testing of Plastics and Elastomers; Determination of Density.*
6. ISO 1675 (1975) *Plastics – Liquid Resins – Determination of Density by the Pyknometer Method.*
7. BS 733: Part 1 (1983) *Pyknometers: Specification.*
8. Payne, N. and Stephenson, C. E. (1974) *Materials, Research and Standards*, **4** (1), 3.
9. BS 3715 (1964) *Concentration Gradient Density Columns.*
10. ISO 60 (1977) *Plastics – Determination of Apparent Density of Material that can be poured from a Specified Funnel.*
11. BS 2782 Method 621A (1978) *Determination of Apparent Density of Moulding Material that can be poured from a Funnel*; Method 621B *Determination of Apparent Density of Moulding Material that cannot be poured from a Funnel.*
12. DIN 53466 (1984) *Testing of Plastics; Determination of Bulk Factor and Apparent Density of Moulding Materials.*
13. ASTM D1895 (1969) *Standard Test Methods for Apparent Density, Bulk Factor and Pourability of Plastic Materials.*
14. ISO 61 (1976) *Plastics – Determination of Apparent Density of Moulding Material that cannot be poured from a Specified Funnel.*
15. ISO 1068 (1975) *Plastics – PVC Resins – Determination of Compacted Apparent Bulk Density.*
16. BS 2782 Method 621D (1978) *Determination of Compacted Apparent Bulk Density of PVC Resins.*
17. ISO 171 (1980) *Plastics – Determination of Bulk Factor of Moulding Materials.*
18. BS 2782 Method 621C (1983) *Determination of Bulk Factor of Moulding Materials.*
19. ISO 4591 (1979) *Plastics – Film and Sheeting – Determination of Average Thickness of a Sample and Average Thickness of a Roll, by Gravimetric Techniques (Gravimetric Thickness).*
20. ISO 4592 (1979) *Plastics – Film and Sheeting – Determination of Length and Width.*
21. ISO 4593 (1979) *Plastics – Film and Sheeting – Determination of Thickness by Mechanical Scanning.*
22. BS 2782 Method 631A (1982) *Determination of Gravimetric Thickness and Yield of Flexible Sheet.*
23. BS 2782 Method 632A (1982) *Determination of Length and Width of Flexible Sheet.*

24. BS 2782 Method 630A (1982) *Determination of Thickness by Mechanical Scanning of Flexible Sheet.*
25. BS 870 (1950) *External Micrometers.*
26. ASTM D374 (1979) *Standard Method of Test for Determining the Thickness of Solid Electrical Insulation.*
27. BS 907 (1965) *Dial Gauges for Linear Measurement.*
28. DIN 863 (1977) *Micrometers, External Micrometers, Terms and Definitions, Technical Requirements, Testing.*
29. DIN 878 (1979) *Dial Gauges.*
30. ASTM D1921 (1963) *Particle Size (Sieve Analysis) of Plastics Materials.*
31. BS 3406: Parts 1–4 (1961–3) *Methods for Determination of Particle Size of Powders.*
32. Particle Size Analysis Subcommittee of the Analytical Methods Committee of the Society for Analytical Chemistry (1963) *Analyst*, **88** (144), 156.
33. Irani, R. R. and Callis, C. F. (1963) *Particle Size: Measurement Interpretation and Application*, Wiley.
34. Dallavalla, J. M. (1948) *Micrometrics*, 2nd edn, Pitman.
35. Orr, C., Jr and Dallavalla, J. M. (1959) *Fine Particle Measurement*, Macmillan.
36. BS 410 (1976) *Test Sieves.*
37. ISO 565 (1972) *Test Sieves – Woven Metal Wire Cloth and Perforated Plate – Nominal Sizes of Apertures.*
38. ASTM D1705 (1982) *Particle Size Analysis of Powdered Polymers and Copolymers of Vinyl Chloride.*
39. DIN 53477 (1965) *Testing Plastics; Determination of the Particle Size Distribution of Powdered Thermosetting Moulding Materials by Sieve Analysis.*
40. ISO 4576 (1978) *Aqueous Dispersions of Homopolymers and Copolymers – Determination of Gross Particle Content by Sieve Analysis.*
41. ISO 4610 (1977) *Vinyl Chloride Homopolymer and Copolymer Resins – Sieve Analysis using Air-Jet Sieve Apparatus.*
42. ASTM D1457 (1983) *PTFE Moulding and Extrusion Materials.*
43. Lines, R. W. (1983) *Polymer Paint and Colour Journal*, **173** (4089), 111.
44. Winzon Research (1978) *Plastics World*, **36** (5), 80.
45. Lake Wales Plastic Corporation (1977) *Plastics Technology*, **23** (5), 13.
46. Autech Corporation (1978) *Plastics Technology*, **24** (3), 9.
47. Brunner, M. (1977) *Plastics South Africa*, **6** (5), 39.
48. Barking, H. L. and Getachev, A. (1976) *Kunststoffe*, **66** (12), 807.
49. F. G. Industries (UK) Limited (1976) *Reinforced Plastics*, **20** (10), 279.
50. Lohrbaecher, V., Boes, D. and Schneiders, A. (1974) *Kunststoffe*, **64** (9), 438.
51. Andersen, K. V. (1982) *Kunststoffberater*, **27** (12), 33.
52. Anon. (1982) *European Plastics News*, **9**(12), 40.
53. Autech Corporation (1982) *Modern Plastics International*, **12** (11), 74.
54. Tormala, S. (1982) *ANTEC '82, San Francisco, Calif., May*, 727.
55. Wright, M. G. B. (1985) *Polymer Paint and Colour Journal*, **175** (4155), 810.
56. ISO 468 (1982) *Surface Roughness.*
57. BS 1134 (1972) *Method for the Assessment of Surface Texture.*
58. Anon. (1984) *Rubber World*, **189** (5), 63.
59. BS 2782 Methods 823A and 823B (1978) *Methods for the Assessment of Carbon Black Dispersion in Polyethylene using a Microscope.*

Chapter Eight

Short-term Stress–Strain Properties

8.1 HARDNESS

8.1.1 General Considerations

'Hardness' is not a fundamental property – indeed, the very interpretation of the word is decidedly subjective; the most widely accepted concept is probably that of resistance to indentation, but others widely held include scratch resistance and rebound resilience. The last-mentioned is considered in Ch. 9 as a dynamic property, and scratch resistance in Ch. 10 under 'wear'.

Resistance to indentation, which is invariably measured before fracture occurs, is readily envisaged as some function of rigidity or modulus and, like this property, measurement of hardness is subject to all the effects of temperature, time and other test variables mentioned in §8.2. Phillips and Ramakrishnan[1] have investigated the effect of time of indentation and of recovery on the Rockwell hardness of various thermoplastics. In common with the traditional methods used for metals, hardness measurement of plastics usually takes the form of forcing a standard indentor – often a hardened steel ball – under known load into a flat surface of the material under examination and measuring the resultant degree of penetration. The viscoelastic nature of plastics introduces two complications for, in addition to the dependence of depth of penetration on the time of application of the load, the results of the metals-type test are calculated from the diameter of the indentation (or some other characteristic parameter if the indentor is not sphere-ended) and so the values may well depend on how speedily the diameter can be measured after removal of the loaded indentor. A fairly successful way of overcoming this problem in one type of test is to interpose a piece of carbon paper between the ball indentor and the test surface (carbon side to the latter) and measure the diameter of the carbon impression left after removal of the load.

There are a few attempts in the literature to relate indentation of plastics, but more often of elastomers, to modulus. Theories may be found for instance in Nielsen,[2] Ritchie,[3] Tangorra,[4] Brown[5] and Yeoh[6] who shows that some of the discrepancies in the various hardness/Young's modulus relationships observed can be attributed, at least in part, to different ways in which the modulus is defined.

Suffice it to state here that in addition to the effects of temperature and time and the nonlinear response of strain to stress, we have the fact that, at best, indentation is a complex function of properties including modulus, force and indentor profile, some effects of the latter having been quantified by Crawford and Stephens.[7]

It therefore follows that hardness values according to one method cannot generally be compared with those derived from another, though some *ad hoc* correlations have been published.

Work by Fett *et al.*[8] and by Bowman and Bevis[9] has sought to relate hardness measurements to orientation conditions in a variety of thermoplastics, while Martinez-Salazar *et al.*[10,11] have attempted to relate hardness to structure and morphology in polyethylene and Balta-Calleja[12] to crystalline polymers generally. Selden and Gustafson[13] have attempted to correlate hardness and tensile properties for a number of materials.

Hardness testing of films presents special problems and Sato[14] describes the use of a pendulum hardness tester but notes that it has significant limitations.

General reviews of hardness testing are given by Davies *et al.*[15] and Fenner,[16] and for plastics specifically by Maxwell,[17] Lysaght,[18] Boor,[19] Haldenwanger,[20] Gouza[21] and Livingston.[22] Craft and Whelan[23] have given a very short, simple résumé of a number of hardness tests.

8.1.2 Standard Test Methods

ISO Standards

There are, at present, three standard methods in ISO: ISO 868 (1985)[24] for the measurement of Shore hardness; ISO 2039: Part 1 (1987)[25] for ball indentation hardness; and ISO 2039 Part 2 (1981)[26] for Rockwell hardness.

ISO 868 describes two methods for the measurement of hardness using different shaped indentors. The Shore A method employs a truncated cone as the indentor and should be used for the softer range of plastics, a specimen thickness of at least 6 mm being required; the Shore D method, for harder plastics, employs a sharp cone with slightly rounded end but is also used on specimens at least 6 mm in thickness. Figure 8.1 gives the essential dimensions of these indentors.

It is recommended that measurements are not made less than 12 mm away from any edge.

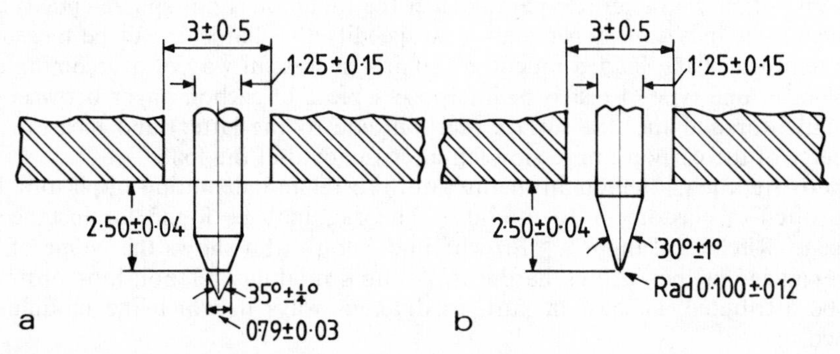

Fig. 8.1 Shore durometer indentors: ISO 868 (dimensions in millimetres). (a) Type A, (b) Type D

As already mentioned and as will be discussed in more detail in §8.2, the viscoelastic nature of the material necessitates that the reading taken after a standard period of time. For these methods this is the set at 15 ± 1 s, although an 'instantaneous' hardness may be estimated after a 1 s application of the durometer spring.

ISO 2039 : Part 1 employs a 5 mm diameter hardened steel ball which is pressed into the specimen under a specified load selected to give an indentation between 0·07 and 0·10 mm for Method A and between 0·15 and 0·35 mm for Method B. The time of application of the load is 30 s before the depth reading is taken and a test piece thickness of 4 mm is recommended. The hardness value is defined as:

$$\text{ball indentation hardness} = \frac{\text{applied load}}{\text{surface area of impression}} \qquad [8.1]$$

ISO 2039 : Part 2 follows the same procedure as laid down in ASTM D785 described more fully later. An annex to the ISO document describes the determination of Rockwell α, which, unlike the other Rockwell scales, correlates closely with the hardness as determined by ball indentation. Fett[27] has shown that

$$H = \frac{(448{\cdot}6)^{1{\cdot}23}}{(150 - R_\alpha)} \qquad [8.2]$$

where H = hardness by ball indentation and R_α = alpha Rockwell hardness number.

British Standards

Five standards exist within BS 2782 for the determination of hardness of plastics: Method 365A (1976)[28] for 'softness number'; Method 365B (1981)[29] for durometer hardness; Method 365C (1986),[30] for Rockwell hardness; Method 365D (1978)[31] for ball indentation hardness; and Method 1001 (1977),[32] the 'Barcol hardness test'.

Method 365B is identical to ISO 868 and Methods 365C and D to ISO 2039 : Parts 1 and 2 and so need not be considered further. Method 1001 uses a Barcol impressor type 934–1 which is identical to ASTM D2583 described below.

BS 2782 Method 365A is in agreement with BS 903 : Part A26 (1969),[33] rubber hardness, which in turn is in technical agreement with ISO 48 (1979)[34] although the method of expressing the result is different. The method consists of measuring, at 23 °C, the increase in depth of indentation of a steel ball 2·5 mm diameter into a flexible material when the force exerted on the ball increases from 0·3 N to 5·7 N. The softness number is defined as the depth of indentation of the ball in 0·01 mm units caused by this increase in force after it has been applied for 30 s. Figure 8.2 shows a typical instrument for carrying out this test.

The specimen should be between 8 and 10 mm thick for the standard test although other thicknesses are permitted provided the thickness is reported along with the softness number observed, an inverse relationship existing between the two. Edge effects are particularly important and no reading must be taken nearer than 10 mm to any edge.

A special note relating to PVC compounds is included as it is known that the hardness of plasticised compounds changes with time after moulding.[35] For such

Fig. 8.2 Dead load softness tester: BS 2782 Method 365A (courtesy H. W. Wallace & Co. Ltd)

materials, conditioning should be carried out for $7 \pm 0{\cdot}2$ days following moulding before the test is undertaken.

USA Standards

Three standard methods are given: ASTM D2240 (1984)[36] which uses Shore A and D durometers; ASTM D785 (1965)[37] which uses the Rockwell scales; and ASTM D2583 (1981)[38] which uses the Barcol impressor.

D2240 is essentially the same as ISO 868 except that the 1 s reading is taken as standard, instead of 15 s, and for the D scale a 3 mm thickness is permitted.

D785 makes use of the well known 'Rockwell hardness tester'; a typical instrument is shown in Fig. 8.3. Brinell hardness (see below) can also be measured on this instrument.

The scales (Table 8.1) overlap to a certain degree – the test increases in severity down the table – but correlation of scales is stated not to be desirable.

The test specimen (sheet 6 mm thick) rests on a flat anvil of at least 50 mm

Fig. 8.3 Rockwell and Brinell hardness tester (courtesy Foundrax Inspection Equipment Ltd)

diameter. Two procedures are described. In Procedure A, essentially the minor load is applied for 10 s and then the major load for 15 s; the hardness reading is taken off the scale 15 s after the major load has been removed, but with the minor loading still operating. In Procedure B, which unlike A is limited to Scale R only, the essential principle is that indentation is recorded 15 s after application of the major load, but with the latter still applied. The test values are quoted as α (alpha) Rockwell Hardness Numbers and are obtained by subtracting the indentation reading from 150.

Table 8.1 Rockwell scales (ASTM D785)

Rockwell scale	Minor load (kg)	Major load (kg)	Indentor diameter (in)	(mm)
R	10	60	0·5000 ± 0·0001	12·700 ± 0·0025
L	10	60	0·2500 ± 0·0001	6·350 ± 0·0025
M	10	100	0·2500 ± 0·0001	6·350 ± 0·0025
E	10	100	0·1250 ± 0·0001	3·175 ± 0·0025
K	10	150	0·1250 ± 0·0001	3·175 ± 0·0025

Further information on the Rockwell test as applied to plastics is given by Lysaght[18] and also by Boor,[19] who includes three scales not included in the above table but applicable to plastics.

D2583 describes the use of the Barcol impressor of Type 934–1. A note refers to Type 935 for softer plastics but its use has not been standardised.

A diagram of the general type of the impressors is given in Fig. 8.4. The indentor is a hardened steel truncated cone, having an angle of 26° with a flat top of 0·157 mm (0·0062 in) diameter. The indicating device has 100 divisions, each representing a depth of 0·0076 mm (0·0003 in) penetration – the higher the reading the harder the material. Test pieces must be at least 1·5 mm ($\frac{1}{16}$ in) thick and large enough in area to ensure a minimum distance of 3 mm ($\frac{1}{8}$ in) in any direction from the point of measurement to the edge of the specimen.

When testing, the housing of the indentor is applied quickly by sufficient hand pressure to ensure firm contact with the test specimen and the highest dial reading is noted; the number of readings necessary ranges upwards from five for homogeneous materials.

German Standards

DIN 53456 (1973)[39] is essentially the same as ISO 2039: Part 1 except that the major load must be selected from 49 N, 132 N, 358 N and 961 N, with a minor load of 9·81 N in all cases. The formula gives

$$H = \frac{0{\cdot}0535F}{h - 0{\cdot}04} \qquad [8.3]$$

where H = the hardness,
h = the penetration (mm),
F = the applied major load.

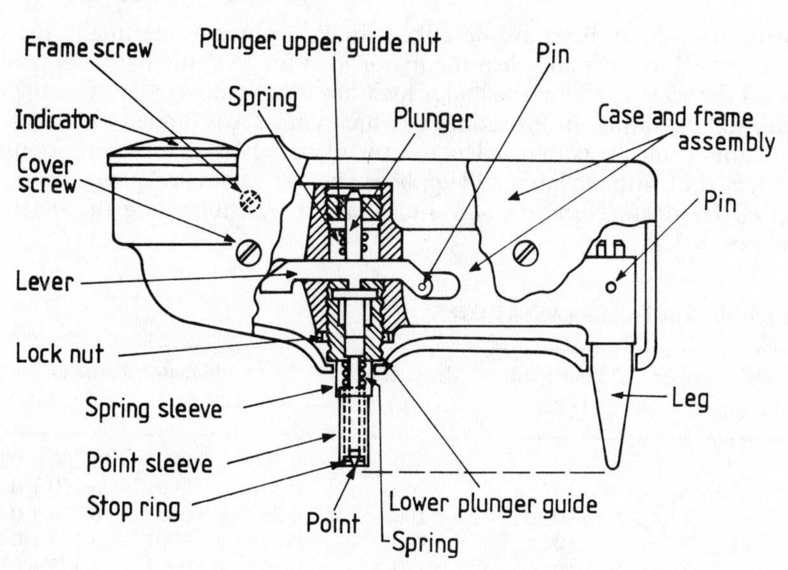

Fig. 8.4 Barcol impressor.[38] ASTM D2583

DIN 53505[40] covers the measurement of Shore hardness.

8.1.3 Other Methods

There are many other arbitrary hardness scales available for rubber and plastics and many which, fortunately, over the years have fallen into disuse. No attempt has been made to produce a comprehensive survey of all these methods and attention is directed below to only the more important.

The Brinell method already mentioned uses a spherical indentor and the hardness number is calculated from the formula:

$$HB = \frac{2F}{\pi D^2[1-(1-(d/D)^2)^{1/2}]} \qquad [8.4]$$

where F = load (kg)
D = diameter of indentor (mm)
d = diameter of impression produced by indentor during period of 15 s (mm).

The indentors used are normally 1, 2, 5 or 10 mm in diameter and loads are selected to give an F/D^2 ratio of 30, 10, 5 or 1; the method was originally designed for metals and therefore an F/D^2 ratio of 5 or 1 is usually needed for plastics materials. The HB value is quoted with the F/D^2 ratio employed in the test.

The fundamental objection to the test for plastics materials is their recovery after removal of the load, and to some extent the indefinition of the indentation during test; the use of carbon paper (§8.1) affords some means of overcoming the former problem. Hounsfield[41] suggests the use of a tensometer to record the depth of penetration of the ball during test, but points out that the general principle is unsound as load is not proportional to depth of penetration; he prefers the use of a paraboloid indentor where the cross-section is proportional to the distance from the apex, i.e. the depth of penetration. He shows plots of load against depth of penetration for Perspex and (low-density) polyethylene which are very nearly linear. The apparatus depicted in Fig. 8.3 has conversion scales relating diameter of indentation to depth of penetration of some of its ball indentors, so that Brinell hardness can be obtained essentially while the load is still applied and the problem of recovery is thus overcome.

Brinell testing of metals is covered by BS 240: Part 1 (1962)[42] and ASTM E10 (1984),[43] the details of which differ in some degree from those given above.

The (Vickers) diamond pyramid test is another standard technique for metals which has been borrowed for hard plastics, from time to time. Its use for metals is described in BS 427: Part 1 (1961)[44] and ASTM E92 (1982).[45] The test employs a right diamond pyramid on a square base, with an apex angle of 136° between opposite facets, the object of which was to simulate one indentor in the Brinell range. The test is conducted in similar manner and the test values are computed as follows:

$$HV = \frac{2F \sin(\theta/2)}{d^2} \qquad [8.5a]$$

where F = applied load (kg)
d = mean diagonal width (mm)
θ = apex angle of the pyramid = 136°.

Thus

$$HV = \frac{1{\cdot}844F}{d^2} \quad [8.5b]$$

A 5 kg load is normally used for plastics.

Campbell[46] has conducted an investigation of the variables encountered in the Vickers test and included acrylics in his assessment of time effects, etc.

The Vickers diamond principle is used in an adaptation of the Wallace microhardness tester. The apparatus is shown diagrammatically in Fig. 8.5.

The loads employed are very small (a fraction of a gram up to 3·5 kg) and the indentations correspondingly so; the test figures necessarily relate only to the surface of the test material, but the test is virtually nondestructive. The table on which the specimen is mounted is raised by the adjusting screw so that the indentor is returned to the exact position it occupied before the major load was applied. The indentor assembly is supported by leaf springs as shown, so that friction is eliminated, and as the principle is to return the indentor to the 'null' position, the leaf springs do not affect the result.

The 'null' position was formerly identified by means of headphones into which two oscillatory circuits were fed beating at nearly the same frequency. One of the circuits contained the capacitor shown and matters were so arranged that the frequencies were identical at the 'null' position, which could thus be readily identified by ear. An electronic 'null indicator' is now used. The wedge system and a suitable dial gauge permit indentation to be measured to 0·000 2 mm, clearly illustrating just how small the total indentation is in the test.

The same apparatus, but with a ball indentor 0·395 ± 0·005 mm in diameter, a contact force of 8·3 ± 0·5 mN and an indenting force increment of 145 ± 0·5 mN,

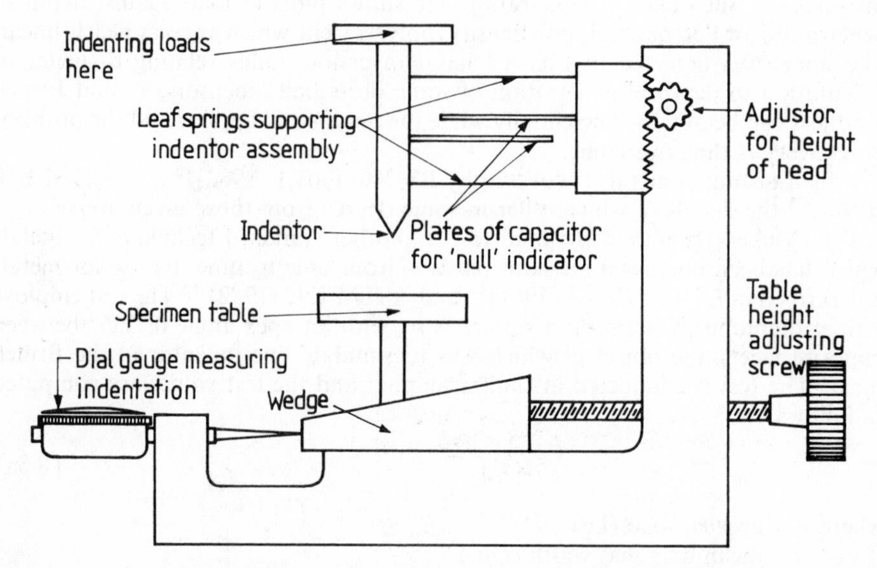

Fig. 8.5 Diagrammatic view of Wallace microhardness tester

is used for the microhardness testing of rubber (BS 903: Part A26 (1969),[33] Method M). The small size of the specimens required allows hardness measurements of small O-rings and irregularly shaped articles.

The results are read directly in 'International Rubber Hardness Degrees' but, depending on a number of variables, may not compare with those obtained from the already mentioned Normal Method, N, of BS 903: Part A26. The microtest is best used as a method in its own right.

Likewise, use of a suitable dial gauge will give a 'microtest' suitable for softness numbers of PVC, but the agreement with BS 2782, Method 365A (see above) is not good and this variation is not standardised.

The microtest for rubber has been standardised at international level and incorporated in ISO 48 (1979).[34]

Before leaving the closely allied subject of rubber hardness, mention must be made of BS 2719 (1975) *Pocket Type Rubber Hardness Meters* (*Methods of Use and Calibration*). One of the instruments described is the Wallace pocket meter, the mechanism of which (Fig. 8.6) illustrates one method of using springs for this purpose. This instrument and the Shore A durometer give very similar results.

Knoop Hardness Number is obtained by using a Knoop indentor on the commercial instrument known as the 'Tukon' tester[47] which is also capable of carrying out the Vickers (136°) diamond test. The Knoop indentor is again a diamond but with the ratio of long and short diagonals approximately 7:1 and, as a result, the depth of indentation is only about one-thirtieth of its length. Recovery problems are less[18] and only light loads need be applied, rendering the liability to shatter of brittle materials less than in the 136° test.

The TNO hardness tester[48] employs an indentor of a sapphire polished into

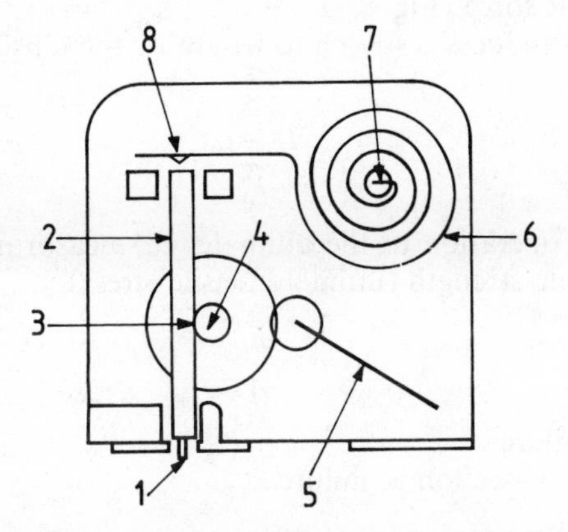

Fig. 8.6 Mechanism of Wallace pocket hardness tester. The indentor (1) is fixed to a shaft (2), having a rack (3) rotating a pinion (4), which transmits movement through a train of gears with hairspring to the spindle carrying the pointer (5). Upward movement of the indentor and shaft is resisted by the clock-type spring (6), fixed at its inner end on the split spigot (7) and having a 'pip' on its free end (8) bearing on the top of the shaft. Spigot (7) is part of a circular plate on the back of the instrument, rotatable by means of a key to loosen or tighten the spring

the form of a Vickers diamond pyramid and uses a capacitive technique for determining depth of penetration.

The Sward hardness test has been suggested for evaluating plastics materials nondestructively.[49]

The indentation testing of flooring materials is considered in detail by Gavan and Wein.[50]

The scleroscope is sometimes described as giving hardness values, but it is essentially a rebound (resilience) test. Likewise, Moh's test and pencil hardness are really for scratch hardness and are accordingly described in §10.2. Boor *et al.*[51] compare results from a number of the above hardness tests (several Rockwell scales, Knoop, Barcol, etc.) with those from scratch hardness (Bierbaum, pencil (Kohinoor) and Moh) and abrasion (ASTM Mar and Taber), examining a wide range of plastics materials. It is concluded, not surprisingly, that they mostly measure different parameters and therefore cannot be expected to give results which compare – even in placing materials in order of merit. Less comprehensive comparisons are given by Maxwell[17] and Lysaght.[18]

From time to time hot penetration tests have been suggested for checking the degree of cure of thermoset plastics[52,53]; the relationships, if any, are essentially empirical and subject to the influence of many compositional variables.

8.2 TENSILE STRESS–STRAIN

8.2.1 General Considerations

The most common type of stress–strain measurement is made in tension, that is by stretching the material. A tensile stress is thus applied, defined for a section of uniform cross-sectional area A_0 by the formula $\sigma_1 = F_1/A_0$ where σ_1 = tensile stress and F_1 = tensile force (Fig. 8.7).

If this tensile stress induces a stretch to length l_1, the tensile strain ε_1 is defined as:

$$\varepsilon_1 = \frac{l_1 - l_0}{l_0} \qquad [8.6]$$

Taking the stressing operation to the ultimate, i.e. measuring the force until the material breaks, tensile strength (ultimate tensile stress)

$$\sigma = \frac{F}{A} \qquad [8.7]$$

where F = force at failure
A = area of cross-section at failure.

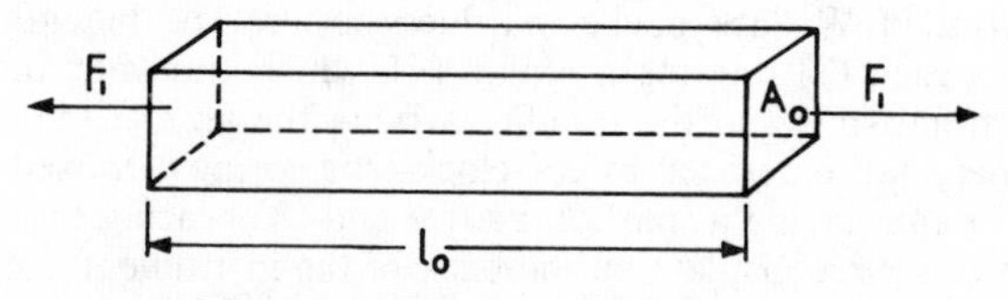

Fig. 8.7 Material in tension

As the material stretches so its dimensions orthogonal to the axis of applied force decrease and thus the area of cross-section decreases. However, for experimental convenience most tensile strengths are based on the original cross-section (A_0) since this is easily measured before the test is started.

Ultimate elongation, or elongation at break, equals $l - l_0$ where l is the length at failure. This is usually expressed as a percentage of the original length:

$$\text{elongation at break} = \frac{l - l_0}{l_0} \times 100 \text{ per cent} \qquad [8.8]$$

According to Hooke's law, for an ideal elastic solid stress is proportional to strain (Fig. 8.8). Even when we ignore the effects of temperature and most of the effects of time, no plastics material comes very close to this ideal. Plastics exhibit a whole spectrum of behaviour which, in qualitative terms, may be expected from such dissimilar materials as soft PVC, polystyrene, nylon and unplasticised PVC.

Carswell and Nason[54] have described five forms of behaviour which they express in the form of stress–strain diagrams (Fig. 8.9). It is immediately obvious that such terms as 'tensile strength' and 'elongation at break' can be misleading or gross oversimplifications even when used with the knowledge that such properties themselves have strictly limited application. For instance, in Fig. 8.9(a), is the correct value for tensile strength that deriving from the stress at break or that at the peak of the graph? Again, what significance has an elongation at break from a behaviour such as that depicted in Fig. 8.9(e), when we have a combination of steady extension with increase in stress, increase in extension with no increase in stress (even a drop in stress) and finally a rapid extension with relatively little increase in stress?

A generalised stress–strain plot is shown in Fig. 8.10.

A Hookean material has already been defined; the constant ratio σ/ε is called the Young's modulus, E. If a given plastics material has a stress–strain relationship which is initially linear, as in Fig. 8.10 up to position L, $E = \sigma_L/\varepsilon_L$.

To this already complex situation must be added the further complication of the viscoelastic nature of the polymer, which causes the material's response to an applied force to vary according to the time-scale and temperature of the experiment. Thus, a tensile test carried out at 500 mm/min on a sample of polyethylene could result in a stress–strain curve of the type given in Fig. 8.9(a), whereas at 5 mm/min the same material could exhibit the behaviour shown in Fig. 8.9(d). Similarly, conducting the test at two different temperatures could convert a hard, brittle plastic into a hard, tough one.

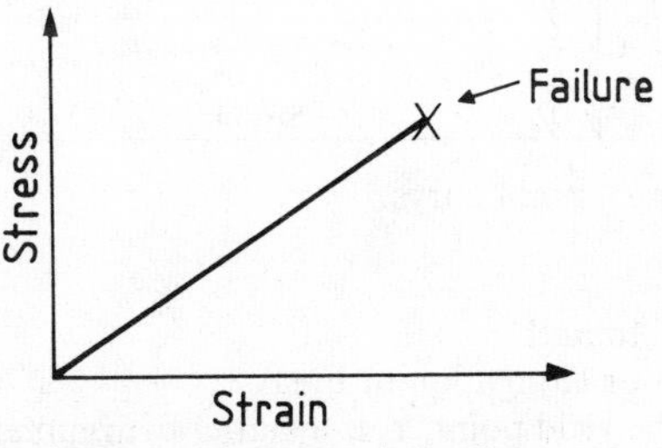

Fig. 8.8 Hookean behaviour

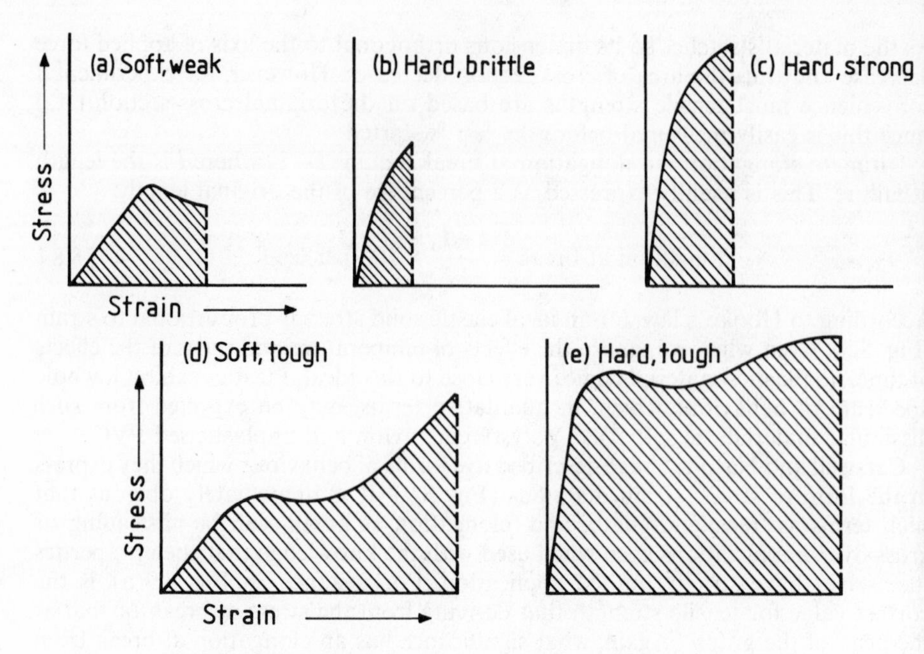

Fig. 8.9 Stress–strain behaviour of various types of plastics[54]

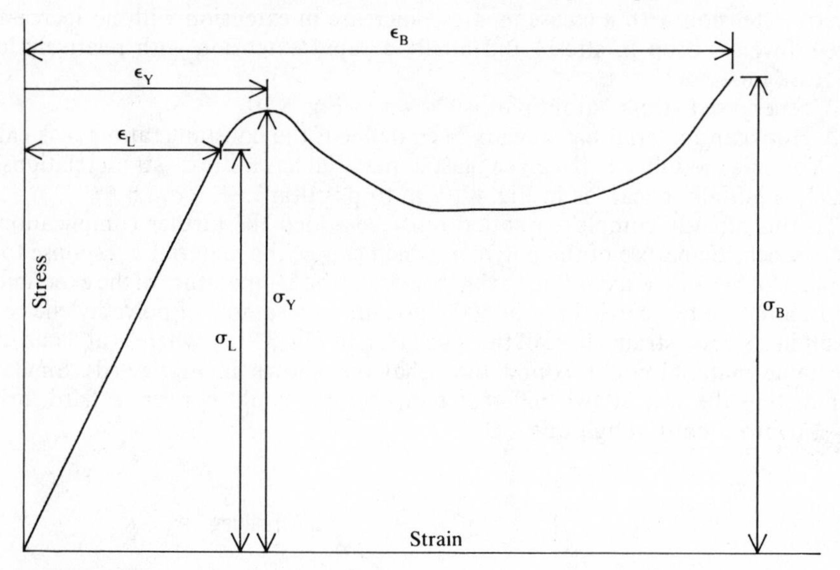

Fig. 8.10 Schematic stress–strain curve
σ_Y = yield stress
ε_Y = elongation at yield
σ_B = ultimate or tensile strength
ε_B = ultimate elongation or elongation at break
The precise location of the yield point, Y, is a matter of argument and, according to different authorities, may be anywhere between Y and L

More detailed descriptions of stress–strain and viscoelastic behaviour can be found in Nielsen[55] and Ferry.[56]

The foregoing serves to illustrate that statements such as 'The tensile strength of poly x is y MPa' are meaningless unless the method of test is also given.

As with processibility tests discussed in Ch. 5, the standard methods for determining the short-term mechanical properties, outlined in this chapter, give only very limited information with regard to the behaviour of the material and are best thought of as instruments of quality control. Data to be used for engineering design purposes must be obtained under conditions which simulate as closely as possible the actual service conditions to be encountered, if grave errors are to be avoided. This will often involve considerable testing, including dynamic mechanical (Ch. 9), fatigue (Ch. 12) and long-term (Ch. 11) property measurements.

8.2.2 Form of Test Piece

In earlier chapters we have talked of pre-conditioning our test specimens (Ch. 4) and even how to mould or machine them (Ch. 3), but there remains the question of the shape and size of pieces on which the measurements are to be performed. It is clear that different tests (for different properties) demand different test specimens – hence the reason for leaving the subject until discussion of each individual method – but it may be less obvious that, for instance, tensile tests on different materials employ specimens of different sizes; the effect of specimen size on tensile strength has been studied by Panov,[57] for instance, and more recently by Hawley *et al.*[58] in relation to the test pieces given in ISO 6239.[59] Again, this complication is due to the wide variety of types of material embraced in the term 'plastics', with quite different ultimate tensile strengths and percentage elongations at break.

To carry out a tensile test, i.e. a stretching test, some form of elongated specimen capable of being gripped at both ends is needed. The simple rectangular bar illustrated in Fig. 8.7 is not suitable. Even if reduced in thickness to facilitate gripping, it is still often unsatisfactory in that fracture can occur anywhere along its length and most probably the point of fracture will be within one or other of the gripped portions, because of the weakening effect of the compression exerted by the gripping chucks (but see below). Thus a failure load may be obtained which is lower than the characteristic tensile value for the material in the specimen cross-section and, additionally, the failure may occur outside the portion of the test piece being examined for elongation (the gauge length). In order to select the portion in which failure takes place and obtain a breaking load on a cross-sectional area unbiased by gripping mechanisms, a 'dumb-bell' specimen is normally employed. For metals this is a fairly accurate description, since the test specimens are usually produced by lathe turning (Fig. 8.11).

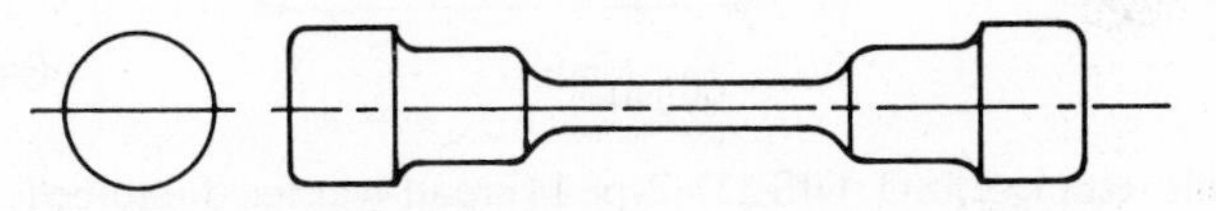

Fig. 8.11 Metal dumb-bell tensile test specimen

Many of the more rigid plastics may be machined on a lathe and so it is possible to produce similar specimens out of cast epoxide or poly(methyl methacrylate), for example; however, more often specimens have to be cut out from (thin) flat sheet or moulded from powder or granules and the orthodox dumb-bell specimen is not used as a standard test piece even when machined from castings. For sheet materials the specimens are 'two-dimensional dumb-bells', retaining the flat profile of the sheet.

Four basic types of test piece are found in the standards literature: the narrow-waisted dumb-bell, the broad-waisted dumb-bell, the 'dog-bone' dumb-bell and the parallel-side strip. These are illustrated in Figs. 8.12–8.15, the dimensions being those contained in ISO/DIS 527 which should be published shortly.[60]

The broader waisted, type 1, dumb-bell is the general purpose test piece for plastics materials. The type 2 test piece finds most use for highly extensible materials and not surprisingly is widely used for rubbers. The type 3 test piece is only employed for thermosetting resins if the type 1 is unsuitable and is little used nowadays. It cannot be used where elongation measurements are required. Finally the type 4 test piece is the parallel strip which, despite the disadvantage pointed out earlier, is found to be the most suitable for reinforced thermoplastic and thermosetting sheets. The problem of fracture in the grips is obviated by bonding end pieces onto the test piece (Fig. 8.15). In addition, the tab ends so formed can be drilled so that the test piece can be bolted into the grips to prevent grip slippage. Plastics films are also frequently tested using parallel strip specimens. There is some debate, however, as to whether parallel strip or dumb-bell shaped test pieces are to be preferred for testing plastics films.

ASTM D1708 (1984)[61] specifies the use of a microtensile dumb-bell which can be used whenever a limited amount of material is available. The use of small test

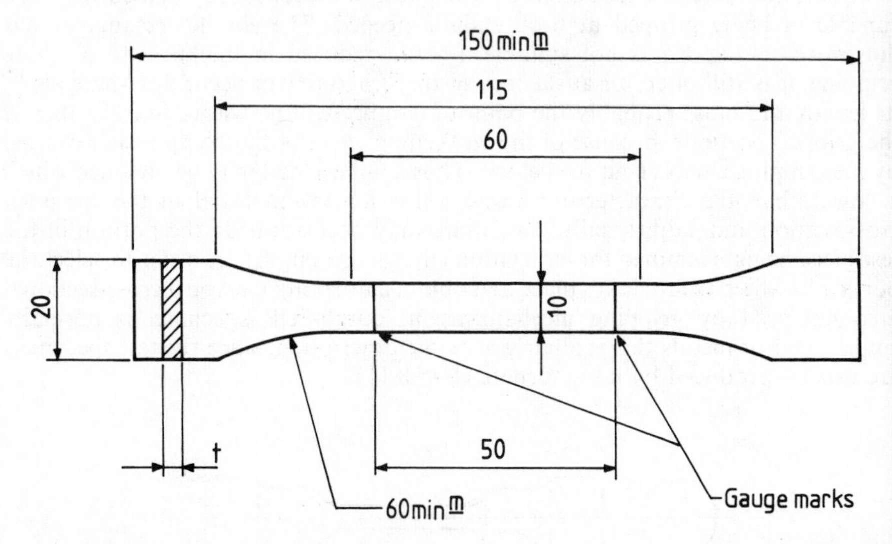

Fig. 8.12 Tensile test piece:ISO/DIS 527 Type 1 (broad-waisted dumb-bell) $t = 1$ minimum, $= 10$ maximum, $= 4$ preferred, $= 4$ for moulded test pieces (dimensions in millimetres)

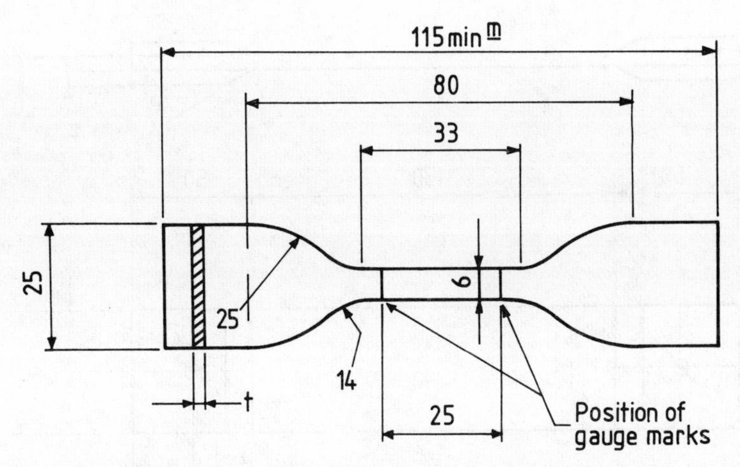

Fig. 8.13 Tensile test piece: ISO/DIS 527 Type 2 (narrow-waisted dumb-bell) $t = 1$ minimum, $= 3$ maximum, $= 2$ preferred (dimensions in millimetres)

pieces, based on scaled-down versions of the current ISO Type 1 dumb-bell, has been referred to.[59]

8.2.3 Apparatus

Testing Machines

The equipment required for carrying out tensile tests can vary from the very simple – adding weights to a freely suspended test piece until it breaks – to the very

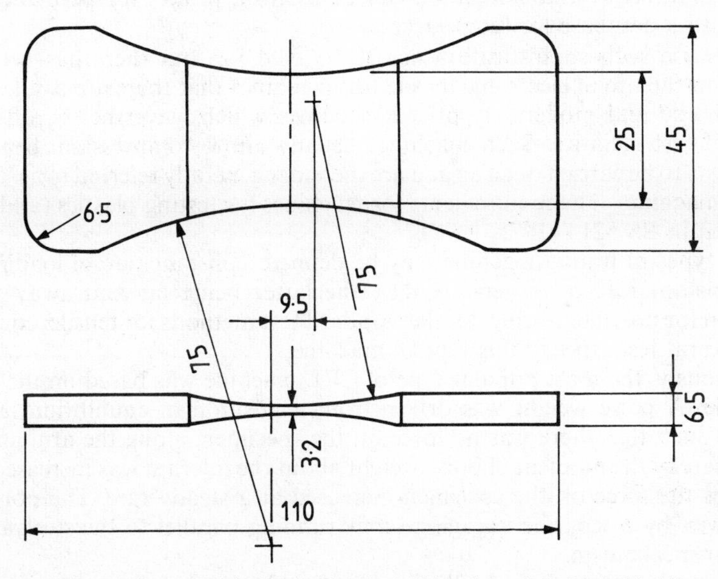

Fig. 8.14 Tensile test piece: ISO/DIS 527 Type 3 ('dog-bone' dumb-bell) (dimensions in millimetres)

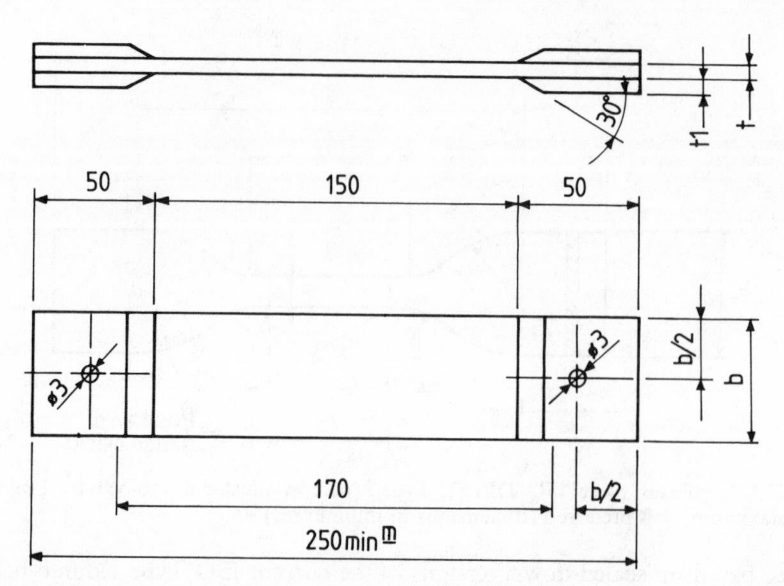

Fig. 8.15 Tensile test piece: ISO/DIS 527 Type 4 (parallel-sided strip) $t = 2$ minimum to 10 maximum, $t_1 = 3$ minimum to 10 maximum, $b = 25$ or 50 (dimensions in millimetres)

sophisticated – fully automatic machines where test pieces are placed in a loading carriage and tested one after the other with no further operator intervention. Built-in microprocessor units can effect initial area compensation and the digital output of strength and elongation can be fed to a printer for permanent record and/or to a database for future access.

Of course such sophistication has to be paid for, but there has been such a growth in the use of electronic tensile test machines that there are available today quite simple and moderately priced machines which, nevertheless, offer a wide range of test facilities. Such machines usually allow compression, bending and shear tests to be carried out also and are therefore generally referred to as 'universal' testing machines. The requirements for machines for testing plastics (and rubbers) are given in BS 5214 (1975, 1978).[62]

Two types of testing machine may be defined; constant rate of loading (CRL) and constant rate of traverse (CRT), the latter being far and away the more common for polymer testing: all the standard test methods for tensile, compression and flexural tests specify this type of machine.

Previously, the most popular type of CRL machine was based on the steelyard principle. A poise weight was driven from a position of equilibrium about the fulcrum such that there was no force on the specimen, along the arm at a steady rate, whereby the moment of poise weight about the fulcrum was increased steadily and thus the force on the specimen increased at a steady rate. The poise weight was driven by a separate engaging screw running parallel to the steelyard, or by a similar mechanism.

With modern electronic feedback circuits it is relatively easy to set up a loop to control the cross-head movement in response to the force increase from the load cell and thus provide a CRL environment as well as CRT on the same machine. Not surprisingly, therefore, the steelyard principle is not used on modern machines. Constant rate of loading conditions are most commonly called for in testing textiles and adhesives.

Constant rate of traverse machines are those where the test is controlled by the straining of the specimen and the resulting force is measured. This is a somewhat crude description because very few machines actually achieve the neat and tidy ideal of constant rate of extension of the specimen. Secondly, the rate of separation of the driven jaws of the test machine is the sum of the deformation of the specimen plus movements in the jaws themselves and in the means of attaching the jaws to the machine, so that the rate of jaw separation, whether constant or not, is related to the specimen deformation (stretching, bending or compression) by a variable and probably unknown amount. Only by taking a gauge length of the specimen (see below), over which straining properties are uniform, and controlling the rate of separation of the jaws with reference to the extension of the gauge length, will true constant rate of extension by achieved. Moreover, many machines fondly thought to be 'CRT' are not this by any manner of means.

The 'dynamometer' or force-measuring device may be found to 'yield' significantly under the load it is measuring and thus contribute, for instance, to the apparent extension of a tensile specimen: machines are termed 'soft' if this yield is appreciable relative to the deformation in the specimen and 'hard' if it is negligible. Finally, the mechanism driving or separating the jaws may alter in performance significantly according to the force it is acting against in the specimen. Again, however, modern load cells are usually designed for high stiffness and feedback circuitry can compensate to a large extent for any tendency of the drive mechanism to slow down with increasing load. So having drawn attention to all these shortcomings it is only fair to comment that with many machines the deviation from consistency of rate of transverse is not so serious as significantly to affect mechanical test results on plastics within practical limits, provided that the load is increased without 'surges' and the time to break is achieved within a defined period of time. Thus many specifications for mechanical tests speak in terms of 'extending the test specimen (or separating the jaws of the machine) at a rate which is essentially constant'.

Grip Systems

Of crucial importance to the successful completion of a tensile test is the grip system used to hold the test piece. This must ensure that the specimen is held firmly to minimise, or preferably prevent, slippage without crushing the tab ends; in addition, the design of the grips or chucks and their mode of attachment to the testing machine must be such as to ensure alignment of the test piece in the direction of strain, without any bending or shearing component.

There are many designs of grips to achieve this aim, some complementary but mostly having special applications according to the nature of the material under test and its physical form, be it strong or weak, soft or hard, wide or narrow, thick or thin. Some designs are described below.

Probably the most widely used is the 'wedge action' type, suitable particularly for rigid flat sheeting (Fig. 8.16). Some control over the friction between the test specimen and the gripping faces of the wedges can be obtained by having different

Fig. 8.16 Wedge action grips (courtesy Instron Ltd)

surface patterns on the wedge faces, varying from serrated patterns to perfectly smooth; smooth faces may be further modified by insertion of emery paper between jaw face and test specimen, with the rough paper surface facing the test specimen. The final choice depends on trial and error.

Also very useful for general purpose work are pneumatically or hydraulically actuated grips: the grip pressure is easily adjusted prior to test and is then maintained at a constant value throughout the test. Grip pressures from a few kilopascals to several tens of megapascals can readily be achieved.

Another type of grip, for the moulded thermoset type of tensile specimen, is illustrated in Fig. 8.17 while for thin brittle films, vice chucks are often suitable (Fig. 8.18).

Very strong laminates and sheet materials are best gripped by strip chucks (Fig. 8.19): the test piece is first drilled with a hole at each tab end to suit the

Fig. 8.17 Grips for moulded test pieces (courtesy Monsanto Ltd)

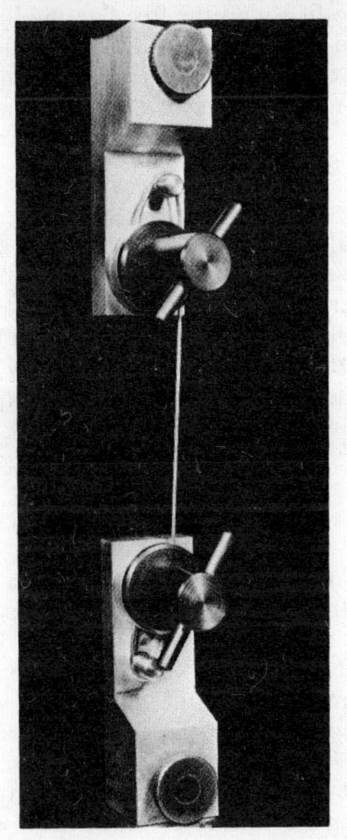

Fig. 8.18 Vice-type chucks (courtesy Nene Instruments Ltd)

Fig. 8.19 Strip chucks (courtesy Monsanto Ltd)

bolts, which are screwed down hard on stout serrated washers which fit into chucks; the chucks are hollow, to fit round the free floating lugs of the test machine and be gripped by pins fitted in the holes shown, matching holes in the lugs. (This method of attaching the chucks to the test machine lugs is widely used.)

Finally, in this by-no-means exhaustive list, may be mentioned self-tightening jaw chucks used widely for soft rubber and plastics sheeting, such as highly plasticised PVC (Fig. 8.20).

Force Measurement

The once universal pendulum-type of force-measuring device has now been all but relegated to a museum piece with the availability of electronic systems that offer more stiffness, negligible error due to inertia or friction and much wider operational range for a given load cell, by varying the amplification factor, than was possible for these early systems. Other advantages are the ability to measure very low forces, ease of providing zero offsets, and ability to convert to digital signals for automatic data processing and archiving on magnetic disc.

BS 5214[62] gives two grades of precision, ± 1 per cent and ± 2 per cent these being measured under static conditions. Performance under dynamic conditions is much more difficult to determine and is not as yet required for this type of testing.

There are a number of ways in which load cells can be constructed with many variations on the basic themes. Two common methods are the proof ring and strain gauge bridge.

A proof ring, shown schematically in Fig. 8.21, depends on the measurement of an extremely small deformation of a stiff but perfectly elastic metal member by a suitable electrical transducer. The proof ring is made from a steel block by cutting a slit and two cylindrical holes, the transducer being screwed into one side of the block and bearing against the opposite site of the slit. Application of a tensile force widens the slit, thus causing the transducer sensor to move and produce an electrical signal proportional to the force.

Fig. 8.20 Self-tightening jaws (courtesy Instron Ltd)

A strain gauge bridge, shown schematically in Fig. 8.22 consists of four resistive elements, three of which are fixed resistors of high stability and the fourth is the measuring element which is rigidly mounted on a plate in the load cell body such that as the force is applied and the load cell deforms, it distorts the measuring element. This alters the resistance of the element and hence produces an out-of-balance signal proportional to the applied force.

Elongation Measurement

The requirements for measuring elongation in highly extensible materials have been dealt with by Brown[5] and BS 5214[62] gives various grades of extensometer with differing degrees of precision. The accuracy requirements range from ± 1 mm down to $\pm 0{\cdot}05$ mm. Such a wide range is clearly required when one remembers that plastics materials vary from plasticised PVC and polyolefin films through to

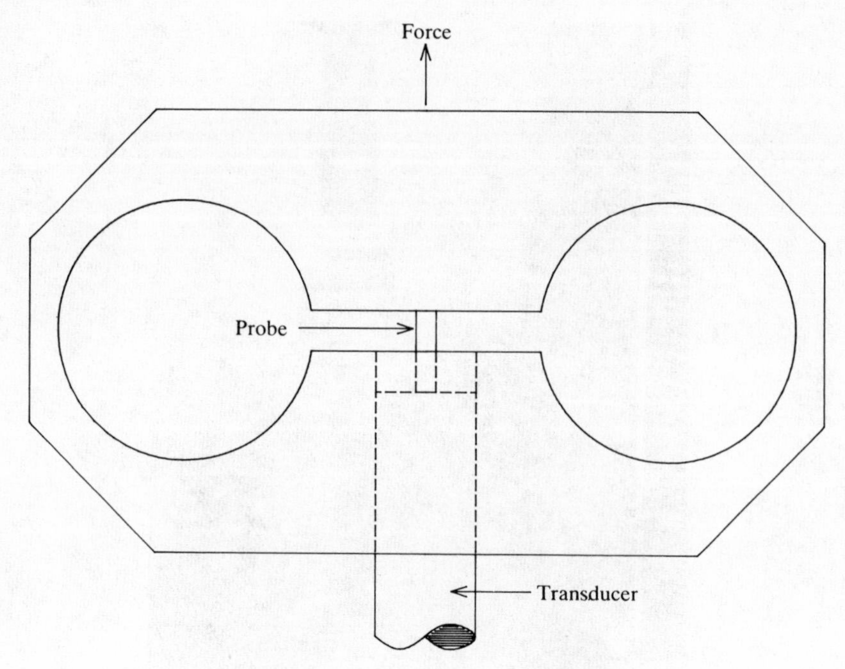

Fig. 8.21 Proof ring (schematic)

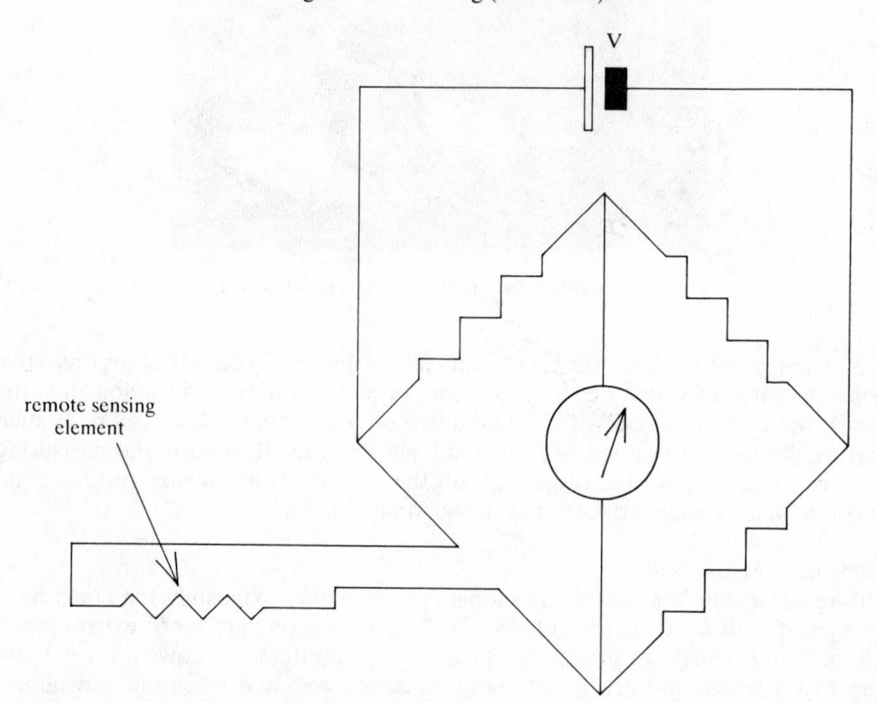

Fig. 8.22 Strain gauge (schematic)

carbon fibre reinforced thermosets, not only ultimate elongation being of interest, but also elongation at yield and Young's modulus.

Extensometers may be of the contact or non-contact type. As the name implies the contact type rely on the physical contact between the extensometer and the specimen to sense the change in length during test. Such devices may use strain gauge elements, LVDTs or potentiometers according to the precision required and can be of relatively simple, robust construction. However, the need for physical contact places conflicting requirements on the design: the clips must exert sufficient force to prevent any slippage during test, yet not so much force as to induce failure at the point or line of contact. These extensometers may also be excluded when tests at elevated or subambient temperatures are required.

The non-contacting extensometer avoids both of these problems since light beams are used to track the movement of contrasting colour gauge marks on the test piece, servomechanisms being used to drive the optical heads in the appropriate direction. The separation of the heads is usually measured electrically via potentiometers or LVDTs. Not surprisingly such systems are the more expensive and can cost as much as the tensile machine to which they are attached.

8.2.4 Definition of Terms

There are a multiplicity of terms used in tensile testing and there is sometimes confusion over the precise meaning of a given term. For example, tensile strength is usually taken to be the maximum tensile stress exhibited during a test, but occasionally is taken as the stress at break; these will not always be the same (see Fig. 8.9(a)).

The following are those that are most commonly found to pertain:

Tensile stress (nominal)	The tensile force per unit area of the original cross-section within the gauge length carried by the test piece at any given moment. The standard unit is megapascal (MPa) = MN/m^2 (meganewton/metre^2) = N/mm^2.
Tensile strength (nominal)	The maximum tensile stress (nominal) sustained by a test piece during a tension test.
Tensile stress at break	The tensile stress which occurs at break of the test specimen.
Yield stress	The tensile stress at which occurs the first marked inflection of the stress–strain curve. Where any increase in strain occurs without any increase in stress, this point is taken as the yield stress.
Offset yield stress	The tensile stress on the stress–strain curve where the curve departs from initial linearity by a specified strain.
Gauge length	The original length between two marks on the test piece over which the change in length is determined.

Strain	The change in length per unit original length of the measured gauge length of the test specimen. It is expressed as a dimensionless ratio.
Percentage elongation	The strain produced in the test piece by a tensile stress, expressed as a percentage of the gauge length.
Percentage elongation at yield	The percentage elongation produced in the gauge length of the test piece at the yield stress.
Percentage elongation at break, or at maximum load	The elongation at break, or at maximum load, produced in the gauge length of the test piece, expressed as a percentage of the gauge length.
Proportional limit	The greatest stress which a material is capable of supporting without any deviation from proportionality of stress to strain (Hooke's law).
Elastic modulus in tension (Young's modulus)	The ratio of tensile stress to corresponding strain below the proportional limit. The stress–strain relationship of many plastics does not conform to Hooke's law throughout the elastic range but deviates therefrom even at stresses well below the yield stress. For such materials the slope of the tangent to the stress–strain curve at low strain is usually taken as the elastic modulus.
Secant modulus	In general, the ratio of stress to strain at any given point on the stress–strain curve.

8.2.5 Standard Test Methods

There are currently two internationally standardised test methods for the determination of the tensile properties of plastics: ISO 527[60]) for testing plastics test pieces at least 1 mm in thickness and ISO 1184 (1983)[63] for test pieces less than 1 mm thick. The test pieces used in ISO 527 have already been illustrated in §8.2.2; the test pieces for ISO 1184 are of types 1 and 2 only, the parallel strip type with a length of at least 150 mm and a width between 10 and 25 mm as given in the previous version (R1184) having been withdrawn.

Nine test speeds are given in ISO 527 as follows:

Speed *A*			1 mm/min ± 50%
Speed *A*1			2 mm/min ± 20%
Speed *B*			5 mm/min ± 20%
Speed *C*			10 mm/min ± 20%
Speed *D*[a]	20	or	25 mm/min ± 10%
Speed *E*			50 mm/min ± 10%
Speed *F*			100 mm/min ± 10%
Speed *G*[a]	200	or	250 mm/min ± 10%
Speed *H*			500 mm/min ± 10%

[a] The two *D* and *G* speeds are allowed because both are in very widespread use throughout the world.

Somewhat more consistently, ISO 1184 allows 2 or 2·5 mm/min for speed *A*1 although it calls this speed *B*.

The British equivalent to ISO 527 is BS 2782 Method 320 (1976)[64] and of ISO 1184 is BS 2782 Method 326 (1977)[65] although these are not identical as the parallel strip is still used in UK for film test pieces. The tensile testing of glass reinforced plastics is specifically covered in BS 2782 Method 1003 (1977),[66] which follows the same basic procedure as Method 320 but with special emphasis on sample preparation and the problems of anisotropy. The measurement of modulus is also given rather more detailed description than is to be found in the more general test method.

Method 320A is for flexible thermoplastic sheet, extrusion and moulding compounds. The test speeds allowed are 100 mm/min and 500 mm/min and the test pieces are stamped from sheet.

Method 320B is for injection-moulded thermoplastic materials including filled and reinforced compounds. The test speeds allowed are 1, 5, 25, 50 and 100 mm/min and the specimens must be injection-moulded.

Method 320C is for rigid thermoplastic and thermosetting sheet; test speeds are the same as for Method B but test pieces are machined from sheet.

Method 320D, as previously mentioned, is for rigid thermosetting moulding materials including filled and reinforced compounds. The test speed is 5 mm/min and the test pieces must be either compression- or injection-moulded.

Method 320E is for fibre reinforced materials incorporating mat, cloth woven rovings. It can be used for pre-pregs. Test pieces are machined from sheet and the allowed test speeds are 1, 5 and 10 mm/min.

Method 320F is for unidirectional fibre reinforced materials, including pre-pregs. The test piece as illustrated in Fig. 8.23 is something of a hybrid between the 'dog-bone' and parallel strip in that it has bonded aluminium end pieces on a strip specimen but the centre section is reduced in thickness from 2 mm to 1 mm. The test speeds are 1 and 5 mm/min.

Method 327A[67] is specific to PTFE. There are a variety of test pieces permitted; turned dumb-bells as in Fig. 8.11, large and small standard 'rubber' dumb-bells, parallel-sided strips and straight cylinders.

Methods 326A, B and C for films differ only in their test speeds, being, respectively, 5, 50 and 500 mm/min. The test pieces are strip specimens of at least 170 mm length and widths between 10 and 23 mm.

The two standard American tests are ASTM D638 (1984)[68] and D882 (1983)[69] and although, as previously mentioned, the test pieces differ dimensionally from European test pieces the test speeds and temperature are in agreement. For D638 the allowed speeds are 1, 5, 10, 50, 100, 500 mm/min according to the test piece selected; for D882 they are selected according to distance between the grips and the required initial strain rate, which can be 10 per cent, 50 per cent or 1000 per cent per minute. A standard test temperature of $23\,^{\circ}\mathrm{C} \pm 2\,^{\circ}\mathrm{C}$ is used throughout.

There is a metric version of ASTM D638, designated D638M which uses test pieces of type 1 and 2 for ISO 527 these being referred to in the ASTM as types M–I, and M–II, respectively. There is also a type M–III which is a nominally one-fifth sized type I, although the scaling used is not uniform.

ASTM D3039 (1976)[70] is specific to oriented fibre composites and uses the

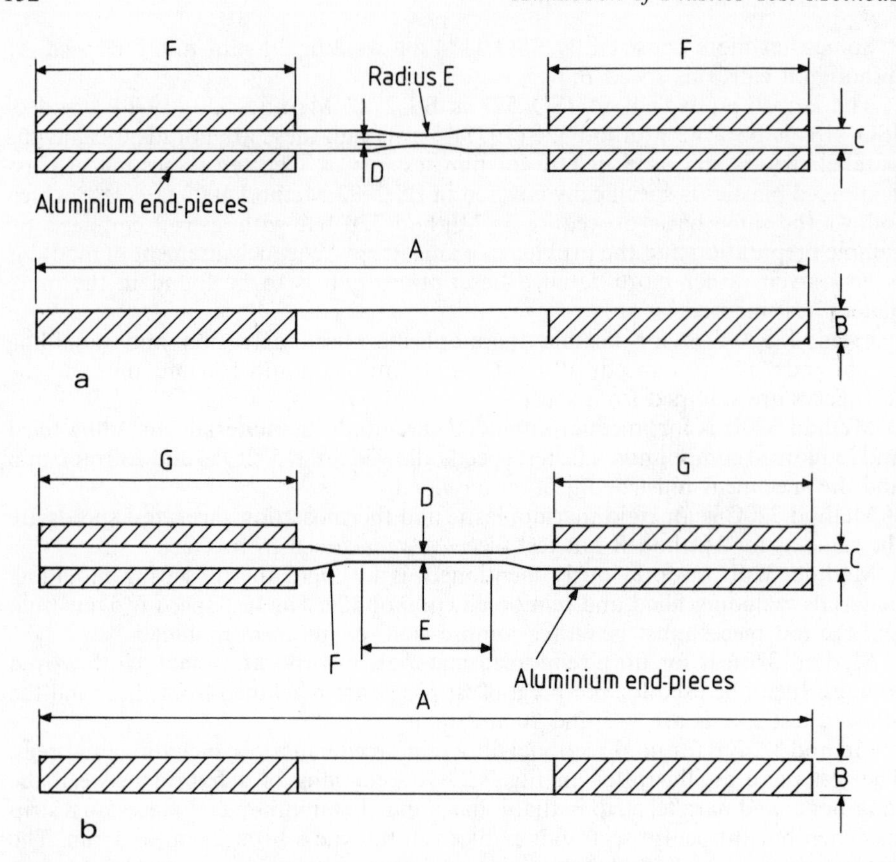

Fig. 8.23 Tensile test piece for unidirectional fibre reinforced materials: BS 2782, Method 320F. (a) Type A test piece; (b) Type B test piece (for modulus measurement only)

strip-type specimen of Fig. 8.15 at a temperature of 23 °C ± 1 °C and a strain rate of 1–2 per cent per min.

ASTM D2289 (1984)[71] covers high-speed (not impact) tensile testing of plastics. The same type of test piece is used as in D638 and D882 but at test speeds of 2500, 25 000 or 250 000 mm/min. Tensile impact testing is standardised in ASTM D1822 (1984) and is dealt with in §8.7.

The standard German test is DIN 53455 (1981 – Testing of plastics: Tensile Test) which includes the testing of films. The standard test speeds are 1, 2, 5, 10, 20, 50, 100, 200 and 500 mm/min.

8.3 COMPRESSION STRESS–STRAIN

8.3.1 General Considerations

Compression stress–strain is the antithesis of tensile stress–strain, both being caused by the application of two equal and opposite colinear forces but acting in the reverse direction to each other.

This reversal of force application, tending to crush the material rather than to stretch it, solves some of the problems inherent in carrying out tensile tests but creates others unique to itself. Thus, while there is no difficulty in gripping the test piece and the deformation of the test piece can be related to the movement of the machine cross-head much more accurately than in a tensile test, there is a greater need to apply a truly axial force. The test piece has a tendency to bend or buckle and friction between the test piece and end plates tends to prevent lateral expansion of the specimen as it is compressed, leading to barrel distortion.

These considerations mean that the parallelism of the test piece – especially of the end faces – and its relative dimensions, as determined by the slenderness ratio (defined later), must be more rigorously controlled than for a tensile test piece.

8.3.2 Test Equipment

There are test machines available which are specifically designed for compression testing but, as mentioned in §8.2.2, the trend today is towards 'universal machines' which allow several types of mechanical test to be applied. These can be used either directly in compression, using a load cell to measure the force, or in the tensile mode using a compression cage (Fig. 8.24) to reverse the test direction on the specimen while maintaining a tensile force on the load cell.

Since such cages may generate frictional forces during test, the compressive forces to be measured must be much greater than the frictional forces if serious errors are to be avoided.

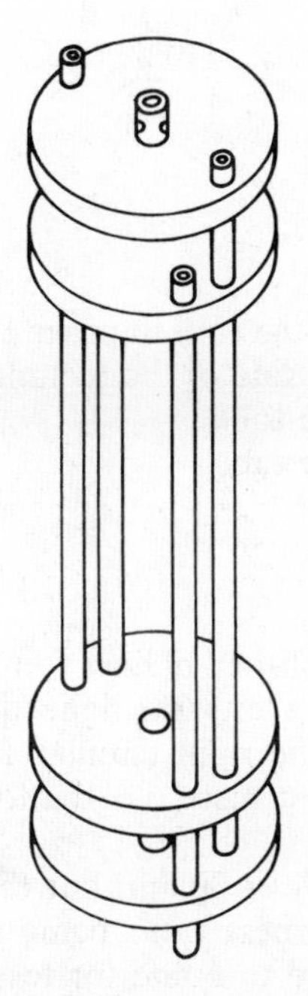

Fig. 8.24 Compression cage for tensile machine

It is worth noting that the compressive forces generated in a compression test are often considerably greater than the tensile forces generated in a tensile test because of the larger dimensions of the test pieces. If a single testing machine is to be used for both types of test, therefore, it may have to be of greater capacity than would be required for tensile testing alone, and, as the forces are high and deformations low, a particularly 'hard' or stiff machine is required.

8.3.3 Definitions

ISO 604 (1987),[72] Sect. 3 defines the various terms used in compression stress–strain tests which for the most part are analogous to the tensile definitions given above in §8.2.4. One term, however, has no counterpart in the tensile test and that is the already briefly mentioned 'slenderness ratio'. This is defined as 'the ratio of the length of a solid of uniform cross-section (column) to its least radius of gyration'. The least radius of gyration, i, is quoted for the commonly used test piece shapes. Thus, for a right square prism,

$$i = \frac{a}{3{\cdot}46} \qquad [8.9a]$$

For a right rectangular prism,

$$i = \frac{b}{3{\cdot}46} \qquad [8.9b]$$

For a right circular rod,

$$i = \frac{d}{4} \qquad [8.9c]$$

For a right circular tube,

$$i = \frac{\sqrt{(D^2 + d_1^2)}}{4} \qquad [8.9d]$$

where a = length of the side of the square prism
b = length of the shorter side of the rectangular prism
D = outer diameter of the tube
d_1 = inner diameter of the tube.

8.3.4 Standard Test Methods

ISO Standards

ISO/DIS 604 shortly to be published defines four types of test piece that may be used for carrying out compressive tests: the right square prism, the right rectangular prism, the right cylinder and the right circular tube. For each of these the test requires that the ends of the test piece, i.e. the load-bearing faces, be parallel to within 0·1 mm.

The preferred test piece height is 30 mm but the allowed range is from 10 mm to 40 mm, the standard slenderness ratio being set to 10. For materials which buckle under test this is reduced to 6 and for tests on sheet to 3, the force being applied in the thickness direction only. The rate of deformation of the test piece is 1 mm/min normally but speeds up to 10 mm/min are permitted for ductile materials.

Where only small amounts of material are available or where standard test pieces cannot be prepared from the particular product available, small specimens may be used as described in an Annex to the standard. Two sizes are permitted: Type 1, 6 × 5 × 3 and Type 2, 12 × 5 × 3 (mm throughout).

British Standards

BS 2782, Method 345A[73] is technically similar to the above ISO standard but allows a greater number of test pieces to be selected: only the Type 4 test pieces conform to ISO.

The Type 1 test piece is a cube of side 12 ± 0·1 mm and is suitable for thermoplastic or thermosetting sheet and for flat compression or injection mouldings, including filled or reinforced compounds, of thickness not less than 12 mm. For thicknesses greater that 12 mm, the test piece is machined on one face only to reduce the thickness to 12 mm.

The Type 2 test piece is similar to the above but used when the sheet or moulding thickness is less than 12 mm. Provided the sheet or moulding is at least 8 mm thick, the height of the test piece is the same as the sheet thickness. For materials less than 8 mm thick, however, test pieces are piled up and machined as appropriate to produce an overall test piece height of 12 mm.

The Type 3 test piece is suitable for casting and laminating resin systems containing no fibrous reinforcement. The test piece is either a cylinder of diameter 10 ± 0·2 mm and height 10 ± 0·2 mm or a cube of side 10 ± 0·2 mm.

The Type 4 test pieces are identical to those of ISO 604.

The Type 5 test piece is for thin sheet materials, glass fibre reinforced laminates and flat injection or compression moulded test pieces. Figure 8.25 illustrates a typical test piece of this type, although parallel-sided test pieces have been found to be satisfactory.

Unlike ISO 604, the test speed is selected according to the type of test piece to be tested.

USA Standards

ASTM D695 (1984)[74] employs the same technique as above but again the test

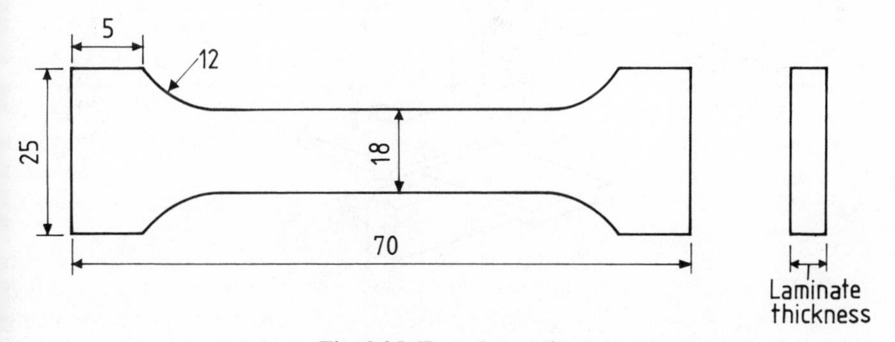

Fig. 8.25 Type 5 test piece

piece dimensions differ, the preferred dimensions being 12·7 mm square by 25·4 mm high for a right prism, and 12·7 mm diameter by 25·4 mm high for a right cylinder.

For rod material the diameter is that of the rod to be tested and the length is such as to produce a slenderness ratio in the range 11 to 15:1.

ASTM D621 (1964)[75] also employs a compression device but is primarily a heat resistance test and is briefly described in Ch. 16.

German Standards

German practice is covered in DIN 53454 (1971)[76] and is very similar to the BS standard in the form and dimensions of the test pieces that may be selected. Test speeds and other conditions are also comparable.

8.3.5 Plain Strain Compression Test

Williams and Ford,[77] using the fact that strain is much easier to measure in compression tests than over gauge lengths, have evaluated the 'plain strain compression test', first developed for metals, as a means of determining the permanent and total deformation curves for plastics up to high levels of strain such as may be encountered in engineering applications. The outline of the experimental set-up is shown in Fig. 8.26.

By this technique, the area under load remains constant. The specimen faces can be lubricated with a number of materials: molybdenum disulphide grease was used by Williams and Ford.

The application of the technique to the examination of polymers at large strains has been further described by Williams[78] and by Narisawa *et al.*[79] and the reader is referred to the first two of these three papers in particular for the full experimental details, mathematics and some results. The technique has not, however, been taken up as a national or international standard.

8.4 SHEAR STRESS–STRAIN

Shear, unlike tensile and compressive forces which are normal to the plane on

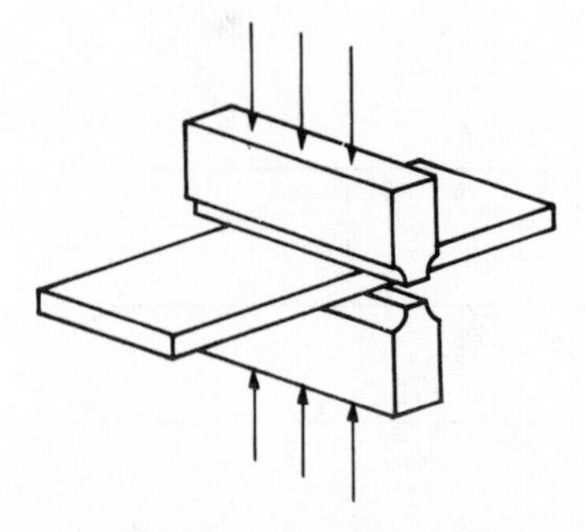

Fig. 8.26 Plain strain compression test[77]

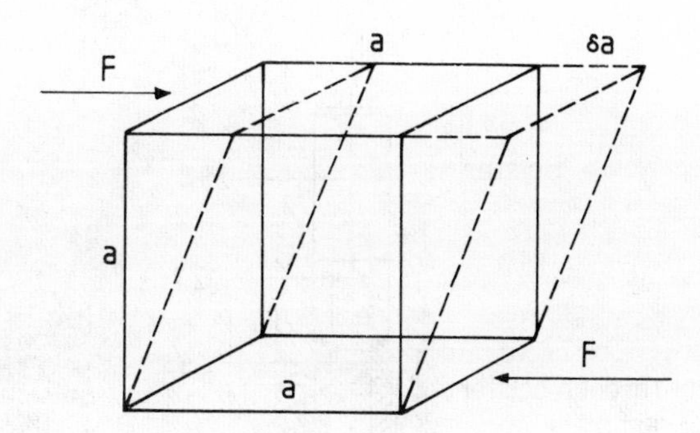

Fig. 8.27 Material in shear

which they act, acts parallel to the plane (Fig. 8.27):

$$\text{shear stress, } \tau = \frac{F}{a^2} \qquad [8.10]$$

$$\text{shear strain, } \gamma = \frac{\delta a}{a} \qquad [8.11]$$

If Hooke's law is obeyed,

$$\text{shear modulus, } G = \frac{\tau}{\gamma a} = \text{modulus of rigidity} \qquad [8.12]$$

Amongst a number of loading systems which give rise to shear forces is the instance where parallel but opposite forces act through the centroids of sections that are spaced an infinitesimal distance apart. This would lead to pure shear but cannot be realised in practice; somewhere near an approximation comes from the punch shear test where a punch bears on a flat sheet of test material resting on a die; the nearer the internal diameter of the die and the external diameter of the punch, the closer the approximation to pure shear. This type has found wide popularity as a standard test for plastics.

BS 2782 Method 340A,[80] for moulding materials, and 340B,[80] for sheet materials, both use the same die and punch assembly (Fig. 8.28).

In Method 340A the test pieces are moulded discs 25·3 ± 0·1 mm in diameter and 1·6 ± 0·1 mm thick, while in Method 340B the test pieces are rectangular bars 6·4 ± 0·2 mm wide, at least 32 mm long and of thickness equal to that of the sheet under test, except where this exceeds 6·35 mm, when the test piece is machined on one surface only to reduce the thickness to 6·10 ± 0·25 mm.

The test is carried out with the tool mounted in a testing machine operated as for compression and must be completed between 15 and 45 s from first application of load. The shear strength for Method A is given by:

$$\text{shear strength} = \frac{F}{\pi D T} \qquad [8.13a]$$

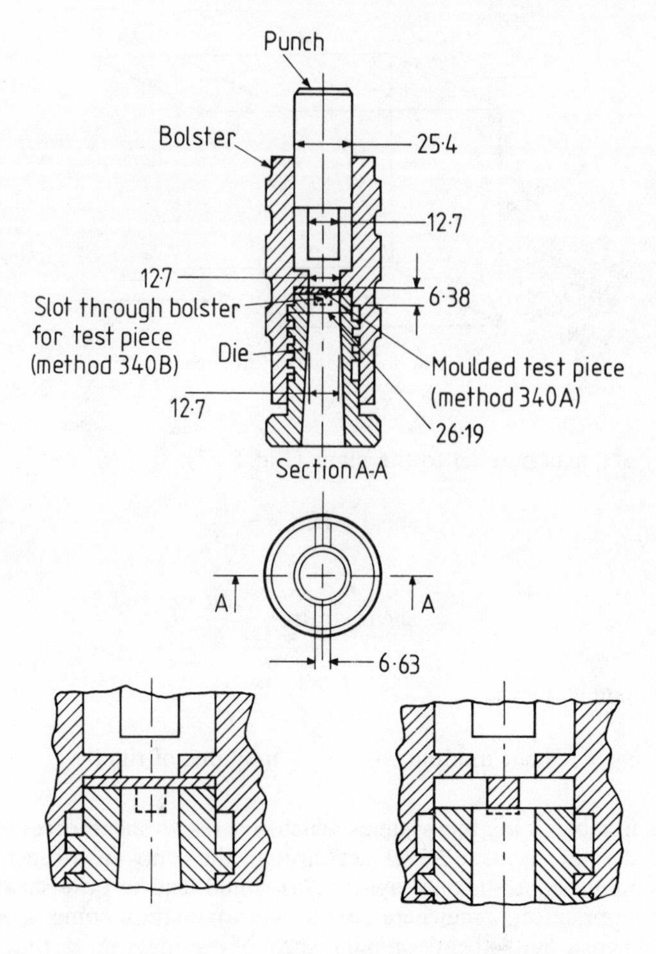

Fig. 8.28 Punching tool assembly: BS 2782 Method 340A or B

and for Method B by

$$\text{shear strength} = \frac{F}{2{\cdot}096BT} \qquad [8.13b]$$

where F = force at fracture
D = diameter of the punch
B = mean width of the test piece
T = mean thickness of the test piece.

The factor 2·096 is required to correct for the curvature of the sheared surfaces in Method B.

ASTM D732 (1984)[81] employs a similar technique but allows any thickness between 0·125 and 12·5 mm to be used and the test piece is drilled centrally to locate a guide pin.

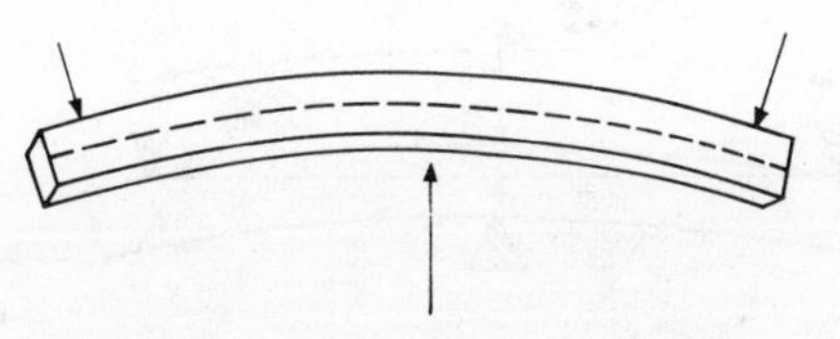

Fig. 8.29 Material in bending

At low temperatures torsional tests giving an apparent shear modulus are most frequently used (Ch. 16).

8.5 FLEXURAL (BENDING) STRESS–STRAIN

8.5.1 General Considerations

When a rectangular beam is bent a continuous change occurs from maximum tensile stress on one surface through the thickness to maximum compressive stress on the other.

In Fig. 8.29 (for a homogeneous isotropic material in pure bending) the top surface is in tension and the under surface in an equal compression; if the tensile and compressive moduli are equal, at the midpoint of thickness the stress is zero, where tension diminishes to zero before compression starts building up, and the dotted line represents this neutral axis.

It is beyond the scope of this book to develop the formulae for bending; suffice it to state that pure bending, i.e. the combination of only tensile and compressive forces, is never achieved and there is always some transverse shear component.

The bending formulae for small deformations given below have been derived assuming pure bending, which is assisted by having a span (length of loaded section) of beam large by comparison with the thickness so that bending approaches a true arc of a circle. Heap and Norman[82] describe flexural testing of plastics in detail and develop equations of more general applicability. For three-point bending of plastics, as in the diagram above, the span–depth ratio should be about 16:1, but this cannot be taken as invariably satisfactory; the appropriate figure for span–depth ratio depends on the characteristics of each individual material.

Three-point Bending

Three-point bending is the type of bending most commonly used in standard tests: for a rectangular beam supported at the mid-point, the flexural stress (or 'maximum fibre stress') is given by:

$$\sigma_F = \frac{3FL}{2bh^2} \qquad [8.14]$$

where F is the force at the mid-point and the other symbols are as shown in Fig. 8.30. At fracture, therefore,

$$\text{flexural strength} = \frac{3F_{\mathrm{m}}L}{2bh^2} \qquad [8.15]$$

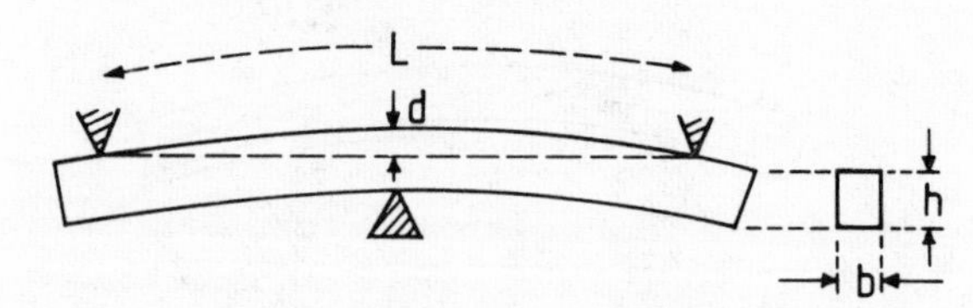

Fig. 8.30 Three-point bending

where F_m is the maximum force recorded. Flexural strength is also known as 'cross-breaking strength'.

A more accurate expression, which takes into account the horizontal component of the flexural moment, is given by

$$\sigma_F = \frac{3FL}{2bh^2}\left(1 + \frac{4d^2}{L^2}\right) \qquad [8.16]$$

The apparent modulus of elasticity in flexure (equal to Young's modulus only as a first approximation) is given by

$$E_b = \frac{L^3}{4bh^3} \times \frac{F}{Y} \qquad [8.17]$$

where F/Y is the slope of the initial linear force–deflection curve.

For a circular rod,

$$\text{flexural strength} = \frac{8F_m L}{\pi D^3} \qquad [8.18]$$

$$\text{modulus of elasticity} = \frac{4L^3}{3\pi D^4}\left(\frac{F}{Y}\right) \qquad [8.19]$$

where F_m, L, F and Y have the same significance as before and D = diameter of rod.

Four-point Bending
For a rectangular beam (Fig. 8.31),

$$\text{flexural strength} = \frac{6F_x L_1}{bh^2} \qquad [8.20]$$

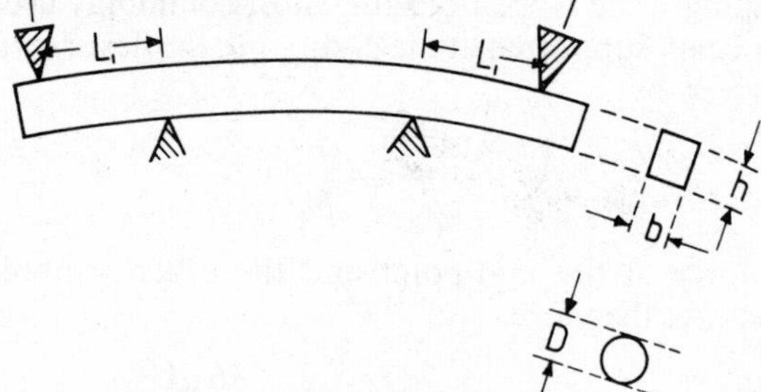

Fig. 8.31 Four-point bending

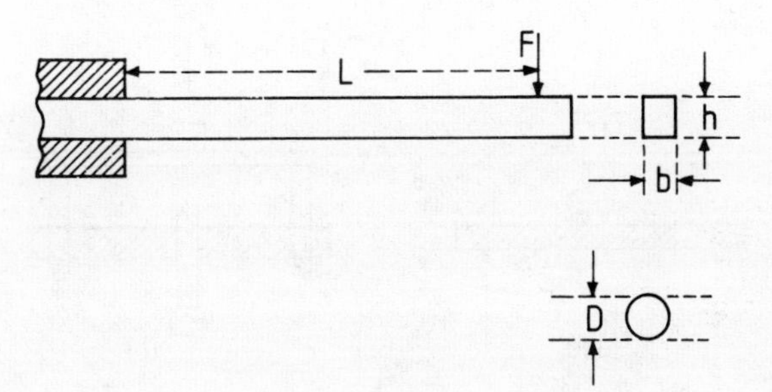

Fig. 8.32 Cantilever bending

where F_x = force on each bearing, i.e. half total force.

For a circular rod,

$$\text{flexural strength} = \frac{32F_xL_1}{\pi D^3} \quad [8.21]$$

where F_x = force on each bearing point, i.e. half total force.

Cantilever Bending

For a rectangular beam (Fig. 8.32),

$$\text{flexural strength} = \frac{6FL}{bh^2} \quad [8.22]$$

For a circular rod,

$$\text{flexural strength} = \frac{32FL}{\pi D^3} \quad [8.23]$$

8.5.2 Definitions

We have seen in the previous section most of the terms used in flexural tests and these are defined in a manner analogous to their counterparts in tensile and compression tests. There is, however, a term which has no counterpart in the other two modes of test. This is the 'flexural stress at the conventional deflection', which is defined in ISO/DIS 178 shortly to be published as the flexural stress at a deflection of 1·5 times the thickness of the test piece. This parameter is particularly useful for materials which do not fracture under test.

8.5.3 Standard Test Methods

Bending tests may be carried out in tensile or compression test machines (Fig. 8.33).

In addition to the usual variables of specimen size and shape, test speed and temperature, there must be added the radius of curvature of the bearing rods, which must not be too sharp to cause fracture. Cross-breaking strength tests are popular because the test specimens are simple and therefore easy to prepare and

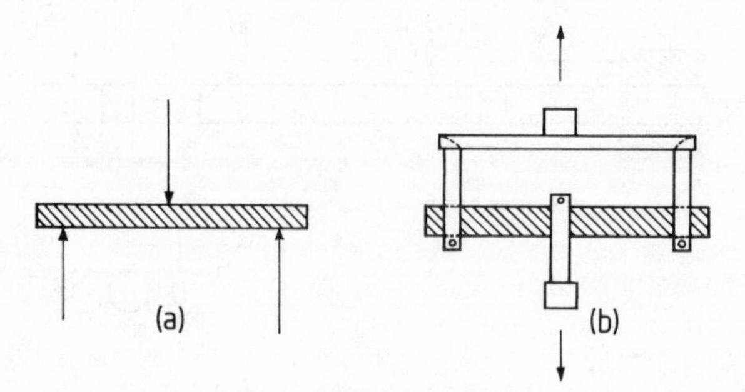

Fig. 8.33 Bending test: (a) in compression machine; (b) in tensile machine

the loads to be measured are relatively low; also gripping problems are eliminated and deflection data more easily obtained. Theoretically, in pure bending, cross-breaking strength and bending modulus are equal to tensile strength and Young's modulus respectively, but because bending is not pure and because many materials tested are not homogeneous and isotropic, these equalities are often very approximate at best.

Again, three-point bending is generally preferred in standard tests, but there are many advocates of four-point tests because in the latter the stress is equal, under practical conditions, over the whole of the span between the inner two supports, in contrast to the local maximum stress which occurs opposite the centre support in three-point bending which is essentially non-pure; clearly errors may result if the material in the region of the centre support is not representative of the whole.

ISO Standards

ISO 178[83] specifies the standard test piece as being at least 80 mm long, $10 \pm 0{\cdot}5$ mm wide and $4 \pm 0{\cdot}2$ mm thick. It does, however, acknowledge that such a test piece may be unsuitable for some materials, in which case the dimensions of the test piece must be selected so that the length is at least twenty times the thickness and the width should be between 10 and 25 mm for normal materials or between 20 and 50 mm for materials containing coarse fillers.

Where the normal sized test specimen cannot be used by reason of lack of material or geometrical constraints of the product available for test, a reduced size specimen is permitted the dimensions of which are: thickness between 1 and 2 mm, width 3 mm and length 20 × thickness.

The span length must be set to between fifteen and seventeen times the test piece thickness for normal testing, although for very thick and unidirectional fibre reinforced test pieces this can be exceeded to prevent delamination due to shear forces and for very thin materials the span-to-thickness ratio can be reduced.

The speed of testing is selected from the same list as used for tensile testing (see §8.2.5) so as to produce within 1 min, a deformation which is closest to the conventional deflection.

British Standards

BS 2782 Method 335A (1978)[85] is technically equivalent to ISO 178 and need not be considered further, while Method 1005 (1977)[85] is technically equivalent in so far as it deals with glass reinforced plastics only.

Method 336B of BS 2782 (1978)[86] employs the cantilever bending mode; the apparatus is illustrated in Fig. 8.34. Test pieces 70·0 mm long, 25·4 mm wide and 1·50 mm thick are moulded under the conditions appropriate to the material and a hole between 2·00 and 2·02 mm in diameter is drilled centrally through each, 50·8 mm from one end. A test piece is clamped as illustrated and the loading arm, without the 50 g weight, is attached. The cursor reading against the mirror scale is recorded after 5 mm, the 50 g weight is attached to the cursor and the new scale reading recorded, again after 5 min. The deflection in bend is calculated from the difference in these two scale readings.

There are two other methods within BS 2782 which, while not being flexural tests in the same sense as the above two methods, nevertheless employ the three-point loading principle.

The first measures the stiffness of plastics film and is detailed in Method 332A (1976).[87] The test measures the force required to push a sample of film into the rectangular gap formed between two metal plates (Fig. 8.35).

The second, detailed in Method 341A (1977),[88] gives a measure of interlaminar shear strength of composite materials containing fibrous reinforcement. The main difference between this method and method 335A above is that the span-to-thickness ratio is reduced from 16:1 to only 6:1, thereby increasing the shear forces so as to induce shear failure – a tensile type failure (illustrated in the standard) is considered unsatisfactory. The shear strength is calculated according to the expression:

$$S = \frac{0{\cdot}75F}{bh} \qquad [8.24]$$

where the symbols are as before.

USA Standards

The principal test method is given by ASTM D790 (1984),[89] which lays down two methods and two procedures. Method I uses a three-point loading system and Method II a four-point loading system, wherein the two load points are equally spaced from their adjacent support points and the distance between the load points is one-third of the support span. Procedure A is followed when the material under test fractures at small deflections and Procedure B is for materials undergoing large deformations during test. A range of support span-to-depth ratios is allowed according to the nature of the material.

ASTM D747 (1970)[90] is somewhat similar to BS2782, 336A although the apparatus and technique are more complex.

German Standards

The standard test method is detailed in DIN 53452 (1977)[91] and is very similar to the ISO standard, using three-point loading and the same, preferred, test piece with other dimensions being permitted if required. A four-point bending test is also given, however, in DIN 53435 (1983)[92] which uses the Dynstat apparatus.

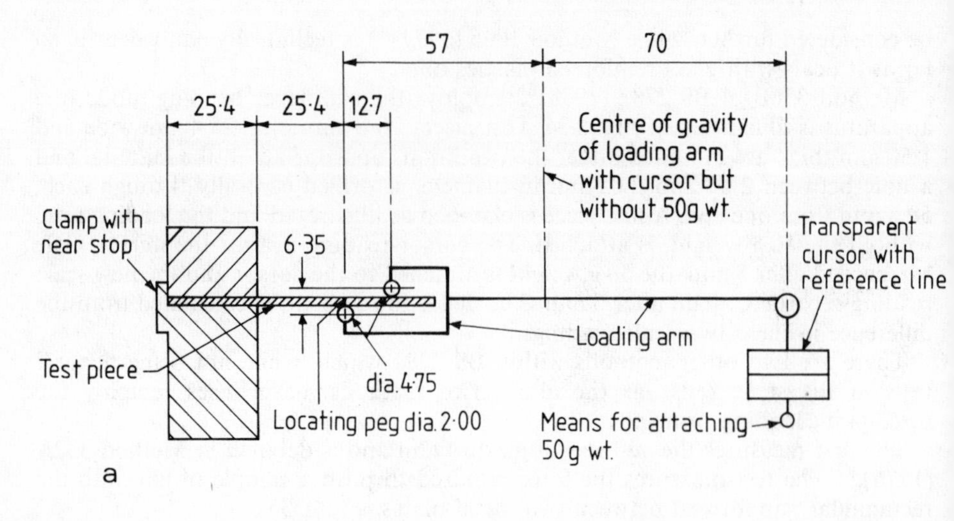

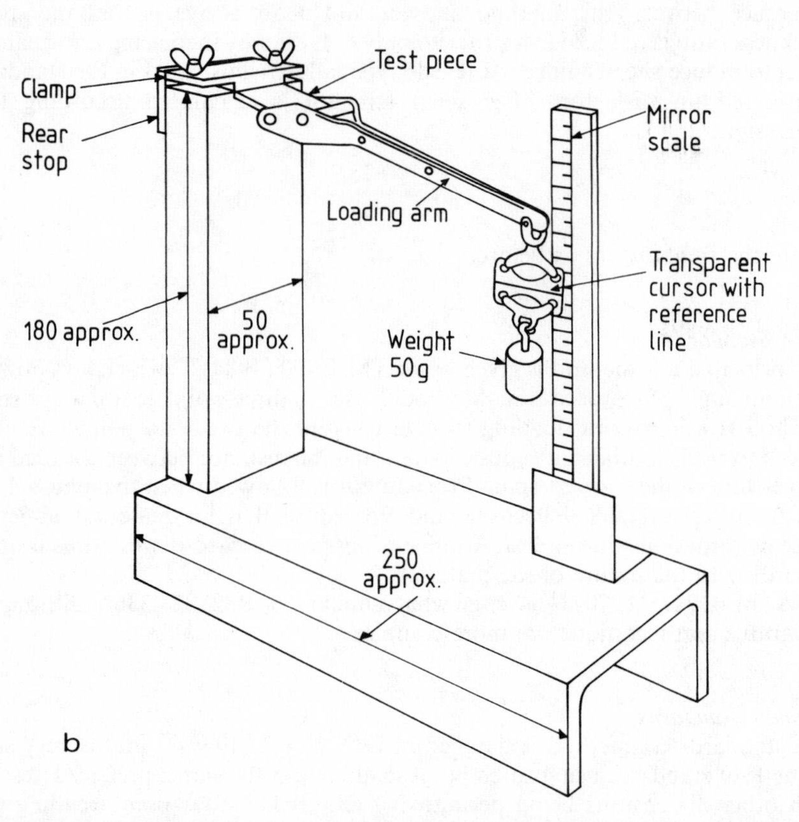

Fig. 8.34 Apparatus suitable for use with BS 2782 Method 336A (dimensions in millimetres)

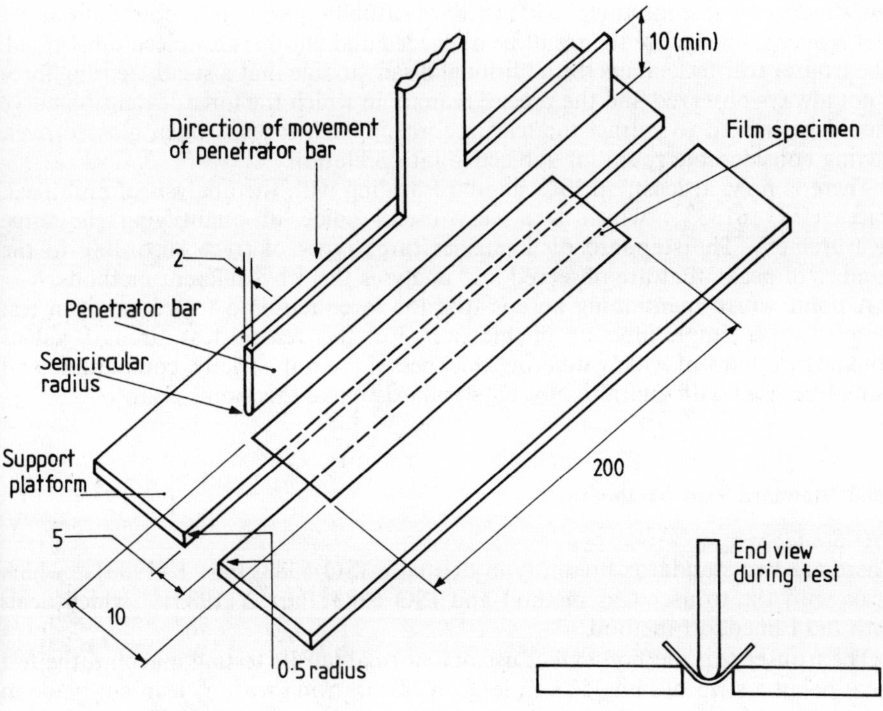

Fig. 8.35 Apparatus for determining stiffness of film: BS 2782 Method 332A

8.6 TEAR TESTS

8.6.1 General Considerations

Tear tests on plastics materials are invariably applied to plastics films, unlike their counterparts for rubbers which are more often applied to standard test sheets of 2 mm nominal thickness. It is easy to see why this should be so, since plastic film finds such widespread use in the packaging industry; obviously puncture resistance (§8.7) and tear resistance are significant parameters to be determined in assessing the suitability of a film material for a particular application.

Although the tests themselves are generally straightforward enough, the reproducibility is often rather poor and usually more test pieces are tested than is normal for, say, a flexural test. Films are also frequently more anisotropic than is the case with, for example, injection-moulded test pieces, and mono- or bi-axially oriented film is common. The properties along and across the machine direction can differ significantly, therefore, and both should be determined if incorrect conclusions are to be avoided.

A common difficulty encountered with tests of the 'trouser tear' and Elmendorf type is that the tear does not propagate in the direction of the test piece length but wanders off at some angle to it. Standards usually specify the amount of wander that is acceptable before the result be discarded and another test piece substituted. The trouser tear test suffers the additional disadvantage that a steady tearing force is not always observed and the precise manner in which the force–extension curve should be treated to extract the tearing force in such cases is often glossed over, leaving considerable room for subjective interpretation.

There is now, at least, an ISO standard dealing with the analysis of multipeak traces (ISO 6133[93]) which is a very useful guide in quantifying the force measurement. The standard distinguishes three types of trace according to the number of peaks that are observed and analyses them by different methods.

A point worth mentioning here is that the force required to tear a given test piece is not a simple function of thickness. For this reason tear strength values obtained on films of widely differing thicknesses are not, strictly, comparable and should be used with caution. Not all standards make this point clear.

8.6.2 Standard Test Methods

ISO Standards

There are two standards presently in existence ISO 6383: Part 1 (1983)[94] which deals with the trouser tear method and ISO 6383: Part 2 (1983)[94] which deals with the Elmendorf method.

The trouser tear method makes use of a normal tensile testing machine, the test piece being a strip of film 150 mm long by 50 mm wide with a clean cut made in one end to a distance of 75 mm, the cut being central to the width of the test piece. The two 'trouser legs' are held in the two jaws of the testing machine and a grip separation rate of 200 or 250 mm/min is employed. After an initial rise in the force–extension curve, a plateau region is generally observed as the tear propagates. The force of this plateau area divided by the film thickness is defined as the tear strength. The trouser tear test piece is now preferred to the angle tear test piece which also makes use of a tensile testing machine.

The Elmendorf apparatus (Fig. 8.36) consists of a stationary jaw and a movable jaw carried on a pendulum, preferably formed by a sector of a wheel or circle, which is free to swing about a substantially frictionless bearing. The test piece is of the 'trouser tear' type and is held in the two jaws. When the sector release mechanism is operated the potential energy of the pendulum causes the cut in the film to propagate along the film's length and the energy absorbed by this process is read directly off the calibrated scale on the pendulum; the scale is so positioned that, with no film sample present, a reading of 0 is obtained when the pendulum is released. The scale reading is converted to tear strength via the known calibration factor for the apparatus, the length of the tear produced and the sample thickness.

British Standards

These also cover the trouser tear and Elmendorf methods, BS 2782 Method 360B (1980)[95] being of the former type and Method 308B (1970)[96] of the latter. These methods are technically equivalent to the ISO standards described above but are not identical in all respects.

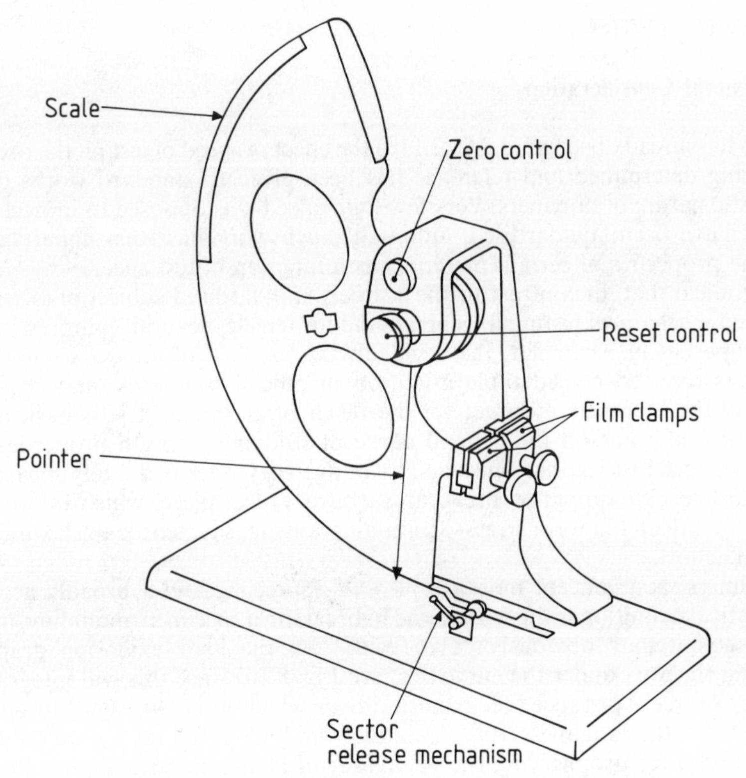

Fig. 8.36 Tear testing apparatus of the Elmendorf type

USA Standards

ASTM has no fewer than four test methods covering the tearing properties of sheet and film. ASTM D1004 (1966)[97] uses the angle test piece now withdrawn in the UK. This test piece measures the resistance of the film to tear initiation and is tested at 51 mm/min, the maximum recorded force being divided by the film thickness to give the tear resistance. This is essentially the same test piece as is used for elastomers in ASTM D624 (1981).[98]

ASTM D1922 (1967)[99] is similar to BS 2782 Method 360A, being of the Elmendorf type, although the dimensions of the apparatus are somewhat different. D1938 (1967)[100] is a trouser tear test but the dimensions of the test piece differ from those in the UK. The test speed is the same, however, at 250 mm/min, as is the treatment of data.

ASTM D2582 (1967)[101] attempts to combine the effect of tear initiation and propagation in one test. A weighted pointer is released from a given height and it falls down a carriage on to a film, clamped in a curved holder. The holder and pointer are positioned in such a way that the point first punctures and then tears the film, its kinetic energy being absorbed by doing so. After the pointer has come to rest, the length of the cut produced is measured from a suitably positioned scale and the tear resistance is calculated from this, the weight of the pointer assembly and the drop height.

8.7 IMPACT TESTS

8.7.1 General Considerations

Mention has already been made (§8.2.1) of the effect of speed of test on the property value being determined and reference has been given to standard works on the viscoelastic nature of polymers. Very few high-speed – as opposed to impact – test methods have been standardised, although much work has been undertaken to assess the properties of certain materials as a function of test speed.[102–116]

It is ironical that, in contrast to the scarcely standardised subject of (scientific) high-speed mechanical testing in general, and for tensile, flexural, compressive and shear properties in particular, the decidedly *ad hoc* field of impact strength and testing has received considerable attention in official standards, materials data sheets and the literature at large; yet the result of an impact test is basically no more than one point on the general curve of studying strength properties as a function of speed of testing. The one advantage they offer is a ready measure of the actual energy required to break an (arbitrary) test piece, which information can only be calculated from stress–strain diagrams in, say, tensile or flexural tests with some effort.

'Toughness' is a concept most people can appreciate and a broadly accepted mathematical definition is the work done in breaking a specimen, moulding, article, etc. As such it may obviously be derived from the load–extension graph by computing the area under the curve (e.g. see Fig. 8.10) since this will integrate all the units of force × distance (i.e. extension over which force operated) to give the total work, i.e. the toughness. As calculated from high-speed tests, such data may be quite useful for prophesying the behaviour of plane face articles, but the area under a conventional tensile or bending test curve is of very limited value because in practice we are mainly interested in toughness under conditions of rapid deformation, i.e. where an article is dropped or has something dropped on it – hence the concept of impact resistance or 'strength' and the introduction of impact tests, or shock resistance and its assessment. The subject is vast and at least one whole book has been given over to it[117] with a whole chapter (II) devoted to plastics and rubber. Davis *et al.*[118] and Fenner[119] give general descriptions of conventional impact testing. Among more recent general papers are included those of Galli[120] dealing with methods, an article in *Plastics Industry*[121] covering the effects of ductile–brittle transitions, molecular parameters, ageing, moisture and impact modifiers on impact strength of thermoplastics and of Savadori[122] which is orientated towards fracture mechanics.

Standard impact tests for plastics divide themselves into: (a) those which use instruments where a pendulum of known energy strikes a specimen of defined size and shape; and (b) those where weights or other impactors are allowed to fall freely through known heights onto specimens and the impact strength is computed from the minimum combination of height and weight required to cause fracture. Tests of type (a) are further subdivided into cantilever (Izod) and supported beam (Charpy) variants, using specimens of flat plane face or more often containing a moulded or machined notch in order to assess sensitivity to weakening by notches, and tensile impact using dumb-bell-shaped test pieces. Data from tests of type (a), and particularly the Izod version, are open to much criticism; they and the other important standard tests will be described first, after which the practical significance or otherwise of the results is discussed briefly.

8.7.2 Pendulum Impact Tests

Izod Test

The basic principle of the Izod test is to allow a pendulum of known mass to fall through a known height and strike a standard specimen at the lowest point of its swing, and to record the height to which the pendulum continues its swing. If the striking edge of the pendulum is sited to coincide with the centre of percussion of the pendulum, the bearings of the pendulum are frictionless and there is no loss of energy to windage, then the product of the mass of the pendulum and the difference between the fall distance and the height it reaches after impacting the test specimen is the impact 'strength' of the latter. (It will be seen that in fact, whatever property is determined by this test, it is certainly not a 'strength' as so often it is called – even in official standards). The result is usually expressed in joules and, since there is by and large no simple or reliable method of reducing the values to a property figure independent of the specimen cross-section and of the distance between the specimen support(s) and the point of striking (see below), the test values must be referred to the original specimen only; sometimes they are reduced to unit width across the specimen, and sometimes to the area behind the notch. The test may be carried out on plain rectangular bars, but most often a carefully defined notch is moulded or machined into the face to be struck. The reason for this is that impact tests are often regarded as a means of assessing the resistance of a material to shock where notches or 'stress raisers' generally are present; the ratio of the impact strengths unnotched/notched can be regarded as a measure of the notch sensitivity of a material.

The ISO test method ISO 180 (1982)[123] in which the energy is normalised to the area behind the notch.

The velocity of the striker on impact has been standardised at 3·5 ± 10 per cent m/s with impact energies of 1·0, 2·75, 5·5, 11·0 and 22·0 J. Four test pieces are permitted, as detailed in Table 8.2, with Type 1 as the preferred size.

Two types of notch may be cut in the test piece (moulded notches are not permitted); both form an angle of 45° but the radius at the tip of notch is 0·25 mm for Type A (preferred) and 1·0 mm for Type B. A Type C test piece is formed by reversing the Type A or B test piece to produce, essentially, an 'unnotched'

Table 8.2 Standard Izod test bars (ISO 180)

Specimen type	Length (mm)	Dimension *y* (mm)	Dimension *x* (mm)
1	80·0 ± 2	10·0 ± 0·2	4·0 ± 0·2
2	63·5 ± 2	12·7 ± 0·2	12·7 ± 0·5
3	63·5 ± 2	12·7 ± 0·2	6·4 ± 0·3
4	63·5 ± 2	12·7 ± 0·2	3·2 ± 0·2

specimen. The notch is always cut into the y dimension and is of depth $y/5$ so that the x dimension becomes the notch length.

The designation of x and y dimensions is especially helpful when referring to tests on laminated test pieces, which may be tested in edgewise and/or flatwise directions (Fig. 8.37). Reference to thickness and width, as hitherto, in relation to testing laminates has always led to confusion over which dimension is which.

The height to which the pendulum rises after impact is recorded by an idler pointer on a scale which reads directly the energy absorbed.

BS 2782 now contains two Izod tests: Method 350 (1984)[124] is identical to ISO 180; while Method 306A (1970)[125] has been retained for a limited period to allow users of 'British' Izod testers time to purchase the new 'ISO/ASTM' machines. The most obvious difference between the two standards is in the impact velocity, which for the older standard is 2·44 m/s, and in the different impact energies of 1·36, 4·07 and 13·6 J. The test pieces have dimensions conforming to Types 2 and 3 above but the Type A test piece must have a moulded notch, cut notches being confined to the Type B and C test pieces. Only the Type B notch, of 1 mm radius, may be used in Method 306A. Laminated products may be tested but in this case the notch is produced across the laminations by drilling and then opening out the drill hole with a saw cut (Fig. 8.38).

ASTM D256 (1984)[126] also follows the same basic principles as the above methods and is closely aligned to the ISO method. The ASTM method has retained the units, energy per unit length of notch, however, and in Method C, which is used for materials with impact strengths below 27 J/m of notch, the toss factor has also been retained. This purports to allow for the energy required to toss the broken half of the test piece after fracture has occurred, although there has been

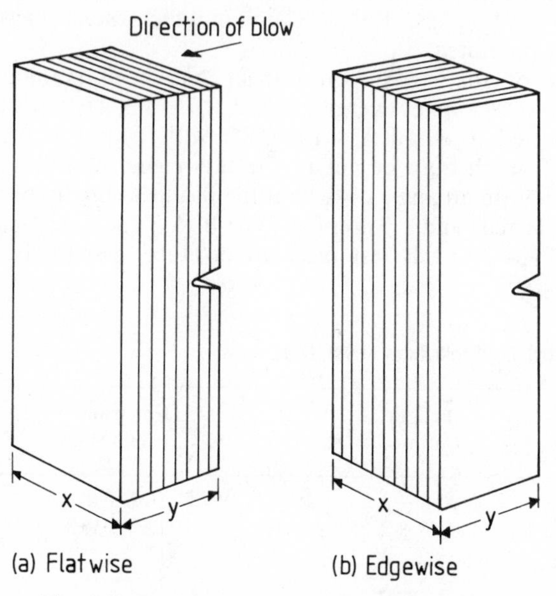

Fig. 8.37 Izod impact testing laminated sheet

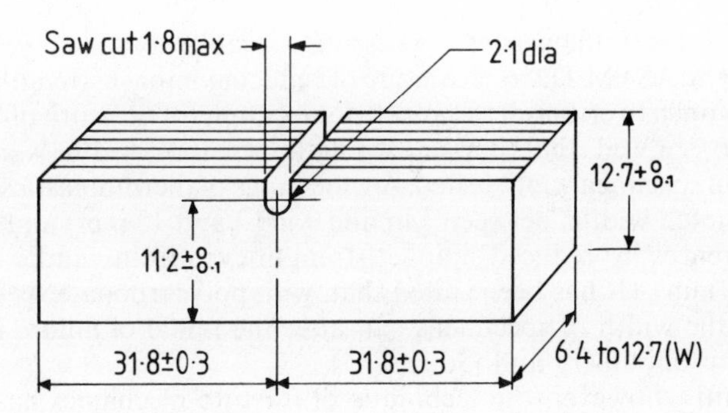

Fig. 8.38 Izod impact resistance test piece – machine laminate: BS 2782 Method 306

considerable criticism of the technique. The toss factor is estimated by replacing the broken half of the test piece on the clamped portion and striking it a second time with the pendulum released from a height corresponding to that to which it rose following the breakage of the test piece. The 'energy of toss', which is subtracted from the impact energy, is then the difference between this reading and the free swing reading obtained from this height.

The Izod test for all its popularity and widespread use, particularly in materials data sheets, has long been recognised as being far from ideal when it comes to correlating the results it gives to real service performance. For instance, Telfair and Nason[127] have analysed the factors comprising 'impact strength' as measured by the Izod test as follows:

(a) Energy to initiate fracture of specimen.
(b) Energy to propagate fracture across specimen.
(c) Energy to deform the specimen plastically.
(d) Energy to throw the broken end of the test specimen.
(e) Energy lost through vibration of the apparatus and its base, and through friction.

Value (a) is probably the one of real interest, though the addition of (b) and (c) has obvious practical significance; (d) is the 'toss factor' referred to in Method C of ASTM D256 (1984) and the significance of the correction in certain cases is well illustrated by Ritchie[128] in a general discussion on the Izod impact test: with low impact strength materials the 'toss factor' can be the major proportion of the measured value. Factor (e) requires a rigid design of pendulum arm in particular and a massive foundation.

Ritchie also mentions further factors which may influence results and lead to errors, illustrating each with results from practical investigations:

(i) Variation of clamping pressure in vice.
(ii) Failure to strike the specimen squarely.
(iii) When notches are machined, the state of the cutter and actual cutting technique used.

The effect of such variables as temperature and notch radius on the impact strength of polyolefins in particular has been reported by Horsley and Morris,[129] who have also examined the influence of processing variables on falling weight impact strength.[130] The ASTM D256 technique of reducing impact strength to unit width of specimen is open to criticism as a result, for example, of the work of Wolstenholme *et al.*[131] who showed that impact strength per unit width is not invariably independent of specimen width tested. A wide range of thermoplastics was examined at specimen notch widths between $\frac{1}{8}$ in and $\frac{1}{2}$ in (3 and 13 mm) and many showed marked differences in 'reduced' impact strength even from widths between $\frac{1}{4}$ and $\frac{1}{2}$ in (6 and 13 mm). (It has been stated that, with polycarbonate, as little as 0·01 in difference in the width of specimens can alter the mode of failure from tough to brittle, see also Shoulberg and Gouza.[132])

More recently, however, the technique of fracture mechanics has been applied to Izod (and Charpy) methods[133,134] with the view of deriving a true material parameter which is independent of the test conditions. And, while this has not been an unequivocal success, it is certainly a step in the right direction.

Charpy Test

Although the Charpy test is similar to the Izod test, in that both are flexural impact tests performed by a pendulum striking a bar-shaped test piece, there are a number of significant differences between them and no general correlations relating the data obtained from each have been developed. In the Charpy Test, the test specimen, supported as a horizontal beam, is broken by a single swing of a pendulum, with the line of impact midway between the supports and, in the case of notched specimens, directly opposite the notch. The Charpy method has long been preferred in Europe.

ISO 179 (1982)[135] is the international test method and a range of impact energies at two striking velocities are permitted as shown in Table 8.3 with standard specimens given in Table 8.4.

It is important to note that the x and y dimensions are reversed here compared

Table 8.3 Characteristics of Charpy pendulum impact testing machines

Impact energy (J)	Velocity at impact (m/s)	Maximum permissible frictional loss (%)	Permissible error after correction (J)
0·5	2·9 (±10%)	4	0·01
1·0	2·9 (±10%)	2	0·01
2·0	2·9 (±10%)	1	0·01
4·0	2·9 (±10%)	0·5	0·02
5·0	2·9 (±10%)	0·5	0·02
2·7	3·8 (±10%)	0·5	0·02
7·5	3·8 (±10%)	0·5	0·05
15·0	3·8 (±10%)	0·5	0·05
25·0	3·8 (±10%)	0·5	0·10
50·0	3·8 (±10%)	0·5	0·10

Table 8.4 Specimen types, dimensions and distances between supports for the Charpy test

Specimen Type	Length l (mm)	Dimension y (mm)	Preferred value of dimension x (mm)	Distance between lines of supports (mm)
1	80 ± 2	$10 \pm 0{\cdot}5$	$4 \pm 0{\cdot}2$	60
2	50 ± 1	$6 \pm 0{\cdot}2$	$4 \pm 0{\cdot}2$	40
3	120 ± 2	$15 \pm 0{\cdot}5$	$10 \pm 0{\cdot}5$	70
4	125 ± 2	$13 \pm 0{\cdot}5$	$13 \pm 0{\cdot}5$	95

with the Izod test, the notch being cut into the x dimension for Charpy but into the y dimension for Izod. (Compare Figs. 8.37 and 8.39.)

The reason for this reversal is that the Charpy test is normally performed in the flatwise direction but the Izod test is performed edgewise.

There are three notch types permitted, Types A and B corresponding to the Izod notches precisely, and Type C which is a square section notch of depth one-third the value of the x dimension. The notch width is either 2 mm in the case of the Type 1 and Type 3 test pieces or 0·8 mm in the case of the Type 2 test pieces. The Type 4 test piece is not so notched.

The impact strength is quoted in relation to the area behind the notch (i.e. the product $x_k . y$).

BS 2782 Method 359 (1984)[136] is identical to ISO 179 and so need not be considered further.

ASTM D256[126] Method B gives the USA Charpy method and is essentially the same as the Type 4 of ISO 179.

DIN 53453 (1957)[137] details the German practice of measuring Charpy impact strength. It is very similar to the ISO standard, using the same impact energies,

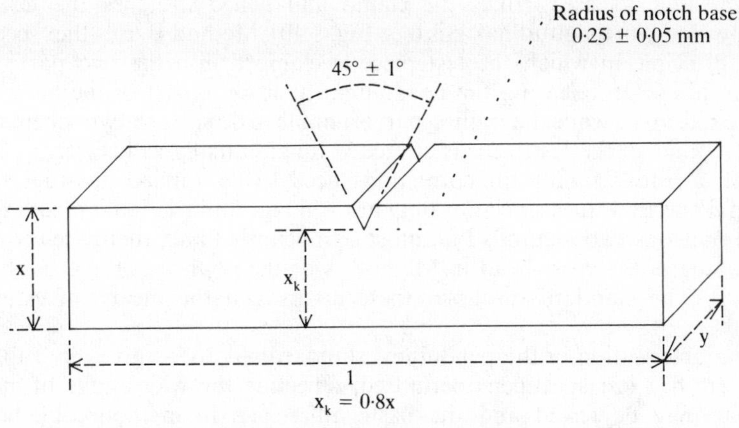

Fig. 8.39 Charpy test piece with type A notch

pendulum velocities, standard test piece dimensions and notch Type C although the tolerances are not always identical. Notches A and B are not specified. (An impact test can also be carried out on the Dynstat apparatus using an unnotched test piece but the clamping geometry is unique, making the test neither a Charpy nor Izod test. This test is detailed in DIN 53435,[92] which also deals with the normal flexural test.)

The instrumentation of the Charpy test with piezoelectric or strain-gauge transducers has been reported in the literature: Charentenay *et al.*[138] for example have proposed a new interpretation of the Charpy fracture energy evaluation in the case of a brittle/ductile fracture mode; Hirose *et al.*[139] have studied the instrumented Charpy tester and claim that modifications to the notch tip radius and energy calculation, etc., lead to acceptable correlations with falling dart impact data; while Merle *et al.*[140] describe the instrumentation of a Charpy impact tester together with damping materials to the striker to reduce mechanical vibration and a collapsible environmental chamber for conducting tests at subambient temperatures.

Tensile Impact Test

Both the tests so far described have used a flexural mode of deformation. However, in service a component may be subject to shock loading in a tensile configuration and standard test methods have been developed to enable data to be gathered from this type of loading. European and American standards have existed for many years but the UK has never standardised on this geometry. Work, however, is at an advanced stage in ISO to produce an international standard and it is more than likely that such a standard would become incorporated into BS 2782 in due course.

ISO/DIS 8256[141] is reasonably well developed and so will be used to describe the technique even though certain minor details have yet to be worked out.

The standard is based on two different methods, A and B, which it advises may not give identical results. Method A uses the 'specimen-in-bed' principle in which the test piece is clamped rigidly in a horizontal plane, one end being mounted on the base of the instrument and the other in a movable (and light) clamp. The pendulum, after release, strikes the clamp and hence stretches the test piece essentially along its longitudinal axis (see Fig. 8.40). Method B uses the 'specimen-in-head' principle in which the test piece is clamped into the two parts of the pendulum head. On releasing the pendulum, the leading part of the head passes between fixed anvils while the trailing part is unable to do so as shown schematically in Fig. 8.41 and so the test piece is stretched longitudinally as before.

In both methods significant corrections need to be applied in order to give meaningful energy values since not only are windage and frictional errors present as in the previous two methods but, most significantly of all, there are errors due to the tossing of the cross-head in Method A or the bounce of the cross-head in Method B. The standard discusses these errors and the means of estimating them.

The characterisation of the pendulums standardised are as given in Table 8.5.

There are five test specimens permitted, reflecting the wide range of material types that may be tested and the basic differences in methodology between Methods A and B: test piece 1 is a dumb-bell shaped specimen of length 80 mm,

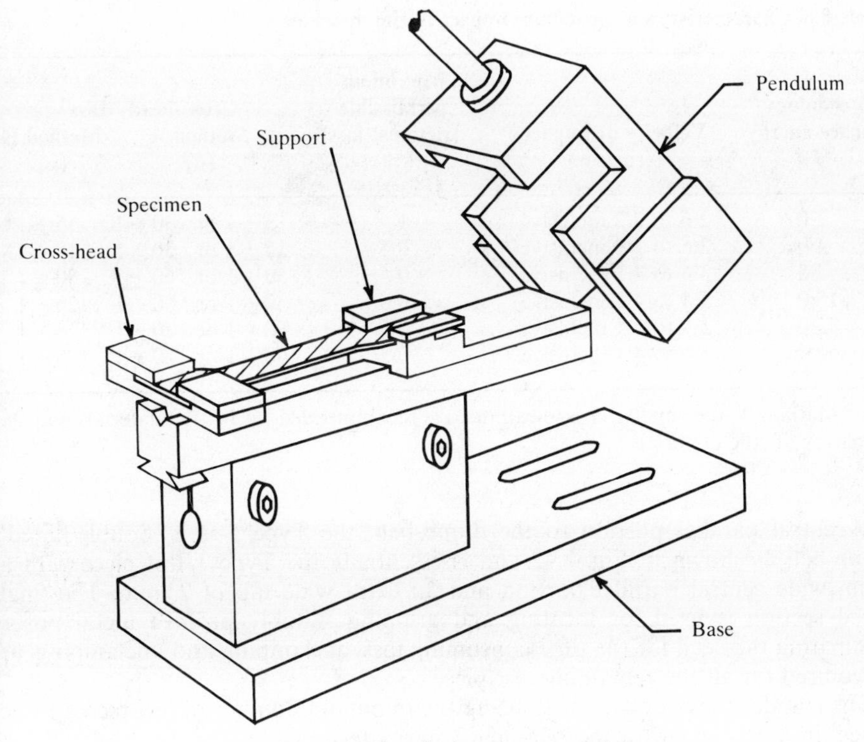

Fig. 8.40 Tensile impact apparatus

overall width 15 mm with a reduced width of 10 mm; the Type 2 test piece is very similar with corresponding dimensions 60 by 10/3; the Type 3 test piece is comparable to the preferred Izod/Charpy specimen but has two 90° notches with 0·5 mm tip radius machined opposite each other across the thickness, so as to leave a 7 mm wide 'land' between them; the Type 4 is the same as Type 2 without

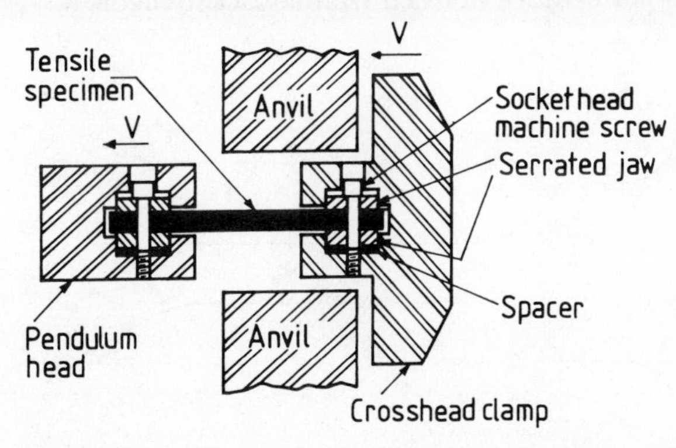

Fig. 8.41 Schematic diagram of tensile impact testing machine: ASTM D1822

Table 8.5 Characteristics of pendulum impact testing machine

Pendulum impace energy (J)	Velocity at impact (m/s)	Maximum permissible frictional loss (%)	Cross-head Mass[a] Method A (g)	Method B (g)
2·0	2·6 to 3·2 inclusive	1	15 ± 1 or 30 ± 1	15 ± 1
4·0	2·6 to 3·2 inclusive	0·5	15 ± 1 or 30 ± 1	15 ± 1
7·5	3·4 to 4·1 inclusive	0·5	30 ± 1 or 60 ± 1	30 ± 1
15·0	3·4 to 4·1 inclusive	0·5	30 ± 1 or 60 ± 1	120 ± 1
25·0	3·4 to 4·1 inclusive	0·5	60 ± 1 or 120 ± 1	120 ± 1
50·0	3·4 to 4·1 inclusive	0·5	60 ± 1 or 120 ± 1	120 ± 1

[a] For Method A, the smaller cross-head mass is recommended for brittle materials and the larger for ductile materials.

any central parallel portion to the dumb-bell; the Type 5 is a 'double-flared' dumb-bell as shown in Fig. 8.42 and is essentially the Type 1 test piece with a 5 mm wide central parallel portion and an extra wide tab of 23 mm. The final flared section is used for locating and gripping suitably profiled metal posts, eliminating the need for the time-consuming task of clamping and unclamping, as is required for all the remaining test pieces.

The standard lays down ten as being the minimum number of test pieces to be tested, in agreement with the Izod and Charpy tests.

The corresponding USA method is given in ASTM D1822 (1984)[142] – and the metricated version D1822M (1984) – this being essentially the same as Method B in DIS 8256. The German standards, conforming closely to Method A of DIS 8256, is DIN 53448 (1977).[143]

Work by Therberge and Hall[144] showed a high degree of correlation between tensile impact values and unnotched 'impact strengths' for glass-filled SAN, polystyrene, polycarbonate, polyethylene, polypropylene, PVC, polyurethane, polysulphone, polyester and nylons-6, -6·6 and -6·10; not surprisingly, the correlation did not exist for notched Izod impact strengths.

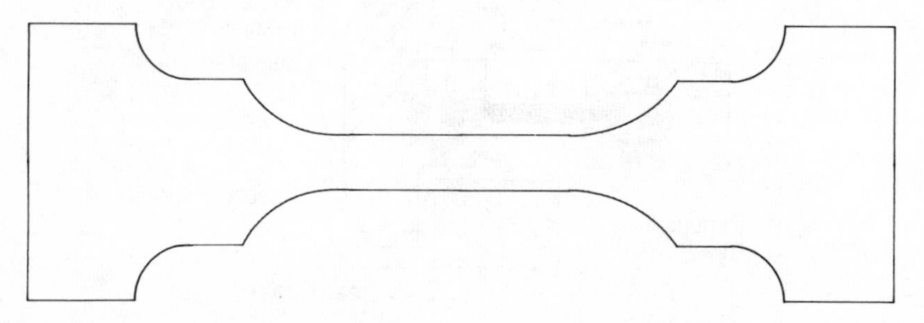

Fig. 8.42 ISO/DIS 8256 Type 5 tensile test piece

8.7.3 Falling Weight Tests

As with the pendulum tests described above, the principle of these tests is very simple. Here, instead of the energy being supplied by a pendulum, falling through an arc, striking the test piece horizontally, it is provided by a mass, usually a sphere or a hemispherical indentor, falling vertically onto the test piece. The test piece is usually a moulded plaque or sample cut from sheet but full mouldings can also be more readily accommodated with this impact methodology than with pendulum machines.

Advances in electronics and in particular the explosive growth in microcomputing power and availability has enabled a significant development in instrumented techniques to occur and it is convenient to subdivide falling weight tests into the non-instrumented or 'traditional' type of test and the newer, more sophisticated, 'instrumental' variants.

Traditional Tests

ISO 6603/1 (1985)[145] lays down the basic framework for this kind of test, there being two test methods given. The preferred method is the staircase method in which a uniform energy increment must be used and the energy is either increased or decreased by this increment after testing each specimen depending on the result (no fail or fail, respectively) observed for the preceding test specimen. At least thirty test pieces are recommended and (as with the second method) no test piece may be struck more than once.

The second method, called the statistical (probit), method tests groups of at least ten test specimens under identical impact conditions. The percentage of failures (or passes) is then plotted against energy on probability paper to determine the average energy to failure. This technique has the advantage that any suitable energy levels may be used – they do not have to differ from each other by any regular increment – but it does require greater numbers of test pieces, the standard quoting forty but in practice it will be seldom less than fifty. Where both brittle and ductile behaviour occurs within a 'homogenous' group this is the mandatory method.

For both methods the energy may be realised by either varying the mass at constant height (preferred) or by varying the height at constant mass. The former is preferred simply because plastics are viscoelastic (cf. §8.1 and elsewhere) and hence velocity-sensitive. In practice, however, this rarely makes any significant difference since the effect is often proportional to log (velocity) and over the range of drop heights used for a given material the change in log (velocity) is small in relation to the statistical scatter of results that are obtained anyway. However, in transition regions this generalisation may not hold.

The standard recognises that 'failure' is a 'movable feast' and lays down five criteria that may be used – without excluding others – from which the most appropriate for a given situation is selected. It then goes on to specify the 50 per cent impact failure point as the characteristic 'mean value' of the test, this being the energy (or mass or height) at which half the test pieces would fail according to the failure criterion adopted.

The standard test specimens are 60 mm diameter or square with a thickness between 1 and 4 mm, 2 mm being preferred. The test piece is supported on a steel cylinder of internal diameter 40 mm and it is struck by a hemispherical dart of diameter 20 mm, with 10 mm also allowed. The preferred drop height for the constant height option is 1 m while for the variable height option the normal operating range is from 0·3 to 2 m.

In the case of films and sheets of thickness less than 1 mm there is not yet published an ISO standard covering these, although a document is in the final stages of preparation; DIS 7765[146] follows very similar lines to those described for ISO 6603/1 with the staircase method of analysis being the only one included. Since the test pieces are not self-supporting, a mechanical, pneumatic or vacuum clamping system must be used to prevent slippage during impact. The test pieces, being 125 mm diameter, are appreciably bigger than the rigid test piece variant as also are the permitted darts (38 or 50 mm diameter). Impact energy is always varied by varying the mass at constant height, the two permitted heights being 0·66 m and 1·5 m.

The current British Standard is BS 2782 Methods 306B and C[147] which differ only in drop height.

The specimen is a 57–64 mm diameter disc or a 57–64 mm square. For moulded or extruded material the specimen thickness is $1{\cdot}52 \pm 0{\cdot}05$ mm, for sheet it is the sheet thickness. At least twenty such specimens are required and no specimen is used more than once.

The apparatus may be of the form shown in Fig. 8.43 (suitable for Method 306B). The specimen support is a hollow cylinder of internal diameter $50{\cdot}80 \pm 0{\cdot}05$ mm, external diameter not less than 57·2 mm and length not less than 25·4 mm. The axis of the cylinder coincides with the line of fall of the striker and a soft shock-absorbing disc is placed inside the cylinder to rest on its base. The specimen may be clamped onto the support.

The striker has a hardened hemispherical striking surface, $12{\cdot}7 \pm 0{\cdot}05$ mm diameter (the 'ball'), and is fitted with a carriage to take weights so that a specified series of increments of energy may be obtained if the striker is released from a height of 610 ± 2 mm (Method 306B) or 305 ± 1 mm (Method 306C) above the upper surface of the specimen. The striker is allowed to fall with or without guides; in the former instance the fall must be substantially without friction. The striker can conveniently be supported electromagnetically and released by switching off the current.

To carry out the test a 'trial run' is undertaken first. The striker is loaded in such a way that the product of weight and fall height is equal to the expected impact strength. If, on release of the striker, the specimen does not fracture, or cracks on one surface only, the result is recorded as 'unbroken'; if the specimen breaks or cracks or tears through from one surface to the other, the result is recorded as 'broken'. Thereafter:

1. If the result was 'broken', a second specimen is tested with an impact energy less than the first by a specified amount S. If this second specimen is 'broken', a third is tested with an impact energy S^* less than that applied to the second – and so on until a specimen does not break.
2. If the first result was 'unbroken', a second specimen is tested with an impact energy greater by amount S obtained as before. As in (1), this is repeated, this time until a specimen does break.

The 'test run' is now carried out, using the remaining test specimens, the energy

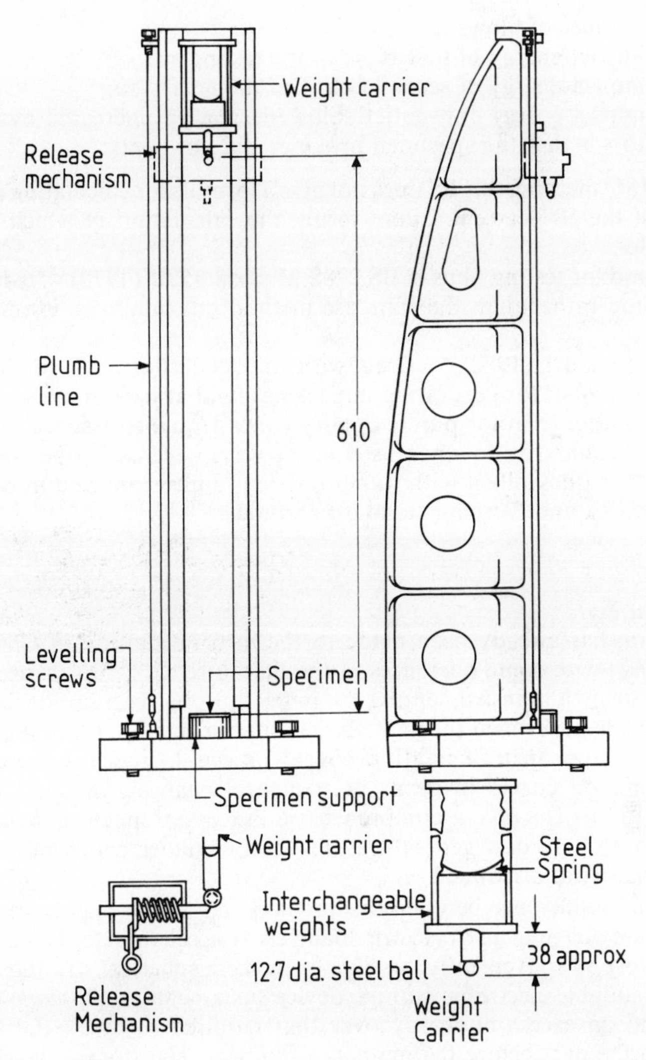

Fig. 8.43 Falling weight impact apparatus: BS 2782 Method 352D (dimensions in millimetres)

of the blow applied to any specimen being *Y* more or less than that on the previous specimen, respectively, according to whether the latter remained unbroken or was broken, with *Y* being specified in the BS method. The test run thus uses a maximum of eighteen specimens (since at least two of the twenty must be required for the 'trial run'), but must not use less than twelve.

The impact strength is calculated as follows:

$$\text{impact strength} = \frac{1}{21 - M}(Y_{m+1} + Y_{m+2} + \ldots + Y_{21}) \qquad [8.25]$$

where m = number of blows in trial run
Y_{m+1} = impact energy of first blow of the testing run
Y_{m+2} = impact energy of second below of the testing run
Y_{21} = impact energy of twentieth blow, decrease or increased by Y according to whether the specimen broke or did not break.

Unlike the ISO method, this BS does not give a formula for calculating the standard deviation of the 50 per cent failure result, the calculation of which is also very much simpler.

The method for testing films is BS 2782 Method 352D (1979)[148] which uses the probit method rather than the staircase method but otherwise is very similar to DIS 7765.

BS 4618 Sect. 1.1 (1972)[149] deals with impact testing in a general way and describes techniques for generating impact results that have some significance for design data rather than for purely quality control/quality assurance purposes as all the above standards do when used in the ways covered by the standards.

Similar techniques, albeit with rather different dimensions and impact energies, are used in USA and German standard methods.[150–152]

Instrumented Tests

Brief mention has already been made to the instrumentation of Charpy impact testing (§8.7.2) with some references in the literature[138–140] to its use, but by far the greatest growth in instrumenting the impact test has been in the falling weight methodology, as is testified to by the large outpouring of papers on all aspects of the technique from instrumentation considerations to the interpretation of the force/deformation curves in terms of structural changes for specific materials. References 153 to 169 give a comprehensive but by no means exhaustive list; of these 153 to 157 are of a general nature, the remainder being more specifically product or material orientated.

The essential difference between this method and the non-instrumented variant is that a strain guage or piezoelectric load cell or accelerometer is mounted in the striker, as close as conveniently possible to the hemispherical striking surface, and linked to a suitable electronic storage device such as an oscilloscope or transient recorder and an excess of energy over that required to break the test piece is imparted to the dart before the moment of impact. The storage device is usually linked to a microcomputer so that the data may be analysed and manipulated in various ways.

As the force is continuously monitored throughout the impact event it is possible to derive, in addition to the energy to failure – which is all that can be achieved with the traditional method, a measure of stiffness from the initial slope of the force/deformation curve, the yield parameters and post-yield behaviour and in the case of composites any pre-failure cracking. The energy value up to any point on the curve may be readily computed having been identified by the person conducting the test and so the technique is of value from fundamental studies of fracture processes via fracture mechanics through to QC testing of a particular product.

There is also the advantage that very few test pieces (as few as five in some cases) are required to produce a meaningful result compared with the laborious staircase or probit methods but, of course, the capital outlay is appreciably greater than for the simple and unsophisticated alternative.

At the time of writing there is only one fully fledged standard on the instrumented test for plastics which is DIN 53443: Part 2 (1984)[170] although work is in hand at ISO to produce an international standard, ISO 6603: Part 2. This is still at a relatively early stage and there are some fundamental disagreements on details of instrumentation and data analysis. Nonetheless many essential details are established and it is worthwhile briefly to note these.

It is normal to use a large excess of energy in the dart assembly so that the change in velocity of the striker as it penetrates the specimen is kept low; up to 20 per cent is permitted in the draft ISO at present. The frequency response of the striker/load cell must be sufficiently high to avoid damping of the signal and the natural frequency of the force measuring system should be greater than $5/t_f$ where t_f is the minimum time to failure that is required to be measured. Similarly the total band width of the amplifier train should be greater than $16/t_f$. Piezoelectric devices, because of their high stiffness and hence inherently high natural frequency, are very popular as load cells.

The geometry of the test pieces, supports and darts are, in the main, the same as those found in Part 1 of the standard but an interesting addition is the use of the flat-ended striker develped in Holland and reported by Mooij.[171] Despite initial caution over its inclusion, an international round-robin showed that this geometry gave at least as consistent results as the hemisphere that everyone has become accustomed to but, not surprisingly, produces quite different absolute values.

A similar development is in hand for the impact testing of films as Part 2 of ISO 7765.

REFERENCES

1. Phillips, P. J. and Ramakrishnan, N. R. (1978) *Polymer Engineering and Science*, **18** (11), 869.
2. Nielsen, L. E. (1962) *Mechanical Properties of Polymers*, Reinhold, p. 220.
3. Ritchie, P. D. (1965) *Physics of Plastics*, Illiffe, p. 185.
4. Tangorra, G. (1966) *Rubber Chemistry and Technology*, **39** (5), 1520.
5. Brown, R. P. (1986) *Physical Testing of Rubbers*, Elsevier, Ch. 8.
6. Yeoh, O. H. (1984) *Plastics and Rubber Processing Applications*, **4** (2), 141.
7. Crawford, R. J. and Stephens, G. (1985) *Polymer Testing*, **5** (2), 113.
8. Fett, T., Nothdurft, W. and Racke, H. H. (1973) *Kunststoffe*, **63** (2), 107.
9. Bowman, J. and Bevis, M. (1977) *Colloid and Polymer Science*, **255** (10), 954.
10. Martinez-Salazar, J., Pena, J. G. and Calleja, F. J. B. (1985) *Polymer Communications*, **26** (2), 57.
11. Martinez-Salazar, J. and Balta-Calleja, F. J. (1985) *Journal of Materials Science Letters*, **4** (3), 324.
12. Balta-Calleja, F. J. (1985) *Advances in Polymer Science*, **66**, 117.
13. Selden, B. R. and Gustafson, C. G. (1984) *Antec '84, New Orleans*, 573, 011, SPE.
14. Sato, K. (1984) *Journal of Coatings Technology*, **56** (708), 47.
15. Davies, H. E., Troxall, G. E. and Wiskoch, C. T. (1964) *The Testing and Inspection of Engineering Materials*, 3rd Edn, McGraw-Hill, Ch. 7.
16. Fenner, A. J. (1965) *Mechanical Testing of Materials*, Philosophical Library, Ch. 8.
17. Maxwell, B. (1955) *Modern Plastics*, **32** (9), 125.
18. Lysaght, V. E. (1948) *Materials and Methods*, **27**, 84.

19. Boor, L. (1960) *ASTM Bulletin*, **244**, 43.
20. Haldenwanger, H. (1961) *Kunststoffe*, **51** (2), 82.
21. Gouza, J. J. (1966) in *Testing of Polymers*, vol. 2 (Ed. J. V. Schmitz), Interscience, Ch. 7.
22. Livingston, D. I. (1967) in *Testing of Polymers*, vol. 3 (Ed. J. V. Schmitz), Interscience, Ch. 7.
23. Craft, J. and Whelan, A. (1983) *British Plastics and Rubber*, September, 65.
24. ISO 868 (1985) *Plastics – Determination of Indentation Hardness by Means of a Durometer (Shore Hardness)*.
25. ISO 2039: Part 1 (1987) *Plastics and Ebonite – Determination of Hardness by Ball Indentation Method.*
26. ISO 2039: Part 2 (1981) *Plastics – Determination of Rockwell Hardness.*
27. Fett, T. (1972) *Material Prüfung*, **14** (5), 151.
28. BS 2782 Method 365A (1976) *Determination of Softness Number of Flexible Plastics.*
29. BS 2782 Method 365B (1981) *Determination of Indentation Hardness by means of a Durometer.*
30. BS 2782 Method 365C (1986) *Determination of Rockwell Hardness.*
31. BS 2782 Method 365D (1978) *Determination of Hardness by Ball Indentation Method.*
32. BS 2782 Method 1001 (1977) *Measurement of Hardness by Means of a Barcol Impressor.*
33. BS 903: Part A26 (1969) *Methods of Testing Vulcanised Rubber – Determination of Hardness.*
34. ISO 48 (1979) *Vulcanised Rubbers – Determination of Hardness (Hardness between 30 and 85 IRHD).*
35. Bessant, K. H. C., Dilke, M. G., Hollis, C. E. and Millane, J. J. (1952) *Journal of Applied Chemistry*, **2** (9), 501.
36. ASTM D2240 (1984) *Rubber Property – Durometer Hardness.*
37. ASTM D785 (1965) *Rockwell Hardness of Plastics and Electrical Insulating Materials.*
38. ASTM D2583 (1981) *Indentation Hardness of Rigid Plastics by Means of a Barcol Impressor.*
39. DIN 53456 (1973) *Testing of Plastics – Indentation Hardness Test.*
40. DIN 53505 (1973) *Testing of Elastomers – Shore A and D Hardness Testing.*
41. Hounsfield, L. H. *Commercial Testing*, Part II, *Plastics*, Tensometer Ltd.
42. BS 240: Part 1 (1962) *Method for Brinell Hardness Test – Testing of Metals.*
43. ASTM E10 (1984) *Test Method for Brinell Hardness of Metallic Materials.*
44. BS 427: Part 1 (1961) *Method for Vickers Hardness Test – Testing of Metals.*
45. ASTM E92 (1982) *Test Method for Vickers Hardness of Metallic Materials.*
46. Campbell, R. F. (1967) *Materials, Research and Standards*, **7** (10), 443.
47. *Plastics and Rubber Weekly* (1982) 14th August, 950, 6.
48. Grodzinski, P. (1953) *Plastics, London*, **18** (194), 312.
49. Roberts, J. and Steel, M. A. (1966) *Journal of Applied Polymer Science*, **10** (9), 1343.
50. Gavan, F. M. and Wein, J. T., Jr (1965) in *Testing of Polymers*, vol. 1, (Ed. J. V. Schmitz), Interscience, Ch. 12.
51. Boor, L., Ryan, J. D., Marks, M. W. and Bartoe, W. F. (1947) *ASTM Proceedings*, **17**, 1017.
52. Bennitt, J. H. and Avenall, C. E. (1952) *Chemistry and Industry*, **39**, 936.
53. McSheehy, W. H. (1964) *Plastics Technology*, **10** (11), 44.
54. Carswell, T. S. and Nason, H. K. (1944) *Modern Plastics*, **21** (10), 121.
55. Nielsen, L. E. (1962) *Mechanical Properties of Polymers*, Reinhold, Ch. 5.
56. Ferry, J. D. (1980) *Viscoelastic Properties of Polymers*, 3rd edn, Wiley.
57. Panov, P. (1969) in *Testing of Polymers*, vol. 4, (Ed. W. E. Brown), Interscience.
58. Hawley, S. W., Tams, P. and Rhodes, A. (1984) *RAPRA Technical Report*, no. 9933.
59. ISO 6239 (1986) *Plastics – Determination of Mechanical Properties by Use of Small Specimens – Tensile Properties.*

60. ISO/DIS 527 *Plastics – Determination of Tensile Properties.*
61. ASTM D1708 (1984) *Tensile Properties of Plastics by Use of Microtensile Specimens.*
62. BS 5214: Part 1 (1975) *Tensile, Flexural and Compression Testing Machines for Rubbers and Plastics*; Part 2 (1978) *Constant Rate of Force Application Testing Machines for Rubbers and Plastics.*
63. ISO 1184 (1983) *Plastics – Determination of Tensile Properties of Films.*
64. BS 2782 Methods 320A to F (1976) *Plastics Tensile Strength, Elongation and Elastic Modulus.*
65. BS 2782 Methods 326A to C (1977) *Determination of Tensile Strength and Elongation of Plastics Films.*
66. BS 2782 Method 1003 (1977) *Glass Reinforced Plastics – Determination of Tensile Properties.*
67. BS 2782 Method 327A (1982) *Determination of Tensile Strength and Elongation at Break of Polytetrafluoroethylene (PTFE) Products.*
68. ASTM D638 (1984) *Tensile Properties of Plastics.*
69. ASTM D882 (1983) *Tensile Properties of Thin Plastic Sheeting.*
70. ASTM D3039 (1976) *Tensile Properties of Fibre–Resin Composites.*
71. ASTM D2289 (1984) *Tensile Properties of Plastics at High Speeds.*
72. ISO/DIS 604 *Plastics – Determination of Compressive Properties.*
73. BS 2782, Method 345A (1979) *Determination of Compressive Properties by Deformation at Constant Rate.*
74. ASTM D695 (1984) *Compressive Properties of Rigid Plastics.*
75. ASTM D621 (1964) *Deformation of Plastics under Load.*
76. DIN 53453 (1971) *Testing of Plastics – Compression Test.*
77. Williams, J. G. and Ford, H. (1964) *Journal of Mechanical Engineering Science*, **6**(4), 405.
78. Williams, J. G. (1967) *Transactions of the Plastics Institute*, **35**(117), 505.
79. Narisawa, I., Ishikawa, M. and Ogawa, H. (1979) *Gakujutsu Bunken Fukyukai*, 265.
80. BS 2782 Method 340A (1978) *Determination of Shear Strength of Moulding Materials*; Method 340B (1978) *Determination of Shear Strength of Sheet Material.*
81. ASTM D732 (1984) *Shear Strength of Plastics by Punch Tool.*
82. Heap, R. D. and Norman, R. H. (1969) *Flexural Testing of Plastics*, Plastics Institute.
83. ISO/DIS 178 *Plastics – Determination of Flexural Properties.*
84. BS 2782 Method 335A (1978) *Determination of Flexural Properties of Rigid Plastics.*
85. BS 2782 Method 1005 (1977) *Glass Reinforced Plastics – Determination of Flexural Properties Three Point Method.*
86. BS 2782 Method 336B (1978) *Determination of Deflection in Bend under an Applied Force.*
87. BS 2782 Method 332A (1976) *Stiffness of Plastics Film.*
88. BS 2782 Method 341A (1977) *Determination of Apparent Interlaminar Shear Strength of Reinforced Plastics.*
89. ASTM D790 (1984) *Flexural Properties of Plastics and Electrical Insulating Materials.*
90. ASTM D747 (1984) *Stiffness of Plastics by Means of a Cantilever Beam.*
91. DIN 53452 (1977) *Testing of Plastics – Bending Test.*
92. DIN 53435 (1983) *Testing of Plastics – Determination of Flexural Properties and Impact Resistance with Dynstat Test Pieces.*
93. ISO 6133 (1981) *Rubber and Plastics – Analysis of Multi-Peak Traces Obtained in Determinations of Tear Strength and Adhesion Strength.*
94. ISO 6383 (1983) *Plastics – Film and Sheeting* – Part 1: *Trouser Tear Method*; Part 2: *Elmendorf Method.*
95. BS 2782 Method 360B (1980) *Determination of Tear Strength of Sheet and Sheeting (Trouser Tear Method).*
96. BS 2782 Method 308B (1970) *Resistance to Tear Propagation of Thin Flexible Sheet.*

97. ASTM D1004 (1966) *Initial Tear Resistance of Plastic Film and Sheeting.*
98. ASTM D624 (1981) *Rubber Property – Tear Resistance.*
99. ASTM D1922 (1967) *Tear Propagation Resistance of Plastic Film and Thin Sheeting by Pendulum.*
100. ASTM D1938 (1967) *Tear Propagation Resistance of Plastic Film and Thin Sheeting by a Single-Tear Method.*
101. ASTM D2582 (1967) *Puncture-Propagation Tear Resistance of Plastic Film and Thin Sheeting.*
102. Dietz, A. G. H. and McGarry, F. J. (1956) in *Symposium on Speed of Testing of Non-metallic Materials*, STP 185, ASTM, p. 30.
103. Supnik, R. H. (1962) *Materials, Research and Standards*, **2** (6), 498.
104. Goldfein, S. (1964) *Modern Plastics*, **41** (12), 149.
105. Sandek, L. (1962) *Plastics Technology*, **8** (2), 26.
106. Silberberg, M. and Supnik, R. H. (1962) *SPE Transactions*, **2** (2), 140.
107. Heimerl, G. L. and Manning, C. R., Jr (1962) *Materials, Research and Standards*, **2** (4), 270.
108. McAbee, E. and Chmura, M. (1963) *SPE Journal*, **19** (1), 83.
109. Holt, D. L. (1968) *Journal of Applied Polymer Science*, **12** (7), 1653.
110. Barnes, C. B., Hawkings, E. L. and Davis, M. V. (1966) *Materials, Research and Standards*, **6** (11), 560.
111. Jones, J. W. (1960) *Journal of Applied Polymer Science*, **4** (12), 284.
112. D'Amato, D. A. (1964) *Journal of Applied Polymer Science*, **8** (1), 197.
113. Patterson, G. D., Jr and Miller, W. H., Jr (1960) *Journal of Applied Polymer Science*, **4** (12), 291.
114. Amborski, L. E. and Mecca, T. D. (1960) *Journal of Applied Polymer Science*, **4** (12), 332.
115. Grimminger, H. and Jacobshagen, E. (1962) *Kunststoffe*, **52** (5), 254.
116. Dietz, A. G. H. and Eirich, F. R. (1960/61/62/64/65) *High Speed Testing*, vols. I, II, III, IV, V, Wiley/Interscience.
117. Späth, W. (1961) *Impact Testing of Materials* (rev. M. Rosner), Thames and Hudson.
118. Davis, H. E., Troxell, G. E. and Wiskocil, G. T. (1964) *The Testing and Inspection of Engineering Materials*, 3rd edn, McGraw-Hill, Ch. 8.
119. Fenner, A. J. (1965) *Mechanical Testing of Materials*, Philosophical Library, Ch. 7.
120. Galli, E. (1982) *Plastics Compounding*, **5**(5), 18.
121. Anon., *Plast. Ind.* (1983) **10** (7), 6.
122. Savadori, A. (1985) *Polymer Testing* **5** (3), 209.
123. ISO 180 (1982) *Plastics – Determination of Izod Impact Strength of Rigid Materials.*
124. BS 2782 Method 350 (1984) *Determination of Izod Impact Strength of Rigid Materials.*
125. BS 2782 Method 306A (1970) *Impact Strength (Pendulum Method).*
126. ASTM D256 (1984) *Impact Resistance of Plastics and Electrical Insulating Materials.*
127. Telfair, D. and Nason, H. K. (1943) *Modern Plastics*, **20** (11), 85.
128. Ritchie, P. D. (1965) op. cit., Ch. 3.
129. Horsley, R. A. and Morris, A. C. (1966) *Shell Polyolefins Engineering Design Data*, Shell International Chemical Co. Ltd/Shell Chemicals UK Ltd.
130. Horsley, R. A. and Morris, A. C. (1966) *Plastics, London*, **31** (350), 1551.
131. Wolstenholme, W. E., Pregun, S. E. and Stark, C. F. (1964) *High Speed Testing*, vol. IV, Wiley/Interscience, p. 19.
132. Shoulberg, R. H. and Gouza, J. J. (1967) *SPE Journal*, **23** (12), 32.
133. Adams, G. C. (1982) *ACS Rubber Division, 121st Meeting, Philadelphia*, paper 30, 23.
134. Vu-Khanh, T. and Charrier, J. M. (1984) Chemical Inst. of Canada, Macromolecular Science Div., *Quantitative Characterisation of Plastics and Rubber Symposium*, Hamilton, p. 114.
135. ISO 179 (1982) *Plastics – Determination of Charpy Impact Strength of Rigid Materials.*
136. BS 2782 Method 359 (1984) *Determination of Charpy Impact Strength of Rigid Materials*

(*Charpy Impact Flexural Test*).
137. DIN 53453 (1975) *Testing of Plastics – Impact Flexural Test.*
138. Charentenay, F. X. de, Robin, J. J. and Vu Khanh, T. (1982) *Impact Fracture Mechanics of Semi-Ductile Polymers Conference, Cambridge*, paper 33.
139. Hirose, H., Kobayashi, T. and Kohno, Y. (1984) *Polymer Testing*, **4** (1), 31.
140. Merle, G., Yong-Sok, O., Pillot, C. and Santereau, H. (1985) *Polymer Testing*, **5** (1), 37.
141. ISO/DIS 8256 *Plastics – Determination of Tensile Impact Strength.*
142. ASTM D1822 (1984) *Tensile Impact Energy to Break Plastic and Electrical Insulating Materials.*
143. DIN 43448 (1977) *Testing of Plastics – Tensile Impact Test.*
144. Therberge, J. E. and Hall, N. T. (1969) *Modern Plastics*, **46**(7), 114.
145. ISO 6603/1 (1985) *Plastics – Determination of Multiaxial Impact Behaviour of Rigid Plastics*: Part 1, *Falling Dart Method.*
146. ISO/DIS 7765 *Plastics – Films and Sheeting – Determination of Impact Resistance – Free Falling Dart Method.*
147. BS 2782 Methods 306B and C (1970) *Determination of Impact Strength* (*Falling Weight Method with Sheet Specimens*).
148. BS 2782 Method 352D (1979) *Falling Weight Impact Resistance of Films.*
149. BS 4618 Sect. 1.2 (1972) *Recommendations for the Presentation of Plastics Design Data – Impact Behaviour.*
150. ASTM D1709 (1975) *Impact Resistance of Polyethylene Film by the Free Falling Dart Method.*
151. ASTM D3029 (1984) *Impact Resistance of Rigid Plastic Sheeting or Parts by Means of a Tup* (*Falling Weight*).
152. DIN 53443: Part 1 (1984) *Testing of Plastics*; *Multiaxial Impact Behaviour*; *Falling Weight Test.*
153. Adams, G. C. and Wu, T. K. (1982) *ANTEC '82 San Francisco, May 10–13*, 898.
154. Gutteridge, P. A., Hooley, C. J., Moore, D. R., Turner, S. and Williams, M. J. (1982) *Kunststoffe*, **72** (9), 543.
155. Zoller, P. (1983) *Polymer Testing*, **3** (3), 197.
156. Bevan, L., Nugent, H. and Potter, R. (1985) *Polymer Testing*, **5** (1), 3.
157. Trubshaw, R. N. (1985) *International Reinforced Plastics Industry*, **4**(6), 16.
158. Hooley, C. J., Moore, D. R., Whale, M. and Williams, M. J. (1981) *Plasticon 81 Symposium 4, Warwick University, PRI Confer. 6125*, paper 19.
159. Casiraghi, T., Castiglioni, G. and Ajroldi, G. (1982) *Plastics and Rubber Processing Applications*, **2** (4), 353.
160. Wardle, M. W. and Tokarsky, E. W. (1983) *Composites Technology Review*, **5**(1), 4.
161. Dao, K. C. (1983) *ANTEC '83, Chicago, May 2–5*, 49.
162. Fernando, P. L. (1983) *ANTEC '83, Detroit, Sept. 20–22*, 161.
163. Crivelli-Visconti, I. (1983) *Macplas Int.*, **11**, 71.
164. Johnson, A. E., Lynskey, B. M. and Forster, J. M. W. (1983) *Composites Plast. Renf. Fibres Verre Text.*, **23** (5), 34.
165. Golovoy, A., Cheung, M. F. and Van Oene, H. (1984) *Reinforced Plastics/Composites '84, New York, Jan. 16–19*, 627.
166. Dormier, E. J., Yarmoska, B. S. and Dan, E. (1984) *ANTEC '84, New Orleans, April 30–May 3*, 294.
167. Szamborski, E. C. and Hutt, R. J. (1984) *ANTEC '84, New Orleans, April 30–May 3*, 884.
168. Meijering, T. G. (1985) *Plastics and Rubber Processing Applications*, **5** (2), 165.
169. Winkel, J. D. and Adams, D. F. (1985) *Composites*, **16** (4), 268.
170. DIN 53443: Part 2 (1984) *Multiaxial Impact Behaviour*; *Impact Penetration Test Combined with Data Processing by Means of Electronic Devices.*
171. Mooij, J. J. (1981) *Polymer Testing*, **2** (1), 69.

Chapter Nine

Dynamic Stress–Strain Properties

9.1 INTRODUCTION

A dynamic test is one in which the plastics material is subjected to a cyclic, usually sinusoidal, deformation with the stress and strain being recorded continuously in most cases. Traditionally, dynamic tests do not include those tests where the prime objective is to assess the life of a component which will be subject to cyclic deformation in service. Such tests are more properly considered as fatigue tests and are discussed in Ch. 12.

Dynamic tests yield data concerning the appropriate modulus of the material, according to the mode of deformation being applied, and, perhaps more significantly, the mechanical damping exhibited by the material under that particular deformation pattern. This information is not obtainable from the 'static' tests detailed in the previous chapter. Such information, while being of immediate value to the design engineer, has also proved to be of immense value in the study of the molecular transitions that occur in high polymers. In order to exploit this, however, the equipment used in such tests must be capable of operating over as wide a frequency and/or temperature range as possible.

This has resulted in the widespread application of these techniques to fundamental stresses and to the generation of design data, but an almost total absence of use for routine and quality control purposes, largely because of the complexity of the apparatus. Until recently only the torsion pendulum test was standardised at national or international level but there has now been some advance which indicates that perhaps dynamic tests for plastics will become more widely used. In the field of rubbers there have been international standards for some time, dealing with general testing requirements as well as particular test methods. These methods and the apparatus have been discussed in detail by Brown.[1]

There are basically two types of dynamic test: the free vibration method in which the test piece is initially deformed, then released and allowed to oscillate without any further input of external energy; and the forced vibration method in which the oscillation of the test piece is maintained by external means. The latter method may also be subdivided into those methods operating at resonance and those operating away from resonance. Before dealing with such methods in more detail, however, it would be useful to outline, briefly, the principles and terms encountered. Ferry[2] provides an excellent detailed text on this subject. Also, ISO 2856[3] gives general requirements for dynamic testing of elastomers.

As mentioned in the previous chapter, the behaviour of a plastic during deformation is not perfectly elastic but depends on the time-scale of the experiment. It behaves, to a first approximation, as a series combination of Voigt and Maxwell elements. The former consists of a perfectly elastic spring in parallel with a perfectly viscous dash-pot and the latter consists of a series combination of spring and dash-pot. The Maxwell element can be ignored in dynamic tests (provided there is no static strain component of deformation) and the behaviour explained by the Voigt element alone.

If a sinusoidally varying stress is applied to the element, the stress at any time is given by

$$\sigma = \sigma_0 \sin \omega t \qquad [9.1]$$

where σ_0 is the maximum applied stress
ω is the angular frequency
t is time.

Because the dash-pot cannot respond instantaneously to the applied stress level, but rather is rate-dependent, the resultant strain lags behind the stress by an amount dependent on the viscous and elastic constants of the element. Thus,

$$\gamma = \gamma_0(\sin \omega t - \delta) \qquad [9.2]$$

where γ_0 is the maximum strain attained and δ is the phase angle (Fig. 9.1).

This use of the simple model above leads to the very useful, if oversimplified, concept of polymer behaviour consisting of two simultaneously acting parts: a perfectly elastic part in which no energy is lost and the applied stress and resultant strain are in phase with each other; and a purely viscous part in which energy is not stored but is dissipated as heat and where the strain vector is 90° behind the stress. This in turn leads to the definition of two moduli, the in-phase or storage modulus given by the ratio of the peak in-phase stress to peak strain and the out-of-phase or loss modulus given by the ratio of the peak out-of-phase stress to peak strain. Reference to Fig. 9.2 shows that

$$M' = \sigma_0'/\gamma_0 \qquad [9.3]$$

$$M'' = \sigma_0''/\gamma_0 \qquad [9.4]$$

where M' is the storage modulus and M'' the loss modulus. (The actual modulus obtained depends upon the mode of deformation: for shear modulus, the symbol

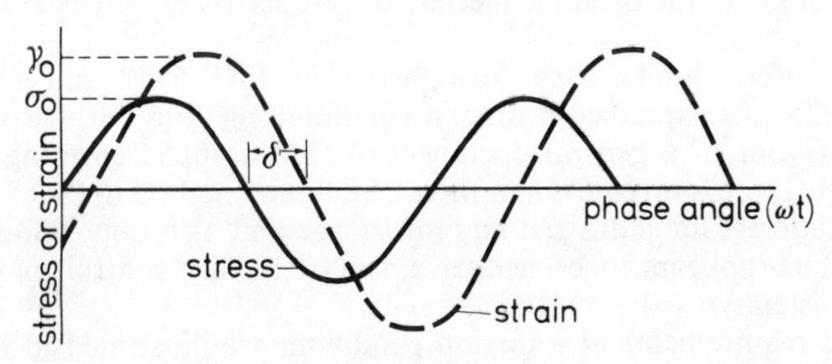

Fig. 9.1 Stress–strain relationship for sinusoidally varying applied stress

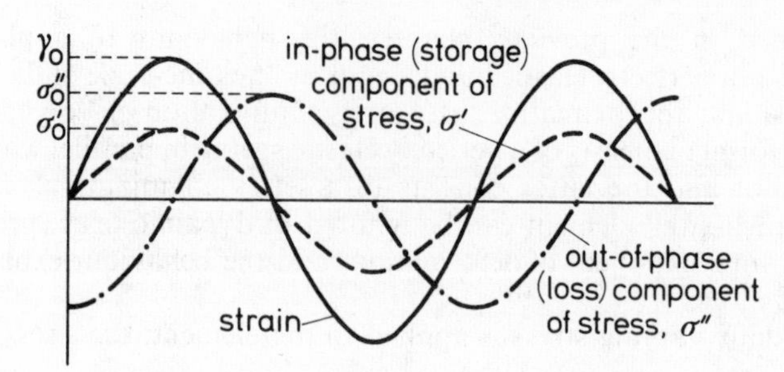

Fig. 9.2 Stress–strain relationship of Fig. 9.1 resolved into storage and loss components

usually employed is G and for extensional or Young's modulus, E.) The complex modulus, M^*, is defined as

$$M^* = \sigma_0/\gamma_0 \qquad [9.5]$$

so that

$$M^* = \sqrt{(M'^2 + M''^2)} \qquad [9.6a]$$

although this is more usually written as

$$M^* = M' + iM'' \qquad [9.6b]$$

where $i = \sqrt{-1}$.

The loss tangent or loss factor is given by $\tan\delta$ and it can be shown that

$$\tan\delta = \frac{M''}{M'} \qquad [9.7]$$

In practice, the amount of deformation that can be applied to a test piece before nonlinear viscoelastic effects become important is very small, although shear tests have a larger linear range than tensile or compression tests. For this and other reasons shear tests are usually preferred.

9.2 FREE VIBRATION METHODS

One of the simplest pieces of apparatus and perhaps the most widely used in the past for the study of the dynamic mechanical properties of polymers is the torsion pendulum.

This has been standardised for plastics in ISO 537[4] and DIN 53445.[5] ASTM D2236[6] also specified a torsion pendulum method but was discontinued in 1985 in favour of a general document, ASTM D4065[7] covering all types of dynamic test. There is no equivalent British Standard method to ISO 537, reflecting the lack of interest for industrial laboratory use and also opposition to the ISO standard which appears to be neither aimed at quality control or fundamental studies consistently.

The basic requirements of a torsion pendulum are illustrated in Fig. 9.3. The test piece is rigidly clamped at one end, while the other end is attached to an

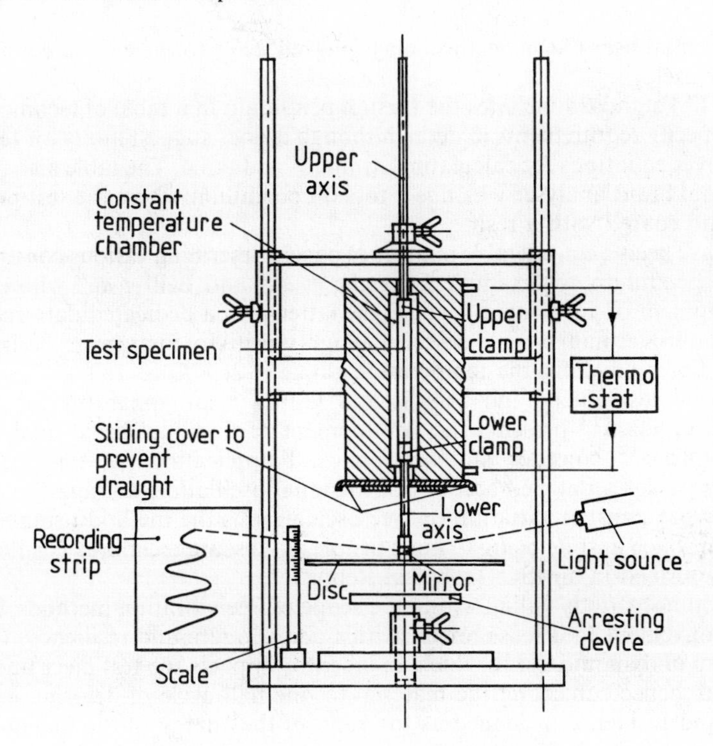

Fig. 9.3 Torsion pendulum apparatus

inertia member which is free to oscillate about the axis of the test piece. The temperature of the test piece is controlled by a thermostated oven and some means of measuring the amplitude of the oscillations without interfering with the damping characteristics of the system is required – an optical system, as illustrated, is frequently used. A variant on this is to counterbalance the weight of the inertia member to minimise the tensile force on the test piece.

ISO 537 suggest that modulus and damping are independent of test piece shape and apparatus – highly unlikely in practice. One test piece size is recommended, a strip 60 mm × 10 mm × 1 mm, but others are allowed. Two variants on apparatus are detailed, one with a fixed upper clamp and one with a counterweight to balance the inertia member. A moment of inertia of almost 30 kg.mm^2 is specified for the former and no figure specified for the latter. The remainder of the apparatus descriptions and procedure is an odd mixture of detail and extremely vague requirements.

The results call for the log decrement and the shear modulus but not tan δ. The wording of how to calculate log decrement is confusing between how it shall be done and how it may be done. Equations are given to calculate the shear modulus (which one is not stated) for each of the two methods which include somewhat complicated correction terms. The derivation of these equations is not given and is by no means obvious.

The potential user of this method may find reference to the equivalent methods for rubber useful.[1]

The ASTM standard includes the torsion pendulum in a table of techniques but does not specify requirements in detail although it does suggest limits for test piece size and gives equations for calculating strain, G' and $\tan\delta$. The table also includes the torsional braid analyser which is a torsion pendulum where the test piece is a textile braid coated with a resin.

There have been a considerable number of papers describing various constructions of torsion pendulum, for example Fritzsche *et al.*[8] and Bell *et al.*[9] who describe the development of automated penduli, the latter with a dedicated data-reduction facility. Aghili-Kermani *et al.*[10] describe a high-sensitivity instrument designed for films and fibres covering the temperature range $-269\,°C$ to $180\,°C$, Seeger[11] a computerised instrument and Pandit and Gupta[12] an apparatus for tubular specimens. Gillham[13] presents the development of torsional braid analysis and Lee and Oliver[14] consider modifications and applications of this technique. Boden[15] provides a detailed account of torsional oscillation testing.

A somewhat unusual variation on free oscillation is the method using Savart's pendulums which despite being to most people an obscure technique has the status of being described in an ISO Technical Report.[16]

A method not strictly falling within the scope of free vibration methods, but one which is convenient to discuss briefly at this point, is rebound resilience. This is a simple form of dynamic test in which an indentor impacts the test piece and is free to rebound, hence subjecting the material to one half cycle of deformation only. The rebound resilience is defined as the ratio of the energy of the indentor after impact to its energy before impact expressed as a percentage and hence, in the usual case where the indentor falls under gravity, is equal to the ratio of rebound height to drop height. It is related to the loss tangent by the expression:

$$R \simeq \exp(-\pi \tan\delta) \qquad [9.8]$$

Thus, by carrying out the experiment over a wide temperature range, the loss tangent can be evaluated and transition temperatures located.

Apparatus can either be in the form of a falling weight or a pendulum. The falling weight is probably the earliest in use and the Shore schleroscope was at one time very widely known. This uses a hemispherical striker in a glass tube, but many apparatuses use a falling ball. The Shore instrument has been quite wrongly attributed with measuring hardness.

Resilience is widely used and standardised for rubbers where pendulum instruments are normally employed and application to these materials has been reviewed.[1] In view of the simplicity of rebound resilience it is perhaps surprising that it has not been used more for plastics, especially in the context of quality control.

Bramuzzo[17] describes what might be termed an 'instrumented resilience test' in which a falling ball impacts a beam on supports. The force/time curve during the contact of the ball and test piece is recorded and the elastic modulus deduced.

9.3 FORCED VIBRATION METHODS

There are several approaches to the measurement of dynamic properties using forced oscillation of the test piece and a great many designs of machine have been used. Nevertheless, such testing has been rarely used for routine material characterisation and quality control of plastics. The basic modes of testing are forced vibration at or near resonance and forced vibration away from resonance. Methods away from resonance can be subdivided into those applying deformation cycles and those applying force cycles. The means of applying the force cycles may be mechanical, in several ways, electromagnetic or servohydraulic.

Mechanical machines are somewhat limited in frequency, the fatigue life may be poor and are generally considered outdated. Electromagnetic vibrators can cover a very wide frequency range, particularly higher frequencies, and are especially convenient for smaller machines and for forced oscillations at resonance. Servohydraulic activation is limited to about 100 Hz in frequency but in all other respects provides the most versatile option, allowing force or strain control, high loads with waveforms other than sinusoidal and the capability to fully characterise complete products. Such machines are complex and expensive which restricts their use.

The term 'dynamic mechanical analyser' (or 'dynamic thermal mechanical analyser') has come into general use relatively recently and is usually taken to mean modest-sized bench-mounted dynamic test machines which are automated to various degrees and allow the measurement of dynamic properties over a range of frequency and temperature. The term is quite loose and may cover free vibration as well as forced vibration instruments. An increasing variety of analysers are now available representing a range of geometries, control systems and degrees of sophistication. Their generic characteristics are their relatively modest cost in relation to efficiency and their suitability for characterising materials as functions of temperature and frequency on a comparative basis rapidly. They are not generally suitable for use with a range of strains and deformation modes and the results may not always be accurate in absolute terms.

Perhaps the original analyser was the Rheovibron developed by Takayanagi[18] which has found widespread use. It is electromagnetically driven, requires only a small test piece, and can operate in tension, compression (although not usually for plastics in this mode) or shear. Its main disadvantages are system resonance, end-effect errors which need correction and (for the non-automated versions at least) labour intensiveness. Kenyon *et al.*[19] describe the automation of the Rheovibron to effect continuous measurements of stress, strain and phase angle as a function of temperature. Massa *et al.*[20] review the capabilities of the instrument and discuss sources of error and equipment modifications. Yee and Takemori[21] and Locati[22] also give details of its use.

The Polymer Laboratories instrument[23] operates in a bending mode and is electromagnetically driven. The du Pont[24] analyser is again electromagnetically driven and the deformation is a form of bending but it operates at resonance. Yokouchi and Kobayashi[25] describe a 'viscoelastometer' operating in tension, and Sternstein[26] illustrates the capabilities of the Dynastat and the Dynalyser.

ASTM D4065 illustrates several forms of dynamic apparatus in a table which is cross-referenced to manufacturers of instruments but gives no details of operation or performance.

Reviews for the measurement of dynamic properties which includes discussions of instruments have been given by Galli,[27] Murayama[28] and Boyer.[29] Dumoulin and Utracki[30] evaluated both dynamic mechanical and dielectric instruments whilst Reed and Duncan[31] critically review a wide range of techniques for obtaining dynamic data including wave propagation (see also Ch. 21) and consider the relations between the complex modulus components.

Malkin *et al.*[32] describe a method using non-sinusoidal vibrations, Farris[33] presents an impulse approach to viscoelastic response, Yee and Takemori[34] consider experimental techniques to obtain both dynamic Young's modulus and Poisson's ratio and Arisawa *et al.*[35] describe a method using a multifrequency excitation signal.

There is one ISO standard covering forced vibrations, ISO 6721,[36] which is a form of vibrating reed apparatus operating at resonance. The method is said to be specially intended for acoustical insulation and it is extremely difficult to see why this should have been singled out for international consideration.

In the vibrating reed apparatus one end of the test piece, which may be of circular or rectangular cross-section, is held in a relatively massive vibrating head while the other end is free to oscillate. The applied frequency of the head is altered via a sine-wave generator and the amplitude of the free end of the test piece is measured by some suitable means. This amplitude varies with the applied frequency and reaches a maximum at the resonant frequency of the reed. Young's modulus for the material can be determined from the expression:

$$E = \frac{38{\cdot}24 d L^4 f_r^2}{D^2} \qquad [9.9]$$

where d = density of the plastic
L = free length of the reed
D = thickness of the reed
f_r = resonant frequency.

The mechanical damping term can be calculated from the width of the resonance peak by the formula:

$$\tan \delta = \frac{\Delta f}{f_r} \qquad [9.10]$$

where Δf is the width of the resonance curve at the half power point (70·7 per cent of the peak height).

ISO 721 covers test pieces either clamped at one end or supported at two vibration nodes, the latter being said to be suitable for plastic coated metal.

Vibrating reed tests are also described by Newman,[37] Atkinson and Eagling,[38] Strella,[39] Scherr,[40] Fielding-Russell and Wetton[41] and more recently by Washburn,[42] who used the technique for studying discontinuities in compression moulded composites and fatigue properties of carbon fibre reinforced phenolics.

REFERENCES

1. Brown, R. P. (1986) *Physical Testing of Rubber*, Applied Science, Ch. 9.
2. Ferry, J. D. (1970) *Viscoelastic Properties of Polymers*, Wiley.
3. ISO 2856 (1975) *Elastomers – General Requirements for Dynamic Testing.*

4. ISO 537 (1960) *Plastics Testing with the Torsion Pendulum.*
5. DIN 53445 (1979) *Testing of Plastics: Torsion Pendulum.*
6. ASTM D2236 (1970) *Dynamic Mechanical Properties of Plastics by means of a Torsional Pendulum.*
7. ASTM D4065 (1982) *Determining and Reporting Dynamic Mechanical Properties of Plastics.*
8. Fritzsche, C., Hoechli, B. and Moser, K. (1975) *Kunststoffe*, **64** (11), 675.
9. Bell, C. L. M., Gillham, J. K. and Benci, J. A. (1974) in *SPE 32nd Annual Technical Conference, San Francisco*, p. 598.
10. Aghili-Kermani, H., O'Brien, T., Armeniades, C. D. and Roberts, J. M. (1976) *Journal of Physics, E*, **9** (10), 887.
11. Seeger, M. (1980) *Rheology*, vol. 2, Plessam Press.
12. Pandit, S. N. and Gupta, V. B. (1981) *Polymer Composites*, **2**, no. 3.
13. Gillham, J. K. (1982) *Development in Polymer Characterisation – 3*, Applied Science Publishers.
14. Lee, W. A. and Oliver, M. J. (1983) *British Polymer Journal*, **15**, no. 1.
15. Boden, H. E. (1984) *Advances in Polymer Technology*, **4** (3).
16. *ISO Technical Report* 4137, 1978.
17. Bramuzzo, M. (1985) *Polymer Testing*, **5**, no. 6.
18. Takayanagi, M. (1963) *Memoirs of the Faculty of Engineering, Kyushu University*, **23**, 41.
19. Kenyon, A. S., Grote, W. A., Wallace, D. A. and Rayford, M. (1977) *Journal of Macromolecular Science, B*, **13** (4), 553.
20. Massa, D. J., Flick, J. R. and Petrie, S. E. B. (1975) *Coated Plastics Preprints*, **35** (1), 971.
21. Yee, A. F. and Takemori, M. T. (1977) *Journal of Applied Polymer Science*, **21** (10), 2597.
22. Locati, G. (1978) *Polymer Engineering and Science*, **10** (10), 793.
23. Wetton, R. E. (1984) *Polymer Testing*, **4** (2–4).
24. Gill, P. S., Lear, J. D. and Leckenby, J. N. (1984) *Polymer Testing*, **4** (2–4).
25. Yokouchi, M. and Kobayashi, Y. (1981) *Journal of Applied Polymer Science*, **26**, no. 12.
26. Sternstein, S. S. (1981) *Polymer Preprints*, **22** (1).
27. Galli, E. (1984) *Plastics Compounding*, **7**, no. 4.
28. Murayama, T. (1980) *Rheology*, vol. 3, Plenum Press.
29. Boyer, R. F. (1981) *Organic Coatings and Plastics Chemistry*, vol. 44.
30. Dumoulin, M. M. and Utracki, L. A. (1984) *Chem. Inst. of Canada Symposium, Hamilton, June 21–22, Proceedings*, p. 68.
31. Read, B. E. and Duncan, J. C. (1981) *Polymer Testing*, **2** (2).
32. Malkin, A. Y., Begishev, V. P. and Mansurov, V. A. (1984) *Polymer Science USSR*, **26** (4).
33. Farris, R. J. (1984) *Journal of Rheology*, **28** (4).
34. Yee, A. F. and Takemori, M. T. (1982) *Journal of Polymer Science: Polymer Physics Edition*, **20** (2).
35. Arisawa, H., Hayakawa, R. and Wada, Y. (1981) *Reports on Progress in Polymer Physics in Japan*, vol. 24.
36. ISO 6721 (1983) *Determination of Damping Properties and Complex Modulus by Bending Vibration.*
37. Newman, S. (1959) *Journal of Applied Polymer Science*, **2** (6), 333.
38. Atkinson, E. B. and Eagling, R. F. (1959) *SCI Monograph*, no. 5, Society of Chemical Industry, p. 197.
39. Strella, S. (1956) *ASTM Bulletin*, **124**, 47.
40. Scherr, H. J. (1966) *Materials, Research and Standards*, **6** (12), 614.
41. Fielding-Russell, G. S. and Wetton, R. E. (1979) *Plastics and Polymers*, **38** (135), 179.
42. Washburn, R. M. (1974) *SAMPE Quarterly*, **6** (1), 22.

Chapter Ten

Friction and Wear

10.1 INTRODUCTION

Traditionally, laboratory measurements of friction and wear have been made as independent, almost unrelated, topics despite the fact that in the 'real world' there is a very close link between the two. However, this is not so surprising when one considers that generally it is either the one or the other property, rather than both simultaneously, that is of real significance in a given situation. A clear exception to this is the use of plastics in bearings applications where both parameters are of crucial importance. Also, of course, when high resistance to wear is required, friction is but one of the many parameters which determine that resistance.

The measurement of abrasion resistance is encountered rather more frequently than friction measurement and this is reflected in the number of standards documents dealing with these subjects.

10.2 FRICTION

10.2.1 General Considerations

The basic 'laws' of friction have been known for several centuries and are usually stated as follows:

1. The frictional force, F, opposing motion is proportional to the normal force, N, between the contacting surfaces, i.e.

$$F = \mu N \quad [10.1]$$

   where the constant of proportionality, μ, is called the coefficient of friction.
2. The coefficient of friction is independent of the apparent area of contact.
3. The coefficient of friction is independent of the velocity between the contacting surfaces provided this is not zero.

These three 'laws', although rarely obeyed by polymers, still form the basis on which the frictional properties of these materials can be studied. James[1] details the effect of various parameters on the frictional behaviour of plastics and rubbers and how these are related to the three classical laws and to modern theories of friction. The most important parameters, as regards methods of test, are described briefly below.

No matter how smooth a surface may appear to be, it is, at the microscopic level, rough. This means that the actual area of contact between two faces is considerably less than the apparent area and it is the real contact area which determines the frictional force, experiments having shown a proportional relationship between frictional force and real area of contact. For materials exhibiting a definite yield stress, the real contact area between two faces is proportional to the normal force, from which the first friction law naturally follows. Metals are generally in this class. For substances which deform elastically the situation is more complex and, in general, frictional force is proportional to normal force raised to a power varying between two-thirds and one. Only where it is unity is the coefficient of friction constant and for all plastics materials the coefficient decreases as normal force increases. This implies that if a friction measurement is to be made on a plastics material for other than routine inspection purposes then the test should be undertaken using a variety of normal loads.

Standard textbooks on physics often refer to 'static' and 'dynamic' friction, although truly static friction is an impossibility since relative motion between contacting surfaces is unavoidable if frictional force is to be measured. The normal meaning of static friction, therefore, is the force required just to induce motion, i.e. the frictional force at very low speed. Dynamic friction on the other hand usually means the force required to maintain a constant, perceptible, relative motion between the surfaces. Clearly, therefore, velocity is a factor in determining the frictional force, although for many materials the variation in the coefficient of friction over many decades of velocity is so small as to be negligible – hence the third law of friction. For polymers, however, this is not so and variations in the coefficient of friction may be as much as three- or four-fold over a few decades of velocity. Unfortunately the effect of velocity cannot be isolated from that of temperature, but clearly an adequate characterisation of the frictional behaviour of a plastic cannot be carried out without examining the effect of velocity.

A phenomenon which sometimes occurs is that of 'stick–slip', in which the two surfaces, or small parts thereof, monentarily cease their relative motion causing the frictional force to rise until a critical value is reached at which point the surfaces slip past each other with the frictional force decreasing to some lower value before the cycle is repeated. For stick–slip to occur the coefficient of friction must fall as velocity increases. The frequency of this stick–slip cycle is often in the audible frequency band giving rise to the familiar squealing sound sometimes heard as surfaces move past each other. The occurrence of stick–slip in service, therefore, can be of great nuisance value.

Obviously atmospheric conditions and the presence of lubricants and dust influence the friction, so test conditions should be adjusted to simulate service conditions as far as is practically possible.

10.2.2 Methods of Measuring Friction

At present there are relatively few standard test methods for measuring friction. At international level one standard has recently been published and another, dealing with terms and definitions for sliding contact, is at the level of a DIS (DIS 6601).

The published standard, ISO 8295,[2] describes the measurement of coefficients of friction of film up to 0·2 mm in thickness either against itself or some other surface. Double-sided adhesive tape is used to secure the test film to a sled of given mass (200 g) and given area (4000 mm^2) which is then drawn over the second surface at a constant velocity (100 mm/min). The frictional force is measured by means of a suitable load cell – which can be attached to the cross-head of a tensile testing machine. For dynamic friction the attachment is rigid but for static friction a relatively 'soft' spring couples the sled to the cell and the first peak force recorded by the cell is used to determine the static friction coefficient.

BS 2782, Method 824A (1984)[3] describes the measurement of the static and dynamic coefficients of friction for polyethylene film. A piece of film is held on a horizontal bed by means of a vacuum-operated clamp and another, similar, piece of film is attached to the underside of a sled of sufficient mass to give a contact pressure of 490 N/m^2 when placed on the bed. The sled is driven at 800 mm/min over the bed and the forces required to initiate relative motion and to maintain constant motion are recorded. ASTM D1894 (1978)[4] for plastic film and sheeting in general, is very similar although a wider variety of apparatus is allowed (Fig. 10.1).

A method for measuring static friction only is the inclined plane method given in BS 3424 (1973).[5] In this the test piece is fastened as before to a sled which rests on an initially horizontal bed. The bed is tilted at a given rate until the sled just starts to slide, at which point the drive mechanism is switched off, usually via a microswitch, and the angle of inclination is read off a suitable scale. The static coefficient of friction is then equal to the tangent of this angle.

Despite the paucity of standard test methods a large number of apparatuses have been constructed to carry out friction measurements. Gough[6] describes the construction of an apparatus consisting of four levers supporting a test platform at its corners in such a way that each lever is pivoted at a point that maintains neutral equilibrium in the mechanism. The test surface is drawn over the platform and the coefficient of friction is given by the tangent of the angle of displacement of the levers. Bowers *et al.*[7] describe the use of a surface grinder modified to allow frictional measurements to be made and James[8] describes a similar technique using a lathe bed. Frictional 'drag' fed to the secondary winding of a differential transformer via a leaf spring has been used by Lauer and Friel[9] in a machine they claim has sensitivity over a wide range of values. The effect of velocity on friction has been investigated by Westover and Vroom[10] who describe their apparatus as a 'variable speed frictionometer'. The use of this apparatus has now been standardised in ASTM D3028 (1972).[11]

A piece of apparatus developed at the Rubber and Plastics Research Association and described by James and Newell[12] allows frictional measurements to be made over a wide range of velocities and normal loads. The apparatus (Fig. 10.2) is fixed to a normal tensile testing machine, giving a very stiff system which enables small forces to be measured accurately. It can be used inside an environmental chamber so that the effect of temperature can be investigated. The test piece geometry is readily changed, allowing tests to be performed on finished products or parts of products. James and Newell, in addition to describing the RAPRA/Daventest friction tester in detail, also provide an excellent source of references for friction test methods and theory.

An apparatus which is of value in measuring the friction of large surfaces in service – flooring or polymeric sports surfaces, for example – is the skid tester

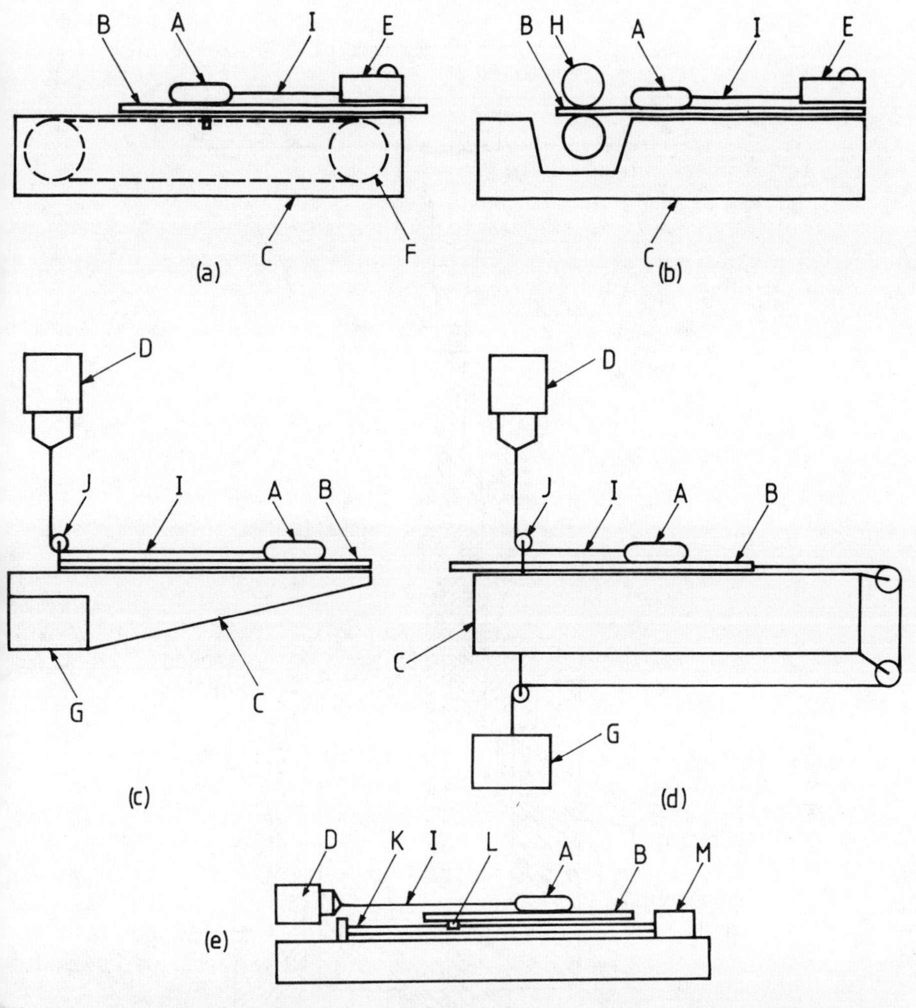

Fig. 10.1 Apparatus for measuring coefficient of friction of film: ASTM D1894: A, sled; B, plane; C, supporting base; D, strain gauge; E, spring gauge; F, constant-speed chain drive; G, constant-speed tensile tester cross-head; H, constant-speed-drive rolls; I, nylon monofilament; J, low-friction pulley; K, worm screw; L, half nut; M, hysteresis, synchronous motor

developed by the Road Research Laboratory.[13] This is a pendulum device which, when released, brushes against the surface to be measured; the energy absorbed is read from the resultant swing of the pendulum in a manner analogous to the pendulum impact testers described in Ch. 8. Giustino and Emerson[14] describe the instrumentation and computerisation of this device to give more accurate and consistent results obtained less laboriously.

In 1975 the British Standards Institution published Section 5.6 of BS 4618[15] dealing

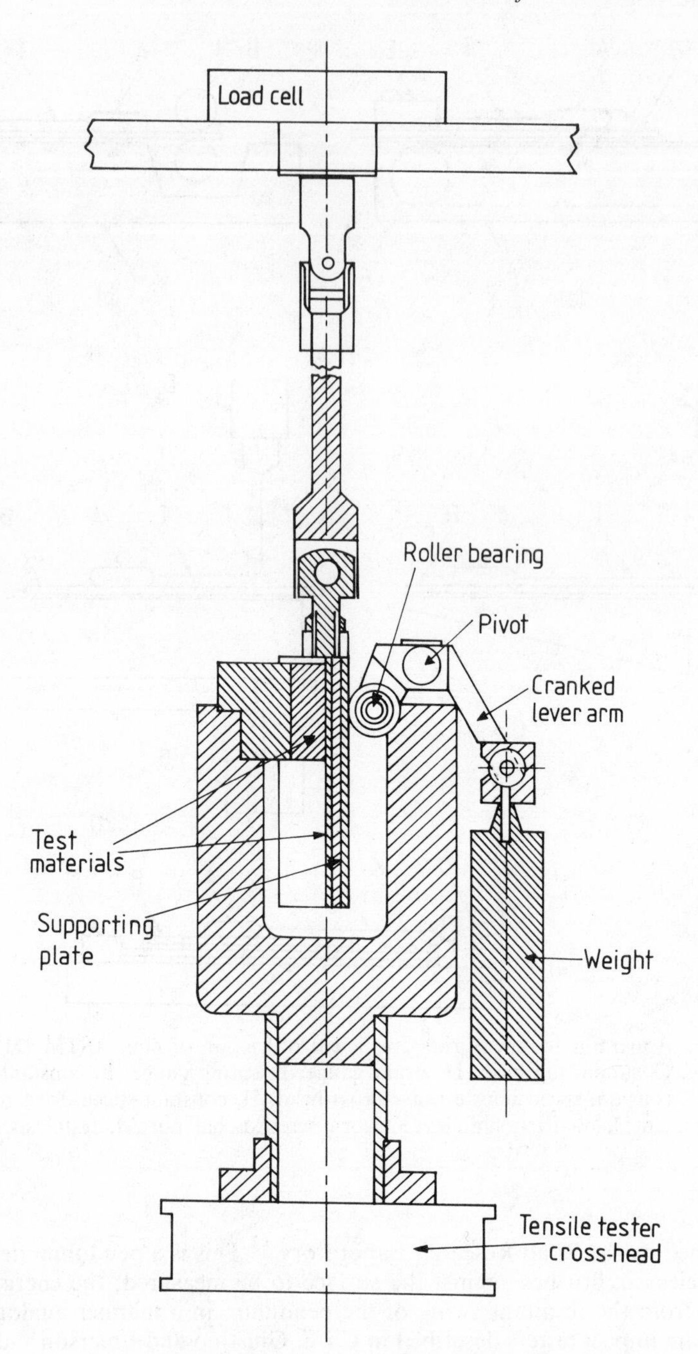

Fig. 10.2 RAPRA/Daventest friction apparatus

with the presentation of plastics design data. Section 5.6 is a guide to sliding friction and is a useful introduction to the subject in the context of plastics materials.

10.3 WEAR

10.3.1 General Considerations

The practical requirements of knowing how well a material will stand up to abrasive wear has been felt for centuries, and probably the oldest 'standard' method is the use of Moh's scale, whereby the ability of the material to resist surface damage is assessed by progressively scratching the surface with numbered minerals of increasing hardness. The material's resistance is expressed as the number of the mineral which just causes surface damage. A similar test is the pencil hardness test which uses the range of draughtsmen's pencils as the abradants. Bierbaum's hardness test, which employed a diamond scribe, was once proposed as a standard method in ASTM but, like the previously mentioned methods, has since been discontinued, probably due to lack of reproducibility. It is clear that the above methods simultaneously measure mar resistance, which is a wear mechanism, and surface hardness (see §8.1), which is not, so that they perform neither function satisfactorily – this may be the more telling reason for their disappearance.

The mechanisms of wear are complex and highly interactive; Lancaster[16] in his review concentrates on the three major processes: abrasion, adhesion and fatigue, the first two being the subject of more recent work by Czichos.[17] More detailed descriptions are given in volumes 5A and B of the Polymer Science and Technology series[18] and a more recent review of friction and wear mechanisms in polymers has been produced by Briscoe and Tabor.[19]

The very complexity of wear processes has given rise to a profusion of test apparatus – and standard methods – while at the same time rendering the quantitative correlation between methods non-existent in general,[20–22] although for a limited range of materials it may be possible to develop approximate interrelationships between test methods. Even more regrettable is the fact that different methods do not always rank materials in the same order, although it must be stressed that in this case the agreement between methods is considerably improved.[23]

The foregoing points clearly indicate that in order for meaningful data to be generated in the laboratory, the tests that are applied must simulate the real service conditions as closely as possible.[24] Even then, the experimenter must usually settle for merely ranking the materials of interest rather than deriving quantitative figures for them.

10.3.2 Standards Test Methods

ISO Standards

There is at present no standard test method for abrasion of plastics although the rotating cylindrical drum abrader, frequently known as the 'DIN abrader' because of its origins, has now been standardised for rubbers as ISO 4649[25] and can be used for many plastics.

With this abrader, the test piece is held in a chuck and traversed over a rotating drum covered with a sheet of the abradant. In this way there is a large area of abradant exposed to the test piece so that the wear of the abradant is slow and uniform. The main advantages of this abrader are the small test piece required – only 16 mm diameter by, at least, 6 mm thick, and the ease of operation which makes it ideal for quality control purposes.

It does, however, lack the versatility of the Taber Abraser, described below, even if the abradant and load are changed from those specified in the standard.

At the time of writing, TC61 had produced its first draft document dealing with the Taber Abraser but as this is at such an early stage, the final form of the standard is likely to be changed considerably and it would be inappropriate to detail it further here other than to note that it is a very general document of principles rather than detailed test methodology.

Another general document, ISO/DIS 6601,[26] already referred to in §10.2.2 is more a work on terminology than a test method.

British Standards

The only abrasion test within BS 2782 at the time of writing is Method 310B (1970),[27] the apparatus for which is illustrated in Fig. 10.3. The test measures print adhesion on PVC sheet. Test pieces 229 mm × 51 mm are used and first washed with a soap solution, rinsed, dried and conditioned. The number of cycles to cause visible damage of the print is recorded.

BS 3424 Method 16 (1985)[28] for coated fabrics specifies an abrasion tester called the 'Rubfastness Tester', which employs a reciprocating table upon which the test piece is placed and a loaded metal knife which rubs against the test piece.

BS 3794: Part 2 (1982)[29] Clause 6 for decorative laminates employs the well known Taber Abraser, described under 'USA Standards' (Fig. 10.4).

For this test the abrasive wheels are calibrated against standard zinc plates. This is a change in procedure from the earlier 1973 version which used standard laminates as the source of reference to normalise the variability of the wheels.

USA Standards

ASTM D1044 (1982)[30] covers measurements of abrasion of transparent plastics using the Taber Abraser with 'Calibrase' CS–10F abrading wheels. The basic principle is the rotation, at a prescribed speed, of test discs 100 mm diameter, or

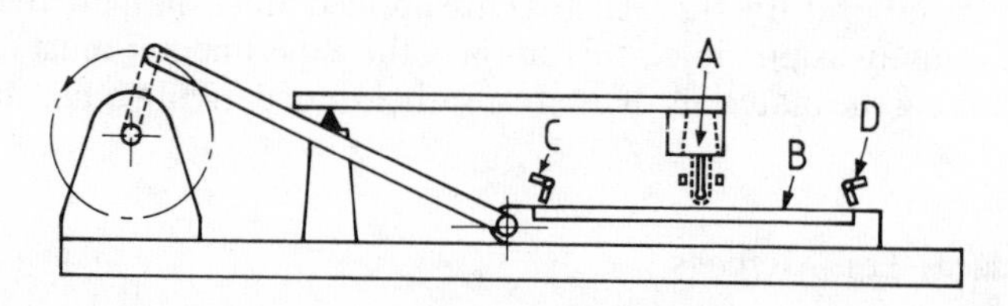

Fig. 10.3 Print adhesion test: BS 2782 Method 310B. The abrading member (A) is a brass peg the centre line of which is at right angles to glass plate (B) over which the test surface is clamped by grips (C, D). The brass peg exerts a force of 9 N on the test piece, through a piece of carefully specified bleached cotton fabric secured round the peg. The rate of reciprocation is 15 ± 2 cycles (each of two strokes) per min

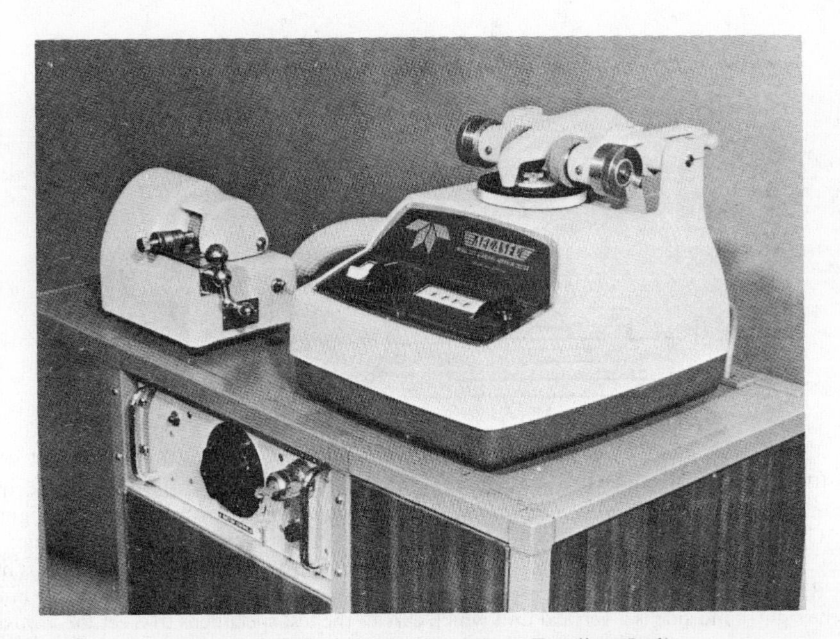

Fig. 10.4 Taber Abraser (courtesy Funditor Ltd)

100 mm squares, under the abrading wheels freely resting thereon with loads acting on them of 250, 500 or 1000 g.

To obtain abrasion the line joining the centres of the wheels is offset from the axis of rotation of the table. Abrasion resistance is measured by the amount of transmitted light which is diffused by the abraded track, using the ASTM D1003 Test for *Haze and Luminous Transmittance of Transparent Plastics* (Ch. 14).

The more conventional use of the Taber Abraser is by determining the weight loss of the test disc or square induced by the abrading wheel under a specified load and as a result of a specified number of revolutions – with a particular type of the four variants of resilient wheels ('Calibrase') or four variants of hard wheels ('Calibrade'). This form of test is included in test LPS–106 of the National Electrical Manufacturers' Association standard for decorative laminated plastics. Great care must be exercised in the use of the wheels: for example, they must be kept clean at all times – an air jet is normally used to blow off abraded material during the test – and the flexible wheels have a very limited life. A study of the reliability of the Taber method has been made by Hill and Nick[31] who conclude that it is best to take the mean of a considerable number of test runs; they also correlate results with the hardness of the (flexible) test wheels.

A great advantage of the Taber Abraser is its versatility, in that the severity of abrasion can be altered by using different types of wheel under different loads. The test can also be conducted in the presence of a liquid medium since the normal table can be replaced by a lipped table to hold the fluid. Its major disadvantage, particularly significant for testing finished products, is the large area of flat specimen required.

Two methods of testing the abrasion resistance of plastics are included in

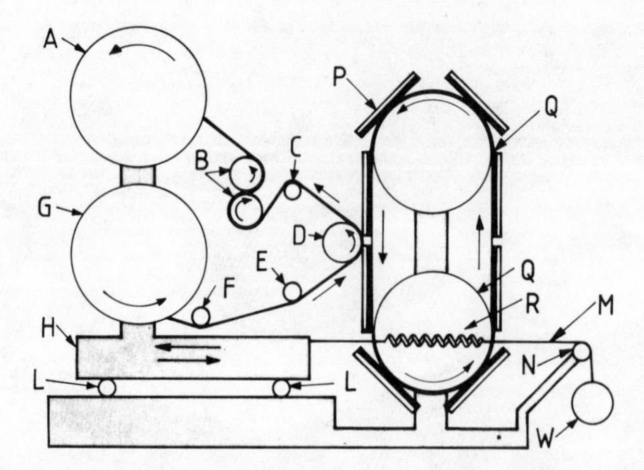

Fig. 10.5 Bonded abrasive abrading machine: ASTM D1242 Method B: A, take-up drum; B, constant-speed-driven rolls; C, abrasive tape; D, steel contact roll; E, F, slotted guide roll; G, abrasive tape supply drum; H, carriage; L, roller bearings; M, spring cable; N, pulley; P, specimen mounting plates; Q, constant-speed-driven pulleys with continuous link belt; R, spring; W, dead weight. Essentially the machine comprises two independent units. On the right-hand side is a vertical unit which carries the test specimens to meet the abrasive tape (C), which rotates in the opposite direction and is kept in contact by being mounted on carriage (H) on roller bearings (L), and tensioned by weight (W) through spring (R)

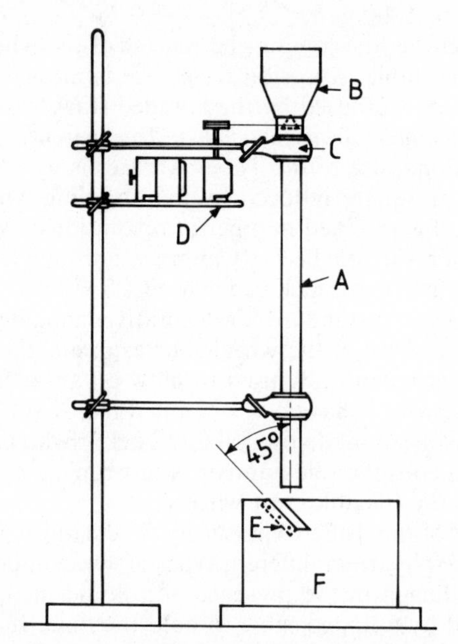

Fig. 10.6 Mar-resistance abrader: ASTM D673. Abrasive grit is allowed to fall onto the test specimen resting in groove (E) from a hopper (B) rotated to produce a controlled rate of feed. The grit falls down (A) onto the specimen and then into receptacle (F)

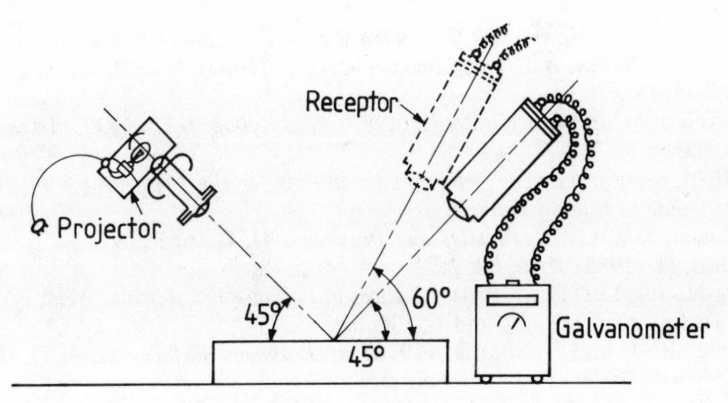

Fig. 10.7 Mar-resistance gloss meter: ASTM D673. The photoelectric receptor is set at 45° and the galvanometer reading is taken (l_1); the reading at 60° is then taken (l_2). Gloss (per cent) = 100 $(l_1 - l_2)/l_2$

ASTM D1242 (1956).[32] Method A uses loose abrasive applied under defined conditions to the test surface and results are expressed as volume loss (from weight loss and relative density). Method B uses a 'bonded abrasive abrading machine' (Fig. 10.5) which also gives results in volume loss.

The resistance of glossy plastics to abrasion may be measured by loss of gloss caused by the abrasive action of impacting carborundum grit (Fig. 10.6); such is the principle of ASTM D673 (1970).[33] The gloss before and after abrasion is measured with a gloss meter (Fig. 10.7).

Wiinikainen[34] has described a 'round robin' of scratch resistance and abrasion tests on transparent plastics materials.

German Standards

DIN 53754 (1977)[35] describes the Taber geometry while DIN 53516 (1977)[36] describes the so-called 'DIN Abrader'.

REFERENCES

1. James, D. I. (1973) *RAPRA Members Journal*, July, 170.
2. ISO 8295 (1986) *Plastics – Film and Sheeting – Determination of the Coefficients of Friction.*
3. BS 2782 Method 824A (1984) *Determination of Coefficients of Friction of Plastics Film.*
4. ASTM D1894 (1978) *Static and Kinetic Coefficients of Friction of Plastics Film and Sheeting.*
5. BS 3424 Method 12 (1973) *Methods of Test for Coated Fabrics – Determination of Surface Drag.*
6. Gough, V. E. (1953) *Journal of Scientific Instruments*, **30**, 345.
7. Bowers, R. C., Clinton, W. C. and Zisman, W. A. (1954) *Modern Plastics*, **31** (6), 131.
8. James, D. I. (1961) *Journal of Scientific Instruments*, **38** (7), 294.
9. Lauer, J. L. and Friel, P. J. (1957) *Review of Scientific Instruments*, **28** (4), 294.
10. Westover, R. F. and Vroom, W. I. (1963) *SPE Journal*, **19** (10), 1093.
11. ASTM D3028 (1972) *Kinetic Coefficient of Friction of Plastics Solids and Sheeting.*

12. James, D. I. and Newell, W. G. (1978) *RAPRA Members Report*, no. 20.
13. Giles, C. G., Sabey, B. E. and Cardew, K. H. F. (1964) *Road Research Technical Paper*, no. 66, HMSO.
14. Giustino, J. M. and Emerson, R. J. (1983) *ACS Rubber Division 123rd Meeting, Toronto, May*, paper 76, 31.
15. BS 4618: Section 5.6 (1975) *Recommendations for the Presentation of Plastics Design Data: Guide to Sliding Friction.*
16. Lancaster, J. K. (1973) *Plastics and Polymers*, **41**, October, 297.
17. Czichos, H. (1983) *Wear*, **88**, 27.
18. Lieng-Huang, Lee (Ed.) (1974) *Advances in Polymer Friction and Wear*, Polymer Science and Technology, vols. 5A and B, Plenum.
19. Briscoe, B. J. and Tabor, D. (1978) in *Polymer Surfaces* (Eds. D. T. Clark and W. J. Feast), Wiley, p. 1.
20. Harper, F. C. (1961) *Wear*, **4** (6), 461.
21. Anon. (1961) *Wear*, **4** (6), 479.
22. Anon. (1964) *Wear*, **1** (3), 302.
23. Frick, O. F. V. (1969) *Wear*, **14**, 119.
24. Anderson, J. C. and Williamson, P. K. (1984) *Polymeric Materials Science and Engineering*, **50**, 388.
25. ISO 4649 (1985) *Rubber – Determination of Abrasion Resistance using a Rotating Cylinder Drum Device.*
26. ISO/DIS 6601 *Plastics – Friction and Wear by Sliding – Terms and Definitions.*
27. BS 2782 Method 310B (1970) *Adhesion of Print on Flexible PVC Sheet.*
28. BS 3424 Method 16 (1985) *Methods of Test for Coated Fabrics – Determination of Colour Fastness to Wet and Dry Rubbing and Determination of Resistance to Printwear.*
29. BS 3794: (1982) *Decorative Laminated Sheets Based on Thermosetting Resins*; Part 2: *Methods of Determination of Properties.*
30. ASTM D1044 (1982) *Resistance of Transparent Plastics to Surface Abrasion.*
31. Hill, H. E. and Nick, D. P. (1966) *Journal of Paint Technology*, **38** (494), 123.
32. ASTM D1242 (1956) *Resistance of Plastics Materials to Abrasion.*
33. ASTM D673 (1970) *Mar Resistance of Plastics.*
34. Wiinikainen, R. A. (1969) *Materials, Research and Standards*, **9** (12), 17.
35. DIN 53754 (1977) *Testing of Plastics: Determination of Abrasion by the Abrasive Disc Method.*
36. DIN 53516 (1977) *Determination of Abrasion.*

Chapter Eleven

Creep, Relaxation and Set

11.1 INTRODUCTION

We have seen in earlier chapters how the viscoelastic nature of a plastics material renders its short-term and dynamic mechanical stress–strain behaviour considerably more complex than that of more perfectly elastic materials such as metals. Here and in the following chapter the long-term effects of applied stress and strains on plastics materials are examined. For the design engineer the long-term behaviour of the material is often of considerably greater importance than the short-term behaviour, since distortion of a component under load (creep), loss of applied force (relaxation) and component failure under loads less than the 'static' load to failure (fatigue) can all seriously reduce the useful life of the component.

Consideration of fatigue testing, which almost invariably results in the catastrophic failure of the item under test, is deferred to Ch. 12, while creep and stress relaxation are discussed below. Set, which is seldom encountered in the literature on plastics, is mentioned briefly at the end of the chapter.

11.2 CREEP

11.2.1 General Considerations

The principle of creep testing is very simple: a fixed load is applied to a test piece and the deformation of the test piece as a function of time is recorded. Parameters such as temperature and humidity, where the latter is known to affect the creep rate (e.g. for polyamides), are kept constant. In practice, of course, the situation is much more involved and great care must be exercised if reproducible results are to be obtained. The deformation-measuring device must be capable of high resolution, in order to measure the (usually) small strains involved, while at the same time having a high stability to long-term thermal or electrical drift. It must not interact significantly with the test piece, thereby modifying the creep behaviour. The test piece should also be well characterised chemically and physically since the time-scale of a creep test allows the individual viscoelastic nature of each test piece to exert itself to the full and slight differences between test pieces, which might not be detectable in short-term tests, may become significant in a long-term test. Effects such as an increase in crystallinity for a crystallisable polymer, post-curing in a thermoset, or different moulding stresses in a thermoplastic, can alter the creep behaviour of the material significantly. All this means that the test environment is normally controlled rather more rigorously than is usual for short-term tests and temperature control is often required to be within ±1 °C or better. Humidity control at ±2 per cent may also be required for materials which are sensitive to water content.

11.2.2 Definitions

The following definitions are taken from BS4618: Section 1.1 (1970):[1]

Creep strain	The total strain, which is time-dependent, resulting from an applied stress or system of stresses.
Creep modulus	The ratio of applied stress to creep strain.
Creep lateral contraction ratio	The ratio of lateral strain to longitudinal strain measured simultaneously in a creep experiment (also known as Poisson's ratio).

The above definition of creep strain is the one currently favoured in plastics engineering practice. There is, however, a 'scientific' definition of creep strain which subtracts the instantaneous, or elastic, strain from the total strain to give the creep strain. Such a definition follows naturally from the phenomenological models of the Kelvin and Voigt type mentioned in Ch. 9 and used to approximate the viscoelastic behaviour of polymers, but in practice the truly elastic, i.e. instantaneous, deformation cannot be separated from the time-dependent deformation.

11.2.3 Standard Test Methods

Creep tests may be carried out in tension, compression, flexure, shear or torsion, although only the first three have been standardised, tests in tension being the most popular.

ISO Standards

There are currently two standards published with one general method for compressive creep of rigid cellular plastics at the stage of a draft.

ISO 899 (1981)[2] deals with tensile creep and uses standard tensile specimens as described in ISO 527 (Section 8.2).

ISO 899 describes the apparatus requirements (which are rather more exacting in terms of axiality of load application than for the corresponding short-term test method) and type of grip system that may be used. The loading system must be designed to apply a load within ±1 per cent of the desired load and be of such a construction that if a test piece breaks the energy released is not transmitted to any of the other test pieces. The extensometer must be capable of measuring to ±1 per cent of the total strain anticipated. Contactless, optical systems are recommended especially for creep rupture tests and the comment is made that electrical resistance gauges should only be used for short-duration tests and then only when perfect adhesion to the test material can be guaranteed.

The test pieces are as specified for the determination of tensile properties but the variation in cross-sectional area within the gauge length should not exceed 2 per cent and dimensional measurements should be made with equipment having an accuracy of at least 0·01 mm. It is recommended that an allowance be made for any electrical and thermal drift of the strain-measuring device by means of the use of a control test piece which is prepared and tested under identical conditions to the real test piece, except that no load is supplied to it.

The loading of the test piece is very important and this should be carried out smoothly and continuously within a period of from 1 to 5 s from the start of loading, zero time for the experiment being taken at the point of full loading. At no time during loading must the stress exceed the level at zero time. The strain is recorded either automatically (preferably) or manually at approximately equally spaced logarithmic time intervals.

The experiment is usually carried out over a range of applied stress since the creep modulus is not independent of stress level except at low stress levels (in the linear viscoelastic region) and it is frequently convenient to derive isochronous stress–strain curves from the strain–time curves obtained under different stresses. Figures 11.1 and 11.2 illustrate this operation. Temperature, too may be considered as a parameter to be examined especially if the data are to be used for design purposes, in which case the whole experiment is repeated at several temperatures.

Creep recovery tests may also be carried out by removing the load, in the same manner in which it was applied, after some suitable time and then following the decrease in strain as a function of time from the instant of unloading.

ISO 6602 (1985)[3] describes the method of determining flexural creep and is an amalgamation of ISO 178 (Section 8.5) for short-term flexural testing with ISO 899 for the long-term requirements of the creep test using the same accuracies and tolerances for dimensions, loads, strains and times as the corresponding values in ISO 178 or ISO 899 as appropriate.

British Standards

BS 4618: Section 1.1 (1970)[1] is the relevant UK document in this context. It is considerably more detailed than ISO 899 and is an excellent source of information

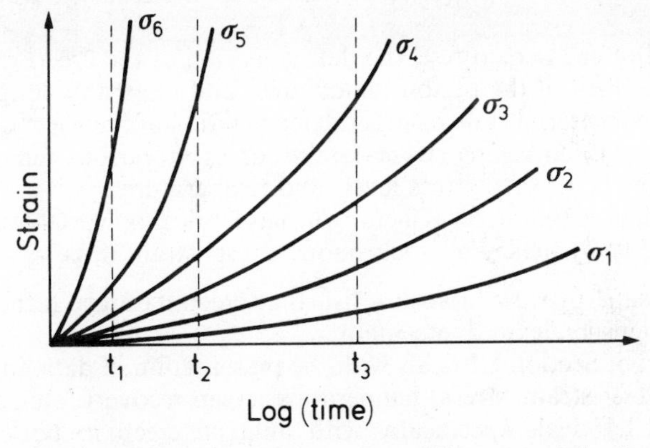

Fig. 11.1 Typical creep curves at different stress levels

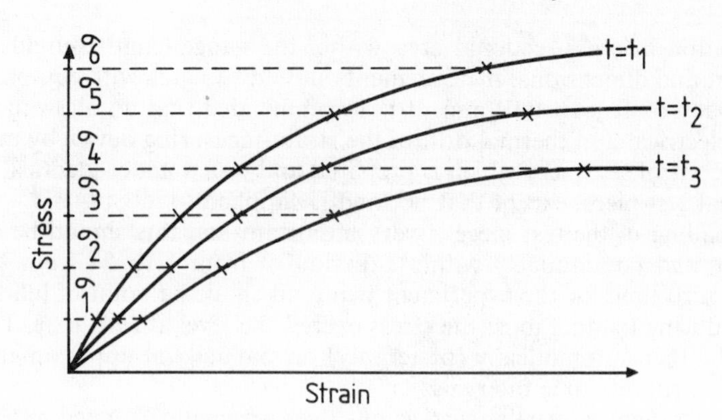

Fig. 11.2 Isochronous stress–strain curves

on the philosophy and procedures for creep testing of plastics. Creep rupture, however, is not included within its scope.

This document makes the point not made in ISO 899 that the strain response immediately after the application of the load depends upon the precise loading path taken and recommends that strain measurements made within a period equal to ten times the loading period be ignored, although in justice to ISO 899 the minimum recommended time measurement is 1 min after application of the load and so this point is covered, albeit by default.

Section 1.1 (of BS 4618) describes an experiment whereby the isochronous stress–strain curve (for a short time) may be derived from a single test piece, but warns that for data suitable for engineering design purposes from fifty to a hundred test pieces may need to be tested. The minimum programme for characterising a nonlinear viscoelastic material is given as:

(i) the determination of the isochronous stress–strain curve at a particular temperature in the range 19° to 24 °C at some specified short time such as 60 s or 100 s;
(ii) for the same temperature and material state as in (i), the determination of at least four creep curves at different stress levels, preferably to duration at least one year;
(iii) at the end of each creep test, the determination of the creep recovery for at least 10 per cent of the period under stress and preferably longer;
(iv) for the same material state as in (i), determination of the effect of temperature by means of isochronous stress–strain curves at various temperatures and creep curves at selected stress levels and temperatures;
(v) determination of the effects of factors such as fabrication variables, environment and humidity by means of isochronous stress–strain tests.

It is not surprising, in view of such a list, that creep data are rather less readily available than tensile, flexural, etc., data!

The final part of Section 1.1 deals with the presentation of data and gives creep, isochronous stress–strain, stress–temperature, creep recovery, etc., curves.

Subsection 1.1.1 deals specifically with uniaxial creep in both tension and compression, giving criteria of suitability for test pieces and apparatus as well as describing test conditions and experimental programmes. A point worthy of note is the classification of the strain-measuring device according to precision: Category A has a minimum detectable increment of 0·002 per cent strain, Category B 0·008 per cent and Category C 0·02 per cent. The requirements of axiality of the test machine are discussed in relation to these three categories.

Subsection 1.1.2 deals with flexural creep in three-point loading. The advantages of using flexural creep tests are:

1. In the linear viscoelastic region, i.e. at low strains, data of high accuracy are more readily obtained than in tension and compression.
2. The apparatus can be simpler.
3. Data can be obtained that are directly relevant to flexural deformation in structures.

These three considerations lead to three categories of testing, the requirements of the first two of which are described in detail in the standard. Category 1 is for the measurement of flexural creep modulus to better than ±3 per cent, Category 2 for measurement poorer than ±3 per cent and Category 3 where the degree of accuracy depends on the purpose for which the data are required (and therefore must be stated in the presentation of the data) but where the general principles of the previous categories are still applied.

Subsection 1.1.3 deals with the measurement of Poisson's ratio, which is one of the four fundamental parameters required to describe uniquely the deformation behaviour of an isotropic, linear, viscoelastic material (the other three being the bulk, Young's and shear moduli). As only two of these four parameters are independent for a given material, the measurement of Poisson's ratio and tensile creep modulus, which can be carried out simultaneously on the same apparatus, enables an isotropic linear viscoelastic material to be fully characterised. Two strain gauges making measurements in the longitudinal and transverse directions are required for this test and the same classification of strain gauge precision as described in Subsection 1.1.1 also applies.

USA Standards

The standard test method is ASTM D2990 (1977),[4] which covers the measurement of creep in tension, compression and flexure. Creep rupture of plastics is also included.

The experimental procedures follow much the same pattern as the ISO and BS standards previously mentioned, although the test temperature control is only to within ±2 °C which is rather surprising. For creep rupture tests it is recommended that a minimum of seven stress levels be selected in order to give failure in approximately 1, 10, 30, 100, 300, 1000 and 3000 h. The standard points out that for simple material comparisons, such as for data sheets, a useful parameter to determine is the stress to give 1 per cent strain at 1000 h. This is obtained from the 1000 h isochronous stress–strain curve, usually by interpolation. The effect of temperature on the material is readily shown via this parameter by carrying out the test over a suitable temperature range.

A creep rupture test specifically designed for plastics pipe is ASTM D1598 (1981)[5] in which a length of pipe is pressurised by means of liquid or gas, as appropriate, and the time to failure is recorded.

German Standards
There are two test methods for creep testing in Germany, these being DIN 53444 (1968)[6] for tensile creep, as described above, and DIN 54852 (1984)[7] which deals with flexural creep but includes a four-point loading variant in addition to the normal three-point method.

11.2.4 Practical Systems

Having looked at the standard test methods available, it is appropriate to examine briefly some practical systems that have been used to produce creep data. More detailed information on creep, its measurement and the effects of molecular parameters on it is given by Nielsen[8] and Ferry[9] and also by Gittus,[10] who provides a particularly detailed account of creep in materials generally, rather than plastics alone. An excellent exposition of creep testing which examines in some detail the pros and cons of various geometries and the means by which dimensional change can be measured over extended time scales is given by Turner.[11]

The simplest possible experimental technique is described by Findley[12] who studied creep in cellulose acetate specimens at $25 \pm \frac{1}{2}$°C and 50 ± 2 per cent r.h. at periods up to 7000 h (about 10 months). The stresses involved ranged between 3·45 MPa and 18·6 MPa (based on original cross-sectional area) and these were achieved by the application of simple dead loads of from 140 to 700 N to the specimens. Under these conditions extensions in the range from 1 to 45 per cent were produced, and measurements of adequate accuracy were obtained with a travelling microscope, which was used to determine the distances between pairs of marks clamped to the specimens. As would be expected the specimens decreased in thickness significantly during the progress of the tests, and it was necessary to spring load the clamps to accommodate this reduction (see Fig. 11.3).

Results of this experimental work are shown in Fig. 11.4. As can be seen, this technique is capable of demonstrating the creep characteristics of this material

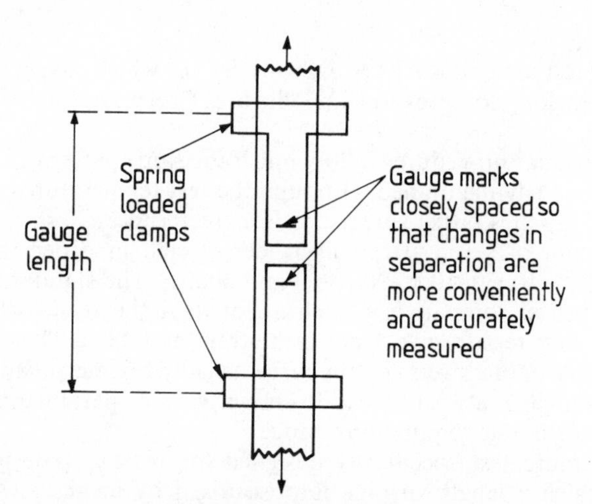

Fig. 11.3 Arrangement of gauge marks[12]

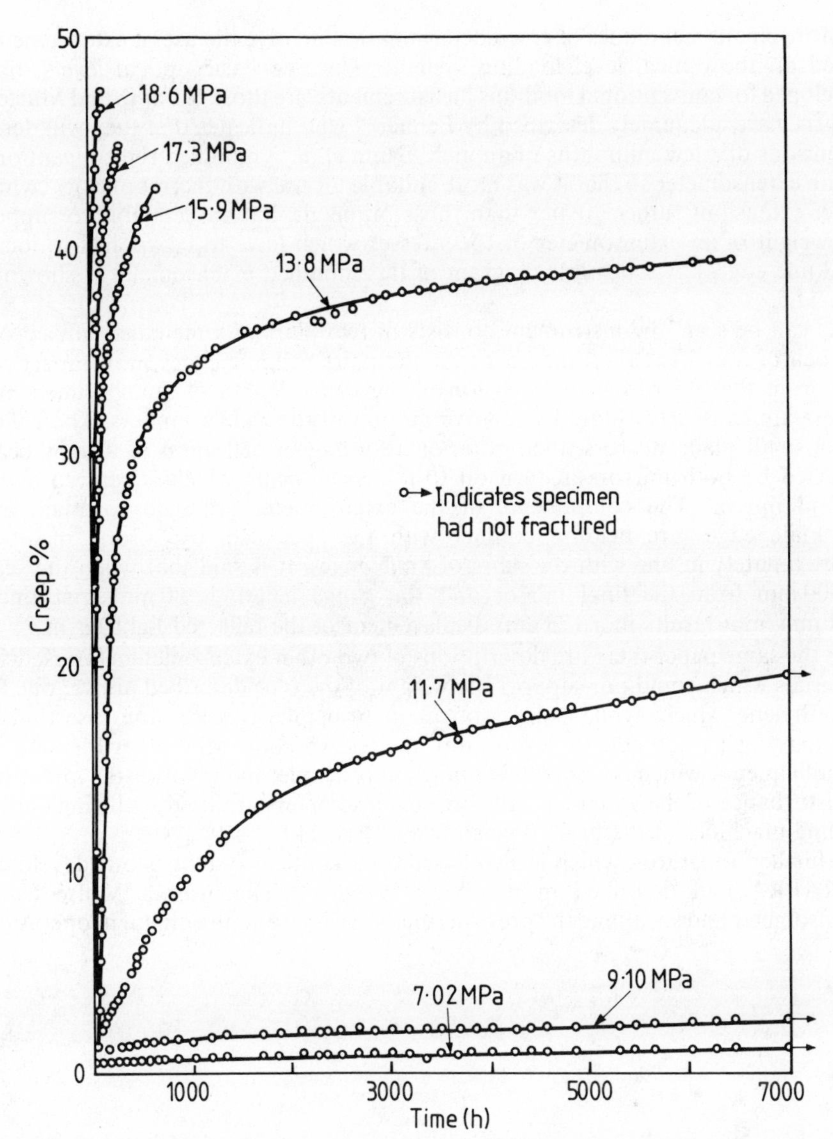

Fig. 11.4 Creep–time diagram of cellulose acetate at 25°C.[12] (reproduced by permission of McGraw-Hill Inc.)

over a range of stresses and deformations which might be relevant to consumer goods and non-critical applications, though the low creep region, of interest for serious engineering projects, is less adequately covered.

Improvements in low strain sensitivity can be made by using strain gauges bonded to the test piece, provided adequate precautions are taken to minimise the many sources of error.[13]

More elgant techniques of creep determination involve the use of extensometers based on the optical level loading systems. The two basic optical levers, both developed for conventional modulus measurements, are those of Lamb and Martens and they are adequately described by Fenner[14] who indicates that they will detect extensions of a few millionths of an inch, Dunn *et al.*[15] modified the conventional Lamb extensometer so that it was more suitable for use with thermoplastics (which suffer extensions rather greater than those of metals, but are less able to support the weight of an extensometer or the stresses which have to be applied to locate the knife edges). A schematic diagram of this modified extensometer is shown in Fig. 11.5.

As can be seen, the instrument consists of two pairs of knife edges affixed one to each of four light rigid members. As the pairs of knife edges move apart one pair from the other, due to extension of the gauge length of the specimen, two rollers are caused to rotate by relative motion of the rigid members. The rollers carry small plane mirrors, their rotation resulting in deflection of a light beam reflected by both mirrors in turn on to a circular scale which is centred at the second mirror. The components of the extensometer are held together, and the knife edges are held in contact with the specimen, by sprung members approximately in line with the pairs of knife edges. It is said that when the scale is 500 mm from the final mirror and the gauge length is 80 mm a strain of 0·01 mm/mm results in a 128 mm displacement of the reflected light beam.

In the same paper there are descriptions of two other extensometers for use with materials which could not support the weight of the type described above: one for polyethylene which, while being optical in principle, requires the insertion of locating pins (which actuate levers) into the test specimen; the other for films or monofilaments, which is completely optical in principle, and involves the minimum of disturbance of the specimen. All three extensometers are used with single-lever loading machines of the basic type shown in Fig. 11.6.

A further apparatus, which is also based on an optical system, is one developed at RAPRA and described in detail by Wright.[16] This uses a Moiré fringe extensometer and so, unlike the previous methods, is essentially digital in operation.

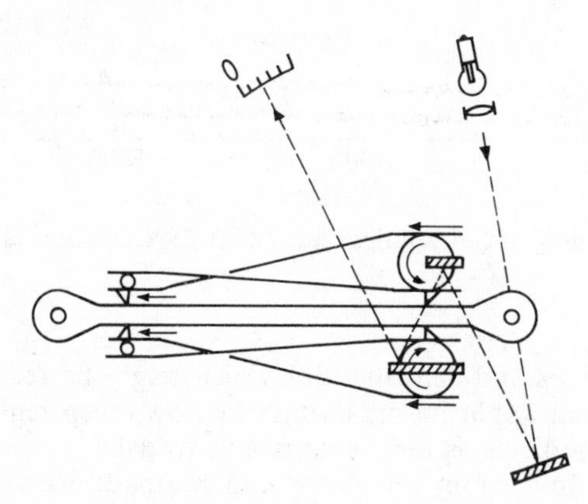

Fig. 11.5 Schematic diagram of modified Lamb extensometer.[15] (reproduced by permission of *British Plastics*)

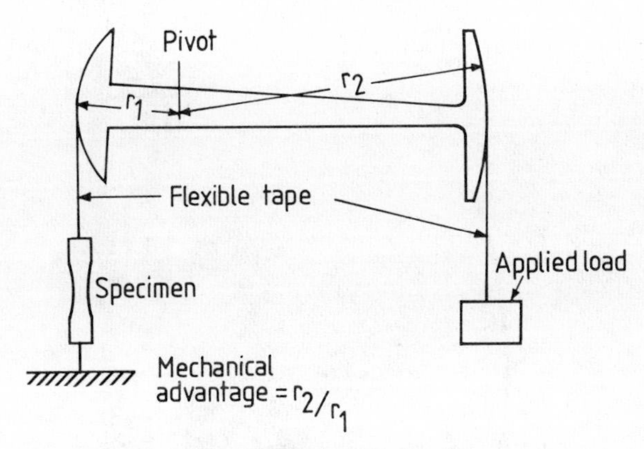

Fig. 11.6 Typical single-lever loading system

This results in a number of advantages over analogue extensometers, in that it has long-term stability, no calibration is required, a 20 per cent strain limit with no loss in resolution is achievable and strain is the independent variable, i.e. the strain is automatically recorded at equal strain increments rather than equal (or logarithmic) time increments. The Moiré fringes are generated via two different gratings inclined at a small angle to each other. One grating is attached via a suitable carriage to one gauge mark of the test piece and the other grating to the other gauge mark. As the test piece extends, the Moiré fringe pattern moves over the surfaces of suitably positioned photocells which are connected to a counting circuit; as each fringe is detected, the time is recorded by means of a data logger. The apparatus, of Class A sensitivity (BS 4618), is illustrated in Fig. 11.7.

Special difficulties arise if it is necessary to determine creep at constant stress – that is, if it is necessary to reduce the applied load in step with the reduction in area of the specimen as it extends. Andrade[17] stressed specimens by means of an appropriately shaped weight (hyperboloid of revolution) which was partially immersed in a liquid. As the specimen extended and decreased in cross-section the load was decreased by buoyancy of the liquid. Ward and Marriott[18] used a single-lever machine, but the circular arc at the 'dead-weight' end of the lever in the conventional machine (as shown in Fig. 11.6) was replaced by a specially profiled cam. In computing the shape of the cam it was assumed that Poisson's ratio was 0·5, which is approximately true for many plastics. No allowance appears to have been made for the fact that elongation and section reduction of the parts of the specimen immediately adjacent to the grips will be less than occurs in the central part of the specimen.

The application of the creep load has been given considerable attention. The investigator is required to apply the load rapidly (to avoid the theoretical embarrassment of a stress which increases over a significant period of time) and reproducibly but without shock or vibration. Findley[12] supported his dead weights on planks, blocked up in such a way that they could be used as levers to lower the weights quickly but gently until they were supported by the specimen. He then 'immediately' recorded extension and time. Dunn *et al.*[15] describe a system for

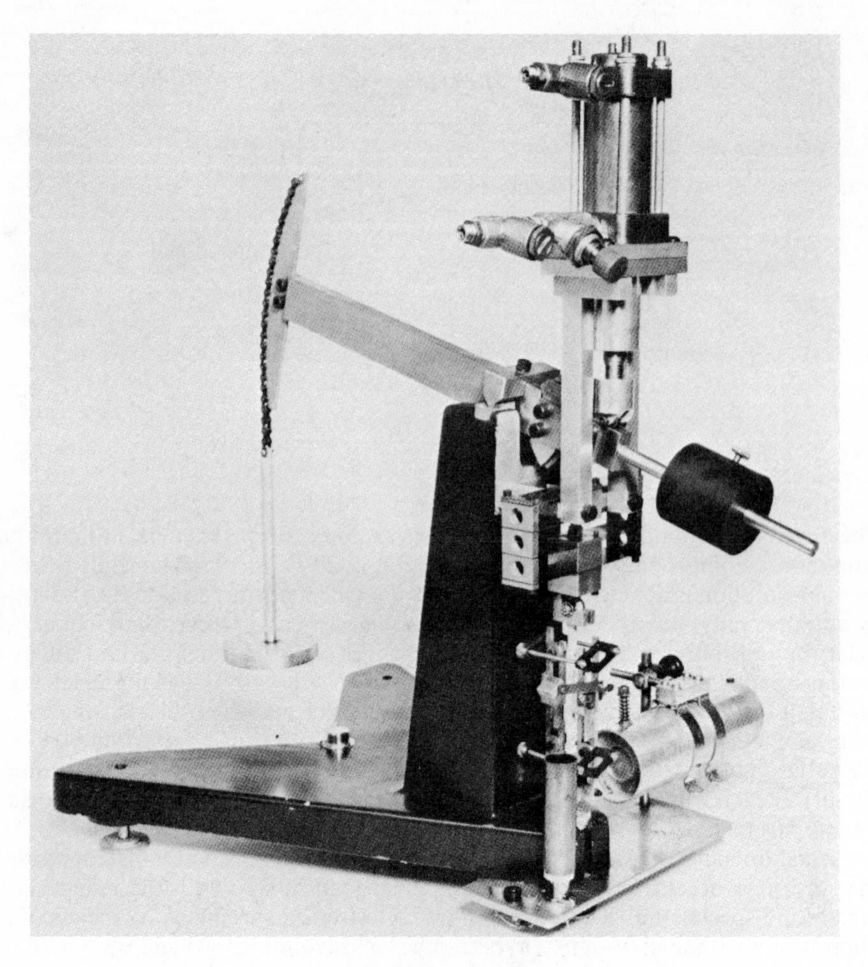

Fig. 11.7 RAPRA–Hampden creep apparatus

automatic shock-free loading which can apply loads up to 30 kg without vibration, in times considerably below $\frac{1}{2}$ s. The RAPRA apparatus uses a hydraulically actuated piston which applies the load in approximately 1 s.

While the tensile mode of creep testing has received the major share of the experimenter's attention, the other types of geometry have not been without their adherents.

The examination of specimens in compression is subject to the difficulties which apply to compression tests in general, described in Ch. 8. Of these, the tendency of the specimens to buckle is the most serious, as this would obviously invalidate the measurements of small creep deformations. Dunn *et al.*[15] and Thomas[19] suggest the use of a hollow cylindrical specimen to prevent or reduce buckling, while Jones *et al.*[20] suggest the use of a square-section specimen which is almost totally enclosed in a close-fitting tube (the method is unsuitable for large deformations). In all three cases the intention is to use a specimen which is sufficiently slender to be stressed without the use of excessively large loads. Decrease in specimen length is assumed to be directly related to the relative movement of the compression platens (which is recorded by dial gauge) in the work reported by Jones *et al.* and Dunn *et al.*, although the latter do indicate that this assumption is of doubtful validity and recommend the use of extensometers; Thomas shows that, for some materials at least, the error associated with the use of platen-mounted dial gauges is unacceptably large.

Little published work appears to have been carried out in flexure, although Hammant and Roberts[21] describe a novel technique in which the centre-loading nose has the same radius as the radius of curvature of the test piece, the load being rapidly applied via electromagnets. More recently Allen[22] reports the use of a LVDT transducer to monitor deflections of high-performance epoxy resins under three-point and four-point loading in water at various temperatures. A pneumatic cylinder was used to provide the bending force over the range from 45 to 1800 N.

Torsional creep is described in some detail by Link and Schwarzl[23] the torque being applied by a three-phase asynchronous motor and the torsional angle measured by a laser beam reflecting from a mirror on to an electro-optical device. Further work on torsional creep has been published by Gacougnolle *et al.*[24] and Shankar *et al.*[25]

Finally, it is perhaps worth mentioning that a number of more commonplace tests, such as hardness tests (load applied for perhaps 30 s), indentation tests on floor tiles, etc., are really elementary creep tests, in so far as they require the application of a constant load to a specimen and the measurement of the resulting deformation. Indeed, Gonzalez[26] has taken this to its logical conclusion and modified a Rockwell hardness tester to set up a screening test for the torsional creep of plastics. Correlation between the degree of penetration and the (measured) torsional creep appears to be good in eleven out of the thirteen materials examined, the two exceptions being reinforced plastics.

11.3 STRESS RELAXATION

In stress relaxation tests, instead of holding the stress (or, more correctly, the load) constant and measuring strain as a function of time, the strain is fixed and the stress is measured. By contrast to creep tests, stress relaxation is seldom measured in plastics materials and this is reflected in the paucity of standard test methods. This situation is reversed for elastomers, as detailed by Brown.[27]

Nielsen[8] describes a simple, basic test equipment for the measurement of stress relaxation, and indicates that simple instruments are suitable for use with low modulus (and, by inference, highly extensible) polymers (Fig. 11.8). Rigid polymers present much greater difficulties, and for accurate work the apparatus must be very stiff because otherwise its deformation may be comparable to that of the polymer. Nielsen cites a relaxation experiment in which a rigid specimen, 1 in long, stretched by 0·001 in. The stress-measuring device (and the apparatus) must not deform by more than 1 in $\times 10^{-5}$ (2·5 mm $\times 10^{-4}$) if stress measurement accurate to 1 per cent is to be achieved.

A more detailed description of a specific instrument is given by Curran *et al.*[28]

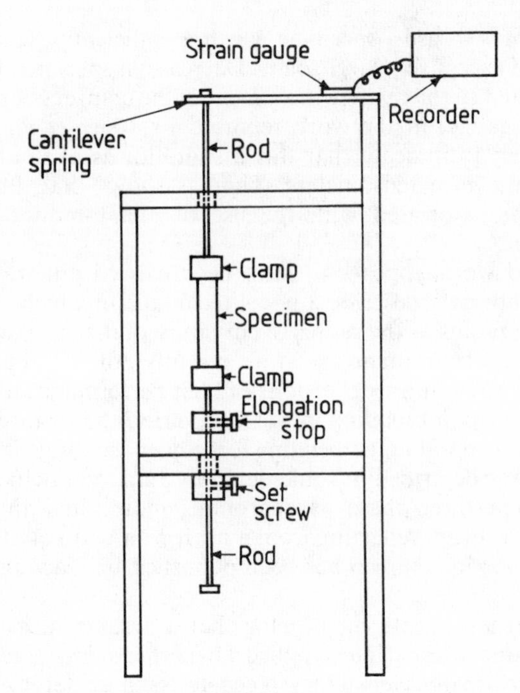

Fig. 11.8 Simple apparatus for measuring stress relaxation[8]

This consists of a rectangular aluminium frame, which is several orders of magnitude more rigid than the test specimens, with the specimen stretched along its longer axis. The requisite stress is applied to the specimen via one grip which is positioned by a keyed micrometer screw passing through one short frame member, and the (initial) strain in the specimen is indicated during the setting up process by a clip-on extensometer, which is subsequently removed. The stress is recorded by a group of four resistance strain gauges which are bonded to a thin strip steel tensile member, which restrain the lower grip. Typical results are shown in Fig. 11.9. This method is suitable for tests involving liquid environments, and a suitable environmental cell is described by the authors.

For the most accurate work this technique would not be acceptable because there are two mechanisms whereby the strain in the specimen could change during the course of the test. The strip steel load weighing member will decrease in length, if only marginally, as the stresses in it (and the specimen) decrease. More significantly, no attention is paid to the changes in strain distribution within the specimen as the test progresses. If one considers a typical dumb-bell specimen it is obvious that the stress will be greater in the narrow portion of the length. Under certain conditions of stress, this narrow portion will creep, permitting the larger sectioned parts of the specimen to retract, and thus invalidating the concept of constant strain. This type of mechanism is also possible in parallel-sided specimens, as their stress distribution is far from uniform because of the disturbance caused by the gripping members, although in this case a simple commonsense analysis is not possible.

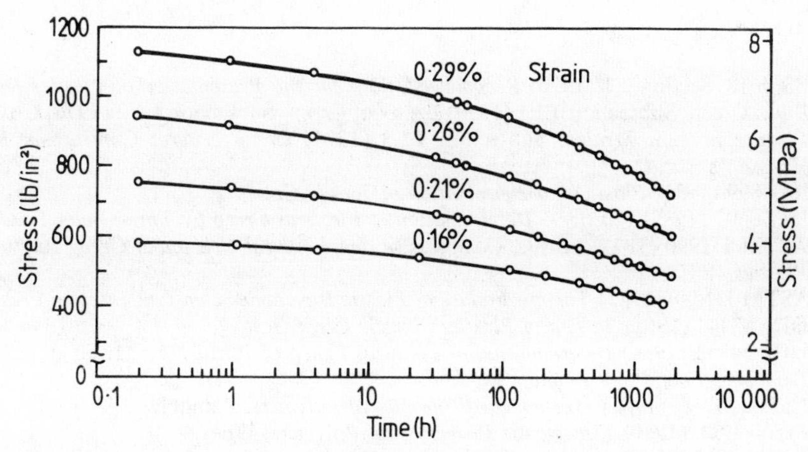

Fig. 11.9 Typical stress relaxation data for 'multiphase' thermoplastics.[28] (reproduced by permission of *Modern Plastics*, McGraw-Hill Inc.)

More recently an apparatus suitable for testing plastics has been described by Kaufmann and Hofmann.[29]

The only standard method for the measurement of stress relaxation in plastics is given in ASTM D2991 (1971).[30] This recommended practice avoids giving detailed test procedures but does indicate environmental and test piece tolerance as well as giving guidance on the measurement of load and strain, and the presentation of data.

ASTM D2552 (1969),[31] which deals with stress rupture, is more properly regarded as a fatigue test and is discussed in Chs. 12 and 17.

11.4 SET

Set is the percentage residual deformation remaining in a test piece after the application of a given reference deformation for a given time followed by a given period of recovery after removal of the load required to maintain that deformation. It is, therefore, essentially a stress relaxation followed by a creep recovery test.

This is not a measurement which is ordinarily carried out on plastics materials although it is commonly used in the rubber industry, partly because the apparatus is very simple and partly because set appears at first sight to be the important parameter for judging sealing efficiency. Stress relaxation is, in fact, the more significant in this context.

Set is dealt with in detail by Brown[27]; briefly a 'normal' compression set test consists of compressing three discs of the material by 25 per cent in a rigid clamp, keeping this at perhaps 70 °C for 22 h, releasing the clamp and then measuring the thickness of the discs after a 30 min recovery period. Compression set is then given by:

$$\text{Compression set} = \frac{(\text{initial thickness}) - (\text{recovered thickness})}{(\text{initial thickness}) - (\text{compressed thickness})} \times 100 \text{ per cent}$$

Clearly, this is not a suitable test for rigid materials.

REFERENCES

1. BS 4618:Section 1.1 (1970) *Recommendations for the Presentation of Plastics Design Data*: Creep. Subsection 1.1.1 (1970) *Uniaxial Creep*; Subsection 1.1.2 (1976) *Creep in Flexure at Low Strains*; Subsection 1.1.3 (1974) *Creep Lateral Contraction Ratio (Poisson's Ratio)*.
2. ISO 899 (1981) *Plastics – Determination of Tensile Creep.*
3. ISO 6602 (1985) *Plastics – Determination of Flexural Creep by Three-Point Loading.*
4. ASTM D2990 (1977) *Tensile, Compressive and Flexural Creep and Creep Rupture of Plastics.*
5. ASTM D1598 (1981) *Time to Failure of Plastic Pipe under Constant Internal Pressure.*
6. DIN 53444 (1968) *Testing of Plastics: Tensile Creep Test.*
7. DIN 54852 (1984) *Determination of Flexural Creep of Plastics by Means of a Three Point and Four Point Loading Method.*
8. Nielsen, L. E. (1965) *Mechanical Properties of Polymers*, Reinhold.
9. Ferry, J. D. (1980) *Viscoelastic Properties of Polymers*, Wiley.
10. Gittus, J. (1975) *Creep, Viscoelasticity and Creep Fracture in Solids*, Applied Science.
11. Turner, S. (1983) *Mechanical Testing of Plastics*, George Godwin.
12. Findley, W. N. (1942) *Modern Plastics*, **19** (12), 71.
13. Staff, C. E., Quackenbos, H. M. and Hill, J. M. (1950) *Modern Plastics*, **27** (6), 93.
14. Fenner, A. J. (1965) *Mechanical Testing of Materials*, Philosophical Library.
15. Dunn, C. M. R., Mills, W. H. and Turner, S. (1964) *British Plastics*, **37** (7), 386.
16. Wright, D. C. (1971) *RAPRA Bulletin*, **25** (6), 133.
17. Andrade, E. N. da C. (1910) *Proceedings of the Royal Society*, *A*, **LXXIV**, 1.
18. Ward, A. G. and Marriott, R. R. (1948) *Journal of Scientific Instruments*, **25** (5), 147.
19. Thomas, D. A. (1969) *Plastics and Polymers*, **37** (131), 485.
20. Jones, E. D., Koo, G. P. and O'Toole, J. L. (1966) *Materials, Research and Standards*, **6** (5), 241.
21. Hammant, B. L. and Roberts, J. (1967) *Engineering Materials and Design*, December, 1922.
22. Allen, R. C. (1982) *SAMPE Quarterly*, **13** (3), 1.
23. Link, G. and Schwarzl, F. R. (1985) *Rheologica Acta*, **24**, 211.
24. Gacougnolle, J. L., Peltier, J. F. and de Fouquet, J. (1982) *ASTM*, *STP*, 765.
25. Shankar, N. G., Bertin, Y. A. and Gacougnolle, J. L. (1984) *Polymer Engineering and Science*, **24** (11), 921.
26. Gonzalez, H. (1976) *Journal of Testing and Evaluation*, **4** (1), 21.
27. Brown, R. P. (1986) *Physical Testing of Rubbers*, Elsevier.
28. Curran, R. J., Andrews, R. D. and McGarry, F. J. (1960) *Modern Plastics*, **38** (3), 142.
29. Kaufmann, W. and Hofmann, D. (1969) *Kunststoffe*, **59**, 173.
30. ASTM D2991 (1971) *Standard Recommended Practice for Testing Stress Relaxation of Plastics.*
31. ASTM D2552 (1969) *Environmental Stress Rupture of Type III Polyethylenes under Constant Tensile Load.*

Chapter Twelve

Fatigue

12.1 INTRODUCTION

Fatigue is the name given to the process whereby a component fails after some period of time throughout part, or the whole, of which it has been exposed to loads of lesser magnitude than the static ultimate load value. As such it should include creep and stress relaxation failures but these are normally thought of as separate from fatigue which ordinarily, though not exclusively, is associated with the action of fluctuating stresses. Creep and stress relaxation has been dealt with in the previous chapter.

The phenomenon of fatigue bedevils probably the whole field of materials science: virtually all materials – natural and synthetic – suffer from its effects. Apart from metal fatigue,[1] which must be known to almost everyone, wood,[2] concrete,[3] bituminous materials[4] and, of course, plastics[5-9] all come under the umbrella of this pervasive effect. The last reference[9] is specifically devoted to the question of fatigue in plastics and contains a comprehensive review of the literature up to 1980. It has been estimated[10] that some 80–90 per cent of all service failures of equipment are due to fatigue.

Even for the traditional engineering materials such as metals, for which fatigue has been a recognised problem for over a century, the fundamental causes of failure are still not fully understood, with all too familiar and often tragic consequences. It should come as no surprise to learn, therefore, that for the newer plastics materials the fatigue processes are only just beginning to be understood and that fatigue data are very sparse. A considerable amount of work is being done to fill this gap, and even though the technique of fracture mechanics is yielding more material-specific rather than geometry-specific data, it is unlikely that plastics design engineers will be provided for many years to come with fatigue data comparable in quality or quantity to those available to engineers using more traditional materials.

Despite the volume of work being undertaken to map the fatigue properties of plastics, very little is carried out to any nationally, much less internationally, recognised test method. Since only two such methods, one American and one German, exist at present specifically for plastics materials, this is understandable. This is in marked contrast to fatigue testing of elastomers for which a number of standard test methods exist, as described in detail by Brown.[11] Fatigue testing of metals is also well covered in the standards literature; in particular ISO R373

(1964),[12] BS 3518: Part 1 (1962)[13] and ASTM E206 (1972)[14] contain much valuable information relating to terminology, data handling and test methods. Test methods and procedures developed for metals can be, and have been, adapted for use with plastics but care must be exercised since metals are low damping, high thermal conductivity materials. Deformation frequencies for plastics testing must, therefore, be much less than for metals testing, as a rule.

The fatigue properties of plastics are often considerably influenced by the medium in which the plastics material is situated. A classic example of this is the well known phenomenon of environmental stress cracking. This effect and the test methods for studying it are considered in Ch. 17.

12.2 TECHNIQUES IN FATIGUE ANALYSIS

As mentioned above, there is little in the way of nationally standardised methods to guide the experimenter in the choice of methodology to adopt and not surprisingly, therefore, a wide range of experimental techniques has arisen. It is not possible to give an exhaustive review of methods here, but the following will serve as a guide to the sort of systems that have been successful in given instances. The reader who is interested in understanding more of the trade-offs that usually have to be made in deciding between the various geometries to choose from is recommended to read Ch. 8 of Turner's book on testing.[6]

In many cases it is sufficient simply to know how long, in the case of static fatigue, or how many cycles, in the case of dynamic fatigue, the test piece will survive without fracturing under the applied stress or strain environment being examined. For this, a simple counter or timer of suitable range can be linked to the apparatus in such a way that it stops working when the test piece breaks. The RAPRA–Hampden creep testing machine, for instance, has been used for measuring creep rupture times in polycarbonate[15] and Gotham[16] describes a simple biaxial fatigue test in which air is used to cyclicly pressurise a cylindrical chamber capped by the test piece. Failure of the latter prevents further pressurisation taking place; this condition being detected by a pressure switch which then no longer activates a counter.

However, there are instances where the progress of the fatigue process is at least as important as the ultimate failure, if not more so. This is the case when an understanding of the fatigue process and the mechanisms of failure are of importance as, for example, in the design of a product or component.

The progressive degradation of the materials' ability to withstand stress may be monitored by measuring the residual strength left in the material after a given stress–strain history which has stopped short of producing failure. Obviously residual strength and residual life before failure are directly related. Stiffness changes may also be used to monitor the fatigue process but although this has the advantage of being non-destructive, there is no clear link between the stiffness loss and the expected life other than by purely empirical evaluations. The theory behind these cumulative damage methods is well presented by Hashin[17] for composites for which it represents a particularly valuable technique. Takemori and Morelli[18] report on residual impact strength of an unmodified polycarbonate under impact fatigue conditions.

For following the development of a growing crack in the polymer, optical methods are a natural choice. This may be done using a microscope of suitable power if the type of specimen is suitable, but more sophisticated methods have also been developed. Schinker *et al.*[19] for example use an interferometric method with a stroboscopic light source to 'fix' the interference pattern as it changes during the cyclic loading and so enable it to be photographed. Fanucci and Mar[20] on the other hand report the use of an out-of-plane Moiré fringe system which, they claim, enabled them to detect underlying delaminations in six-ply graphite/epoxy laminates. Hahn and co-workers[21] have used infrared microscopy utilising the heat generated in the crack tip to follow the progress of the crack.

Mai and Kerr[22] report an interesting use of a fine electrically conducting grid deposited over the area of interest so that as the crack grows it breaks strands of the grid thereby altering the resistance in finite increments. A later development[23] shows this technique to have progressed to the stage where automatic logging and data analysis has taken much of the manual labour and tedium out of the fatigue test.

A technique that has grown rapidly over the past few years is that of acoustic emission, and there are hundreds of papers dealing with this methodology, usually in relation to simple fracture processes. One paper specifically devoted to fatigue life prediction is that by Dijauw and Fesko[24] who make it clear that while the technique has very exciting potential as a non-destructive method, it requires considerable skill and sophisticated equipment to record and interpret data in a meaningful way for fatigue applications.

Before leaving the topic of technique to examine specific types of fatigue experiment, mention must be made of fracture mechanics. Although not a 'technique' in the sense of those previously described, but rather the application of physical mathematics to the energy balance that must be maintained as a crack propagates, it has had a very significant influence on the design of experimental methods and on our understanding of the fracture process and hence a better appreciation of the fatigue problem. As its name implies it is applicable to the generality of fracture processes and is not confined to the fatigue environment. It has been applied, for instance, to such divergent topics as impact strength and environmental stress cracking.

The interested reader is recommended to study the recent book by Williams[25] which gives a detailed exposition of the technique as applied to polymeric materials although it is not a book for the mathematically faint-hearted!

The fracture mechanics method is a development of the ideas first established by Griffiths[26] in the early part of the century who showed that for perfectly elastic solids (his work being done on inorganic glasses) the failure stress is proportional to the square root of the energy required to create the new surfaces as a crack grows and inversely with the square root of the crack, or flaw, size. For polymers especially, this simple theory is complicated by the plastic as well as the elastic deformations that arise. Nonetheless using an elastic system, the energy balance equation shows that the crack propagation may be defined by the energy release rate, G, which is a material parameter (i.e. is independent of the test geometry). An alternative route to the analysis is to consider the stress field around the growing crack and this analysis produces a material characteristic known as the stress intensity factor, K_I, which again is geometry-independent. It can be shown that

$$G = K_I^2/E \qquad [12.1]$$

where E is the Young's modulus for the material and

$$K_I = p\sqrt{\pi a} \qquad [12.2]$$

where $2a$ = crack length
and p is the stress over an infinite, uniformly loaded plate.

Since real test pieces are not infinite in extent various correction factors have been derived for particular types of geometry such as centre notches, double-edge notches, single-edge notches, three-point bending, cantilever bending, etc.

Where significant energy dissipation mechanisms are occurring the above (relatively) simple mathematical analysis can no longer be applied and recourse must be made to more sophisticated forms of analysis such as the J-integral method and the use of crack opening displacement.

Before the development of fracture mechanics, testing consisted of determining the effects of variables on the fatigue life of unnotched specimens, and presenting these in the form of stress versus number of cycles to failure – the familiar *S–N*-curves of the type shown in Fig. 12.1. Indeed, this is still the way the nationally standardised methods are used. However, since failure under this regime arises from inherent flaws, there is always a considerable scatter in the data necessitating the use of a large number of tests.

The use of pre-cracked specimens and the measurement of crack growth rates to characterise fatigue behaviour significantly reduces the labour involved and hence is growing in popularity. Undoubtedly this will filter through to the standards arena. The data are usually presented in the form of graphs of crack growth rate per cycle, da/dN, against applied stress intensity factor range, ΔK, it having been found for metals that this gives reasonably simple power law relationships over extended ranges. Also it is the range of load variations which controls growth rate in many cases.

The form of relationship is conveniently expressed by the Paris law:

$$\frac{da}{dN} = A\Delta K^n \qquad [12.3]$$

where A and n are constants, n generally being in the range from 2 to 6 and frequently around 4. Some polymers follow the Paris law very well, for example

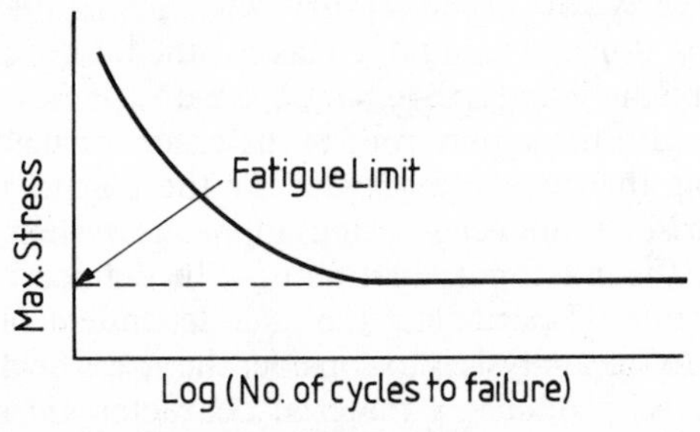

Fig. 12.1 Typical *S–N* fatigue curves

uPVC, whereas others, like PMMA, do not and instead are strongly influenced by the ratio of maximum to minimum stress in the test being conducted.

Fatigue tests may be conveniently subdivided into those operating under a cyclic stress–strain environment and those operating under static environments, the so-called static rupture tests.

12.3 CYCLIC TESTS

A large number of cyclic deformation patterns is possible. As stated above the data are frequently presented as the number of cycles to failure at a given maximum stress level (Fig. 12.1), the fatigue life and fatigue limit being derived from such plots.

The fatigue life, or endurance, is the number of cycles to failure at a specified stress level. The greater the applied stress, the fewer cycles needed to break the test piece. Generally, the fatigue life of a polymer is reduced by increasing the temperature.

The fatigue limit, or endurance limit, is the stress condition below which the material may endure an infinite number of stress cycles without failure. For most real materials the fatigue limit is not as clear-cut as depicted in Fig. 12.1 and estimates based on statistical procedures are all that can be obtained. For many polymers the fatigue limit is between 25 and 30 per cent of the static tensile strength.[27]

The deformation imposed may be tensile, compressive, flexural or shear or any combination of these, generally with sunusoidally varying stress, since this is the easiest to generate, although saw-tooth and square-wave functions may be obtained without undue difficulty. Tests may be carried out with constant amplitude of cyclic stress or constant amplitude of cyclic strain. The latter method has the advantage of ease of setting up but also has disadvantages: stiff materials are subjected to greater stresses than soft (i.e. low-modulus) materials – it is, therefore, a more severe test for such materials; furthermore, once cracks develop the imposed stresses lessen, so that often a considerable number of cycles is required to break the test piece. In constant stress tests, on the other hand, once cracks start to develop failure usually follows very quickly.

Further variants on the above methods are to superimpose the oscillating stress (or strain) on to a static stress (or strain) or to increase the amplitude of the cyclic stress (or strain) with time, as employed by Lazar.[28]

One popular test method, borrowed from the field of metals testing, is to rotate a cylindrical test piece about its axis of symmetry while the ends are constrained by bearings which are symmetrically misaligned. Figure 12.2 shows a schematic drawing of the apparatus. Each elementary volume of the test piece is subjected to a stress which varies sinusoidally between similar maxima of tension and compression. The frequency of test is readily altered, as is the overall applied stress. Lazar[28] describes tests made on PMMA, polystyrene and nylon using this technique while Romualdi *et al.*[29] describe a similar technique, but with the additional refinement that the test pieces can be subjected to a superimposed tensile load, so that alternating stresses varying from complete reversal to pure tension can be developed by varying the magnitudes of the axial and bending stresses.

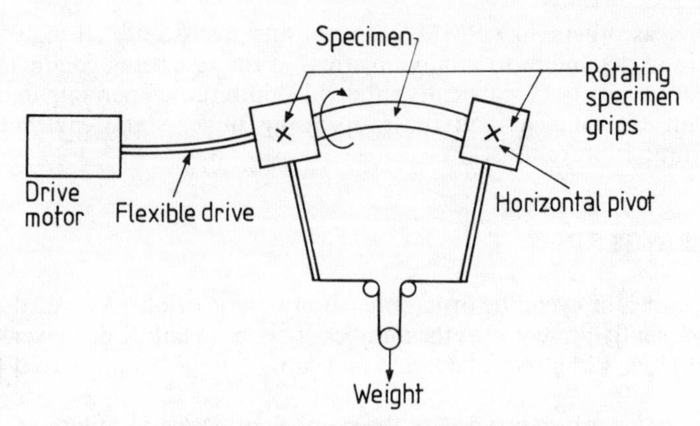

Fig. 12.2 Basic design of rotational fatigue apparatus

Findley *et al.*[30] developed another machine suitable for use down to temperatures of − 196 °C.

Axially applied tensile and compressive stresses have been considered by Lazar[28] and Boller[31] and also by Owen,[32] who describes a tension fatigue testing machine in some detail.

Flexural tests (without rotation) can be performed on rectangular bars and may be carried out in any of the modes available to static flexural tests (§8.5). Thus Gotham[5] and Cessna *et al.*[33] employed cantilever bending, while Lavengood and Gulbransen[34] used three-point loading and Hutchinson and Benham[35] used four-point loading. More recently Newaz[36] has used four-point loading, constant amplitude tests to evaluate the influence of the matrix material on the fatigue performance of unidirectional composites. While flexural tests are somewhat easier to perform than tensile–compression tests, since the problems of axiality are less severe, the data obtainable are more limited, and so reinforced thermosetting resins are more usually tested in tension–compression because of the technical superiority of this mode. For thermoplastics (reinforced and unreinforced), flexural tests are often adequate.[6]

Lhymn[37] reports on the use of impact fatigue tests to characterise the behaviour of glass reinforced polyphenylene sulphide.

Another approach to fatigue testing, which is particularly applicable to flexible materials, involves the repeated severe bending of thin sheet specimens. An appropriate technique is described by Carey[38] and the De Mattia fatigue tester is readily used for plastics-coated fabrics, although its use is more common in the rubber industry.[11]

12.3.1 Standard Test Methods

USA Standards

ASTM D671 (1971)[39] utilises the constant amplitude of force approach mentioned above. It describes terms relevant to the subject and indicates that the fluctuating stresses and strains need not be of constant amplitude or frequency as they can result from random variations. Two test pieces are described, to be selected according to the test piece thickness and stress range over which the measurements are to be made. Both specimens are of triangular form, with a rectangular cross-section, thus providing a uniform stress distribution over their respective test spans. Details of machining these rather awkwardly shaped test pieces are given, as sample preparation techniques can significantly influence the fatigue life. Having pointed out the importance of frequency of testing, the standard goes on to describe the apparatus which, in fact, only operates at a fixed frequency of 30 Hz. The setting up procedure also seems somewhat involved. Brief descriptions of producing S–N curves and of deriving fatigue life and fatigue limit data are given.

One of the major disadvantages of this apparatus is its inflexibility in terms of test frequency, which is considered by some workers to be too high for the majority of plastics materials and may explain why it is seldom used – in the UK at least. As already mentioned, plastics are relatively high damping, low thermal conductivity materials, which means that repeated straining of an article leads to a temperature rise generated within and throughout its body. If the stress–strain cycle is sufficiently rapid, significant heating of the article can occur, possibly leading to the glass transition boundary being crossed (see §16.4) and inducing thermal failure.[40,41] While the existence of this phenomenon is important to note, its quantitative effect is highly frequency-dependent and test frequency becomes an important parameter if the phenomenon is to be adequately evaluated. Where the thermal effect is to be minimised, and cooling of the test piece is inadequate since the heat is generated within the body and not on the surface of the test piece, much lower frequencies – of the order of a few hertz – must be employed. This contrasts very sharply with metals testing, where frequencies in excess of 30 Hz are common, although it has been observed[42] that carbon fibre reinforced plastics can be tested at over 100 Hz without significant heating.

German Standards

The German test method, DIN 53442 (1975),[43] by contrast, employs the constant amplitude of deformation mode in flexure and is more of a general test document than the ASTM standard – the test frequency is not specified and the comment is made that fatigue effects are frequency-dependent and that frequency should, therefore, be one of the parameters investigated. The apparatus briefly described and illustrated schematically in the standard is capable of operation over a range of deformation amplitudes and at non-zero mean strain if required. If this latter mode is selected, the standard explains, relaxation (see Ch. 11) of the test piece will mean that a non-constant mean stress level will be experienced by the test piece during the test. Dumb-bell-shaped test pieces are used, similar to tensile dumb-bells but with the parallel-sided gauge section omitted.

One interesting difference from the ASTM standard is the requirement that surface temperature be recorded as a function of number of cycles as well as the stress levels (measured by means of a suitable load cell attached to the non-driven end of the test piece). Thus, in addition to the normal S–N curve, a plot of surface temperature against $\log N$ is also constructed.

The notes at the end of the standard explain that, while fatigue testing of plastics follows the same lines as fatigue testing of metals, seldom can a machine designed for use with metals be used on plastics without suitable modification.

12.4 STATIC STRESS RUPTURE

Stress rupture tests are of a form similar to the long-term continuous stress tests described in Ch. 11. They are, in fact, creep tests divested of the onerous task of determining the associated (usually) small elongations. Specimens are subjected to a stress, often defined as some fraction of their ultimate short-term failing stress, and the time which elapses between application of load and ultimate failure is recorded. For short-term tests it is usually sufficient for the specimens to be observed by an operator with a stop watch or process timer; in the other extreme, a calendar may be sufficient! Tests of intermediate duration – days, weeks or months – may involve the use of more elaborate devices to ensure that the time of failure is recorded with sufficient accuracy. Stress rupture test results are influenced by temperature, vibration, etc., as are creep and stress relaxation tests.

Examples of this type of test occur only infrequently in standard specifications. It is perhaps appropriate that two 'stress rupture' tests described in British Standards are relevant to plastics pipes, as it was necessary to justify the use of plastics in this field by extensive tests of this nature. BS 3505 (1968)[44] and BS 3506 (1969)[45] both describe a 'long-term hydraulic test'. A specimen of pipe is maintained at a constant (internal) pressure so selected that the specimen will burst after between 1 and 10 h. Throughout the test the internal pressure is kept constant within ± 2 per cent and the specimen is immersed in a water bath maintained at 20 °C $\pm$ 1 °C. The time to burst is recorded. A second specimen is tested in a similar fashion, except that the applied pressure is selected to give a burst after between 100 and 1000 h. The minimum wall thickness and mean outside diameter of each specimen are determined before the test, and the circumferential stress in the specimen is derived from

$$P = \frac{2\delta t}{D - t} \qquad [12.4]$$

where P = pressure to be applied (bar)
δ = circumferential stress (bar)
t = minimum wall thickness (mm)
D = mean outside diameter (mm).

As a quality control procedure on a pipe production unit, the pipe manufacturer is required to record data for all pipes produced to the above-mentioned specifications, grouped according to size and class (pressure rating) of pipe.

Subsequently, circumferential stress is plotted against time on a log–log scale, and a regression line drawn through all the production data is extrapolated to yield a 50-year (438 000 h) value for circumferential stress. Even on a logarithmic scale this is a large extrapolation, but in this case data exist[46] that support its validity for unplasticised PVC. Both BS 3505 and BS 3506 require that the 50-year value of burst (circumferential) stress should exceed 230 bar (23 MPa) for 7 in (178 mm) diameter and smaller pipes, and 260 bar (26 MPa) for larger pipes. Work on predicting the 50-year burst stress for polyethylene pipes for water and for gas distribution has also been reported by Greig[47] and by Gray[48] *et al.*

Subsection 1.3.1 of BS 4618 (1975)[49] deals with aspects of static fatigue failure, describing experimental procedures for deriving data suitable for use as design criteria.

A stress rupture stress method similar to those in BS 3505 and BS 3506 is described in ASTM D2239 (1983).[50] ASTM D1598 (1981)[51] also applies. Another ASTM method, D2552 (1969)[52] is worthy of mention. This describes a standardised method for submitting dumb-bell specimens to continuously applied forces, while the specimens are immersed in a surface active agent which is maintained at a constant temperature. The time taken for specimens to fracture is recorded, and specimens showing ductile failures involving 'a noticeable degree of uniform cold drawing' are not considered in the assessment of results. The end-point of a given test (at one test–temperature combination) is defined as the 'probable time required for 50 per cent of the specimens to fail in a brittle manner' (see also Ch. 17). Finally, when considering specifications it should be remembered that the guidance given in ASTM D2990 (1977)[53] and DIN 53444 (1968),[54] directed specifically towards creep tests (§11.2), is almost all equally applicable to 'stress rupture' tests.

For most situations apparatus and techniques used for creep tests are equally applicable to stress rupture measurements. Carey[55] describes a simple laboratory apparatus in which horizontally mounted tensile specimens (of polyethylene) are stressed by a simple multiplying lever. Elongations of up to about 20 per cent are possible in the apparatus, and the specimen is immersed if required, in a temperature-controlled liquid. Numerous results are shown graphically. Millane[56] has examined a series of glass reinforced plastics laminates, based individually on melamine, epoxide, phenolic, silicone and polyester resins. As an essential preliminary, the flexural strengths of groups of three specimens 6 in long and $\frac{1}{8}$ in thick (152 × 3 mm) were determined at a 5 in (127 mm) span. Subsequently single specimens of similar size were stressed in the same configuration, under various applied loads, in air or water at various temperatures. Stress levels were selected to give flexural stress–rupture time curves extending to 1000 h; examples are shown in Fig. 12.3.

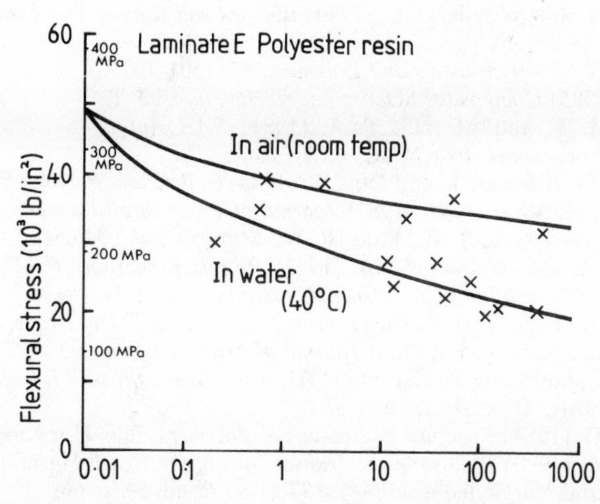

Fig. 12.3 Typical stress rupture data for glass cloth reinforced laminates. (reproduced by permission of *British Plastics*)

Millane's paper describes some practical aspects of the requisite apparatus including an electromechanical timing device. However, simple elapsed time counters are probably the most suitable equipment for this duty, and it is relatively simple to arrange for a counter to be switched off at time of fracture of the specimen.

Static stress–rupture measurements are very often used as a measure of resistance to environmental stress cracking (§17.3.3).

REFERENCES

1. Forrest, P. G. (1962) *Fatigue of Metals*, Pergamon.
2. Lewis, W. C. (1946) *ASTM Proceedings*, **46**, 814.
3. Ople, F. S. and Hulsbos, C. L. (1966) *Proceedings of the American Concrete Institute*, **63**, 59.
4. Pell, P. S. and Taylor, I. F. (1969) in *International Conference on Civil Engineering Materials, Southampton University*.
5. Gotham, K. V. (1969) *Plastics and Polymers*, **37**, 309.
6. Turner, S. (1983) *Mechanical Testing of Plastics*, Iliffe.
7. Nielsen, L. E. (1965) *Mechanical Properties of Plastics*, Reinhold.
8. Schultz, J. M. (1977) *Treatise of Materials Science and Technology*, vol. 10, Part B, Academic Press, p. 599.
9. Hertzberg, R. W. and Manson, J. A. (1980) *Fatigue of Engineering Plastics*, Academic Press.
10. Bennett, J. A. and Quick, G. W. (1954) *NBS Circular* 550, September.
11. Brown, R. P. (1986) *Physical Testing of Rubbers*, Elsevier, Ch. 12.
12. ISO R373 (1964) *General Principles for Fatigue Testing of Metals.*
13. BS 3518: Part 1 (1962) *General Principles (of Fatigue Testing of Metals)*; Part 5 (1966) *Guide to the Application of Statistics.*
14. ASTM E206 (1972) *Definition of Terms Relating to Fatigue Testing and the Statistical Analysis of Fatigue Data.*
15. Gotham, K. V. and Wright, D. C. (1984) *Plastics and Rubber Processing Applications*, **4**, 43.
16. Gotham, K. V. (1969) *Plastics and Polymers*, **37** (130), 309.
17. Hashin, Z. (1985) *Composites Science and Technology*, **23**, 1.
18. Takemori, M. T. and Morelli, T. A. (1981) *SAE Automotive Plastics Durability Conference Proceedings, Troy, Miss., 63 Tr. Ro*, 67.
19. Schinker, M. G., Konczol, L. and Doll, W. (1984) *Colloid and Polymer Science*, **262**, 230.
20. Fanucci, J. P. and Mar, J. W. (1982) *Journal of Composite Materials*, **16**, 94.
21. Hahn, M. T., Hertzberg, R. W., Lang, R. W., Manson, J. A., Michel, J. C., Ramirez, A., Rimnac, C. M. and Webber, S. M. (1982) *PRI Deformation, Yield and Fracture of Polymers, 5th International Conference, Cambridge*, paper 19.
22. Mai, Y. W. and Kerr, P. R. (1984) *Journal of Materials Science Letters*, **3**, 971.
23. Mai, Y. W. and Kerr, P. R. (1985) *Journal of Materials Science*, **20**, 2199.
24. Dijauw, L. K. and Fesko, D. G. (1981) SAE Automotive *Plastics Durability Conference Proceedings, Troy, Miss., 63 Tr. Ro.*, 57.
25. Williams, J. G. (1984) *Fracture Mechanics of Polymers*, Ellis Horwood, Ltd.
26. Griffiths, A. A. (1920) *Philosophical Transactions of the Royal Society*, **A221**, 163.
27. Lazan, B. J. and Yorgiadis, A. (1944) *ASTM STP*, no. 59, p. 66.
28. Lazar, L. S. (1957) *ASTM Bulletin*, **220**, 67.
29. Romualdi, J. P., Chang, Chiao-Lio and Peck, C. F., Jr (1954) *ASTM Bulletin*, **200**, 39.

30. Findley, W. N., Jones, P. G., Mitchell, W. I. and Sutherland, R. L. (1952) *ASTM Bulletin*, **184**, 53.
31. Boller, K. (1964) *Modern Plastics*, **41** (10), 145.
32. Owen, M. J. (1967) *Journal of the Plastics Institute*, **35**, 353.
33. Cessna, L. C., Levens, J. A. and Thomson, J. B. (1969) *Polymer Engineering and Science*, **9** (5), 339.
34. Lavengood, R. E. and Gulbransen, L. B. (1969) *Polymer Engineering and Science*, **9** (5), 365.
35. Hutchinson, S. J. and Benham, P. P. (1970) *Plastics and Polymers*, **38** (134), 102.
36. Newaz, G. M. (1985) *Composites Science and Technology*, **24**, 199.
37. Lhymn, C. (1985) *Journal of Materials Science Letters*, **4**, 1221.
38. Carey, R. H. (1955) *ASTM Bulletin*, **206**, 52.
39. ASTM D671 (1971) *Flexural Fatigue of Plastics by Constant Amplitude of Force.*
40. Riddell, M. N., Koo, G. P. and O'Toole, J. L. (1966) *Polymer Engineering and Science*, **6**, 363.
41. Constable, L., Williams, J. G. and Burns, D. J. (1970) *Journal of Mechanical Engineering and Science*, **12**, 20.
42. Owen, M. J. and Morris, S. (1970) *SPE Conference*, Section 8E.
43. DIN 53442 (1975) *Testing of Plastics: Fatigue Test in the Field of Bending Strain of Flat Specimens.*
44. BS 3505 (1968) *Unplasticised PVC Pipe for Cold Water Services.*
45. BS 3506 (1969) *Unplasticised PVC Pipe for Industrial Uses.*
46. Richard, K. and Diedrich, G. (1955) *Kunststoffe*, **45** (10), 429.
47. Greig, J. M. (1981) *Plastics and Rubber Processing and Applications*, **1** (1), 43.
48. Gray, A., Mallinson, J. N. and Price, J. B. (1981) *Plastics and Rubber Processing and Applications*, **1** (1), 51.
49. BS 4618: Subsection 1.3.1 (1975) *Recommendations for the Presentation of Plastics Design Data: Static Fatigue Failure.*
50. ASTM D2239 (1983) *Standard Specification of Polyethylene (PE) Plastic Pipe (SIDR–PR) Based on Controlled Inside Diameter.*
51. ASTM D1598 (1981) *Time to Failure of Plastic Pipe under Constant Internal Pressure.*
52. ASTM D2552 (1969) *Environmental Stress Rupture of Type III Polyethylenes under Constant Tensile Load.*
53. ASTM D2990 (1977) *Tensile, Compressive and Flexural Creep and Creep Rupture of Plastics.*
54. DIN 53444 (1968) *Testing of Plastics: Tensile Creep Test.*
55. Carey, R. H. (1958) *Industrial Engineering and Chemistry*, **50** (7), 1045.
56. Millane, J. J. (1960) *British Plastics*, **33** (5), 199.

Chapter Thirteen

Electrical Properties

13.1 INTRODUCTION

The tests described in this chapter are principally material tests, although some tests for semi-finished products such as sheet, film and tape are covered. Tests on cables and finished products are not specifically included as these become rather specialised, although the principles of the tests mentioned in this chapter are often applicable to products.

Reference is made to tests drawn up by the International Electrotechnical Commission (IEC) since the ISO does not draw up electrical standards when adequate IEC standards exist.

No attempt is made to describe the detailed phenomena underlying the electrical tests. The theory of electrical behaviour is described by Baird[1] and Blythe[2] (general), and by Daniel[3] (dielectric properties).

Fairly elementary descriptions of dielectrics and dielectric processes, mostly with special reference to polymeric materials, are given by Moullin,[4] Swiss and Dakin,[5] Hoffman,[6] Vail,[7] Sharbaugh[8] and Devins and Sharbaugh.[9] These references are interesting, not least for the variety of approaches to the subject adopted by the authors.

Broader descriptions of electrical insulation properties and test methods considered mainly from the application viewpoint are given by Mason,[10] Stark[11] and Baker.[12]

A number of useful chapters dealing with electrical properties are to be found in more general books on polymers.[13–22]

13.2 WHY MEASURE ELECTRICAL PROPERTIES?

The purpose of electrical insulation may generally be described as the prevention of unwanted contacts between electrical conductors at different potentials.

A simple example such as two mounted terminals can usefully serve as an illustration. The insulation must maintain the separation under a wide variety of environmental hazards which include humidity, temperature, vibration, radiation, the presence of gases, moisture and other contamination. It must be sufficiently strong physically to withstand the mechanical forces which may be exerted by the conductors and at the same time must prevent any significant flow of current between the conductors.

No insulator is perfect, but what constitutes a significant flow of current depends upon the application. A few microamps in an electrical supply circuit would be negligible, but in an electrometer input terminal could ruin any measurement. In every case the current must be sufficiently small to eliminate secondary effects such as temperature rise, mechanical stress and electrochemical action. A significant current may arise in various ways.

When a small or moderate d.c. voltage is applied across an insulator, a current flows. To minimise this the material must have an adequately high resistivity (§13.3). The current may be predominantly through the volume of the material, when the property involved is the *volume resistivity*. On the other hand, the current may be predominantly through a surface layer of different composition (e.g. due to adsorbed materials) and the property involved is then the surface resistivity. Since *surface resistivity* is often related to contamination, it is not generally a material property, although material properties may influence contamination and hence surface properties.

It is sometimes useful, especially in testing products, to measure the *insulation resistance*, which is the resistance between two defined electrodes, no differentiation being made between the passage of current along the surface and that through the bulk of the material (§13.4).

There are uses of plastics where the passage of a d.c. current is desirable, e.g. for the dissipation of static electrical charges, for heating, or in plastics potentiometers, and special tests are required for *antistatic* and *conductive* plastics (§13.5).

The ability of an insulator to 'resist' the passage of direct current is not necessarily synonymous with its ability to resist alternating currents; whereas in the former case the resistivity of the material is of prime importance, in the latter the *power factor* and *permittivity* (§13.6) are more relevant, since they determine the power loss occurring in alternating fields, which loss may vary enormously with applied frequency. These two properties are collectively known as *dielectric properties*. The power factor is due to the orientation of dipoles and the movement of charged ions in an alternating field. These motions, together with the movements of electrons with respect to such dipoles and ions and also to atoms, and of adjacent atoms with respect to each other when the field is applied, contribute to the permittivity.[1] In many insulation applications power losses are important (and undesirable) and thus low values of both power factor and permittivity are desirable. Since capacitance is proportional to permittivity, and in most cases the capacitance of a given conductor–insulator system is an unwanted effect which must be tolerated, the capacitance must be kept low by using low permittivity material. In screened cables, for example, of small diameter and hence low thickness of dielectric, it is fairly common practice to use dielectrics of cellular plastics materials to reduce the capacitance, and such materials may have effective permittivities approaching unity, the value for air. The problem of unwanted capacitance becomes more acute with increasing frequency.

The major field of application for dielectric materials where a high permittivity is the chief requirement is in capacitors. Here the emphasis is continually towards reduction in physical size and, since the capacitance of a given electrode–dielectric system is proportional to the permittivity of the material (capacitance being defined as the ratio of stored charge to applied voltage), the importance of a high permittivity need hardly be stressed. As far as plastics materials as capacitor dielectrics are concerned, the available range of permittivity is approximately 2–20, so that the possibilities of size reduction by choice of permittivity alone are limited; these possibilities are reduced even further when power factor is considered, since virtually all the low power factor materials have permittivities at the lower end of the range. Capacitance is also inversely proportional to the dielectric thickness and so the only real direction of advance at present is in the use of thinner films; plastic foils of thickness 1 μm or less have been developed and used to a limited extent – limited because the problems of handling and manipulating such material are severe; some work is in progress on the vapour deposition of polymeric dielectric films *in vacuo*.

Power factor and permittivity are seldom significant in 250 V power frequency cables, but become important when the frequency and/or voltage are raised.

It is sometimes desirable that the product of power factor and permittivity be high, so that the energy supplied by a high-frequency dielectric field may be converted into heat. In dielectric or microwave heating, it is possible to produce more rapid heating throughout a body than would be possible by heat conducted from the surface. In a related application, synthetic resin adhesives used in furniture may be cured by selective heating of the glue lines. The use of water-soluble adhesives such as those based on urea formaldehyde eases the problem, since the permittivity of water is approximately 80 and it also has a high power factor. High frequency fields are also used to weld, for example, two sheets of PVC together, where very rapid heating–cooling cycles can be attained. Another rather more specialised application is in energy-absorbent materials for use at microwave frequencies.

So far, we have discussed only weak field phenomena, and those properties which are fairly well defined and constant for given environmental conditions and the tests themselves do not cause any irreversible alteration to the insulation. Under the influence of relatively strong fields, however, plastics materials may be permanently altered, either in form or character, and, under continual stress, degradation may occur and ultimately lead to failure.

The *electric strength* of a material (§13.7) is a measure of the ability of a material to withstand a high voltage, usually in the presence of discharges controlled by the rounded edges of electrodes, and is important for high voltage applications.

The tests for *resistance to surface discharges, tracking and arcs* (§13.8) relate to surface rather than bulk breakdown and emphasise the effects of discharges of various types on the surface of the test piece.

In the vicinity of high voltage conductors ionisation of the air occurs due to the high local electric stresses, and insulation subjected to prolonged exposure to the consequent discharges may fail though degradation or erosion. *Discharge resistance* is therefore of practical importance in high voltage bushings and insulators, which often utilise epoxide resins and other thermoset materials.

Tracking resistance is important even at low voltages since the presence of an aqueous contamination can cause localised stresses and result in the formation of a conductive carbonised track across the surface.

Arcs may be produced in many electrical devices, even at low voltages, and we need therefore to know how materials will withstand contact with an arc. In *arc resistance* tests, an arc produced between two surface-mounted electrodes is allowed to play along the surface of the insulation so that the predominant mode of failure is thermal degradation. However, the failure is accelerated by the surface carbonisation that occurs, resulting in current actually flowing within the surface layers of the material. This property is important in high-current switchgear and circuit breakers, where arcs occur on making and breaking contacts.

The wider use of plastics as electromagnetic interference screening materials has resulted in the recent development of methods of measurement of their *EMI screening efficiency* (§13.9).

Two other electrical tests, one for the *corrosion of copper* by adhesive tapes and one for the determination of *conductivity of water extract* are mentioned briefly in §13.10.

13.3 RESISTIVITY OF INSULATING PLASTICS

The methods used are different from those used for conductive and antistatic plastics which are described in §13.5.

13.3.1 Definitions and Qualifications

Volume resistance is the quotient of a direct voltage applied between two electrodes in contact with opposite faces of a specimen and the current between the electrodes after a specified time of application of the voltage, excluding current along the surface, and ignoring possible polarisation effects at the electrodes.

Volume resistivity is the volume resistance calculated to apply to a theoretical cube with unit side. It is the quotient of the nominal voltage gradient (applied voltage divided by thickness) and current density, although the true voltage gradient may differ from the nominal value because of contact effects.

Surface resistance is the quotient of a direct voltage applied between two electrodes on a surface of a specimen and the resulting current between the electrodes after a given time of application of the voltage, ignoring possible polarisation effects.

Surface resistivity is the surface resistance calculated to apply to a theoretical square area: the size of the square is immaterial. It is the quotient of the nominal voltage gradient and the current per unit width of current path.

Both volume and surface resistivities may depend markedly (often by several orders of magnitude) on such factors as the time of application of the voltage, the temperature and humidity of the atmosphere, and the moisture content of the test piece. They also depend on the nature and geometry of the electrodes and the magnitude of the applied voltage. The changes of surface resistivity may follow very rapidly (in seconds) upon any humidity changes. In addition there may be slow irreversible changes (especially of surface resistivity) due to chemical changes.

In explaining various facts it is often useful to use the terms *conductance* and *conductivity*, the reciprocals of resistance and resistivity, respectively.

It is important to note that in any surface resistivity test, even though a guard plate is used, some of the current passes (between the two electrodes on one surface) through the volume of the material. Thus if the surface layer is not of lower 'resistance' than the bulk or is sufficiently thin, the measured value of 'surface resistivity' is substantially a function of the volume resistance, and the 'surface resistivity' has no significance. This was discussed by Geppe.[23] With most standard tests a useful guide is to ignore surface resistivity results when the volume resistance is not greater than 10 per cent of the measured surface resistance or, what amounts to a similar criterion in most cases, where the volume resistivity in ohm cm is not numerically greater than the calculated surface resistivity in ohms. Samples which are thin (relative to the distance between the two electrodes on the same surface), are anisotropic, or have high contact resistances, may give meaningless surface resistivity results.

13.3.2 Units, Ranges and Precision of Values

The basic SI unit of volume resistivity is the ohm metre (Ω m), derived from $\Omega\,m^2/m$. In practice a submultiple of the SI unit, the Ω cm, is more commonly used.

The volume resistivity of common insulating plastics covers a span of some ten decades between, very approximately, 10^8 and $10^{18}\,\Omega$ cm; with the conventional size of specimen in sheet form, the corresponding resistances to be measured lie between $10^6\,\Omega$ and $10^{16}\,\Omega$. At the upper extremity of this range, above approximately $10^{17}\,\Omega$ cm, lie the non-polar plastics, PTFE, polyethylene, polystyrene, etc., and measurements on these materials are considered sufficiently difficult (and the results perhaps of rather academic interest only) to be excluded from our present terms of reference. Dorcas and Scott[24] and Sazhin and Skurikhina[25] give further details.

The SI unit of surface resistivity is the ohm, Ω, derived from Ω m/m. It is sometimes referred to as 'Ω per square'.

Because of the variation of resistance of a given test piece with test conditions and because of non-uniformity of the same material from test piece to test piece, determinations of surface and volume resistivity are rarely reproducible to closer than ± 10 per cent and are often much more widely divergent (the ratio of the maximum and minimum values may occasionally be as high as 10). This has led some people to quote the logarithm of the resistivity, which is very useful in tabulation and graphical presentation, but clumsy in textual references. As a general guide to the accuracy expected from industrial materials, two materials would not be considered very significantly different unless their resistivities differed by at least half an order of magnitude. Measurements on at least two test pieces are considered essential.

13.3.3 Measurement Techniques

Only the measurements of resistance up to approximately $10^{15}\,\Omega$, which is the upper limit of commercially available resistance meters, will be considered, the normal lower limit of such instruments being of the order of $10^6\,\Omega$. As surface resistance values for plastics are in general of the same order as, or lower than, the volume resistance for standard specimens, what follows applies to both properties. The techniques given below are suitable for materials with volume resistivities above about $10^8\ \Omega$ cm. For more conductive materials, see §13.5.

The usual measurement technique is best illustrated by considering the simple case of a two-terminal resistance, for example a resistor, before turning to the additional complications involved with measurements on insulation.

In general the unknown resistance is connected in series with a current-measuring instrument across a high potential supply. The potential is high to provide the necessary sensitivity and is commonly of the order of 100–500 V d.c. The current-measuring instrument has historically been a sensitive galvanometer, but is now more commonly a valve or solid state circuit which measures the small voltage drop across its input resistance. This resistance is related to (may actually be) a known fixed value resistor, which may be switched to provide a wide series of ranges. The current is the voltage developed divided by the input resistance. The resistance of the 'unknown' is usually calculated by dividing the applied voltage by the current, since the voltage drop across the measuring instrument is generally less than 1 per cent and may be as low as 0·001 per cent of the applied voltage.

Multi-range megohmmeters are available which have built-in voltage sources and indicate the resistance directly. Factors which must be considered in purchasing a megohmmeter are considered in §13.3.4.

When measurements on insulating materials are considered it is usual in specification tests to use a sheet specimen in the form of a disc or a square. A two-terminal specimen consisting of such a disc with a suitable electrode applied to each surface would be simple to measure, but the current passing through the body of the material would be augmented by current passing over the surface and round the edge of the specimen, so that it would be impossible to separate the volume conductance from the total measured conductance. To obviate this difficulty a third electrode, called a 'guard ring', is placed concentrically around one of the existing electrodes (which is then known as the 'guarded electrode' and which is reduced in size and separated from the guard electrode). By modified circuitry the desired volume conductance can now be measured direct, and further, by a simple interchange of connections, the surface resistance can be measured also, using the same specimen. The electrodes, current flows and theoretical circuits are shown in Figs. 13.1–13.3, the symbols R_v, R_s and R_g representing the volume, surface and guard resistances, respectively, and R the input resistance of the measuring instrument.

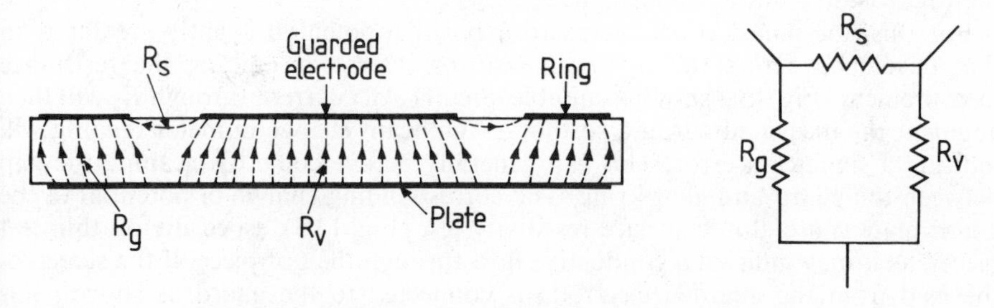

Fig. 13.1 Electrode arrangement and equivalent circuit

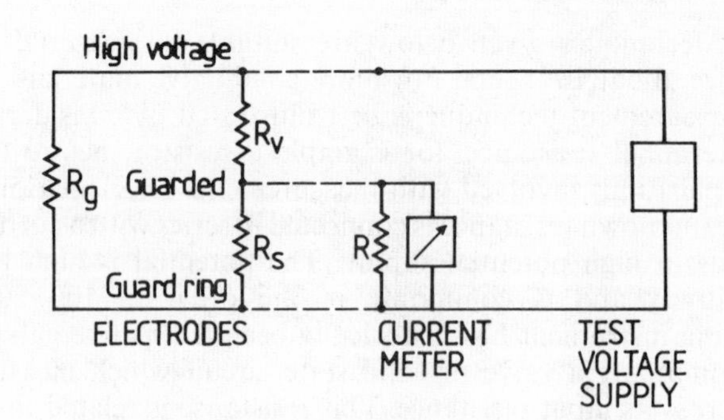

Fig. 13.2 Volume resistivity – theoretical circuit

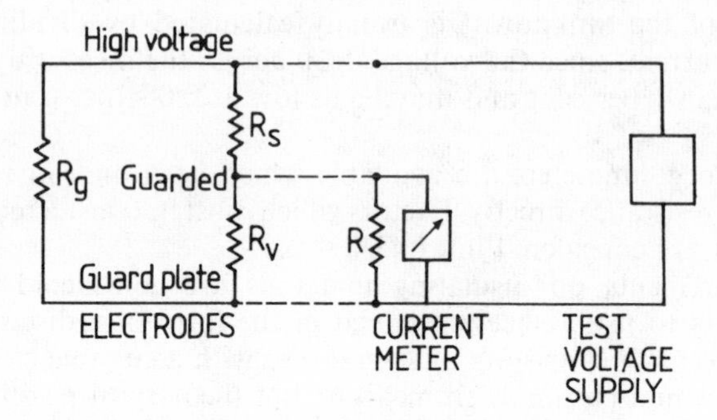

Fig. 13.3 Surface resistivity – theoretical circuit

In the circuits of Figs. 13.2–13.3 the input resistance of the current-measuring instrument is shunted, respectively, by the surface resistance in one case and the volume resistance in the other.

Suppose volume resistance is being measured (Fig. 13.2). In many cases R_s is of the same order as R_v and if the voltage developed across the measuring instrument is small compared with the test voltage, R is much less than R_s and the current through R_s is negligible. Very occasionally R_s may be of the same order as R or even much less than R, so that much of the current which should go through the measuring instrument is diverted. It is therefore useful to be able, at the end of the test, to carry out a 'leakage check'.

For this, the guard is connected to a positive potential slightly greater than that developed across the current-measuring instrument during the resistance measurement (Fig. 13.4 shows a suitable circuit). Any current through R_s will then augment the measured current, so that a significant change in meter reading will indicate a significant error. This can generally be overcome by cleaning the gap between the guard and guard ring. The corresponding change of potential of the guard plate is useful in a surface resistivity test (Fig. 13.5), especially on thin test pieces, as it may indicate a conductive flow through the test piece. If the screen of the lead from the guarded electrode is connected to the guard as shown, this procedure also checks the adequacy of the insulation on this lead. (It will be

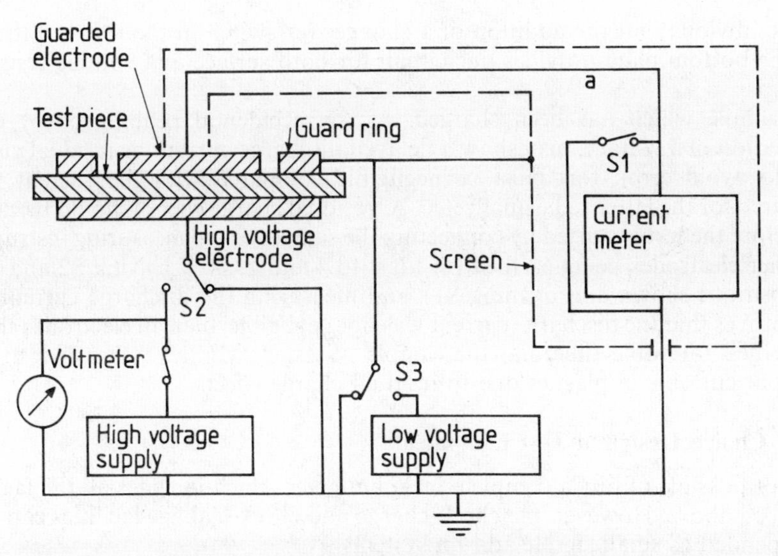

Fig. 13.4 Volume resistivity – practical circuit

necessary to disconnect the screen on this lead at the connection to the current-measuring instrument if a facility for this check is to be incorporated.)

The resistance between the high-voltage electrode and the guard electrode in both cases is a shunt across the supply voltage and therefore has no effect on the measurement, unless it lowers the applied voltage.

The circuits of Figs. 13.4 and 13.5 have been shown separately for clarity but it

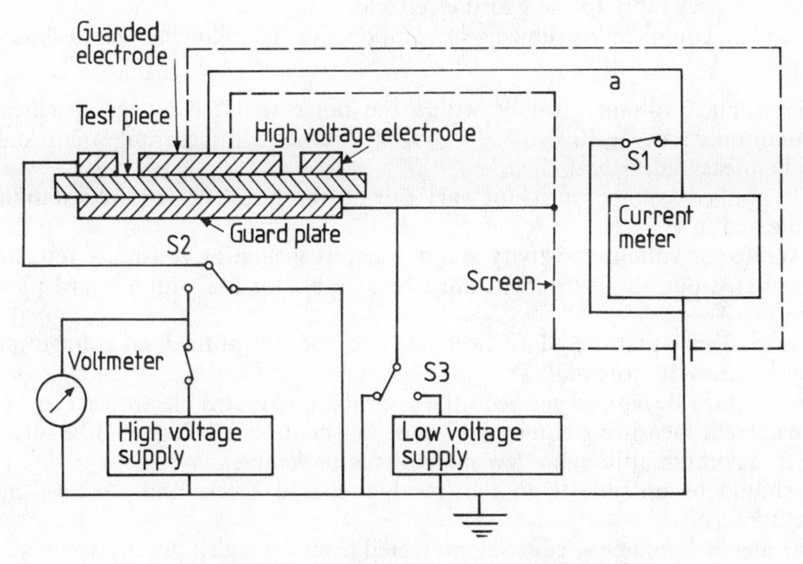

Fig. 13.5 Surface resistivity – practical circuit

will be obvious that the addition of a changeover switch in the leads to the ring and the bottom plate provides one circuit for both surface and volume resistivity tests.

A sample which has been charged, e.g. by accidental rubbing or by direct application of a voltage, may show a delayed discharge current or residual current and, to avoid error, this must be negligible compared with the current to be measured for the latter to be unaffected. A 'residual charge check' should be carried out before the test is started by connecting the sample to the measuring instrument, all other electrodes being earthed (in Figs. 13.4 and 13.5 by leaving S2 and S3 in the positions shown and opening S1), and measuring the discharge current. If it later proves that the discharge current was not negligible, then all electrodes should be earthed for some time and the sample re-tested. Munick[26] has investigated transient currents in plastics due to residual charge effects.

13.3.4 Choice, Design of Test Equipment

It is not possible to buy a complete megohmmeter which includes all the facilities described above. However, the check-charge and check-leakage facilities can often be appended as small simple add-on systems.

The alternative is to buy a power supply unit and a picoammeter and to assemble the control circuitry. If this is done, care must be taken that all the circuit (leads, plugs, sockets, switches) connected with the guarded electrode (a, Fig. 13.5) has an insulation resistance much above the input resistance of the current-measuring instrument and that this situation is not nullified by dirt or careless assembly. It is also essential that the above-mentioned parts of the circuit are adequately screened in a rather special sense. Not only must they be free of any electrostatically induced external influences, but any leakage path over or through any piece of insulation on this part of the circuit must terminate on part of the circuit connected to earth or preferably to the guard electrode.

In buying complete instruments or components, the following points should be checked.

1. The applied voltage must be within the range required by the specification (commonly 100 V $\pm$ 10 V or 500 V $\pm$ 50 V) over the complete range of resistances to be measured.
2. The applied voltage must not vary during the test. The required stability is indicated in §13.3.6.
3. If surface or volume resistivity and not merely insulation resistance tests are to be carried out, the instrument must be suitable for use with a guard plate or ring.
4. It is necessary that the guard electrode and one side of the high voltage supply can be at earth potential.
5. The voltage developed between the guarded and guard electrodes (i.e. across the current-measuring equipment) must be small. One volt should be regarded as a maximum although a few millivolts is preferable.
6. It should be possible to fit the 'residual-charge check' and 'leakage check' facilities.
7. The meter should be adequately protected from damage if the electrodes should short-circuit or a conductive test piece is accidentally introduced. This can be

done by ensuring that the current-measuring instrument will withstand the full test voltage. This may not be possible on the higher current ranges, but a rapidly acting current-limiting system on the voltage supply can deal with this.

8. In addition to using the above check list, the purchaser should ascertain from the supplier that equipment is suitable in other ways for tests to the required specifications.

A suitable screened cable with high insulation resistance must be used for the connection to the guarded electrode.

Ordinary screened cable can produce quite large amounts of electrical noise when flexed, but non-microphonic types are available in which the noise level is considerably reduced by the insertion of a semiconducting layer between the dielectric and the screening braid. It is essential that this layer be removed at the cable ends, preferably for at least 10 mm.

In the author's experience, the measurement of high resistivity values is beset with difficulties. To avoid uncertainty about the functioning of the measuring apparatus it is useful to have a set of high resistances to hand, ranging from 10^6 to 10^{13} Ω. Such resistances are now available with accuracies of 2 per cent or better, even for the higher values, but care should be taken with the selection, since the voltage coefficient is large. This is normally negative, so that a nominally 10^{13} Ω component can have a value much lower at 500 V than the manufacturer's reported value, usually measured at a few volts only. The manufacturer should be consulted for details of the voltage coefficient of his resistors. France[27] gives details of measurement of high value resistors.

Hitchcox[28] has written extensively on high resistance measurements and the use of electronic techniques for extending the upper limits of measurement.

13.3.5 Electrodes

Many types of electrode are available. Amongst the most widely used are the following: graphite, conductive paints, metallic foils applied with a thin layer of viscous liquid, vacuum-deposited metallic coatings and conductive rubber sheets. Most of these require solid metal backing plates to contact the electrodes proper, but whichever type is used, intimate contact with the specimen is essential and the choice will be dictated by environmental conditions.

Some specifications include mercury electrodes but the use of these is now deprecated because they present a toxic hazard. They should certainly not be used for tests at elevated temperatures. Colloidal graphite suspensions painted onto the test specimen are often successful, but with some materials adhesion is poor and (especially with flexible plastics) they tend to part from the surface. Colloidal graphite in alcohol wets most plastics better than the water suspension which is often specified and may be useful for plastics not affected by alcohol.

Conductive paints, often based on a suspension of metallic silver in a suitable resin or resin–solvent system, have the virtue of a high conductivity, but care should be taken that the solvent does not adversely affect the specimen; residual solvent can produce odd results. All the brushed-on coatings require care in ensuring that the guard ring gap is not encroached upon.

Thin metallic foils (about 25 μm thick) may be applied with a thin layer of petroleum jelly, silicone grease or a mixture of poly(ethylene glycol), water, a wetting agent and an electrolyte. The foils, rolled onto the wetted surface very forcibly with a metal roller, are easily removed and therefore useful where other tests may be required to be made on the same specimen. Their effectiveness depends on the thickness of 'adhesive' being extremely small so that, despite its fairly high resistivity, its actual volume resistance is relatively low. This resistance makes foils generally unsuitable for low-resistivity materials and also, since it is difficult to avoid contamination of the guard gap, they are not recommended for surface resistivity measurement. Gold leaf could be used, since it may be applied without grease.

Conductive rubber sheets have been used for rapid comparative measurements on materials, but fairly high contact pressure is required and conductive rubbers tend to become less conductive with age. Too high a pressure may change the conductivity of the rubber before the contact resistance is adequately reduced. The contact resistance may be variable and much higher than the resistance of the rubber electrode.

The dimensions of the electrodes for volume resistivity are not critical, provided that the guard ring gap is not too large. A popular size, calling for a 100 mm test piece approximately 3 mm thick, consists of a guarded disc electrode 50 mm in diameter, a guard ring of 60 mm internal diameter and 80 mm external diameter, and a bottom (applied voltage) disc electrode at least as large as the outer diameter of the guard ring. A larger electrode area and a thinner test piece are advantageous for material of very high resistivity. Some volume resistivity standards regard the area through which the current flows as being that of the guarded electrode, while others use the area of the circle having a diameter equal to the mean of the inner guard ring diameter and the guarded electrode diameter. The difference between these is small (21 per cent) for the electrodes described compared with errors generally outside the control of the operator. In the USA (see ASTM D257 (1978)[29]) not only is the mid-diameter used, but an additional fringing correction is also applied, based on the dimensions of the electrodes and the thickness of the test piece, where a high precision is required! The resistivity is given by $AR_v/t\,\Omega\,\text{cm}$ where A is the area (cm^2), t the thickness (cm), and R_v the measured volume resistance (in Ω) between centre and lower electrodes.

If D and d are the diameters corresponding to the outside and inside, respectively, of the electrode gap, and provided that $(D-d)$ is small compared with D,

$$\text{surface resistivity} = \pi \frac{D+d}{D-d} R_s \qquad [13.1]$$

where R_s is the surface resistance measured between the two upper electrodes.

Note the absence of length dimensions in the surface resistivity expression, which is thus stated in ohms. The units of volume resistivity are still sometimes misquoted as Ω/cm^3 or Ω/cc, which are dimensionally incorrect and misleading.

13.3.6 Effect of Time of Application of Voltage

All resistances have an associated capacitance. Thus, when a voltage is applied, a relatively large current will flow to change this capacitance and it will be necessary to short-circuit the current-measuring instrument for a short period after applying the voltage to prevent overloading it. After this time, the current will still fall with increasing time of application of voltage (a delayed charging current or alternatively one due to the sweeping of mobile ions to the electrodes) over a period which may be weeks. To standardise in reporting values of resistivity, it is customary to adopt the value attained 1 min after the initial application of voltage, and strictly such values should be reported as *apparent resistivity* (*one minute value*) though this is seldom done. These values for polar materials will generally be within an order or two of the true infinite time resistivity. Reddish[30] has shown the errors that can occur with plasticised PVC compounds, where the one-minute value is predominantly due to polarisation current and may be useless as a criterion even for comparison of such compounds.

Another effect of the capacitance of the test piece is that any variation of the applied voltage will give rise to a current equal to the product of the capacitance and the rate of change of voltage. For example, if the test piece has a capacitance of 100 pF (10^{-10} farads) and the voltage changes at $-0{\cdot}001$ V/s, a current of -10^{-13} A will flow. If the resistive current being measured were only 10^{-13} A, this would give rise to a zero current and an apparently infinite resistance.

The voltage supply must therefore have extremely good short-term stability. A bank of lead–acid cells is probably ideal but generally impracticable. A well stabilised mains-driven power supply is normally used, but an adequate time should be allowed for warming up of the power supply unit. Dry cell batteries should not be used because they can drift at a relatively constant rate giving rise to a steady capacity current. If there are occasional large fluctuations of the mains supply so that some significant changes in output of a stabilised supply occur, these generally give rise to a fluctuating capacity current which is easily distinguished from the current to be measured. When fluctuations occur, the test must be repeated at a more opportune time.

13.3.7 Effect of Temperature

Resistivity is a temperature-dependent property and for non-metallic materials it invariably decreases with increasing temperature, unlike the behaviour of many metals. The decrease is always relatively large and of the order of several decades when the increase is from room temperature to, say, 100 °C. Sensitivity to temperature change is possibly more marked for volume resistivity than surface resistivity, but it is advisable always to consider the possible effects of humidity and temperature variations on both properties. It is important in measurements at any temperature to ensure that temperature is maintained very constant during the test. Fluctuations in temperature produce changes in measured current which cannot be distinguished from the normal time-dependent charging current, and may cause errors greatly in excess of those due to the temperature dependence itself.

It is common to plot log resistivity against the reciprocal of absolute temperature, $1/T$, in common with many other phenomena, and frequently the result is a straight line for the long-time steady values. A precaution in determinations at elevated temperatures is to measure with increasing temperature and to ensure that the specimen is fully discharged and short-circuited before changing the temperature. It is possible to produce a residual charge on a specimen by cooling with the voltage applied, which charge may only disappear after a very long time interval.

It is always advisable finally to cool and re-test, to ensure that initial and final measurements at room temperature agree, any difference suggesting a change in cure, for example, or some other non-reversible temperature effect.

At subambient temperatures an additional hazard is caused by the possible condensation of moisture on insulators, and this can be avoided by using an evacuated system or one which can be mounted in a vacuum chamber placed in a low-temperature bath. Alternatively the electrodes can be swept continuously with very dry nitrogen.

13.3.8 Effect of Humidity

Resistivity is often very dependent on humidity, differences in which may cause orders of magnitude changes, dependent upon the type of plastic. Not only is surface resistivity particularly dependent upon ambient humidity, but in some cases it responds in fractions of a second, as may be observed by breathing on the sample during test. On the other hand, the time taken for equilibration of the volume resistivity of some materials (especially thick test pieces) with the atmosphere may be so long (e.g. months) that an acceptable period of conditioning before test can at best only minimise the effects of test piece history. The importance of ambient conditions upon the measurement of resistivity is underlined by the requirements of most specifications, which generally call for tolerances of ±5 per cent in relative humidity and ±2 °C for the pre-conditioning period, usually of a minimum 16 h duration. In view of what has been said about the speed of response to humidity changes, some official specifications quite inadequately specify the conditions during measurement, which should be as rigorous as those of the pre-conditioning cycle where surface resistivity is concerned.

13.3.9 Standard Test Methods

Resistivity test methods are described in IEC 93 (1980),[31] which has been dual numbered as BS 6233 (1982).[32] Methods similar to some of those given in these documents are described in BS 2782 (1982), Methods 230A and 231A,[33] which specify standard test conditions of 23 °C and 50 per cent r.h. In the BS 2782 methods the electrodes specified for surface resistivity are graphite, metal film or conductive silver paint. In addition, for volume resistivity, metal foil or conductive elastomeric electrodes are permitted. In each case rigid metal backing plates are required. Both methods 230A and 231A contain information on the methods of eliminating some errors in the measurement.

Surface resistivity tests are sometimes required to be made after immersion in distilled water, but in that case the precise conditions between immersion and testing may influence the result. Volume resistivity measurements are seldom required to be made at elevated temperatures, but BS 3815 (1964)[34] is an exception and a test at 90 °C is involved. Here colloidal silver or graphite electrodes are stipulated. BS 5823 (1979),[35] which is equivalent to IEC 345 (1971),[36] describes methods for surface and volume resistivity measurements at very high temperatures.

The blanket ASTM specification covering resistivity measurements is ASTM D257 (1978).[29] Electrode sizes are not stipulated for sheet material (nor even the shape, since round, square or rectangular types are permitted). In the case of surface resistivity the gap width between guard ring and centre electrode is made approximately equal to twice the specimen thickness. The test voltage is usually 500 V applied for 1 min as in the British test. The ASTM specification gives a great deal of additional information about the techniques and hazards of measurement and also covers methods for tubing.

BS 4618 (1975)[37] prescribes the presentation of volume (Section 2.3) and surface (Section 2.4) resistivity data. The recommendation to pre-condition the test piece at not more than 1 per cent r.h. for the study of effect of temperature is necessary to minimise changes in moisture content as the temperature is raised, but it may be queried whether the data obtained are then relevant to design data. The test methods quoted are those in IEC 93, but the limit below which these become of doubtful use (given in BS 4618 as $10^3\ \Omega$ cm) should be about $10^8\ \Omega$ cm (§13.5).

ISO 1325 (1973)[38] for thin sheet and film uses the methods of IEC 93 for volume resistivity, but with a test voltage of 100, 10 or 1 V according to the thickness of the material, and the electrodes may be of evaporated metal, silver paint or metal foil. A document covering the same subject is under consideration by IEC and ISO 1325 may be withdrawn. ASTM D2305 (1982)[39] gives a volume resistivity test for thin films. Conductive silver electrodes of 25·4 mm in diameter are painted on each side of a test piece; no guard ring is required for films less than 0·12 mm thick. Evaporated metal electrodes may alternatively be used. The test voltage is 100 V d.c. applied for 1 min.

A test for volume resistivity of sleeving is given in IEC 684–2 (1984).[40] This uses an inner electrode of copper rod or wires of specified dimensions and an outer electrode of high conductivity metal paint 200 mm long with guard rings.

13.4 INSULATION RESISTANCE

13.4.1 General Considerations

Insulation resistance is the resistance between defined electrodes applied to a defined piece of sheet, rod or tube or to a product. It, or rather its reciprocal, *insulation conductance*, is a summation of a surface conductance and a volume conductance, although frequently predominantly the former. Insulation resistance is a function of the shape of the sample and the electrode system and its value is useless without reference to the test method. In practice, the utility of the test is confined to comparative measurements on a few materials of similar type, e.g. synthetic resin bonded paper or other laminates. In such cases the test is commonly used to assess the effect of moisture penetration.

The unit of insulation resistance is the ohm (Ω) and values may range from a few megohms or less to more than $10^{15}\ \Omega$. Most of what has been said in §13.3 is true of insulation resistance, although different electrodes, generally of solid metal, are used. In addition, in the case of insulation resistance there is no guard electrode and the resistance is measured using only two electrodes. There is therefore no guarded electrode/guard leakage problem, although the procedure of raising the potential of the screen of the lead connected to the current-measuring instrument may be used to check the adequacy of the insulation of the lead.

13.4.2 Standard Test Methods

The methods given in BS 2782 (1983)[41] are based partly on IEC 167 (1964)[42] and partly on ISO 2951 (1974),[43] although the ISO standard referred to rubber only. The first three methods in BS 2782 are primarily for rigid materials and involve immersion of the test piece in distilled water for 24 h before the test. Upper and lower limits are placed on the interval between the end of immersion and the start of the test in an attempt to avoid errors due to variable degrees of drying. There are two methods for rigid rod and tube, one using a 25 mm length test piece with plane parallel ends clamped between flat metal plate electrodes under a pressure of 35 kPa and another using taper pin electrodes. The remaining method for rigid sheet also uses taper pin electrodes. Two further methods which do not involve water immersion are prescribed primarily for flexible materials. The first uses stripes of conductive paint as electrodes and is applicable to sheet, rod or tube, while the second, for thin materials, specifies metal bar clamps as electrodes. The ASTM D527 (1978)[29] insulation resistance tests are somewhat similar to the BS ones but still include nut-and-bolt (or binding post) electrodes spaced 32 mm apart in a line. The method using bar clamp electrodes may be used for testing the insulation resistance of thin films as in ASTM D2305 (1982),[39] where it may additionally be used to detect the tendency of the films to corrode metals.

For tests on adhesive tapes BS 3924 (1978)[44] prescribes bar clamp electrodes (see Figs. 13.6 and 13.7). In ASTM D1000 (1982)[45] the test piece, clamped in the electrodes, is pre-conditioned at 23 °C, 96 per cent r.h. for 18 h and the resistance between pairs of electrodes is measured at 100–130 V after 15 s. The specially built humidity chamber described in BS 3924 seems inadequate and the method of guarding is inferior to that in BS 2782.

Tests on insulating sleeving are usually restricted to electric strength and insulation resistance. In BS 2848 (1973)[46] insulation resistance is determined on specimens with a copper wire or tube insert, three foil electrodes being wound externally and connected so that end leakage is avoided. The test is conducted either at 50 per cent r.h. 23 °C, or after three cycles of accelerated damp heat conditioning. IEC 684–2 (1984)[40] specifies a similar test but, in place of the heat cycling, includes a test at high temperature and another after 4 days at 40 °C and 93 per cent r.h.

13.5 TESTS FOR ANTISTATIC AND CONDUCTIVE PLASTICS

13.5.1 General Considerations

Electrostatic charges may be unbalanced whenever an insulating material is separated from another (or from a conductor) leaving an excess or deficit of electrons on the insulating material. The problem of electrostatic charging is a serious one in several industries.[47] Many situations involving rubbers and plastics, where 'static' can be built up to such an extent that a spark occurs, have an inherent fire hazard because of the presence of flammable solvents or explodable

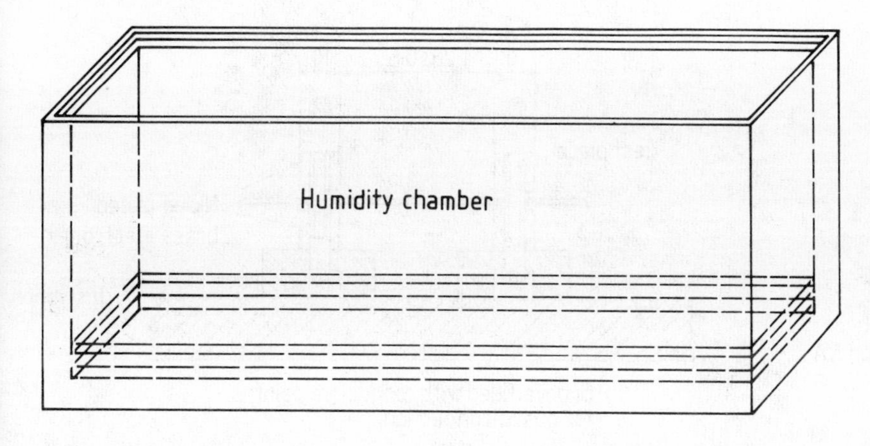

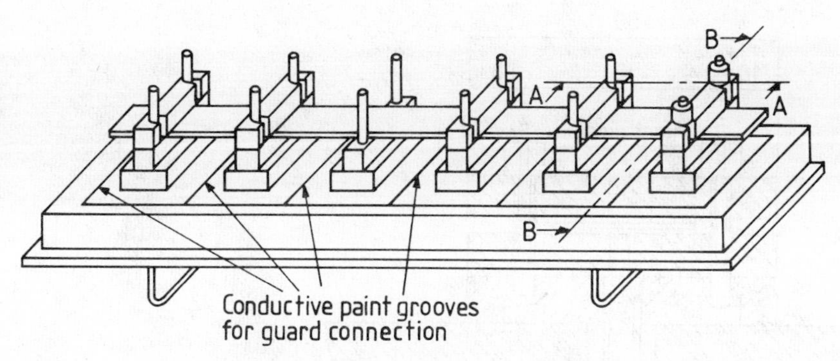

Fig. 13.6 Bar clamp electrode: BS 3924

gas mixtures. Mining, the petroleum industry and hospital operating theatres are examples. The packaging industry has an interest in the problem of dust and dirt pick-up on plastics which is aggravated by electrostatic attraction due to surface charges. It also has sometimes a problem in handling of charged plastics films.

A review by Morris[48] covers the theory of electrostatic charging in some detail, with an extensive bibliography.

The use of the terms 'conductive', 'conducting' and 'antistatic' is very confusing. The first two are interchangeable, although under most circumstances the first is more accurate. 'Antistatic' originally referred to articles which would dissipate static charges without being sufficiently conductive to cause electric shock or a fire hazard if they contacted live electric mains. This distinction is no longer valid in relation to specifications; for the purposes of this book the three terms may be regarded as synonymous.

It is theoretically possible for two materials to be 'electrically matched',[49,50] and thus cause no charge separation; however, each of these may charge if separated from a third (or if separated from each other when surface conditions have changed as a result of a change in atmospheric conditions). Thus the only sure way to

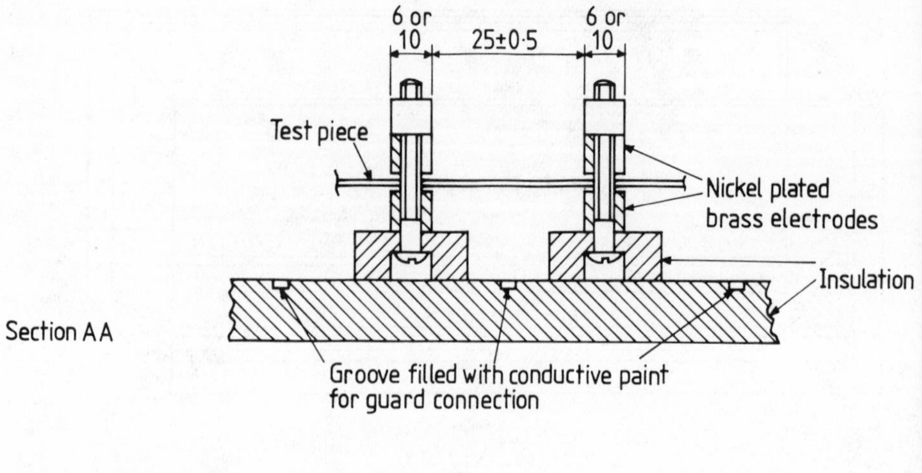

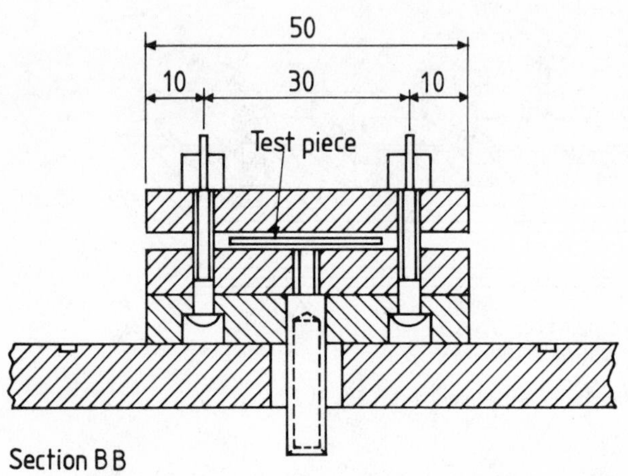

Fig. 13.7 Detail of bar clamp electrode: BS 3924 (dimensions in millimetres)

prevent the build-up of static charge is to make all materials at least somewhat conductive, so that separated charges leak away sufficiently fast. In the author's opinion, this makes measurements of charge developed by (say) rubbing, which are notoriously variable, invalid as specification tests. It is also questionable whether charge decay measurements, which are limited in the useful range of times (compared with the wide range of resistivities which can be measured), give any useful information which cannot be gained from a suitable resistivity test. Beesley and Norman[51] showed that a good correlation could be obtained between decay time and surface resistivity for a range of commercial PVC films (Fig. 13.8). However, all three types of measurement have been used and will therefore be outlined.

Conductive coatings may be used on plastics for electromagnetic screening. Klouda[52] describes screening measurement techniques using a signal generator

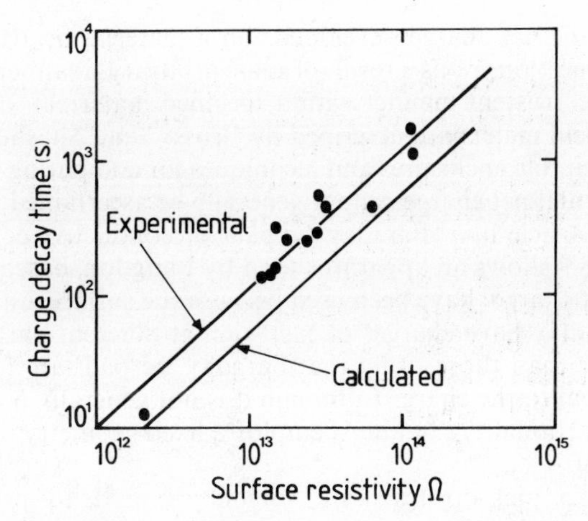

Fig. 13.8 Relationship between charge decay time and resistivity[51]

and a receiver. He indicates that, for initial selection of coatings for screening, a resistivity test can be used.

In applications other than antistatic ones, e.g. plastics potentiometers, a resistivity test on the material or a resistance test on the product (using its own terminals as electrodes) is again the obvious choice.

Further details on types of test for conductive and antistatic materials are given by Norman,[47] Ch. 8.

13.5.2 Effects of Temperature, Humidity and Strain

There are two general types of conductive plastics. The first is one which is filled with conductive filler particles, such as carbon black or metal powders. Such plastics generally have resistivities from about 10^7 Ω cm down to about 10^{-2} Ω cm. In this range they are relatively insensitive to temperature, humidity and voltage, but if they are flexible they are sensitive to strain history.

The second type contains additives known as 'antistatic agents', which are organic molecules that diffuse to the surface and either dissociate to give mobile ions or are hydrophilic and collect a film of moisture on the surface. These are sometimes highly sensitive to humidity and temperature but little is known about their sensitivity to voltage, and they are insensitive to strain.

Antistatic agents are capable of giving resistivities down to 10^7 Ω cm in plasticised PVC, and can then be used for some of the antistatic purposes where a hazard is involved. Antistatic agents in other plastics tend to give resistivities from about 10^{10} Ω cm upwards and are suitable then only to combat 'nuisance static', i.e. to prevent dust pick-up and, with thin films or fibres, handling problems.

13.5.3 Charge Measurements

In charge measurements a high impedance instrument such as an electrostatic field meter or an electrometer is used to measure the *net* charge developed (or a figure proportional to the net charge developed) on a material or article under service or production conditions or as a result of some arbitrary treatment such as rubbing in a (hopefully) consistent manner with a specified material.

Electrostatic field meters are described by Cross[53] and Shashoua.[54] Langdon[55] gives details of suitable enclosures and techniques for comparing fabricated articles and films. The actual net charge cannot generally be ascertained without enclosing the test piece or article in a 'Faraday ice pail' or conductive container of known capitance. Fig. 13.9 shows an apparatus used by Langdon, based on this principle.

The words 'net charge' have been used because the surface, especially of a newly moulded object, may have charges of each sign at different parts of its surface.[56]

Methods other than those using the 'Faraday ice pail' are used to measure a figure proportional to the charge (although it is not generally possible to state the constant of proportionality), but must employ a fixed geometry of apparatus inside an earthed screen.

Dust or smoke pick-up tests are sometimes used to provide qualitative information.

There appears to be no specification test for the direct measurement of surface charges on plastics materials, probably because of the difficulty (or even impossibility) of providing a reproducible method of charging which bears any resemblance to common charging mechanisms.

One logical application for measurement of charge developed was described by Moreno and Gross[57] in connection with the corona charging of electrets (solid dielectrics possessing a persistent dielectric polarisation) for use in electret microphones. Their method permits a measurement while the electret is being charged or discharged.

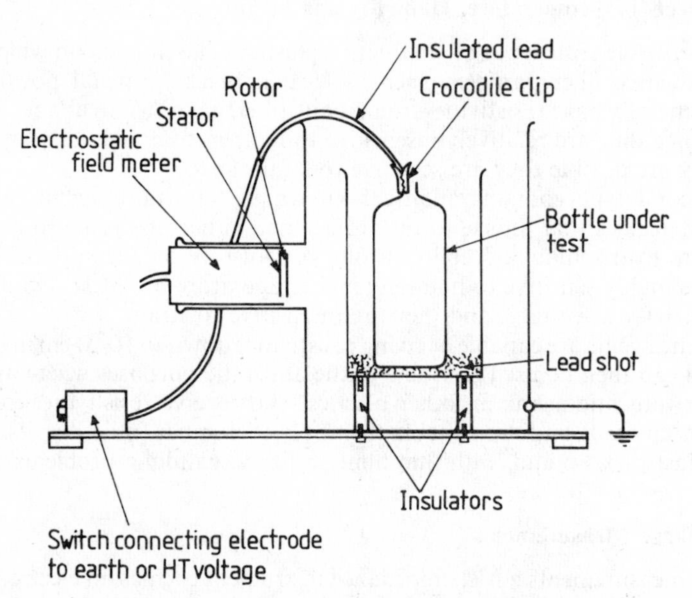

Fig. 13.9 Antistatic test for moulded articles[55]

13.5.4 Tests for Rate of Charge (or Discharge)

Three methods of test are described in BS 2782 (1976)[58] for use on film materials.

Method 250A (see Fig. 13.10) is based on a method described by Langdon[55] and uses a long strip of film threaded over a series of rollers. The rollers A and B are initially connected to a source of 5000 V and, after a given period, are connected to earth; the time of fall of the reading of the field meter to 50 per cent of its maximum value is measured.

Method 250B is suitable only for very flexible films. The equipment is rather like a gold leaf electroscope, where the test piece replaces the gold leaf. A voltage (10 kV) is applied to the test piece support and the time taken for the sample ends to diverge to an angle of 90° is recorded.

Method 250C uses a test piece which is clamped between 'brims' on two coaxial 'top hat' cavities. In one cavity is an electrode which may be connected to 300 V or to earth. In the other is an electrode connected to an electrometer. When the first electrode is switched from 0 to 300 V, the electrometer reading rises sharply to a maximum and then decays. The time for it to decay to half the steady reading which would be obtained in the absence of the film is recorded.

The useful range of the above three methods is generally from 1 to 1000 s, the lower value being often limited by instrumental time constants in Methods A and C and by the mechanical time constant of the sample in Method B, although the standard implies that 0·1 s is attainable. Readings of greater than 1000 s may be related to other effects that the sample properties.[54]

The above tests are suitable for ensuring there will be no 'nuisance static'. Where 'hazardous static' is a problem, much shorter discharging times must be attained and resistance or resistivity measurements are required (§13.5.5).

13.5.5 Resistance and Resistivity Tests

The resistivity tests for materials having resistivities below about 10^8 Ω cm differ from those described in §13.3 for the following reasons:

1. The high voltage gradients used for insulating rubbers would cause more conductive materials to heat up or even ignite.
2. Contact resistance effects, while possibly less than for insulating materials are practically of greater significance.
3. With materials of resistivity below about 10^7 Ω cm, the effects of humidity are usually negligible.
4. The resistivity of some conductive materials is highly sensitive to strain.

Materials having resistivities above about 10^8 Ω cm should be tested by the methods described in §13.3. For materials containing antistatic agents a surface resistivity test is more appropriate.

Resistance tests for antistatic and conductive products are in essence insulation resistance tests (§13.4), although with such products most of the conductance is usually a volume conductance. A series of test methods (which involve applying electrodes except where metal parts of the product are used as electrodes) using a conventional ohm-meter is described in ISO 2878 (1978)[59] for rubber and almost identically in BS 2050 (1978)[60] for flexible polymeric materials generally. In most

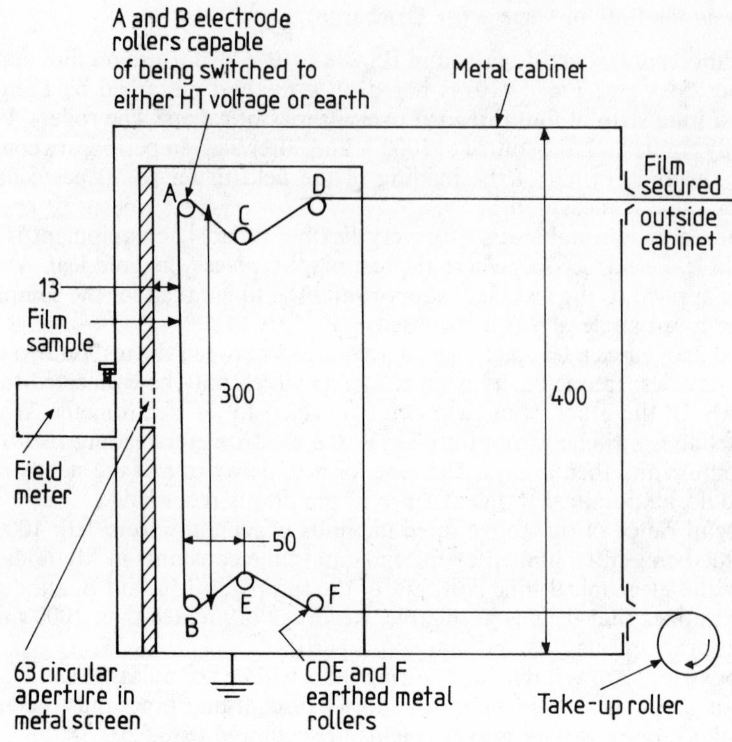

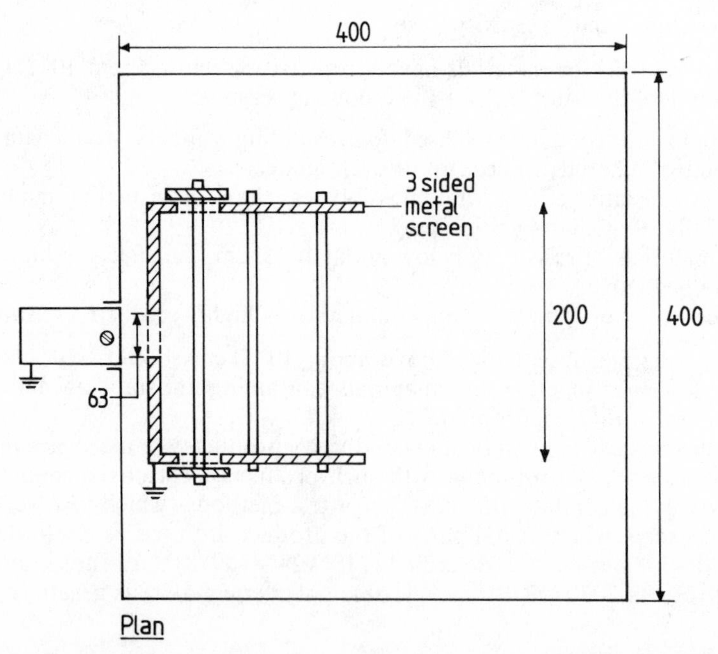

Fig. 13.10 Antistatic test for film: BS 2782 Method 250A (dimensions in millimetres)

cases the electrodes consist of a mixture of poly(ethylene glycol), water, wetting agent and electrolyte, backed by brass electrodes. The test is made with an instrument which has an open circuit voltage of 500 V, but whose output resistance limits the maximum power dissipation in the product. A number of British Standard product specifications use methods similar to those in BS 2050, sometimes with the omission of the electrolyte.

The test for conveyor belts given in ISO 284 (1982)[61] and the technically equivalent BS 3289 (1982)[62] uses a similar poly(ethylene glycol) mixture but without the electrolyte; if the belts are too rough, foil electrodes have to be used between the glycol and the backing plates.

BS 2044 (1984)[63] and ISO 3915 (1981)[64] describe essentially the same potentiometric test for the measurement of resistivity. In this method a current is passed between two electrodes at the ends of the test piece and the voltage drop is measured between two electrodes at intermediate positions. This test method eliminates the contact resistances at the two ends of the test piece. The contact resistances of the potential electrodes may be very high and it is therefore essential that the current supply and measuring circuit be isolated (to better than $10^{12}\,\Omega$) and that the input impedance of the voltage-measuring instrument be greater than $10^{11}\,\Omega$.

Blythe[65] described a potentiometric test using small-area probe electrodes on a semi-infinite surface. 'Correction divisors' to convert the results on test pieces of finite geometry were devised by Uhlir.[66] A more general account of four-point probe methods which takes into account anisotropic resistivities was given by van der Pauw.[67]

13.6 POWER FACTOR AND PERMITTIVITY

13.6.1 General Considerations

The measurement of power factor and permittivity and the related parameters outlined in §13.6.2 (often collectively known as *dielectric properties*) may need to be carried out over a wide range of frequencies from a few Hz to several tens of GHz, but most measurements are made between 50 Hz and 100 MHz.

The range of permittivities which may be encountered is from just over 1 for foams to 20 or more. Power factors may range from 0·0002 for polyethylene to 0·1 or even higher for a poor dielectric. (Note that a poor dielectric is a good material for dielectric heating!)

Materials with anisotropic structures may have anisotropic dielectric properties.

13.6.2 Definitions and Equivalent Circuits

The *relative permittivity* of an insulating material is the ratio of the capacitance of a capacitor, in which the space between (and around) the electrodes is entirely and exclusively filled with the insulating material in question, to the capacitance of the same configuration of electrodes in a vacuum (or to an accuracy of 0·05 per cent in air).

The test methods used are such that the words 'and around' can be ignored and we need only consider that part of the test piece between the electrodes, with a correction for the fringe field at the edge if necessary.

The word 'relative' in the context of tests on plastics is usually dropped for convenience, although the unqualified term 'permittivity' has another meaning. There is no cause for confusion since the magnitudes expressing relative permittivity are in the range from 1 to about 80 (dimensionless) whereas absolute permittivities are of the order of 10^{-11}–10^{-9} F/m.

The *dielectric constant* is the same as relative permittivity and the terms capacivity and specific inductive capacity, found in American usage, also have the same meaning.

The *loss angle* (δ) of an insulating material is the angle by which the phase difference between applied voltage and the resulting current deviates from $\pi/2$ radians when the dielectric of the capacitor consists exclusively of the dielectric material. It is sometimes expressed in microradians.

The *dissipation factor*, often known as *tan δ* or *loss tangent*, is the tangent of the loss angle.

The *power factor* is the ratio of the power loss in watts dissipated in the material to the product of the r.m.s. voltage applied and the r.m.s. current passing through the material. It is equal to the cosine of the phase angle, i.e. to sin δ.

It is common practice in dielectric work to use the tangent of the loss angle rather than the sine, since many of the theoretical equations for dielectrics are then simpler. Additionally, the equations relating to measuring circuits invariably give simpler relations to the tangent. With good dielectrics (tan δ less than 0·1) the difference between power factor and tan δ is less than 0·5 per cent. Thus the habit which has grown up among some people dealing with dielectrics of using the word 'power factor' when 'tan δ' is implied does not generally lead to significant error.

The *loss index*, formerly called *loss factor*, is the product of the relative permittivity and tan δ. For a fixed geometry of capacitor subjected to a fixed frequency and voltage, the power dissipated is proportional to the loss index.

The power factor has been defined in terms of energy dissipation, and the dissipation factor in terms of the loss angle, but in considering measurement at frequencies up to 10^8 Hz it is often convenient to consider equivalent circuits, i.e. series or parallel arrangements of a resistor and a capacitor (Fig. 13.11) which *at a single frequency* have the same electrical characteristics as the capacitor formed by the electrodes and the dielectric material in question.

It must be emphasised that for any dielectric the equivalent values of c and C

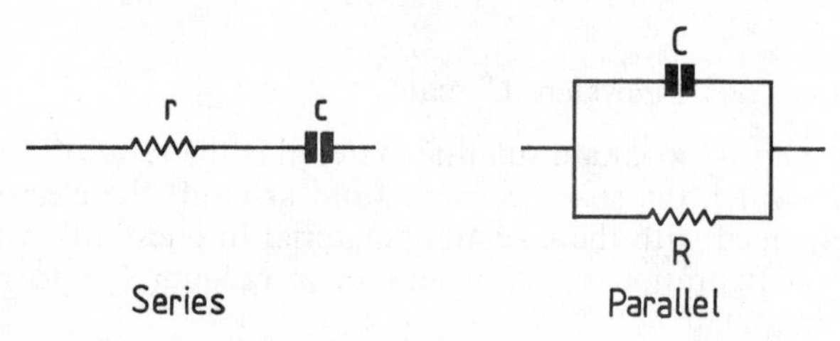

Fig. 13.11 Equivalent circuits

change slowly with frequency while the values of r and R change rapidly, often approximately inversely proportional to frequency. For good dielectrics, the values of c and C are nearly equal while the values of R and r are very different. The choice of the equivalent circuit is usually made on the basis of the measuring circuit employed, since in general the measuring circuit components are so arranged as to give the appropriate capacitance and power factor in terms of either the series or the parallel equivalent.

The condition that the series and parallel equivalents are identical at a given frequency is that their impedance (ratio of current to voltage) is the same with respect to both magnitude and phase. Using complex notation ($j^2 = -1$), the impedances of the series and parallel equivalents are, respectively,

$$Z = r - \frac{j}{\omega c} \quad \text{and} \quad Z = \frac{R}{1 + \omega^2 C^2 R^2} - \frac{j\omega C R^2}{1 + \omega^2 C^2 R^2} \qquad [13.2\text{a,b}]$$

where $\omega = (2\pi \times \text{frequency})$ and the units of capacitance and resistance are, respectively, farads and ohms. Equating real and imaginary parts leads to the following relations:

$$r = R\frac{\tan^2\delta}{1 + \tan^2\delta} \quad \text{and} \quad c = C(1 + \tan^2\delta) \qquad [13.3\text{a,b}]$$

The equations for the series and parallel circuits above lead, respectively, to

$$\tan\delta = \omega c r \quad \text{and} \quad \tan\delta = \frac{1}{\omega C R} \qquad [13.4\text{a,b}]$$

For values of $\tan\delta = 0{\cdot}1$ or less, eqns [13.3a,b] reduce within 1 per cent to

$$r = R\tan^2\delta \quad \text{and } c = C \qquad [13.5\text{a,b}]$$

To illustrate the significance of these equations, consider the example of a low-loss test piece of polyethylene with a capacitance of 10 pF and loss tangent of 0·0001. At $\omega = 10^4$ rads/s (frequency = 1600 Hz, approximately), the series equivalents would be $r = 1000\ \Omega$ and $c = 10$ pF and the parallel equivalents $R = 10^{11}\ \Omega$ and $C = 10$ pF. At $\omega = 10^7$ rad/s ($f = 1{\cdot}6$ MHz, approximately), the equivalents would be $r = 1\ \Omega$, $R = 10^8\ \Omega$, both capacitances again being 10 pF. The small value of r at the higher frequency also illustrates the importance of keeping any lead resistances in the measuring circuit small, since if they are included in the measurements the loss tangent will be augmented accordingly by an amount $\omega c r'$, r' being the extraneous resistance and c the measured capacitance. At the lower frequencies the reverse situation applies and it is important to ensure that no extraneous resistance in parallel with the specimen is included in the measurement unless it is extremely high compared with the specimen equivalent parallel resistance. The latter resistance should not be confused with the d.c. resistance of the specimen or component and it cannot be measured by d.c. methods.

13.6.3 Effects of Temperature, Frequency and Humidity

The effects of temperature and frequency are interrelated. With non-polar materials the changes of properties with frequency and temperature are small. With polar polymers very large changes may occur. For example, with plasticised PVC the power factor may change by a factor of two over a 10 °C range in temperature or a tenfold change in frequency, and the permittivity may change by 20 per cent over similar ranges.[68]

With polar materials (or even impure non-polar materials), especially at low frequencies, the effect of humidity can be large, e.g. a thirtyfold increase in loss index and a 20 per cent increase in permittivity may be produced by the absorption of 0·5 per cent of water between a dry condition and 75 per cent r.h.[69]

13.6.4 Electrodes

As in resistivity measurements, a sheet dielectric material with an electrode applied to each surface would be simple to measure, but the capacitance between the electrodes due to the test piece would be augmented by extraneous capacitance around the edge of the test piece either wholly or partly in air, which would be difficult to assess and allow for. A third electrode, placed concentrically around one of the existing electrodes (which is diminished in area) and as close to it as possible without touching, serves as a guard ring. When this is used with appropriate circuitry it reduces the uncertainty of the measured capacitance to an extremely small amount, since the effective area of the guarded electrode very nearly corresponds to that based on the mid-gap radius. Usually the requirement of the measuring circuit is that the guarded and guard electrodes are maintained at exactly the same potential and phase during measurement to ensure that the capacitance and associated loss between them do not contribute to the measured values; alternatively, the capacitances of the guard/guarded electrode and the guard/non-guarded electrode may, with some circuits, be connected so that they do not affect the balance or accuracy.

Three-terminal measurements are more accurate than two-terminal ones, but are mainly limited to bridge measurements and are thus rather impracticable above about 1 MHz, although there is an increasing tendency for radio frequency bridges to be designed to accommodate them. Above 10 MHz two-terminal systems are standard, and corrections may need to be made for edge effects.

Consideration will now be given to the means by which contact is established between dielectric test pieces and the measuring instrument. The importance of understanding the principles and techniques involved cannot be overstated, since probably more difficulties and errors are associated with electrodes than with the measurement itself. Much of what has been said in §13.3.5 about electrodes in connection with resistivity measurements is applicable here also, with the recognition that the requirements are probably more stringent with dielectric measurements, since greater accuracy can be obtained and is usually desired.

In what follows, essentially only sheet test pieces are considered, but the techniques are commonly applicable to tubes with appropriate changes in the permittivity formulae.[70]

All electrodes are designed either to ensure as intimate a contact as possible between the test piece and the measuring apparatus, or alternatively and rather paradoxically, to ensure a precisely defined air or liquid gap between the test piece and the electrodes, for which accurate allowance can be made. Amongst the *contacting electrodes* appear conductive paints, vacuum-deposited metallic films, mercury, conductive rubbers, metal foils applied with grease and colloidal graphite. Although each of these types may have some utility under particular circumstances, the most widely used is probably that involving metal foils, and possibly the best is that utilising metal films deposited *in vacuo*. Conductive paints and graphite suspensions tend to give relatively high-resistance films, which cause spuriously high loss values on low-loss materials especially at high frequencies, and the solvent base may attack the material. Conductive rubbers have relatively high resistance and are thus only suitable for 'lossy' materials, but have the virtue of being speedy and 'clean' to apply. Mercury presents a toxic hazard and requires carefully designed clamps and considerable care in avoiding air bubbles. Vacuum deposition of metallic films such as gold or aluminium ensures intimate contact; appropriate masks enable the electrode diameter to be defined accurately.

Tin or aluminium foils applied with silicone grease or petroleum jelly and rolled on very hard with a roller are convenient and simple. The foils are usually 0·03–0·05 mm thick and the rolling process produces an extremely thin film of grease, which must have a capacitance at least two orders greater than the test piece to reduce the measurement error to 1 per cent. This is not difficult to achieve with low permittivity test pieces 2 mm thick or so, but errors may become very large with high permittivity or thin test pieces ($<0{\cdot}5$ mm, say) and these foils are not generally suitable for measurements on films. Gold foil does not require grease but it is difficult to handle; on the other hand, it does provide very intimate contact and can be used even on thin films. It should be emphasised that no one electrode material or technique is superior to all others and necessarily gives the 'correct' answer. The only procedure with a doubtful result is to substantiate it by at least one other completely independent method.

All the above electrodes need metal backing plates of the same size as the electrodes. The use of flat plates alone (under a high normal pressure) is mentioned in ASTM D150 (1981).[71]

If d_o is the diameter mid-way in the gap between a circular guarded electrode and its guard ring the permittivity is given by

$$\varepsilon = \frac{144tC}{d_o^2} \qquad [13.6]$$

where t = test piece thickness (mm)
C = capacitance (pF).

For two-terminal electrode systems, when the test piece is at least equal in diameter to the electrodes, the permittivity is given by

$$\varepsilon = \frac{144t(C - C_e)}{d^2} \qquad [13.7]$$

where d = electrode diameter (mm)
t = test piece thickness (mm)
C = capacitance of test piece/electrodes (pF)
C_e = edge capacitance of electrodes (pF).

The edge correction may be obtained either from the Kirchhoff relation[72,73] or by calibration with test pieces of known permittivity.

The power factor is not related to the electrode dimensions.

A difficulty with most contacting electrodes is that any pre-conditioning of the test piece must be carried out before their application, but this difficulty does not arise with *non-contacting* electrodes. With *air gap* systems the electrodes are spaced apart and the test piece is introduced leaving a deliberate gap. Two measuring techniques are then available: *fixed gap* and *fixed capacitance.* In the first, the capacitance and loss tangent of the system are measured with the test piece in and then with the test piece removed, the electrode spacing being left untouched throughout. The characteristics of the material are given by

$$\varepsilon = \frac{t_s}{t_s - t_o[1-(C_o/C_i)]} \qquad [13.8]$$

$$\tan\delta = \tan\delta_i\left(1 + \varepsilon\frac{t_o - t_s}{t_s}\right) - \varepsilon\frac{t_o}{t_s}\tan\delta_o \qquad [13.9]$$

where t_s = test piece thickness
t_o = electrode separation
C_i = capacitance, test piece in
C_o = capacitance, test piece out
$\tan\delta_i$ = loss tangent, test piece in
$\tan\delta_o$ = loss tangent, test piece out
ε = test piece permittivity
$\tan\delta$ = test piece loss tangent.

For a properly designed electrode system, where $\tan\delta_o$ would normally be negligibly small compared with $\tan\delta_i$,the second term would vanish.

The units of capacitance and distance are arbitrary (but consistent) since only ratios are involved. The equations are rather cumbersome, but their virtue is that the calibration of the electrodes need not be known, since their area is not involved, and no electrode adjustment is required, although this limits the permissible test piece thickness as it is desirable for the sake of sensitivity to keep the gap small. As a guide, the gap should be of the order of from 10 to 20 per cent of the test piece thickness. Smaller gaps tend to cause difficulty due to dust particles and fibres, and with larger gaps the electric field tends to distort and cause errors.

With a fixed capacitance system, measurements are made of capacitance and loss tangent as before, with the test piece in, but the electrode spacing is adjustable and after removal of the test piece the spacing is reduced until the capacitance of the empty electrodes is the same as when the test piece was in place. The equations are then

$$\varepsilon = \frac{t_s}{t_s - t_x} \qquad [13.10]$$

$$\tan\delta = (\tan\delta_i - \tan\delta_o)\frac{t_o}{t_s - t_x} \qquad [13.11]$$

where t_s = test piece thickness
t_x = electrode movement between test piece in and out positions
t_o = electrode separation, test piece out

$\tan \delta_i$ = loss tangent, test piece in
$\tan \delta_o$ = loss tangent, test piece out
ε = test piece permittivity
$\tan \delta$ = test piece loss tangent.

With suitable test pieces a precision of the order of 1 part in 1000 is achievable for permittivity, such precision being required, for example, in the control of polyethylene used in submarine cables. Lynch[74] gives useful information on the use and construction of micrometer air gap electrodes with a full analysis of electrode and test piece imperfections; BS 4542 (1970)[75] is largely based on his work and principally covers the frequency range from 1 to 100 kHz.

In recent years *liquid immersion electrodes* have been developed, with which very precise measurements of permittivity can be made; these are based on the fixed gap technique, but with the gap filled with a suitable non-polar liquid. Principally developed to meet the submarine cable requirement referred to above, a high accuracy is achieved by using a fluid of closely similar permittivity to that of the test piece so that only small changes in the electrode capacitance and loss occur when the test piece is inserted. If there were no change at all in these parameters on insertion, the test piece would have an identical permittivity and loss tangent to that of the liquid.

If the three-plate electrode system described in ASTM D1531 (1981)[76] is used with a liquid whose permittivity is within 0·1 of that of the pair of test pieces,

$$\varepsilon = \varepsilon_l + \left(\frac{\Delta C}{C}\frac{t_o}{t_s}\right) \quad [13.12]$$

$$\tan \delta = \tan \delta_l + (\tan \delta_i - \tan \delta_l)\frac{t_o}{t_s} \quad [13.13]$$

where t_o = electrode separation
t_s = test piece thickness
ΔC = increase in capacitance when test piece inserted (pF) (negative sign if decrease)
$C = 0{\cdot}0177084\, A/t_o$ (pF)
ε = permittivity of test piece
ε_l = permittivity of liquid
$\tan \delta$ = loss tangent of test piece
$\tan \delta_l$ = loss tangent of liquid
$\tan \delta_i$ = loss tangent with test piece inserted
A = area of one face of the centre electrode (mm).

This method of measuring very precise permittivity is mainly limited to materials the permittivities of which can be matched to readily available liquids. ASTM D1531 gives details of the requirements for polyethylene measurement using either benzene or silicone fluid.

13.6.5 Measuring Techniques – General

The measurement of power factor and permittivity, over the wide band of frequencies in which these properties are important, is rendered difficult by the limitations of existing test equipment and the necessarily large variety of methods required to cover this range. Insulating materials are used as dielectrics at commercial frequencies between approximately 50 Hz and 10^{11} Hz and the variety of test methods used reflects the many differing practical techniques necessary to generate and distribute power at different parts of this frequency spectrum. The range up to 10^8 Hz, it is true, can be covered by only two types of measuring equipment, but between approximately 10^9 and 10^{11} Hz a different set of equipment is generally required for each specific frequency of interest, since the waveguides used at these frequencies are of particular dimensions directly related to the wavelength.

The two kinds of technique related to these two frequency ranges (above and below 10^8 Hz) are quite different and are sometimes loosely differentiated by the terms 'lumped circuit' and 'distributed circuit'. In the former case the circuitry involves the use of the three basic electrical components: inductors, capacitors and resistors. A passive circuit (one containing no sources of voltage), however complex, comprising these elements can be represented at a given frequency by an equivalent circuit having single values of inductance or capacitance and resistance. This notion of equivalence is a powerful aid to the solution of such complex circuits. At the higher frequencies, say 10^9 Hz and above, the three basic parameters can no longer be 'lumped' in this fashion and the waveguide behaves as though these parameters were distributed properties, so that solutions can only be found in terms of the equations relating to the associated electric and magnetic field distributions in and around the waveguide.

The power factor and permittivity of an insulator (or a capacitor) are so closely related that in practice many measuring circuits are designed to give both properties simultaneously, and as in any case the determination of one property can only be made with accuracy by adjustment of a circuit component which compensates for the presence of the second property, it is generally expedient to measure the two together.

13.6.6 Techniques for Measurements at Frequencies below 20 kHz

Bridge circuits are invariably employed in the audio frequency range (50 Hz–20 kHz) and beyond up to 100 kHz, but the upper limit is generally of the order of 10 MHz, although bridges suitable for the latter frequency have to be of rather special construction. The basic bridge circuit (Fig. 13.12) consists of four arms, and the simplest case involving four resistances is that of the Wheatstone bridge for measuring resistance. The circuit is completed by the addition of a voltage source across one diagonal of the bridge and a suitable voltage detector across the other. At least one of the arms must be variable and the bridge is balanced when the detector indicates zero voltage; the relation $P/R = Q/S$ then holds. Thus one unknown component can be measured in terms of three known ones.

For a.c. measurements at least one arm other than that containing the test piece must contain a reactive component, that is one having inductance (self or mutual) or capacitance, and two examples will be considered in some detail since most present-day determinations are based on them or their variants. These are the Schering bridge and the transformer ratio-arm bridge.

Dielectric measurements at u.l.f. (ultra low frequency)[30,77–79] are of some interest

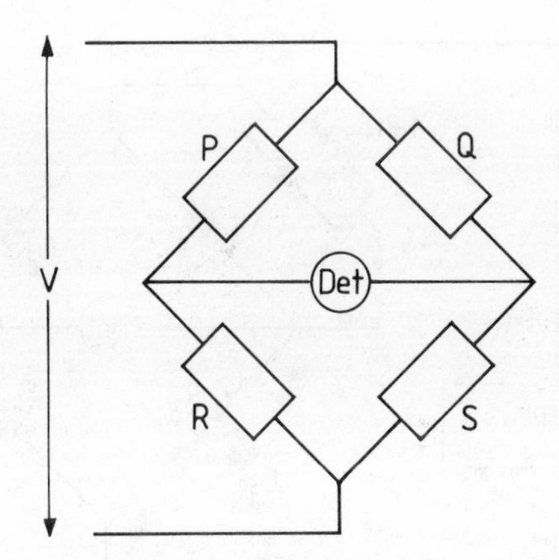

Fig. 13.12 Generalised bridge circuit

in view of the light they throw upon the basic structure of some materials, notably glasses, but are outside the brief of this book. Such measurements have been made down to 0·001 Hz and lower.

A bridge similar to the Schering bridge but suitable for the range from 0·02 Hz to 80 Hz was illustrated by Stoll.[80]

Schering Bridge

The Schering bridge in its simplest form comprises two fixed resistors of equal value and two capacitance arms, one of which is a relatively loss-free variable standard capacitor and the other the unknown, i.e. the test piece. At balance the total capacitance in the unknown arm is equal to the total capacitance in the standard arm, while the loss balance is achieved by a variable capacitor across the resistance arm diagonally opposite the unknown arm, the loss tangent being a function of the capacitance and resistance of this arm and the unknown capacitance.

This arrangement is seldom used in practice, since the residual stray capacitances which are inevitably associated with each arm of the bridge are difficult to assess and allow for. Instead, a substitution method is employed using a modified Schering bridge containing two capacitive arms and two resistive arms (Fig. 13.13), the bridge being initially balanced with the unknown disconnected and then re-balanced with the unknown connected across one capacitive arm containing a low-loss calibrated standard. The diagonally opposite resistive arm has a calibrated loss-balancing capacitor in parallel and the other resistive arm usually also contains an uncalibrated capacitor to facilitate initial balancing. The capacitance and loss are given by the changes in settings of the standard and the loss capacitor, provided that the remaining components are left untouched after initial balance. The virtue of this arrangement is that most of the residual capacitances and losses in the bridge remain constant during the measurement and do not enter into the bridge equations.

In Fig. 13.13, C_1 is the low-loss standard capacitor of known calibration, C_2 a low-loss uncalibrated balancing capacitor, and C_3 and C_4 two capacitors of lesser

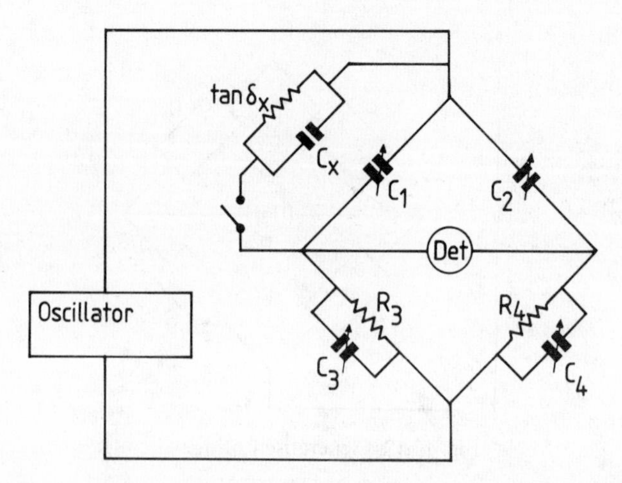

Fig. 13.13 Modified Schering bridge for two-terminal measurements

quality, the latter suitably calibrated. With the unknown C_x (of loss $\tan \delta_x$) disconnected and assuming equal ratio-arms ($R_3 = R_4$), the initial balance gives $C_1 = C_2$ where these symbols denote the total capacitance in arms 1 and 2 including extraneous capacitance. Any difference in loss in arms 1 and 2 will be balanced out by the difference in settings of C_3 and C_4 (again including stray capacitance). The balance equations give

$$C_x = \Delta C_1 \qquad [13.14]$$

$$\tan \delta_x = \omega \frac{C_1'}{10^{12}} R_4 \frac{\Delta C_4}{\Delta C_1} \qquad [13.15]$$

where $\omega = 2\pi \times$ frequency
ΔC_1 = change in capacitance of C_1 (pF)
ΔC_4 = change in capacitance of C_4 (pF)
C_1' = initial setting of C_1 (pF)
$R_4 = R_3$ ratio-arm resistance (Ω).

It should be noted that the residual losses in C_1 and C_2, although small, are finite, but they disappear provided that the loss in C_1 is due to a constant equivalent parallel resistance, which is usually the case for good quality capacitors.

In the two-terminal example given it is important to connect the unknown via a low-capacitance switch as shown, where any alteration in the distribution of the stray capacitances from the specimen electrodes to earth between the two balances has a negligible effect. If the switch were in the upper or high voltage lead, a large stray capacitance to earth could appear across arm 3 on the initial balance, which would disappear on final balance, and the resultant change in the capacitance of this arm would be reflected as a spurious contribution to the loss balance ΔC_4. The values of C_3 and C_4 are often of the order of from 500 to 1000 pF, while C_1 and C_2 are usually 1000 pF or less. The values of R_3 and R_4 are typically from 1000 to 10 000 Ω; R_3 and R_4 are sometimes made unequal, the ratio giving a multiplying factor by which both capacitance and loss ranges may be extended.

Three-terminal electrode systems are accommodated on the modified Schering bridge by the provision of a further two arms, similar to one half of the existing bridge, which form a Wagner earth circuit. There are many forms of this arrangement. That shown in Fig. 13.14 is given in BS 2782 (1982)[81] and shows the screening in detail. It also includes an inductance L_6 which may be required to balance the Wagner earth. The Wagner earth eliminates not only the effect of stray capacitances between the guard and guarded electrodes, but also of those within the bridge and may thus be useful even with two-terminal electrodes.

The main capacitance in the electrode system to be measured is connected across C_1, with the large electrode to the high voltage side of the bridge (or to earth), and the smaller central electrode to b. The guard ring is connected to earth. The main feature of the Wagner earth is that the detector can be switched either across bc, the bridge proper, or to c and earth so that when the detector indicates a null in both positions the points b and c are both at earth potential and the guarding of the electrode system is perfect. The procedure in the substitution method is as follows:

1. With specimen main (high voltage) electrode connected to earth, balance the main bridge using C_2 and C_3, C_1 and C_4 being meanwhile left at their zero settings (corresponding usually to maximum capacitance and minimum capacitance, respectively).

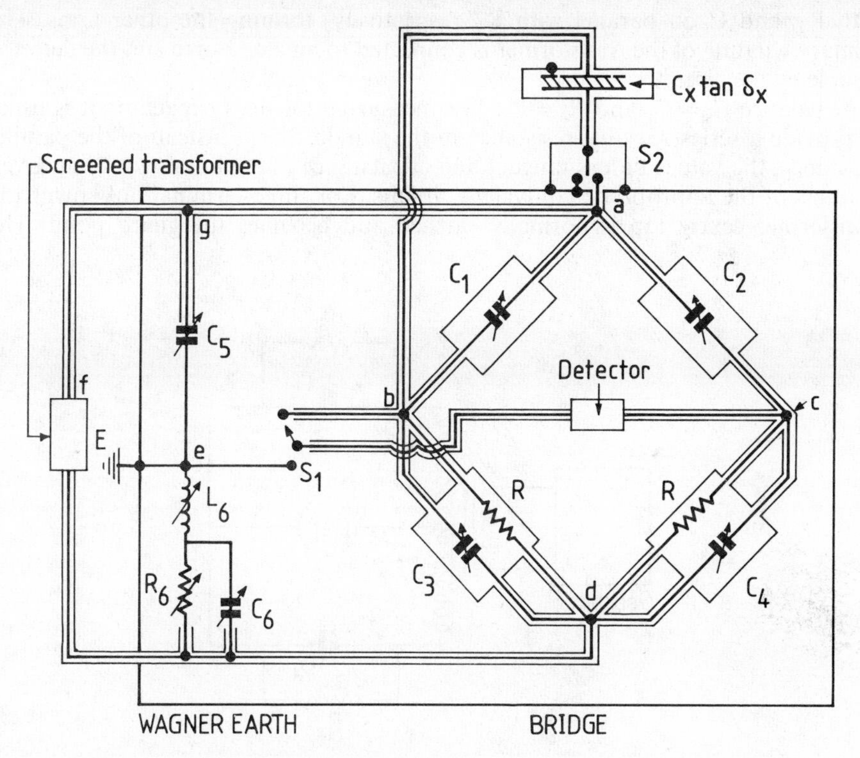

Fig. 13.14 Modified Schering bridge for three-terminal measurements

2. Transfer detector to ce and balance the Wagner earth components C_5 and C_6 (and L_6 if necessary).
3. Alternate detector between bc, balancing C_2 and C_3, and ce, balancing C_5 and C_6, until both positions give a null simultaneously. Thereafter C_2 and C_3 must be left undisturbed.
4. Connect main electrode to a and re-balance main bridge with C_1 and C_4 only (detector at bc) followed by Wagner earth with C_5 and C_6 (detector at ce). Alternate between bridges until both are in balance simultaneously as before.

The loss tangent and capacitance are then derived as for the two-terminal arrangement.

Hague's excellent book[82] on a.c. measurements, although somewhat dated, gives an extremely detailed account of Schering's bridge and many others.

Transformer Ratio-arm Bridge

The other type of bridge circuit which will be considered in detail is the transformer ratio-arm bridge. This has been known for many years but has not been widely used until fairly recently, its development being somewhat delayed until advances in transformer design and materials enabled its virtues to be better exploited.

Figure 13.15 shows such a bridge in its simplest form, with the transformer secondary winding divided electrically exactly into two, forming two of the bridge arms, and the unknown and standard impedances (represented by C_x in parallel with R_x, and C_s in parallel with R_s, respectively) forming the other arms. The primary winding of the transformer is connected to an a.c. source and the detector completes the circuit.

At balance $C_s = C_x$ and $R_s = R_x$. For measuring low-loss capacitors it is usual to provide a series variable resistance in the standard arm instead of the parallel resistance, the values of loss tangent and capacitance of the unknown being given in terms of the appropriate equivalent circuits. For three-terminal unknowns the transformer centre tap is normally earthed and becomes the guard point. The

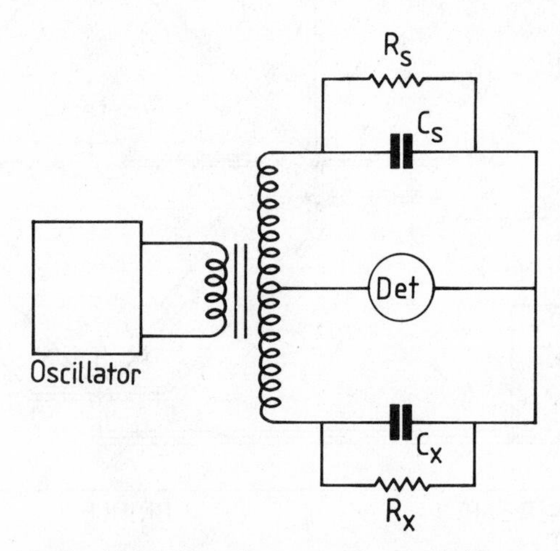

Fig. 13.15 Transformer ratio-arm bridge

capacitances to guard from the remaining two terminals or electrodes of the unknown are thus either a shunt on the detector, or across one half of the transformer winding. The effect is either to decrease the sensitivity of the detector (negligible for capacitances of the order of a few hundred pF) or to impose an additional loading onto the transformer winding. This latter effect is in general small with modern constructional methods and the improved magnetic core materials now available.

The simplicity of the basic circuit conceals the versatility of the transformer ratio-arm bridge and its outstanding advantages are its range and accuracy. In some bridges the single standard capacitor is replaced by a number of standards in decade steps, say 1 pF, 10 pF, 100 pF, etc. The 'standards' side of the transformer secondary is tapped at 10, 20, 30 per cent and so on of the winding and the switches are arranged so that any standard capacitor or parallel resistor can be switched to any ratio. The total capacitance of the 'standard' arm is then given by the sum of the standards multiplied by their respective transformer ratios.

The inherent accuracy of the bridge arises from the precision with which modern transformers can be wound and tapped, combined with the use of precision fixed-value capacitors. In the best bridges the ratios can be made as accurately as 1 ppm or better. The use of a number of relatively cheap fixed precision standard capacitors in place of the expensive variable or decade capacitors required in conventional bridges, results in a considerable economy. The modern transformer bridge has a discrimination of 1 ppm in both loss and capacitance and an acccuracy of 0·01 per cent in the latter variable. Its virtues include a self-guarding facility, simple balancing procedure and freedom from corrections. It can be used over the whole of the frequency range from 50 Hz to 10 kHz.

Automatic self-balancing bridges with digital read-outs based on transformer ratio-arms, the advantages of which hardly need extolling where rapid routine measurements are required, are now available.

13.6.7 Techniques for Measurements at 10 kHz–100 MHz

The frequency range from 10 kHz to 100 MHz corresponds approximately to that of radio transmissions, and telecommunication techniques based upon the tuned (or resonant) circuit are widely used for measurement purposes. The resonant circuit, comprising basically an inductance and capacitance in parallel, has the important property that it is frequency-sensitive, and at one particular frequency its sensitivity to relatively slight changes in either frequency or capacitance is enormously greater than at another frequency. These two features are exploited in the Q-meter and in the susceptance variation methods of measuring dielectric properties, whereby the addition of a capacitance in the form of a dielectric test specimen to the circuit causes a change in resonant frequency, in the first case the change being compensated by frequency adjustment and in the second by capacitance adjustment. The Q-meter will not be dealt with since it is used rather less in the UK than the second method, although it is probably capable of as great an accuracy in its refined forms.[83,84]

Hartshorn and Ward Apparatus

The *susceptance variation* method was originated by Hartshorn and Ward[85] in

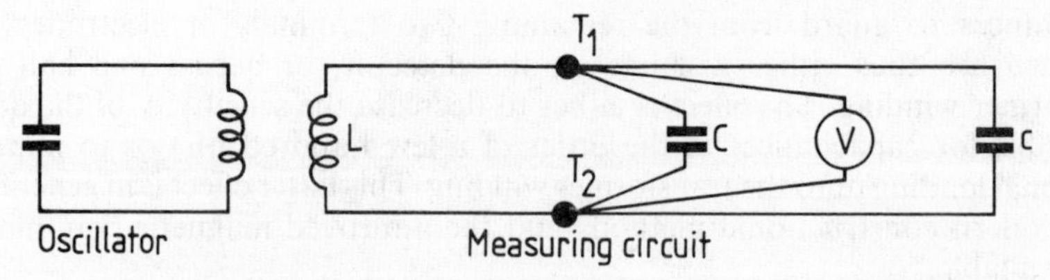

Fig. 13.16 Basic circuit of Hartshorn and Ward apparatus[85]

1936. Refinements by Barrie[86] enable it to be used to measure loss tangents of the order of 0·001 to a few μR (1 μR = 0·000 001).

In this method a resonant circuit (see Fig. 13.16) comprising an inductance, L, in parallel with a variable air capacitor, C, which also forms the electrodes for the test piece is loosely coupled electromagnetically to a stable oscillator. A valve voltmeter with a square law relationship (i.e. one whose output is proportional to the square of its input voltage) is connected in parallel with the circuit and serves both to indicate resonance and to measure the degree to which the circuit is adjusted off resonance during a measurement. The voltmeter output is connected to a reflecting galvanometer, to achieve the high sensitivity required. A second small capacitor, c, is also connected in parallel with the main circuit and is used to produce a given degree of de-tuning in the circuit.

In practice a test piece is placed between the electrodes and the oscillator frequency is adjusted to resonance as indicated by maximum galvanometer deflection θ. The circuit is then de-tuned using capacitor c to produce a deflection of $\theta/2$, first on one side of resonance by increasing the value of c and then on the other side of resonance by decreasing c, the total change of capacitance being denoted by Δc_i (Fig. 13.17). Finally c is adjusted back to resonance. The test piece

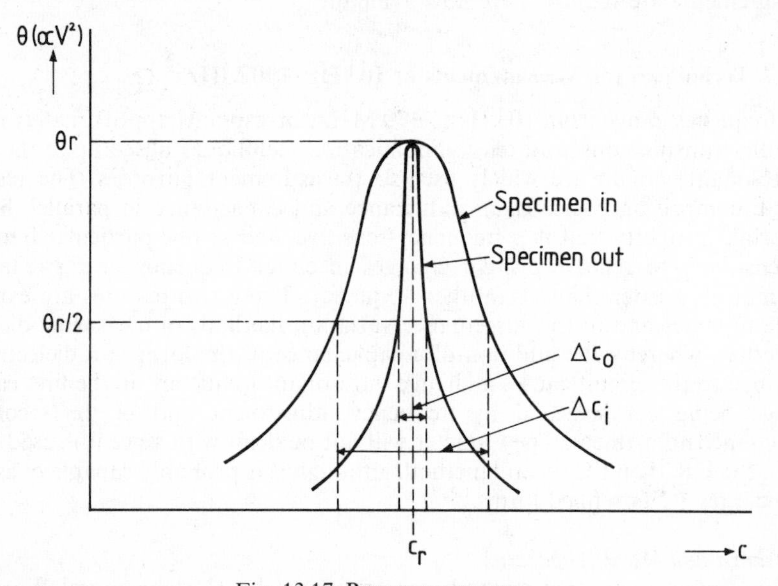

Fig. 13.17 Resonance curves

is then removed and resonance is re-established by increasing the capacitance of C, after which the de-tuning to half deflection is repeated using c, the total change in its value being denoted by Δc_o (less than Δc_i).

The true test piece capacitance (C'_s) is thus given by the changed setting of the main calibrated capacitor (C) after appropriate edge corrections, and the loss tangent is given by the expression

$$\tan\delta = \frac{\Delta c_i - \Delta c_o}{2C'_s} \qquad [13.16]$$

The deflections at resonance for both conditions are shown in Fig. 13.17 as equal, but this is not necessary; for accuracy they should be kept as large as possible and this is achieved by moving the oscillator nearer or away, as appropriate.

For low loss tangents, of the order of 0·0002 or less, the difference $\Delta c_i - \Delta c_o$ becomes too small to measure accurately in this way (for example, with $C'_s = 20$ pF typical values are $\Delta c_i = 0{\cdot}65$ and $\Delta c_o = 0{\cdot}64$ pF at 1 MHz for $\tan\delta = 0{\cdot}000\,25$) and an alternative method based on the ratio of deflections at resonance is used. It can be shown that

$$\frac{\Delta c_i}{\Delta c_o} = \left(\frac{\theta_o}{\theta_i}\right)^{1/2} \qquad [13.17]$$

where θ_o and θ_i are the deflections at resonance corresponding to test piece 'out' and test piece 'in', respectively, *provided that the oscillator is not disturbed during the measurements* (i.e. the coupling is not altered). By re-arranging,

$$\Delta c_i - \Delta c_o = \left[\left(\frac{\theta_o}{\theta_i}\right)^{1/2} - 1\right]\Delta c_o \qquad [13.18]$$

so that the loss tangent becomes

$$\tan\delta = \frac{[(\theta_o/\theta_i)^{1/2} - 1]\Delta c_o}{2C'_s} \qquad [13.19]$$

and for small losses this approximates further to

$$\tan\delta = \frac{\Delta\theta\Delta c_o}{4\theta_i C'_s} \qquad [13.20]$$

where $\Delta\theta = \theta_o - \theta_i$.

Since Δc_o is a constant for a particular apparatus and frequency, by using a fixed value of θ_i a determination is reduced to a measurement of $\Delta\theta$ and C'_s only. For the example given above of a loss tangent of 0·000 25, $\Delta\theta$ would be 25 mm using a metre scale and $\theta_i = 800$ mm so that measurements down to 0·000 01, corresponding to $\Delta\theta = 1$ mm, are quite feasible. Some care in ensuring stability of the deflections is essential at these low levels and Barrie's paper[86] is particularly helpful.

13.6.8 Techniques for Measurements above 100 MHz

At frequencies above 10^8 Hz the two test methods of principal interest are those employing standing wave methods and cavity resonator techniques.

In the standing wave method a waveguide is employed appropriate to the particular frequency range of interest and a dielectric test piece of rectangular or cylindrical section (depending on the type of waveguide), which must be a close fit in the guide, is inserted in one end. The length of the test piece required is a function of its permittivity and the wavelength corresponding to the test frequency and it may be necessary to adjust the length by machining after an initial measurement. The frequency of a generator coupled to the opposite end of the guide is adjusted to resonance, a suitable crystal probe inserted in a slot in the guide serving as a detector. The detector may be accurately positioned along the axis of the guide by means of a vernier slide so that the positions of the standing wave nodes can be evaluated. The standing wave pattern is studied with the test piece both in position and removed, and the permittivity and loss tangent are given by rather complicated equations; these yield a number of solutions and some care is needed in obtaining the correct one.

Further details are given by Roberts and von Hippel,[87] Westphal,[88] Gevers[89] and Dube and Natarajan.[90]

With cavity resonators the test piece is normally a disc, which forms the end of a cavity, coupled to a waveguide. As with the standing wave methods, a generator and crystal detector (in this case usually a square law type) are also coupled to the guide and the frequency is adjusted to resonance. Measurements are now made of the bandwidth of the system, usually at half power points, by means of a fine frequency control (accurately calibrated) in a manner somewhat analogous to that using the Hartshorn and Ward apparatus. The solutions are again complex and tedious to solve. Gevers,[89] Parry,[91] Horner *et al.*[92] and Brydon and Hepplestone[93] describe in detail various cavity resonators.

The two methods described are suitable for the frequency range from approximately 3×10^8 Hz to 3×10^{10} Hz.

13.6.9 Standard Test Methods

IEC 250 (1969),[70] relating to insulating materials in general, gives information on the form of test pieces, types of measuring equipment, electrode materials, testing procedures and calculation of results for measurements from 15 Hz to 300 MHz. It is, however, very general, and national standards for plastics generally lay down restrictions on these factors, although the results are not affected much by the measuring equipment used and sometimes 'other equipment which can be shown to give the same results' is permissible.

In BS 2782 (1982),[81] a Schering bridge with a Wagner earth or a transformer ratio arm bridge is specified for 50 Hz and for from 800 to 1600 Hz. BS 771 (1980)[94] invokes IEC 250 (1969)[70] in Part 1 and BS 2782 in Part 2.

BS 4618 (1970)[95] lays down a method of presentation of permittivity data (section 2.1) and loss tangent data (Section 2.2). These recommendations give advice for selecting temperature intervals, etc., and graphically displaying the results in a clear, standardised way. If these documents become widely accepted, then the present confusion resulting from the use of different methods, frequencies, temperatures, stresses, etc., by manufacturers may yet disappear. As with resistivity, it may be questioned whether the measurements at less than 1 per cent r.h. have any relevance for design purposes.

The blanket ASTM specification covering dielectric measurements from 1 Hz to several hundred MHz is D150 (1981)[71] which, although extremely helpful and informative, is written in quite general terms and makes very few specific recommendations about methods or apparatus. For example, no electrode sizes are even suggested and the only recommendations made are that the guard gap should be as small as possible and the guard width should be at least twice the specimen thickness. No particular type of apparatus is required to be used, but an appendix suggests details of a number of bridge and other circuits. Some details are given of the liquid immersion technique, but this method is dealt with at some length in ASTM D1531 (1981).[76] The cell is a two-terminal type with a double electrode arrangement requiring two specimens. The fluid is either very pure dry benzene or silicone oil of viscosity 1·0 cS, both of permittivity near 2·28. The frequency range covered is between 1000 Hz and 1 MHz; although no one type of apparatus is mandatory, equations are given for direct reading bridges and Q-meters. Figure 13.18 shows the cell.

ASTM D1673 (1979)[96] for cellular plastics, although involving ASTM D150 for test methods, is interesting in that it takes account of the heterogeneity of such materials and their availability in large pieces by permitting specimens of up to 400 mm square by 50 mm thickness to be tested, though at reduced maximum frequency (< 1 MHz). ASTM D1674 (1967)[97] for embedding compounds is also relevant in that a simple two-terminal electrode system suitable for measurements up to 300 °C is described (Fig. 13.19).

For insulating sleeving, IEC 684–2 (1984)[40] specifies electrodes consisting of an inner metal mandrel which 'provides good contact with the bore' and an outer electrode with guard rings of metal foil or conductive metal paint.

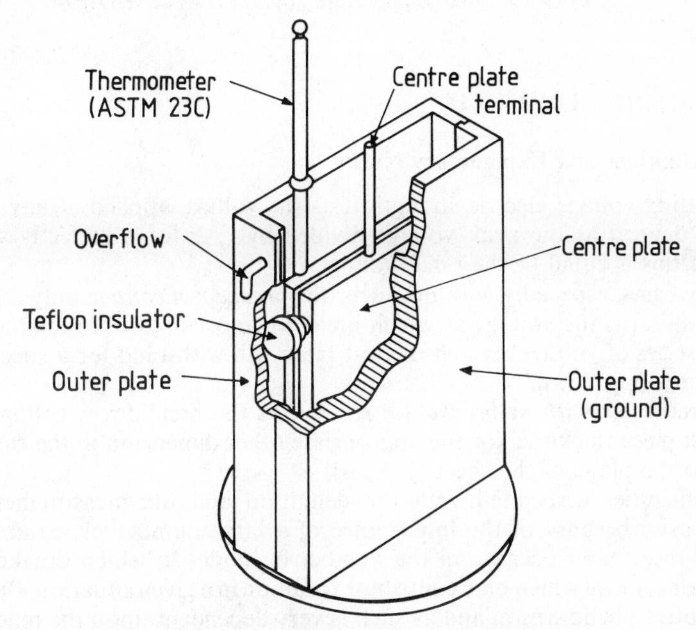

Fig. 13.18 Liquid displacement cell: ASTM D1531

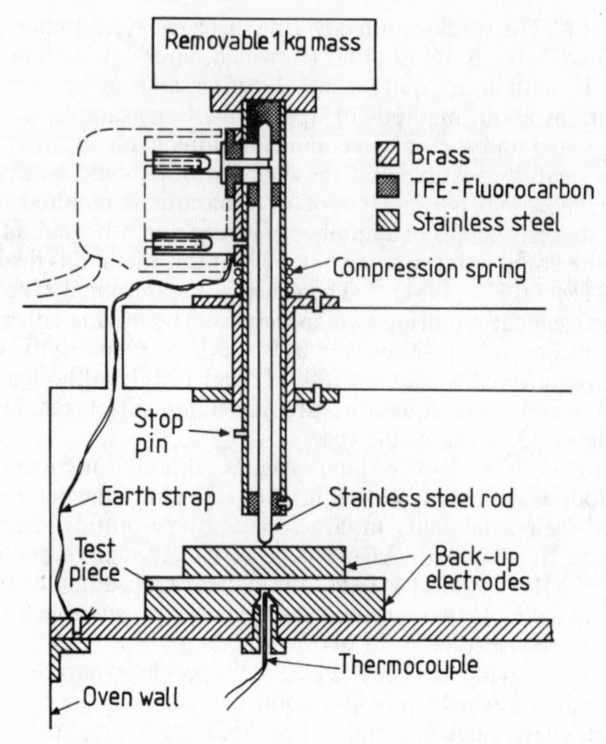

Fig. 13.19 Elevated temperature apparatus: ASTM D1674

13.7 ELECTRIC STRENGTH

13.7.1 Definitions and Explanatory Notes

In alternating voltage electric strength tests the voltage applied at any instant is normally defined as the peak voltage divided by $\sqrt{2}$; for a perfectly sinusoidal waveform this is equal to the r.m.s. voltage.

In some cases, especially with products, the *breakdown voltage* only is recorded. This is either: (a) the voltage at which breakdown occurs; or (b) the highest of a specified series of voltages which the test piece will withstand for a specified time without breaking down.

The *electric strength* or *breakdown strength* is the breakdown voltage divided by the test piece thickness (or the appropriate other dimension in the case of tests parallel to the plane of the sheet (§13.7.6)).

The difficulties associated with the definition and the measurement of the property arise because of the importance of a large number of variables which affect the result, and because of the number of modes in which breakdown can occur, all or some of which may contribute to failure in a given situation. Breakdown is a disruptive phenomenon and as such is very dependent upon the macroscopic, and even the microscopic, structure of the insulation. It may occur through dielectric heating causing an excessive temperature rise (and with many materials, power factor increases with temperature, resulting in further heating and ultimately 'thermal runaway'); the voltage stress may be sufficiently high to accelerate electrons through the material, causing an electron avalanche; gaseous voids inside the material may ionise, causing bombardment of the material and subsequent degradation; the stress may even be high enough to break the chemical bonds in the polymer structure and cause decomposition into less resistant materials; and so on.

Breakdown is a 'weak point' phenomenon and as such it may be instructively compared with, say, a tensile strength failure on a fabric specimen where one thread invariably parts first, thus initiating rupture; similarly breakdown commences where the stress is highest, and one of the major difficulties in electric strength tests lies in the facts that normally the electric stress varies, being greatest near the electrode edges, and that discharges occurring in the ambient medium just outside the electrodes normally initiate breakdown. Irregularities in electrode surfaces or in the test piece due to inhomogeneity affect the results. It is therefore necessary to keep the electrodes clean and free from pitting (resulting from arcing) and to use homogeneous and smooth test pieces.

Under the most carefully controlled conditions, when all spurious discharges are eliminated and all secondary mechanisms of breakdown are avoided, it is possible to achieve for a material a consistent value known as the *intrinsic electric strength*, which is independent of thickness. Rather surprisingly, the range of values obtained for many different homogeneous plastics materials is small and of the order of 0·5–1 MV/mm. Such measurements are of great interest to the theoretician, but in practice industrial measurements are usually made under relatively crude conditions, in order to simulate more closely practical situations, and breakdown can occur by a number of modes, sometimes operating simultaneously.

The tests to be described measure the *industrial electric strength* at power frequencies under discharge conditions (impulse and d.c. tests are specialised and will not be considered). Since breakdown under these conditions is a time-dependent phenomenon, and the rate of voltage application is analogous to rate of loading in the tensile example given earlier, it is important to standardise on the time of voltage application and even the rate of initial voltage increase; so-called 'industrial electric strength tests' are careful to specify such details. The results of such tests are very thickness-dependent in contrast to intrinsic electric strength tests, and the values obtained in general rarely approach even one-tenth of the intrinsic values.

The results are affected by the electrical (and sometimes by the thermal) characteristics of the ambient medium and by the dimensions (and sometimes by the thermal properties) of the electrodes. In addition, the results are dependent upon the frequency and waveform of the applied voltage.

While most materials are tested with electrodes placed so that the electric stress is applied normal to the plane of the sheet, in some cases (principally laminates where electric discharges along the laminae may occur at lower electric stresses than through the thickness) the test is carried out so that the electric stress is applied parallel to the plane of the sheet.

The electric strength is dependent on so many variables that there is no 'correct' value to aim at, and consistency of results between one experiment and another can only be achieved by mutual attention to the test variables. In other words, specifications are essential, and it should not be expected that two apparently fairly similar test methods will give similar results. Comparisons between materials are only strictly valid when experimental conditions are strictly controlled and virtually identical.

A high industrial electric strength does not necessarily mean that the material will withstand long-term degradation processes such as erosion or chemical deterioration by discharges (§13.8) or electrochemical deterioration in the presence of water which may cause eventual failure in service at much lower stresses. On the other hand, if discharges are eliminated in service a higher stress may be withstood.

Thus these tests are only of limited value for design, although they are useful for quality-control purposes.

13.7.2 Schedule of Voltage Application

There are three types of schedule of voltage application commonly used in standards.

In the *rapidly applied voltage test*, the voltage is raised at a uniform rate until breakdown occurs. The rate may be directly specified or be such that breakdown occurs within 10 to 20 s (on average). (In ASTM D149 (1981),[98] 50 per cent of the short-time breakdown voltage may be applied initially and the voltage then raised at a slow uniform rate.) The electric strength is calculated from the voltage at which breakdown occurs.

In the *20 s step-by-step test* a trial specimen is usually tested initially, using the rapidly applied voltage schedule. The starting voltage for the test proper is then taken as 40 per cent of the failure voltage of the trial specimen and the test proceeds in defined voltage steps in, for example, the series (kV) 0·50, 0·55, 0·60, ..., 1·0, 1·1, 1·2, ..., 2·0, 2·2, 2·4, ..., 4·8, and then 10 and 100 times this series up to 100 kV, each step being maintained for 20 s. Here the breakdown strength is generally calculated from the highest of these voltages withstood for the complete 20 s.

In a *proof test* the voltage is raised as rapidly as possible without overshoot to a specified value and then held at this value for a specified period (usually 1 min) and breakdown constitutes a failure.

13.7.3 Ambient Medium

Most tests are carried out in insulating oil at room temperature, but the thermosetting plastics more usually are tested at 90 °C (see BS 771 (1980)[94] for example). Some tests may also be made in air. If this is done at elevated temperatures, the oven required must have an adequate insulating bushing and be sufficiently large to prevent flashover to the oven and significant influence on the electric field in and close to the test piece.

13.7.4 Number of Tests

The number of test results required varies widely with the specification. In some cases only two are required in addition to those for any initial trial tests, while IEC 243 (1967)[99] calls for five with a further five tests being carried out if any result differs from the mean by more than 15 per cent. In the last case the mean of all ten results is calculated. BS 2782 (1970)[100] calls for the mean result of two tests or, for reference purposes, the central value from five tests.

13.7.5 Voltage Supply and Measurement

Safety apart, there are a number of hazards associated with the tester itself, amongst which are errors caused by loading of the transformer due to excessive specimen leakage current or discharge currents, distortion of waveform by poor transformer design, malfunctioning of circuit breakers, and so on. Since it is the peak value of voltage which determines breakdown, whereas many indicating meters are average value instruments (although scaled in r.m.s.), it is essential to check the meter readings against a peak measuring device, such as a sphere gap. BS 358 (1960)[101] is particularly helpful here; 50 mm spheres are appropriate for the range from 8 to 65 kV peak, which is relevant to most of the requirements in plastics specifications. A guide to high-voltage testing techniques is given in BS 923 (1980)[102] and most standards detail requirements of the breakdown tester and voltage measurement methods. These requirements in most standards are very similar.

13.7.6 Standard Test Methods

The methods given in BS 2782 (1983)[100] are substantially in line with IEC 243 (1967).[99] Method 220 requires a rapidly applied voltage and Method 221 a step-by-step application of voltage, but they are otherwise similar in content.

Methods 220A and 221A are for sheet materials. The electrodes consist of solid brass cylinders, the lower one being 75 mm in diameter and 15 mm high and the upper one 25 mm in diameter and 25 mm high. The electrodes have radiused (3 mm) edges to reduce the local stress. The arrangement is shown in Fig. 13.20. The electrodes are immersed in air or insulating oil and the temperature is either 23 °C or 90 °C. The methods for narrow strips and tapes (220B and 221B) use 6 mm coaxial brass rod electrodes with 1 mm radius at the edges, the upper electrode weighing 50 g. An example (see Fig. 13.21) shows five pairs of such electrodes mounted in insulating blocks. Essentially the test is for thin materials tested in air and it is limited to voltages of about 10 kV or less.

Methods 220C and 221C are for rigid tubes. For tubes not exceeding 100 mm in diameter, a closely fitting long inner electrode of rod, tube or foil is used with an outer electrode consisting of a band of foil 25 mm wide. For tubes larger than 100 mm in diameter the inner electrode is a disc of foil 25 mm in diameter and the outer band around the tube is 75 mm wide.

The next three methods in each document cover flexible tubing. Methods 220D and 221D are essentially the same as for rigid tubing of small diameter and the

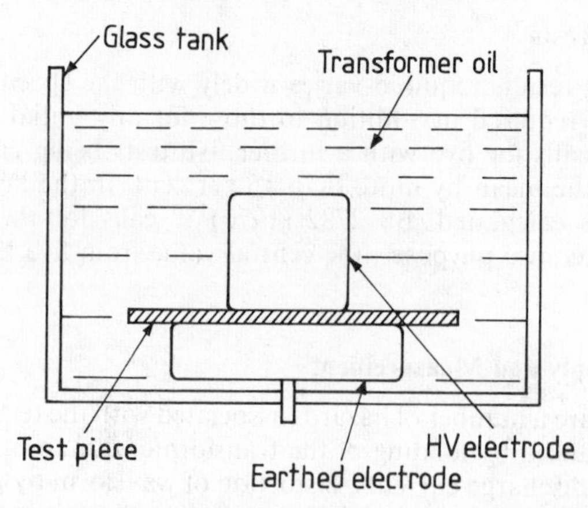

Fig. 13.20 Electric strength electrodes for sheet materials

inner electrode is straight. Methods 220E and 221E use a similar inner electrode but bent into a U-shape, while the outer electrode is of metal shot (Fig. 13.22). A subsidiary test is made to ensure adequate conductivity through the bulk of the shot.

Methods 220F and 221F are for large-bore flexible tubing, where the tubing is cut longitudinally, opened flat and tested as for sheet material. The tests for sleeving given in IEC 684–2 (1984)[40] are similar to some of those given in BS 2782.

The remaining methods of BS 2782 (Methods 220G and 221G) are intended principally for laminated materials whose electric strength may be much lower in the plane of the laminate than normal to it. The test is therefore conducted on test pieces 25 mm wide placed edgewise between metal plates large enough to overlap all round by at least 15 mm (Fig. 13.23). It is important that the test piece has the contacting edges truly machined. To assist the mechanical stability of the arrangement, two or three test pieces may be used to support the electrodes, although of course fresh test pieces are required for replicate tests. For sheet material and for large bore tubing, a test piece about 100 mm long is cut, while for rods and tubes up to 100 mm in diameter (or external side dimension) the complete cross-section is used.

An unusually small electrode system is given in BS 6564: Part 2 (1985).[103] The electrodes are designed so that very small samples can be tested. A test piece 9·5 mm in diameter and 0·75 mm thick is prepared by cutting from a product. The test piece is inserted in the centre of a length of non-rigid electrical grade sleeving (see Fig. 13.24) which must be of such a bore as to grip it tightly; this assembly is mounted in the electrodes (Fig. 13.24) which are then immersed in transformer oil. The test piece is subjected to a proof voltage of 24 kV/mm for 1 min.

Other special kinds of electrodes are used for finished products. For example, BS 2848 (1973),[46] which includes sleeving of PTFE, silicone rubber, PVC and polyethylene, gives an *ad hoc* test in which a test piece is threaded onto a wire or rod of diameter approximately 75 per cent of the bore and wrapped round a metal mandrel. A proof voltage is applied between the mandrel and the wire or rod and

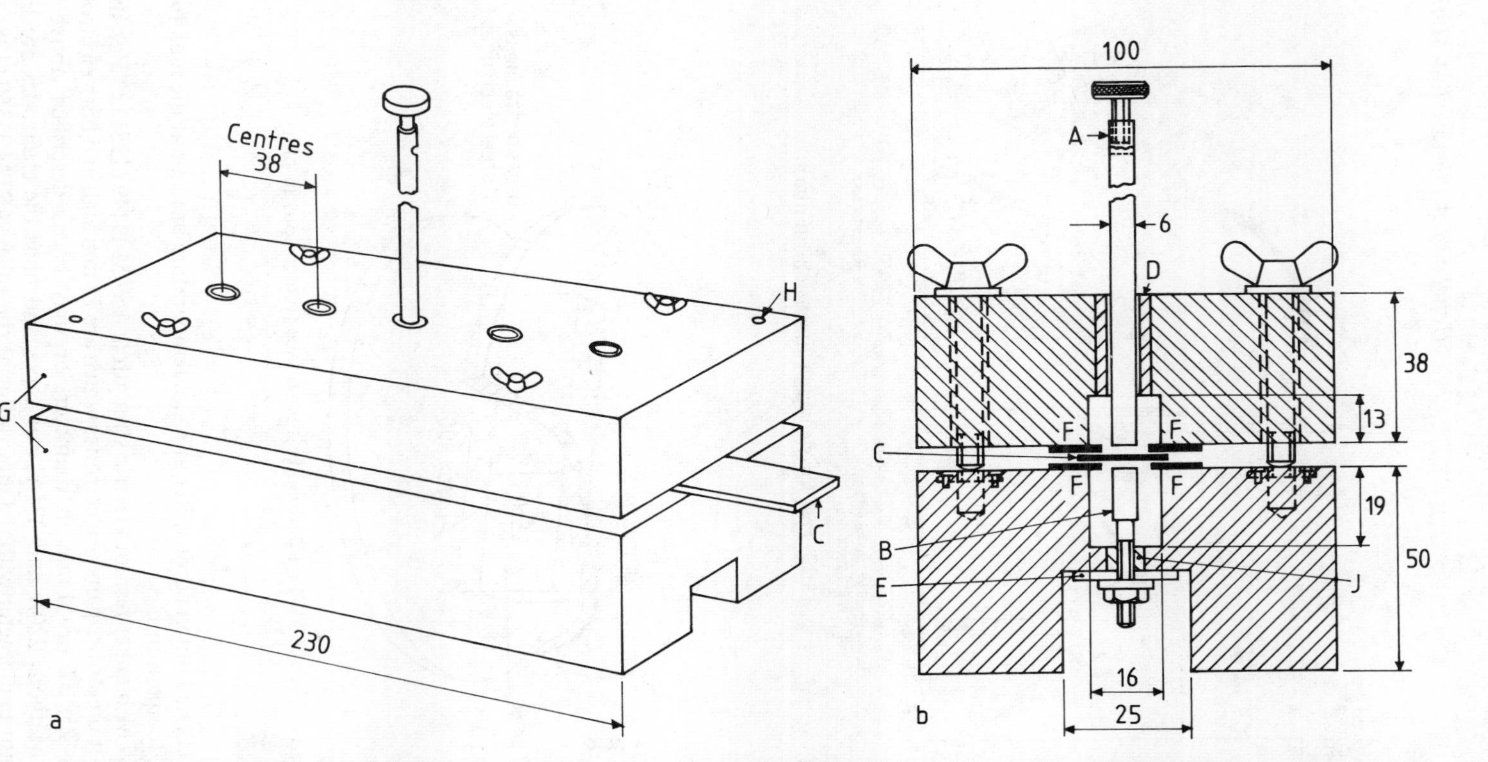

Fig. 13.21 Test for tapes (dimensions in mm): (a) general arrangement of apparatus; (b) section with top slightly raised. A, upper electrode to be an easy fit in bush (D); B, lower electrode; C, test piece; D, brass bush inside diameter just sufficient to clear 6 mm rod; E, brass strip 25 mm wide connecting all lower electrodes; F, strips of suitable insulating material overlapping edges of sample; G, blocks of suitable insulating material, e.g. paper-filled laminate; H, dowel hole; J, brass bushing with internal thread

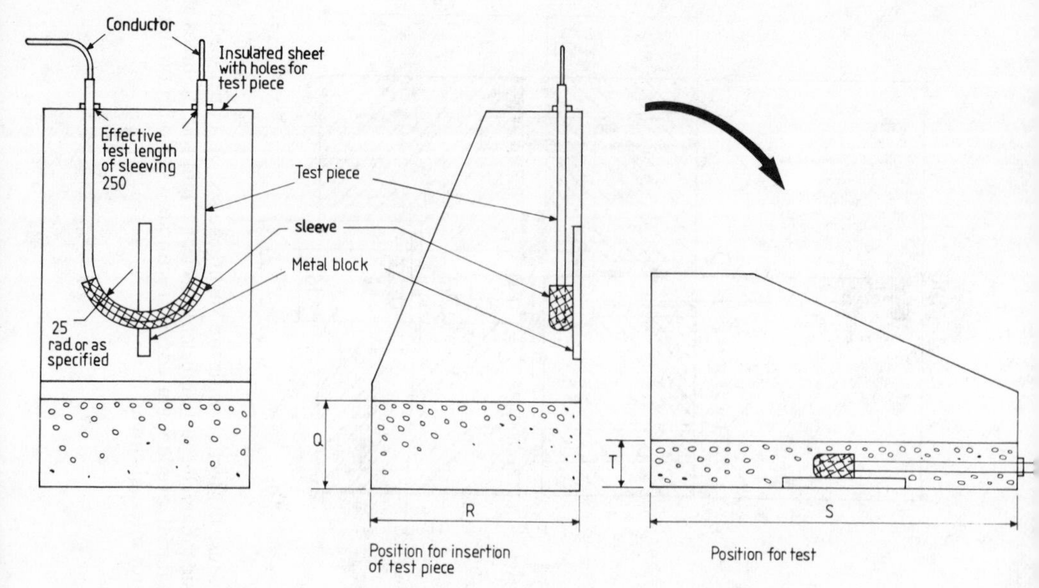

Fig. 13.22 Test for sleeving (dimensions in mm)

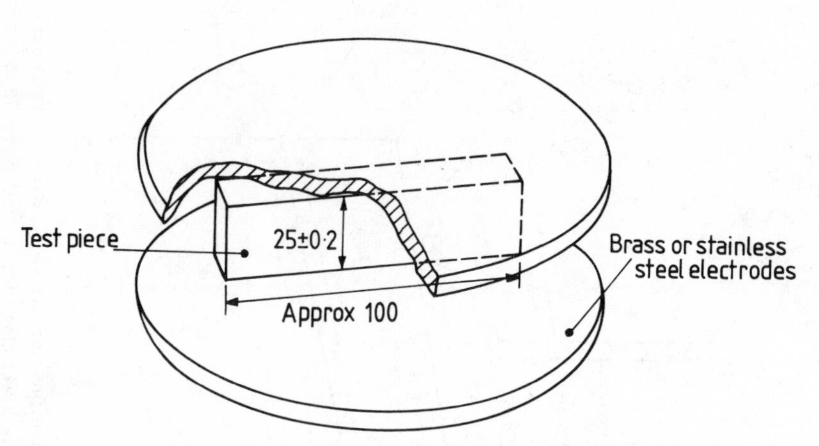

Fig. 13.23 Test parallel to the surface (dimensions in mm)

the test is carried out in air or water at room temperature or in air at the maximum operating temperature.

The ASTM blanket test for electric strength is ASTM D149 (1981).[98] For sheet materials, it prescribes pairs of electrodes which are 25 mm high and both electrodes are either 25, 37 or 51 mm in diameter, or may be as described above for BS 2782: Methods 220A and 221A. For tapes and films the electrodes are long rods 6·4 mm in diameter with 0·8 mm radius at the edges. ASTM D350 (1984)[104]

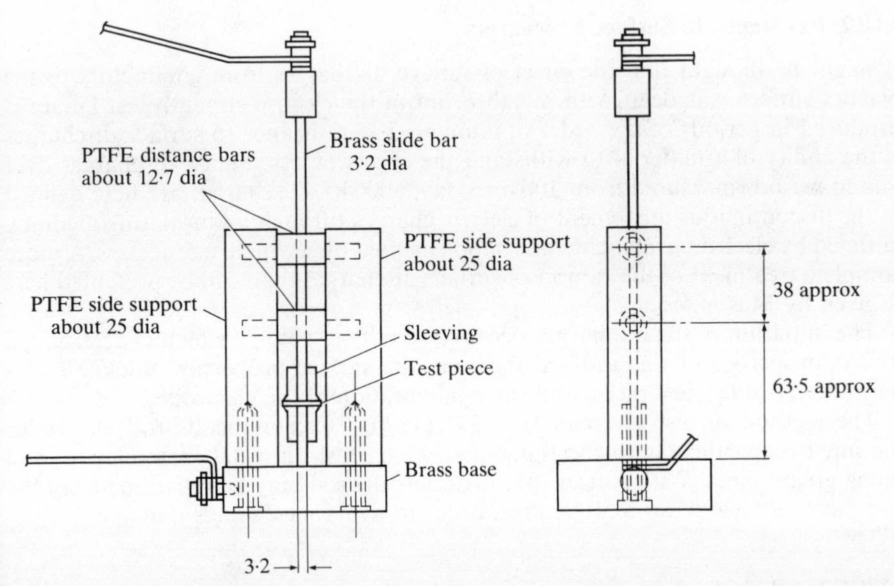

Fig. 13.24 Electrodes for testing small PTFE mouldings: BS 6564 (dimensions in mm)

gives a breakdown voltage test for sleeving on a sample with foil outer and rod or wire inner electrodes.

A test for determining conductive paths in tapes is described in BS 6564: Part 2 (1985).[103] The test piece is laid flat on a polished brass or steel plate and traversed by a roller of polished brass or steel, a voltage of 18 kV/mm being applied between the two metal electrodes. The number of breakdowns is counted.

13.8 RESISTANCE TO SURFACE DISCHARGES, TRACKING RESISTANCE AND ARC RESISTANCE

13.8.1 General Considerations

Tests for resistance to surface discharges, tracking resistance and arc resistance are all concerned with the destruction of the surface of the material by sustained electrical discharges or arcs. The results obtained are very dependent on so many factors that: (a) standard test methods need to be very tightly specified; and (b) the results of these tests, while being an aid to material selection, are not directly applicable to calculating design stresses or creepage distances.

In the surface discharge test there is no deliberate contamination of the surface; in tracking tests, the surface is wet. In both these tests, the discharges do not initially develop into a persistent arc, whereas in the arc resistance test a persistent arc is created initially.

Evans[105] pointed out that tracking tests are notoriously variable and claimed that 'standard deviations' (presumably coefficients of variation) as high as 40 per cent are not uncommon.

13.8.2 Resistance to Surface Discharges

It might be thought that the effect of surface discharges from conductors to the plastics surface was dealt with in §13.7, but in the electric strength test failure is produced in periods of the order of minutes. The resistance to surface discharges is the ability of a material to withstand the effects of less intense discharges over a long period (measured from 100 h to, say, 5000 h). Discharges are here defined as the discontinuous movement of electric charges through an insulating medium, initiated by electron avalanches and supplemented by secondary processes. A more complete treatment of the subject of surface discharges than can be presented here is given by Mason.[106]

The initiation of the discharges occurs at a critical voltage which is determined by the properties of the ambient medium, the surface properties, thickness and permittivity of the test piece, and the configuration of the electrodes.

The method of test given in IEC 343 (1970)[107] recommends rod electrodes mounted perpendicular to the test pieces which rest on a flat plate electrode of much greater area. With soft or thin materials the rod may be separated from the test piece surface by up to 100 μm in order to avoid mechanical damage.

13.8.3 Tracking Resistance

Tracking is defined as the progressive formation of a conductive (carbonised) path across the surface of an insulator by surface discharges.

Since the presence of carbon in the insulator is a prerequisite for this type of failure, it might be expected that plastics in general would be prone to tracking, but this does not follow and some of the polymers which are richest in carbon, such as polystyrene, are non-tracking, although other adverse effects such as erosion occur in this particular case. Tracking is influenced considerably by the presence of surface films of moisture and dirt, etc., and the test methods involve the introduction of conducting media such as salt solutions on the surface in the form of droplets or by dipping, in order to produce an acceleration of the natural processes. Although tracking may occur in a variety of practical situations – two common examples are the failure of domestic light switches and car distributor caps – the problem is more usually associated with areas of high atmospheric pollution, and those near the coast and on board ships where salt spray is a natural hazard.

Numerous tracking resistance tests have been developed in attempts to simulate the results of practical failures in the field and two types of test have received unusually general support. The first is the 'comparative tracking index' (CTI) test, which gives a numerical result only on materials which are not highly tracking-resistant, and gives only a 'does not track' result on a number of plastics. It appears that this shortcoming has resulted in the second test, the so-called 'inclined plane test', which leaves fewer materials undifferentiated.

Comparative Tracking Index (CTI)

The most widely used test is described in IEC 112 (1979),[108] which is dual numbered as BS 5901 (1980).[109] The test involves two chisel-shaped platinum electrodes 4 mm apart pressed on the surface by a force of 1 N (Fig. 13.25). A 48–60 Hz

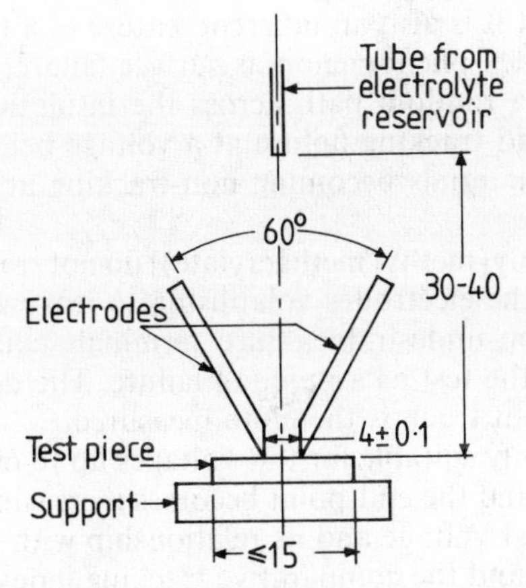

Fig. 13.25 Comparative tracking index test: IEC 112 (dimensions in mm)

voltage with a short-circuit current limit of 1 A is applied between the electrodes and drops of aqueous solution of size 20 mm^3 are allowed to fall between the electrodes at intervals of 30 s. Two solutions are permitted. The preferred solution, A, is less aggressive than solution B and consists of 0·1 per cent NH_4Cl.

The test is performed at a number of different voltage levels and the number of drops to track at each level is noted (Fig. 13.26), tracking being defined arbitrarily by the current in the circuit reaching at least 0·5 A for 2 s.

The CTI is the voltage which the material will just withstand for 50 drops and is quoted, for example, as CTI 350. However, if the material tracks in less than 100 drops at 25 V below the 50 drop value, the voltage corresponding to failure in 100 drops is determined and given in parentheses after the 50 drop voltage. Additionally, the letter M is appended if solution B is used. A proof test with preferred voltages is also given.

Some care should be taken in applying the test. The error of the number of drops to track at a given voltage is high. This partly reflects the inhomogeneity

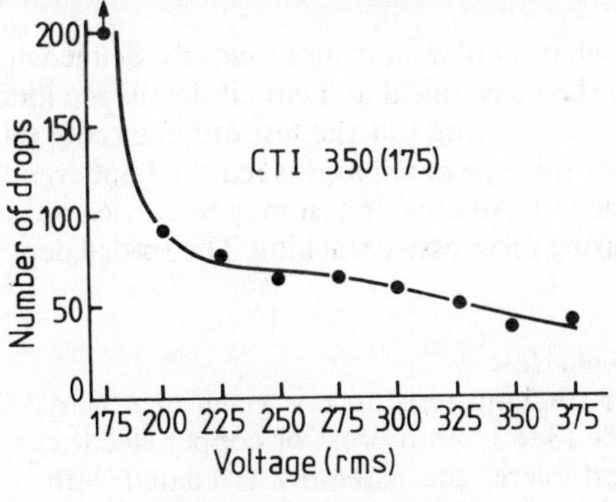

Fig. 13.26 CTI curve

of most filled plastics but it is also an inherent feature of a test which endeavours to measure such a complex phenomenon as surface failure, dependent as it must be on the formation of a random path across the insulation surface. It is often possible to obtain the odd tracking failure at a voltage below the CTI and there are cases on record of materials becoming non-tracking at voltages significantly higher than the CTI.

Some materials (e.g. poly(methyl methacrylate)) do not track, but fail by erosion, or by material between the electrodes volatilising to leave a crater. As in service such behaviour would be undesirable, since terminals could become loosened, erosion is recognised in the test as a mode of failure. The depth of erosion (mm) of test pieces which do not track is therefore measured.

The test described is only suitable for test voltages up to 600 V. Above this level flashovers tend to occur and the end-point becomes increasingly difficult to define. The significance of the test voltage and its relationship with practical use voltages has not been established and the comparative tracking index obtained should not be regarded as a design level below which it is safe to operate. The test given in ASTM 3638 (1985)[110] is based on this method.

Inclined Plane Test

This test is described in IEC 587 (1984)[111] and is dual numbered as BS 5604 (1986).[112] A large flat specimen is inclined at 45° and electrolyte is allowed to flow down its under-surface, forming a bridge between two electrodes pressed against it. The electrolyte stream 'boils' under the influence of the test voltage. In Method A the applied voltage is 2·5, 3·5 or 4·5 kV and is applied for 6 h or until failure occurs. In Method B a voltage of 250 V is applied for 1 h, then 500 V for the next hour, 750 V for the next hour, etc., until failure occurs. The stepwise tracking voltage is the highest voltage withstood for 1 h without failure.

There are two criteria of failure. With criterion A the test is terminated when a current of 60 mA flows; with criterion B (not applicable to Method B) when a track reaches a mark 25 mm from the lower electrode. The advantage of this test is that conditions can be carefully controlled so that steady scintillation occurs, unlike the intermittent and somewhat uncontrolled scintillation which occurs in 'drop' tests.

When no tracking occurs the depth of erosion at the end of the test may be measured.

There is a somewhat similar and more closely defined test in ASTM D2303 (1985).[113] Many of the mechanical and circuit details are identical to the IEC test but the procedures for carrying out the test are markedly different.

Figure 13.27 shows the type of apparatus required but even this does not indicate the complexity of the test. An erosion test may be carried out using a lower voltage (2 kV) than that causing progressive tracking. The eroded depth and eroded volume are recorded.

Dust and Fog Tracking Test

The US method for tracking resistance is given in ASTM D2132 (1985).[114] The principle is that 51 × 13 × 3·2 mm brass or copper electrodes are placed flat upon the surface of a test piece, the assembly is coated with a synthetic dust and subsequently wetted by fine water spray in an appropriate chamber. A relatively

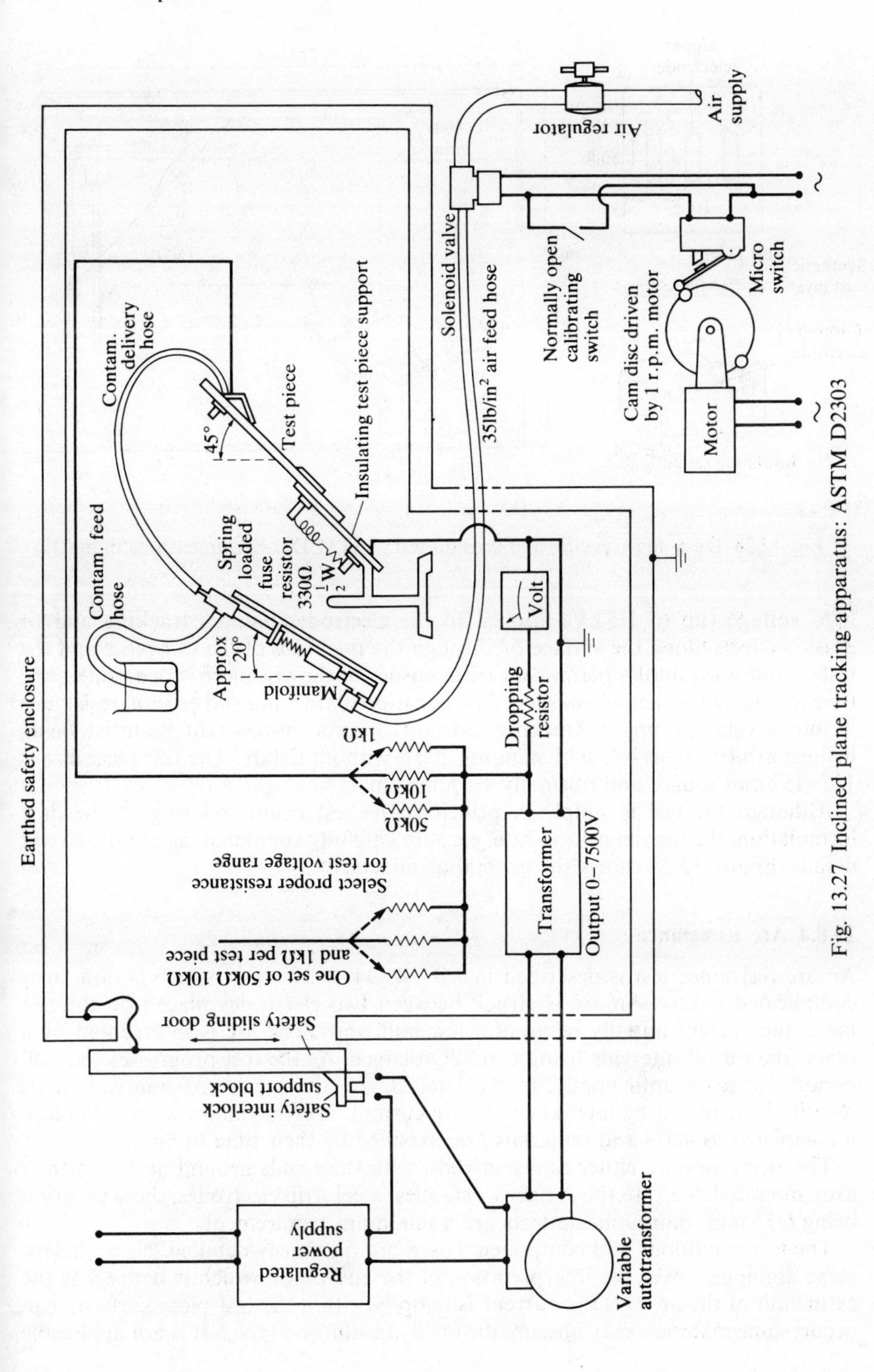

Fig. 13.27 Inclined plane tracking apparatus: ASTM D2303

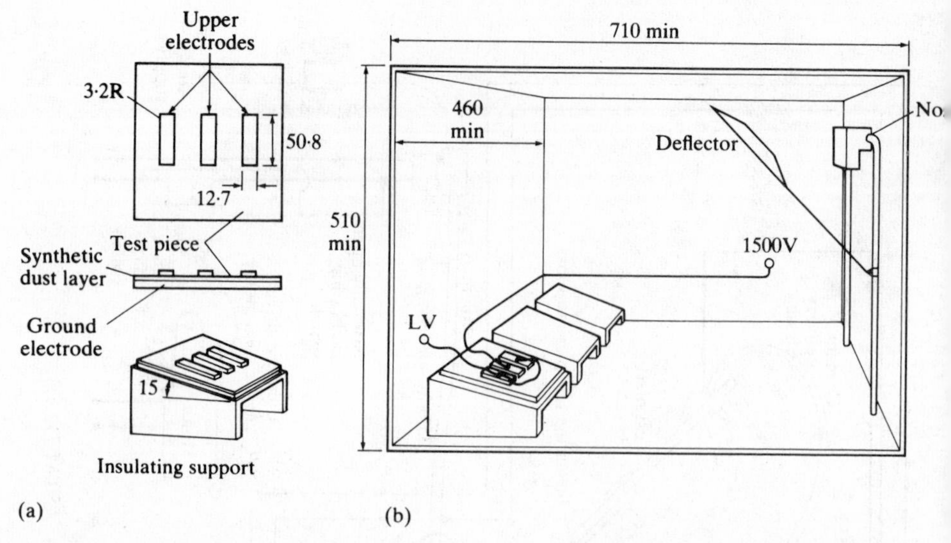

Fig. 13.28 Dust–fog tracking and erosion test: ASTM D2132 (dimensions in mm)

high voltage (up to 1·5 kV) applied to the electrodes induces tracking and/or erosion effects along the surface or through the thickness of the test piece and the test is continued until a permanent track ensues or the test piece is penetrated, the time to such failure being recorded. Classification of tracking and erosion resistance is into several groups, a tracking-resistant or erosion-resistant material being defined arbitrarily as one withstanding 200 h without failure. The test piece size is 127–152 mm square and normally 1·59 mm thick.

Although the test is simple in principle, the test details relating to the dust formulation, the fog deposition rate, etc., are carefully stipulated, as are the circuit details. Figure 13.28 shows the general arrangement.

13.8.4 Arc Resistance

An arc resistance test is described in ASTM D495 (1984)[115] which is of a fairly complicated nature. An arc is struck between two electrodes placed on the test piece, the current initially being of a few milliamps. The arc is interrupted by a timer, the on/off intervals being carefully defined. As the test progresses the 'off' period decreases until finally, in the later stages, the arc is continuous and its severity is increased by increasing the arc current. The maximum time of the test, if completed, is 420 s and materials are classified by their time to failure.

The electrodes are either tungsten rods, with their ends ground at 30° to their axes, mounted at 35° to the surface or stainless steel strip electrodes, the separation being 6·35 mm: quintuplicate tests are a minimum requirement.

The test conditions and components used are rigorously detailed. Nevertheless, some ambiguity over the interpretation of the end-point, which is defined as the extinction of the arc (i.e. the current is entirely within the test piece surface), can occur: some materials may obscure the arc by burning and the test is not applicable to materials that do not produce conductive paths under the action of the arc or which melt and float the conductive residues out of the test area.

Most arc resistance tests are carried out using moderate alternating currents. However, Watanabe *et al.*[116] have developed an apparatus for testing GRP using a unidirectional current up to 1200 A or more (derived from the discharge of a large capacitor) applied from twenty to forty times a minute.

13.9 ELECTROMAGNETIC INTERFERENCE SHIELDING EFFECTIVENESS

Over the past decade, there has been a very large increase in the usage of plastics for cases of instruments (both industrial and domestic) which either emit electromagnetic interference (EMI) or are affected by EMI; additionally earlier cases were generally metallic, providing shielding (screening) against such interference. This has resulted in the demand for plastic EMI shielding materials, which has brought a requirement for the development of techniques for measuring the *shielding efficiency* (SE) of materials, although the SE of a case depends not only on the materials used, but also on the joints and ports in the case. The testing of a complete case (normally carried out on a large open field site by a free space method) will not be discussed here. Generally materials are tested in sheet form by measuring the electric field strength or the power transmission between a transmitter and receiver before and after installation of the sample.

$$\text{SE (in decibels)} = 20\log(A_i/A_t) = 10\log(P_i/P_t) \qquad [13.21]$$

where A is a field strength and P a power, and the subscripts t and i refer to the received signal with and without the sample, or more strictly to the transmitter and incident radiation, respectively. Many references quote the expressions in the brackets inverted, but then proceed to ignore the minus sign!

Electromagnetic waves close to the transmitter (the *near-field region*) behave differently from those at greater distances (the *far-field* region). In practical terms (for small aerials) the near-field region extends to about $\lambda/2\pi$ from the transmitter, where λ is the wavelength of the radiation. In the far-field region the electrical (E) and magnetic (H) components are related, but in the near-field it is necessary to know both the electrical and magnetic components in order to have a complete description of the electromagnetic field. Thus measurements in the near-field region are more difficult to interpret.

Two basic types of equipment[117–119] are used commonly (if this is the right term to use where equipment of any type exists in only a relatively few laboratories). The first is the *dual chamber method*; the sample in this case is a flat sheet in a window between two screening chambers, which are either adjoining or one within the other (a box within a screened room). A small transmitting aerial is set up in one chamber and a small receiving aerial in the other. The ratio of the transmitted signal strengths with and without the sample is determined. For normal laboratory-sized equipment, near-field conditions apply when the frequency is less than about 100 MHz and in this region the results sometimes fluctuate considerably with frequency, while at higher frequencies the results change rather less with frequency.

The second method is the *transmission line method*; in this the signal from the transmitter is passed along a torpedo-shaped expanded section of a coaxial transmission line to a receiver. The transmitted signal is measured with and without a washer-shaped test piece in the 'torpedo'. This equipment gives effectively far-field (plane wave) conditions at all frequencies. The results do not show the violent fluctuations observed with the dual chamber method at low frequencies and tend to give results more consistent with that method at frequencies above about 100 MHz.[118–121]

The results obtained with the dual-chamber method in various laboratories may vary by 80 dB in the near-field region, while the results are much more consistent in the far-field region.

Other methods exist and some interesting techniques were described by Adams and Ondrejka.[122] Even though their minimum frequency was 100 MHz, there were differences of some 25 dB between the results of the various methods.

The improvement in understanding and designing equipment for testing EMI shielding has been so rapid that it appears possible that some of the fluctuations in results obtained 6 or 7 years ago may not be found with more modern equipment.

Bigg *et al.*[121] found good agreement between the results of a dual-chamber test and a free space test on a carbon black loaded moulded box tested at 1 GHz. They also found good agreement at 1 GHz between tests on a moulded plaque and on a box moulded at the same thickness. However, when they tested a composite containing conductive fibres the agreement between the results on the box and the sheet was poorer.

There is a theoretical relationship between SE and resistivity, although this does not always seem to hold, possibly because of errors in both measurements or because there is variability or anisotropy of resistivity in the test sample. For adequate EMI shielding, the resistivity needs to be below about 1 Ω cm, according to Regan.[123] Simon,[119] using a transmission line method and taking great care to minimise contact resistance effects, found in general that SE values were within 5 dB of those predicted by volume resistivity measurements. Bigg *et al.*[121] also found agreement between the calculated and measured SE at 1 GHz for far-field measurements.

13.9.1 Standard Test Method

ASTM Emergency Standard 7 (1983)[117] prescribes a dual-chamber method for the near-field measurements and a transmission line method for far-field measurements. Calibration specimens, consisting of insulating specimens coated with a thin gold film, are used in both methods.

13.10 OTHER ELECTRICAL TESTS

The properties outlined above are considered to be the prime ones: in all applications of plastics involving electrical insulation, some knowledge of one or more of these properties is required to ensure the proper and efficient use of the appropriate material.

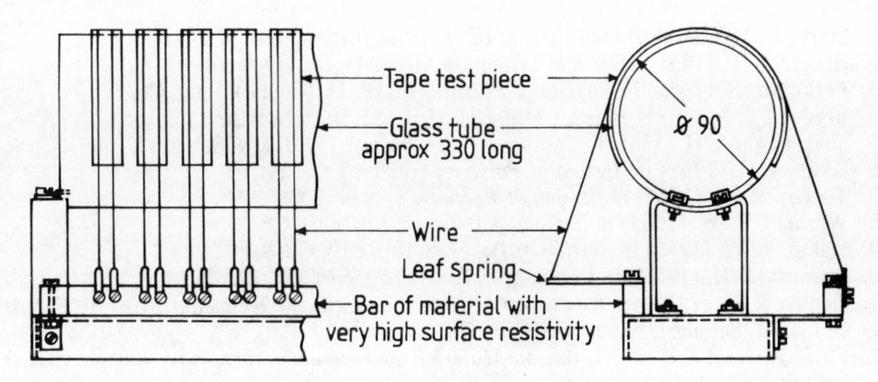

Fig. 13.29 Apparatus for determining electrolytic corrosion: BS 3924 (dimensions in mm)

A more specialised test which does not fit into the above classification is the *copper corrosion test* for adhesive tapes. It is an important requirement of adhesive electrical tapes that the adhesive should not corrode copper, as this is a particular hazard in high-humidity, high-temperature atmospheres. In BS 3924 (1978)[44] a test for electrolytic corrosion involves a length of tape wound round the outside of a large glass tube (Fig. 13.29), with two bare copper wires wound under tension over the top of it and anchored. The wires are connected to a 120 V d.c. supply and the assembly is subjected to a high humidity for 15 days, after which the breaking loads of the positive wire and a blank sample wire are tested. A *corrosion liability factor* (CLF) is then calculated as the percentage (loss in strength of the tested wire divided by the strength of the untested wire). Water-soluble impurities in either the adhesive or the backing tend to be anionic and are therefore attracted to the positive wire, which they attack and weaken.

In practice, ten test pieces are employed, half with wires in contact with the adhesive side, and half on the non-adhesive side of the tape. All positive wires are tested and the standard deviation is reported.

BS 3924 also specifies tests for the determination of *conductivity of water extracts* of, for example, insulating tapes.

REFERENCES

1. Baird, M. E. (1973) *Electrical Properties of Polymeric Materials*, Plastics Institute.
2. Blythe, A. R. (1979) *Electrical Properties of Polymers*, Cambridge University Press.
3. Daniel, V. V. (1967) *Dielectric Relaxation*, Academic Press.
4. Moullin, E. B. (1944) *Journal of the Institution of Electrical Engineers*, Part 1, **91** (48), 448.
5. Swiss, J. and Dakin, T. W. (1954) *Westinghouse Engineer*, **14** (3), 144.
6. Hoffman, J. D. (1957) *Transactions of the British Institution of Radio Engineers*, CP4 (June), 42.
7. Vail, C. B. (1962) *Electro-Technology*, **69** (February), 82.
8. Sharbaugh, A. H. (1962) *Electro-Technology*, **69** (February), 146.
9. Devins, J. C. and Sharbaugh, A. H. (1961) *Electro-Technology*, **68** (February), 104.
10. Mason, J. H. (1962) *Transactions of the Plastics Institute*, **30** (87), 171.

11. Stark, K. H. (1962) *Transactions of the Plastics Institute*, **30** (87), 184.
12. Baker, W. P. (1965) *Electrical Insulation Measurements*, Newnes.
13. Parkman, N. (1965) in *Physics of Plastics* (Ed. P. D. Ritchie), Iliffe, Ch. 6.
14. Brydson, J. A. (1975) *Plastics Materials*, 3rd edn, Butterworth.
15. Sharbaugh, A. H. (1965) in *Testing of Polymers*, vol. 1 (Ed. J. V. Schmitz), Interscience.
16. Scott, A. H. (1965) in *Testing of Polymers* (op. cit.), Ch. 6.
17. Tucker, R. W. (1965) in *Testing of Polymers* (op. cit.), Ch. 7.
18. Warfield, R. W. (1965) in *Testing of Polymers* (op. cit.), Ch. 8.
19. Dakin, T. W. (1965) in *Testing of Polymers* (op. cit.), Ch. 9.
20. Brunton, J. H. (1965) in *Testing of Polymers* (op. cit.), Ch. 10.
21. Kepler, R. G. (1977) in *Treatise on Materials Science and Technology*, vol. 10, Part B (Ed. J. M. Schultz), Academic Press, Ch. 4.
22. Fava, R. A. (1977) in *Treatise on Materials Science and Technology*, vol. 10, Part B (op. cit.), Ch. 4.
23. Geppe, A. P. (1958) *Elektrichestvo*, **3**, 60.
24. Dorcas, D. S. and Scott, R. N. (1964) *Review of Scientific Instruments*, **35** (9), 1175.
25. Sazhin, B. I. and Skurikhina, V. S. (1960) *High Molecular Compounds*, **10**, 1535.
26. Munick, R. J. (1956) *Journal of Applied Physics*, **27** (10), 1114.
27. France, G. (1957) *Electronic Engineering*, **29** (6), 30.
28. Hitchcox, G. (1956) *Journal of the British Institution of Radio Engineers*, **16** (June), 299.
29. ASTM D257 (1978) *Tests for DC Resistance or Conductance of Insulating Materials.*
30. Reddish, W. (1959) *The Physical Properties of Polymers, SCI Monograph*, no. 5, Society of Chemical Industry, p. 138.
31. IEC 93 (1980) *Methods of Test for Volume Resistivity and Surface Resistivity of Solid Electrical Insulating Materials.*
32. BS 6233 (1982) *Methods of Test for Volume Resistivity and Surface Resistivity of Solid Electrical Insulating Materials.*
33. BS 2782 (1982) *Methods of Testing Plastics.* Method 230A, *Determination of Volume Resistivity*; Method 231A, *Determination of Surface Resistivity.*
34. BS 3815 (1964) *Epoxide Resin Casting Systems for Electrical Applications.*
35. BS 5823 (1979) *Method of Test for Electrical Resistance of Insulating Materials at Elevated Temperatures.*
36. IEC 345 (1971) *Methods of Test for Electrical Resistance and Resistivity of Insulating Materials at Elevated Temperatures.*
37. BS 4618 (1975) *Recommendations for the Presentation of Plastics Design Data*; Part 2, *Electrical Properties*; Section 2.3, *Volume Resistivity*; Section 2.4, *Surface Resistivity.*
38. ISO 1325 (1973) *Determination of Electrical Properties of Thin Sheet and Film.*
39. ASTM D2305 (1982) *Testing Polymeric Films used for Electrical Insulation.*
40. IEC 684–2 (1984) *Specification for Flexible Insulating Sleeving*; Part 2, *Methods of Test.*
41. BS 2782 (1983) *Methods of Testing Plastics*; Method 232, *Determination of Insulation Resistance.*
42. IEC 167 (1964) *Determination of Insulation Resistance of Solid Insulating Materials.*
43. ISO 2951 (1974) *Vulcanised Rubber. Determination of Insulating Resistance.*
44. BS 3924 (1978) *Specification for Pressure Sensitive Adhesive Tapes for Electrical Insulating Purposes.*
45. ASTM D1000 (1982) *Testing Pressure-sensitive Adhesive Coated Tapes used for Electrical Insulation.*
46. BS 2848 (1973) *Flexible Insulating Sleeving for Electrical Purposes.*
47. Norman, R. H. (1970) *Conductive Rubbers and Plastics*, Elsevier.
48. Morris, W. T. (1970) *Plastics and Polymers*, **38** (February), 41.
49. Davies, D. K. (1967) in *Static Electrification*, Institute of Physics/Physical Society, p. 29.
50. Morris, W. T. and Norman, R. H. (1972) *RAPRA Bulletin*, **26**, 254.

51. Beesley, J. and Norman, R. H. (1977) *RAPRA Members Report*, no. 5.
52. Klouda, J. C. (1978) *Journal of Cellular Plastics*, **14** (January/February), 33.
53. Cross, A. S. (1953) *British Journal of Applied Physics*, Supplement no. 2, 47.
54. Shashoua, V. E. (1958) *Journal of Polymer Science*, **33** (126), 65.
55. Langdon, S. J. (1964) *Plastics, London*, **29** (322), 43.
56. Bertein, H. (1967) in *Static Electrification*, Institute of Physics/Physical Society, p. 11.
57. Moreno, R. A. and Gross, B. (1967) *Journal of Applied Physics*, **47**, 3397.
58. BS 2782 Method 250 A–C (1976) *Antistatic Behaviour of Film.*
59. ISO 2878 (1978) *Rubber, Vulcanised – Antistatic and Conductive Products – Determination of Electrical Resistance.*
60. BS 2050 (1978) *Electrical Resistance of Conducting and Antistatic Products made from Flexible Polymeric Materials.*
61. ISO 284 (1982) *Conveyor Belts – Electrical Conductivity – Specification and Methods of Test.*
62. BS 3289 (1982) *Conveyor Belting for Underground use in Coal Mines.*
63. BS 2044 (1984) *Methods for Determination of Resistivity of Conductive and Antistatic Plastics and Rubbers (Laboratory Methods).*
64. ISO 3915 (1981) *Plastics – Measurement of Resistivity of Conductive Plastics.*
65. Blythe, A. R. (1984) *Polymer Testing*, **4** (2–4), 195.
66. Uhlir, A. (1955) *Bell Systems Technical Journal*, 105.
67. van der Pauw, L. J. (1961) *Phillips Research Reports*, **16**, 187.
68. Clark, F. M. (1962) *Insulating Materials for Design and Engineering Practice*, Wiley.
69. Norman, R. H. (1953) *Proceedings of the Institution of Electrical Engineers*, **100**, IIA, 41.
70. IEC 250 (1969) *Recommended Methods for the Determination of the Permittivity and Dielectric Dissipation Factor of Electrical Insulating Materials at Power, Audio and Radio Frequencies including Metre Wavelengths.*
71. ASTM D150 (1981) *Tests for A–C Loss Characteristics and Permittivity (Dielectric Constant) of Solid Electrical Insulating Materials.*
72. Electrical Research Association (1958) *Technical Report*, L/S9.
73. Hartshorn, L. (1941) *Radio Frequency Measurements by Bridge and Resonance Methods*, Chapman and Hall.
74. Lynch, A. C. (1965) *Proceedings of the Institution of Electrical Engineers*, **112** (2), 426.
75. BS 4542 (1970) *Method for the Determination of Loss Tangent and Permittivity of Electrical Insulating Materials in Sheet Form (Lynch Method).*
76. ASTM D1531 (1981) *Test for Dielectric Constant and Dissipation Factor of Polyethylene by Liquid Displacement Procedure.*
77. Vince, P. M. (1964) in *IEE Conference on Dielectric Insulating Materials*, London.
78. Scheiber, D. J. (1961) *Journal of Research of the National Bureau of Standards*, **65c**, 23.
79. Barney, W. M. (1961) in *Annual Conference on Electrical Insulation*, NAS–NRC 973, p. 59.
80. Stoll, B. (1985) *Colloid and Polymer Science*, **263** (11), 873.
81. BS 2782 (1982) *Methods of Testing Plastics*; Methods 240A and B, *Determination of Loss Tangent and Permittivity at Power and Audio Frequencies.*
82. Hague, B. (1943) *Alternating Current Bridge Methods*, 5th edn, Pitman.
83. Weston, D. (1962) *Plastics, London*, **27** (302), 105.
84. Hazen, T. (1967) in *Conference on Plastics Telecommunications Cables, Plastics Institute, London, April.*
85. Hartshorn, L. and Ward, W. H. (1936) *Journal of the Institution of Electrical Engineers*, **79**, 597.
86. Barrie, I. T. (1965) *Proceedings of the Institution of Electrical Engineers*, **112** (2), 408.
87. Roberts, S. and von Hippel, A. (1946) *Journal of Applied Physics*, **17**, 610.
88. Westphal, W. B. (1954) *Dielectric Materials and Applications*, Section II A2, MIT

Technology Press/Wiley.
89. Gevers, M. (1955) in *Precision Electrical Measurements, Proceedings of the National Physical Laboratory Symposium*, HMSO.
90. Dube, D. C. and Natarajan, R. (1974) *Journal of Physics, E*, **7**, 256.
91. Parry, J. V. L. (1951) *Proceedings of the Institution of Electrical Engineers*, **98**, III, 303.
92. Horner, F., Taylor, T. A., Dunsmuir, R., Lamb, J. and Jackson, W. (1946) *Journal of the Institution of Electrical Engineers*, **93**, III (21), 53.
93. Brydon, G. M. and Hepplestone, D. J. (1964) in *IEE Conference on Dielectric Insulating Materials, London.*
94. BS 771 (1980) *Phenolic Moulding Materials.*
95. BS 4618 (1970) *Recommendations for the Presentation of Plastics Design Data*; Part 2, *Electrical Properties*; Section 2.1, *Permittivity*; Section 2.2, *Loss Tangent.*
96. ASTM D1673 (1979) *Tests for Dielectric Constant and Dissipation Factor of Expanded Cellular Plastics used for Electrical Insulation.*
97. ASTM D1674 (1967) *Testing Polymerisable Embedding Compounds used for Electrical Insulation.*
98. ASTM D149 (1981) *Tests for Dielectric Breakdown Voltage and Dielectric Strength of Electrical Insulating Materials at Commercial Power Frequencies.*
99. IEC 243 (1967) *Electric Strength of Solid Insulating Materials at Power Frequencies.*
100. BS 2782 (1983) *Methods of Testing Plastics*; Method 220, *Determination of Electric Strength – Rapidly Applied Voltage Method*; Method 221, *Determination of Electric Strength – Step by Step Method.*
101. BS 358 (1960) *Method for the Measurement of Voltage with Sphere-gaps (One Sphere Earthed).*
102. BS 923 (1980) *Guide on High-voltage Testing Techniques.*
103. BS 6564 (1985) *Specification for Polytetrafluoroethylene (PTFE) Materials and Products*; Part 2, *Specification for Fabricated Unfilled Polytetrafluoroethylene Products.*
104. ASTM D350 (1984) *Testing Flexible Treated Sleeving Used for Electrical Insulation.*
105. Evans, R. E. (1982) *Plastics Polymer Science and Technology* (Ed. M. D. Baijal), Wiley.
106. Mason, J. H. (1971) *Insulation Engineer*, **1**, 5.
107. IEC 343 (1970) *Recommended Test Methods for Determining the Relative Resistance of Insulating Materials to Breakdown by Surface Discharges.*
108. IEC 112 (1979) *Method of Test for Determining the Comparative and Proof Tracking Indices of Solid Insulating Materials under Moist Conditions.*
109. BS 5901 (1980) *Method of Test for Determining the Comparative and Proof Tracking Indices of Solid Insulating Materials under Moist Conditions.*
110. ASTM D3638 (1985) *Test Method for Comparative Tracking Index of Electrical Insulating Materials.*
111. IEC 587 (1984) *Test Method for Evaluating Resistance to Tracking and Erosion of Electrical Insulating Materials used under Severe Ambient Conditions.*
112. BS 5604 (1986) *Method of Test for Evaluating Resistance to Tracking and Erosion of Electrical Insulating Materials used under Severe Ambient Conditions.*
113. ASTM D2303 (1985) *Test for Liquid-contaminant, Inclined-Plane Tracking and Erosion of Insulating Materials.*
114. ASTM D2132 (1985) *Standard Test Methods for Dust and Fog Tracking and Erosion Resistance of Electrical Insulating Materials.*
115. ASTM D495 (1984) *Test Method for High-voltage, Low-current, Dry Arc Resistance of Solid Electrical Insulation.*
116. Watanabe, T., Sato, M. and Kuwajima, H. (1982) *Composites*, **13** (1), 24.
117. ASTM ES7 (1983) *Emergency Standard Test Method of Electromagnetic Shielding Effectiveness of Planar Materials.*
118. Bigg, D. M. and Stutz, D. E. (1983) *Polymer Composites*, **4** (1), 40.

119. Simon, R. M. (1983) *Modern Plastics International*, **13** (9), 124.
120. Woodham, G. and Gerteisen, S. (1980) *Thirty-fifth Annual Technical Conference, SPI Reinforced Plastics/Composites Institute.*
121. Bigg, D. M., Mirick, W. and Stutz, D. E. (1985) *Polymer Testing*, **5** (3), 169.
122. Adams, J. W. and Ondrejka, A. R. (1984) *EMI/RFI Shielding Plastics; RETEC SPE (Chicago Section) and SPE (Electrical and Electronic Division).*
123. Regan, J. (1982) *Polymer–Plastics Technology and Engineering*, **18** (1), 47.

Chapter Fourteen

Optical Properties

14.1 INTRODUCTION

For most plastics products it is aesthetic properties such as surface texture, colour, haze, gloss and reflectance which are important. Light transmission and refractive index measurements are needed for only a few products such as lenses, windows and roof lights.

This is unfortunate, for light transmission and refractive index are a good deal easier to define and measure accurately than the other properties mentioned. Surface texture, colour, haze, gloss and reflectance are difficult to define, measure or to correlate with visual experience. Detailed information about these properties should be sought from specialised textbooks[1] and published literature. In this chapter attention will be confined to those instruments and specifications which have relevance to the plastics industry. However, the measurement of colour is common to many different industries and because the methods employed are very similar, the references will be more wide ranging.

Additionally, fibres represent a special case. Here the optical properties of refractive index and birefringence are important not so much for their influence on the appearance of the product but as an aid to fibre identification. Although similar identification techniques are applicable to transparent plastics in general, such tests are not widely used outside the forensic science laboratory. Infrared and ultraviolet absorption are more often used for polymer identification, but these techniques and others such as X-ray scattering are outside the scope of this chapter. Similarly, methods of studying changes in the refractive index of solutions as an aid to following chemical reactions in solution, or methods of molecular weight determination, will not be considered.

Changes of colour on exposure to light or heat are dealt with in Ch. 17 and referred to only briefly here.

BS 4618 Section 5.3 (1971),[2] *Optical Properties*, is a useful guide outlining the properties of refraction, transparency, haze, gloss and light transfer, together with a brief account of the available experimental methods.

ISO 31/V1 (1980),[3] *Quantities and Units of Light and Related Electromagnetic Radiation*, is a table giving the international symbols and units relating to light and other radiations.

Ross and Birley[4] review optical properties and Venable[5] discusses the role of spectrophotometry in the specification and measurement of optical properties of translucent materials. A useful review is that given by Christie and Hunter,[6] who make the observation that when sufficient economic demand arises for more sophisticated instrumentation, new instrument development will keep pace with the need.

Bugash[7] points out that by using a modulated light source and appropriate electronics much of the uncertainty associated with conventional instruments can be eliminated.

Test pieces for optical measurements need to be prepared to a very high standard. Platens of borosilicate glass without a release agent are recommended in order to obtain surfaces of adequate optical quality.

14.2 REFRACTIVE INDEX

14.2.1 Terms and Definitions

The ISO definition[3] of refractive index is given in terms of the velocity of light: 'Refractive index is the ratio of the velocity of electromagnetic radiation *in vacuo* to the phase velocity of electromagnetic radiation of a specified frequency in the medium.[2] Although accurate, this is unhelpful and the ray definition is more useful.

When a ray of light passes from one isotropic medium to another, the sine of the angle of incidence bears a constant ratio to the sine of the angle of refraction (both measured with respect to the normal) for all angles of incidence. This ratio depends not only on the two media concerned, but also on the wavelength of the light used and on temperature.

'Refractive index' is the term used for this ratio when light passes from a vacuum (or from air for less accurate work) into a more dense medium; it is always greater than unity.

Refractive index is dimensionless. For anisotropic materials the state of polarisation of the light (and its direction, where appropriate) must be defined relative to a reference axis in the sample. It is then customary to quote two refractive indices; additionally, the maximum difference between two indices measured in two mutually perpendicular directions is termed the 'birefringence' of the material.

14.2.2 Methods of Measurement

General methods of measuring refractive index, not necessarily confined to plastics, can be found in older textbooks on optics.[8] Perhaps the best known methods are those depending on matching the unknown solid with known liquids, using the Becke line technique[9] or dispersion staining.[10] Other well known methods are those based on the Abbe refractometer[8] and the measurement of the ratio of real depth to apparent depth.[8]

The two methods recommended in ISO 489 (1983)[11] are the Abbe refractometer and the Becke line technique.

The test piece for the refractometer method should be about 12 × 6 × 3 mm with one flat face and one truly perpendicular surface, these two surfaces intersecting along a sharp line without a bevelled or rounded edge. (For anisotropic materials specimens should be made with their polished surfaces parallel and perpendicular to the direction of orientation.) The test piece is attached to the prism of the refractometer with a drop of liquid of refractive index higher than the test piece by at least 0·01 and this liquid should not soften, attack, or dissolve the plastics material. A table of suggested liquids is given in the specification which states that measurement should be carried out at 20 °C ± 0·5 °C using white light. If it were updated it is probable that a mean temperature of 23 °C would be specified. The instrument gives the refractive index for the sodium [D] line to a precision of about 0·001.

ASTM D542–50 (1977)[12] is similar and here the specified conditions are 23 °C ± 2 °C and 50 ± 5 per cent r.h. In cases where the material has a high thermal coefficient of refractive index the temperature has to be accurately controlled at 23 °C. The claimed accuracy of the method is four significant figures and where maximum accuracy is required the use of sodium light is recommended.

DIN 53491 (1955),[13] which gives useful advice on practical details relevant to refractometer measurements, is similar to ISO 489 in specifying a test temperature of 20 °C ± 0·5 °C, but like ASTM D542 recommends the use of sodium light for maximum accuracy which is claimed to be four decimal places. All the specifications give lists of liquids suitable for different plastics.

It is convenient to consider the alternative, less preferred method in ASTM D542 at this point as this also requires a test piece with parallel faces. This method is based on the well known phenomenon that the apparent depth of a transparent object viewed through the thickness is less than the true depth, refractive index being the ratio of real depth to apparent depth. ASTM D542 outlines the technique for carrying out this measurement with a standard microscope having a magnifying power of at least 200 diameters and a means of measuring the relative movement of the objective lens to the sample with an accuracy of 0·025 mm. The claimed accuracy for this method of measurement is three significant figures.

The alternative measurement in ISO 489, the Becke line method, is much more useful in that it can be used with powdered or granulated transparent material or indeed with any small chip of material taken from a larger specimen. A microscope having a magnifying power of at least 200 diameters is required, together with a range of liquids of known refractive index. If an Abbe refractometer is available, but only small chips of material rather than parallel-sided test pieces, then the refractometer can be used to calibrate test liquids for use with the Becke line method. The test pieces should have a thickness significantly less than the working distance of the 8 mm microscope objective and linear dimensions sufficiently small and so distributed that simultaneous observation of approximately equal areas of sample and surrounding field is possible.

The material under test is mounted in a liquid of known refractive index and examined in monochromatic light with the condenser adjusted to give a narrow axial beam. When the test pieces and the liquid have different refractive indices each particle is surrounded by a narrow luminous halo (the Becke line) which moves as focus is adjusted. If the focus is lowered then the Becke line moves towards the medium having the lower refractive index. The test is repeated with particles mounted in other immersion liquids until a match is found or until the index of the test sample lies between two known indices in the series of liquid standards. If the Becke line phenomenon does not appear then the refractive index of the material being examined is equal to that of the immersion liquid. ISO 489 recommends a test temperature of 20 °C ± 0·5 °C, but doubtless if it were up-dated the mean temperature would be changed to 23 °C.

ISO 489 does not recommended any conditioning procedure prior to the test. ASTM D542 specifies that test specimens must be conditioned for at least 40 h at 23 °C ± 2 °C and 50 ± 5 per cent relative humidity. DIN 53491 recommends that hygroscopic materials should be dried for 24 h over anhydrous calcium chloride, phosphorous pentoxide or concentrated sulphuric acid prior to measurement.

Dispersion staining is another useful microscopical method for the determination of refractive index described in detail by McCrone in ASTM STP 348.[10]

In general, methods have changed very little since the nineteenth century so that the older references still have value. Wiley and Hobson's review[14] covers refractometer methods and immersion methods whereas Brown, McCrone *et al.*[15] deals with dispersion staining. Billmeyer,[16] Ellis[17] and Schael[18] deal specifically with polymer films where it must be borne in mind that, as in the case of fibres, birefringence may result from the processing conditions. There is no doubt that when anisotropic materials are considered these are best dealt with by means of the polarising microscope. Suitable techniques are discussed in §14.3 below.

14.3 BIREFRINGENCE

14.3.1 Definition

Birefringence is a measure of optical anisotropy. It is defined as the maximum algebraic difference between two refractive indices measured in two perpendicular directions. Some materials are isotropic until stressed elastically, others have permanent birefringence induced by processing, e.g. drawing or rolling.

14.3.2 Methods of Measurement

The most convenient way of measuring birefringence is by means of a polarising microscope. It would be inappropriate to discuss methods of analysis in detail here, but some idea of the principle will be given.

When a beam of light falls on an isotropic material it is resolved into two components. A polariser is a device which is designed so that one of these components is totally absorbed (or internally reflected in the case of a Nicol prism). Thus light emerging from the polariser is plane polarised. A similar sheet of polarising material mounted with its plane of polarisation at right angles to the first will absorb all of the light. Only if the two directions of polarisation are parallel will light pass through both filters. This second piece of polarising material is known as the 'analyser'. The two arrangements described above are referred to as 'crossed polars' and 'open polars', respectively.

Consider now a birefringent object mounted between crossed polars. The simplest case is that where a rectangular object, having its planes of polarisation parallel to its sides, is mounted with one of these planes parallel to the plane of the incident light (Fig. 14.1(a)). Light from the polariser passes through the birefringent test piece and is absorbed by the analyser, so that the field is dark.

Similarly, if the test piece is rotated through 90° the field remains dark

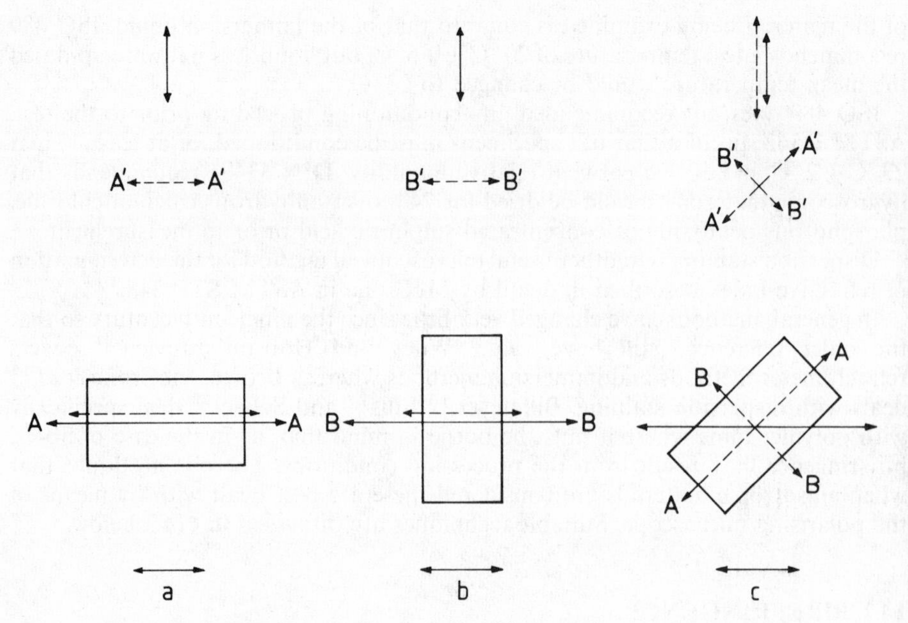

Fig. 14.1 Interaction of plane-polarised light with a birefringent object

(Fig. 14.1(b)). If the test piece is mounted at 45° to the plane of polarisation, however, the object stands out brightly on a dark field. To understand this, reference must be made to Fig. 14.1(c). The original plane-polarised beam may be considered to be resolved into two components A and B corresponding to the principal planes of the test piece. These two components traverse the test piece at different velocities so that the two emergent beams A′ and B′ are out of phase. On reaching the analyser each of the components A′ and B′ may be resolved into components parallel to and at right angles to the analyser direction. Components parallel to the analyser direction pass through and, since these are out of phase, interference results. Because the phase difference varies with wavelength, brilliant interference colours may be obtained. The effect on these colours of retardation wedges or compensators enables an estimate of birefringence to be made. Detailed descriptions of these methods can be found in standard textbooks.[19,20]

If a polarising microscope is available the experimental technique involved is relatively straightforward. Many of the difficulties arise from a lack of understanding of the physical principles behind the method and initial study of the theoretical principles of polarising microscopy is well worth while.

Fibres sometimes present difficulty in that there is a different orientation at the skin and in the core of the fibre. In these cases it must be borne in mind that the Becke line method gives the refractive index of the skin, whereas birefringence measured by a compensator method is that of the fibre as a whole. Care is needed in interpretation and advice should be sought from specialised textbooks or, better still, from specialists in the field.

14.4 TRANSPARENCY

14.4.1 General Considerations

Light transmission, haze and see-through clarity are closely linked properties which together constitute transparency. Aberration of the image formed by any optical system or the merging of points distinct in the original object is known as 'loss of clarity', whereas the scattering of light with consequent loss of contrast is termed 'haze'. Light transmission is the percentage of light transmitted without deviation.

As an illustration of these properties, glazing material for windows and diffusing material for lighting fixtures must both have a high light transmittance, but the former must be free from haze and very transparent while the latter must have maximum diffusion and minimum transparency. Film for packaging must have good see-through clarity, a property which should not be confused with haze as hazy materials often have quite good see-through clarity.

Webber[21] and Miles and Thornton[22] have expressed the view that see-through clarity is a function of light scatter over a very small forward angle, as distinct from the larger angles of scatter usually considered in haze measurements.

Haze and loss of clarity indicate discontinuities or irregularities within or at the surface of the material and these could be filler particles, impurities, minute bubbles or voids, surface roughness, or regions of different refractive indices brought about, say, by crystallisation. Abrasion, weathering, absorption of water and even temperature change can all bring about a change in the haziness of a material, so that test conditions need to be closely specified.

BS 4618 Section 5.3 (1972)[2] gives useful background material and the principles of experimental methods. Other papers covering general principles are those by Venable,[5] Christie and Hunter[6] and Willmouth.[23]

14.4.2 Light Transmission and Haze – Small-scale Test Methods

Light transmission may be defined as the percentage of light transmitted without deviation, and haze as the percentage of incident light which is transmitted with more than a certain specified angular deviation by forward scattering (say, 2.5°). Both properties vary with the geometry of the optical measuring system used which must be carefully specified.

The three principal light sources used in illumination work and photometry are CIE (Commission International de l'Eclairage) Sources A, B and C which correspond, respectively, to incandescent light, noon sunlight and overcast sky daylight. Although methods of measuring light transmission and haze are under consideration by ISO TC61 (*Plastics*) no international standards have yet been agreed and the principal standard in use for small systems is ASTM D1003–61 (1977).[24] This method gives CIE Source C as the preferred light source, but allows Source A as an alternative.

The method was originally devised for low-density polyethylene film, which in practice involves a fairly narrow range of thicknesses. When applied to other thin films, or to sheets or plaques, the method can give misleading answers, so that some caution in its application is called for.

ASTM D1003 allows several different instruments, all based on the use of an integrating sphere. One is based on a spherical hazemeter which is pivotable about a vertical axis through the entrance port, where the specimen is placed (Fig. 14.2). In the normal position the collimated incident light passes straight through the sphere, leaving through the exit port which is closed by an absorbent light trap. Any light which is scattered by the instrument alone (specimen removed) or instrument plus specimen (specimen in) is reflected from the region around the edge of the exit port and finally collected by a photocell positioned at 90° to the beam after multiple reflections from the highly reflective walls of the sphere.

When the sphere is rotated slightly so that the incident light hits the opposite highly reflecting wall of the sphere (which forms a reflectance standard) adjacent to the exit port, a measurement with and without specimen gives a measure of the total transmittance.

The three properties of interest are:

$$\text{total transmittance, } T_t = \frac{T_2}{T_1} \quad [14.1]$$

$$\text{diffuse transmittance, } T_d = \frac{T_4 - T_3(T_2/T_1)}{T_1} \quad [14.2]$$

$$\text{haze (per cent)} = \frac{T_d}{T_t} \times 100 \quad [14.3]$$

where T_1 = photocell output, specimen out, reflectance standard in beam
T_2 = photocell output, specimen in, reflectance standard in beam
T_3 = photocell output, specimen out, reflectance standard out of beam
T_4 = photocell output, specimen in, reflectance standard out of beam

T_1, T_2, T_3, T_4 represent, respectively, the incident light, the total light transmitted by the specimen, the light scattered by the instrument and the light scattered by

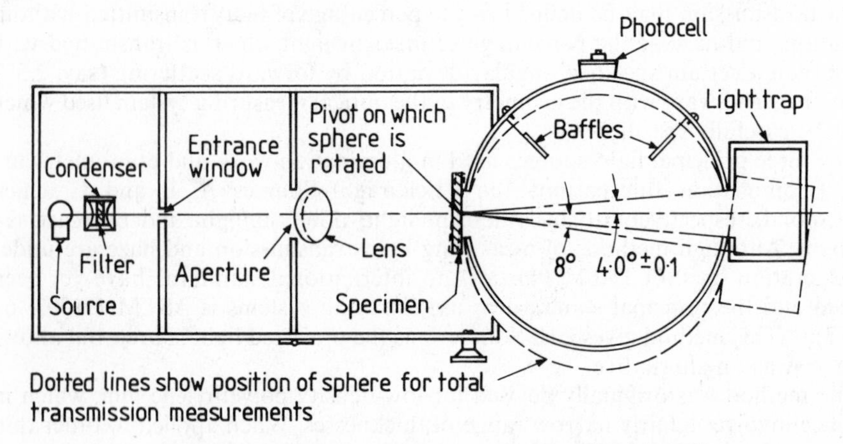

Fig. 14.2 Integrating sphere hazemeter: ASTM D1003

instrument and specimen. The principles of the other instruments are very similar.

The method of ASTM D1003 is also invoked by ASTM D1044 (1985)[25] to measure the diffuse transmittance of specimens after Taber abrasion tests, as a measure of the degree of abrasion (§10.3).

The wider implications of transmittance and reflectance measurements carried out on translucent plastics are discussed by Johnston *et al.*[26]

14.4.3 Light Transmission and Haze – Large-scale Test Methods

A light transmission test for panels 600 mm square appears in two British Standards, BS 4154 (1985)[27] and BS 4203 (1980)[28] and in ASTM D1494–60 (1980)[29] (see also Meyer and Smith[30]). Since it is a test on a finished product rather than a material it will not be described in detail. A photocell measures the transmitted light and the transmittance is given as the ratio of photocell outputs with the specimen in and the specimen out. In BS 4154 an additional test for light diffusion is given, based on a simple slit diffusion photometer, and an appropriate correction is then calculated and applied to the total light transmission. This refinement is not required in BS 4203, presumably because only transparent material is involved.

14.4.4 See-through Clarity

There is no method universally accepted for measuring see-through clarity. The following method, outlined in BS 4618 Section 5.3 (1972),[2] is based on the work of Webber.[21]

Sets of charts (Snellen charts) are viewed through the film under test. Each chart consists of sets of parallel lines which are vertically perpendicular, each group of lines having a different separation. The angle subtended at the eye by the minimum resolvable separation between the lines is noted and compared with the angular resolution without a film. The difference between the two values gives a measure of clarity. The angular resolution with a film present depends on the distance between chart and sample and this must be specified or recorded.

14.5 GLOSS

14.5.1 General Considerations

According to Hunter,[31] ‘Gloss is defined as the degree to which a surface simulates a perfect mirror in its capacity to reflect incident light.’ He goes on to say that ‘gloss is determined by the surface’s geometric selectivity in reflecting light’. In most practical glossmeters specular gloss is the property determined and this is usually taken to be the fraction of light flux reflected in the direction of mirror reflection (the specular direction) when a specimen is illuminated by a parallel light beam.

The standard methods of measuring gloss are numerous. Superficially many of the methods are similar, but small differences in optical geometry can cause large variations in results from one type of instrument to another.

The review of Christie and Hunter[6] usefully outlines different types of instrument.

The greatest difficulty with gloss measurements lies in correlating measurements with visual impressions; this is discussed by Hammond,[32] Dinsdale and Malkin,[33] Hunter[34] and Elm.[35]

Knittel[36] points out that an observer, in assessing visual gloss, turns the sample over using many angles of incidence and reflection, and his judgement is therefore based on a whole series of observations.

Because there is no absolute standard for gloss, existing standards (usually optically flat black glass plates) are assigned values which depend on the refractive index of the glass employed and which may vary from one test to another.

14.5.2 Standard Test Methods

ASTM D523 (1985)[37] is the recommended standard for gloss measurements on plastics (excepting films) and uses a 20° (to the vertical) incident reflected light beam geometry for high gloss materials and 85° geometry for low gloss materials. An intermediate 60° angle is used for inter-comparative purposes and for deciding which of the other angles should be used with a given specimen.

Although the geometry of source and receiver is defined with close tolerances on the angles, the actual instrument to be used is not described in great detail: Figs. 14.3 and 14.4 (taken from the standard) merely indicate the general layout. The primary standard of gloss is a highly polished, plane, black glass surface with a refractive index of 1·567 to which is assigned the arbitrary value of 100 gloss units for each of the three geometries (the reader is warned that elsewhere and in other circumstances different values may be assigned to similar gloss standards).

The specification requires that any measurements should be in accordance with CIE Source C values, but states that spectral corrections need to be applied only to highly chromatic low-gloss specimens.

For films, ASTM D2457–70 (1977)[38] is based on ASTM D523 for the 20° and 60° angle tests, but the third angle is 45°, not 85°. The 45° test is based on a

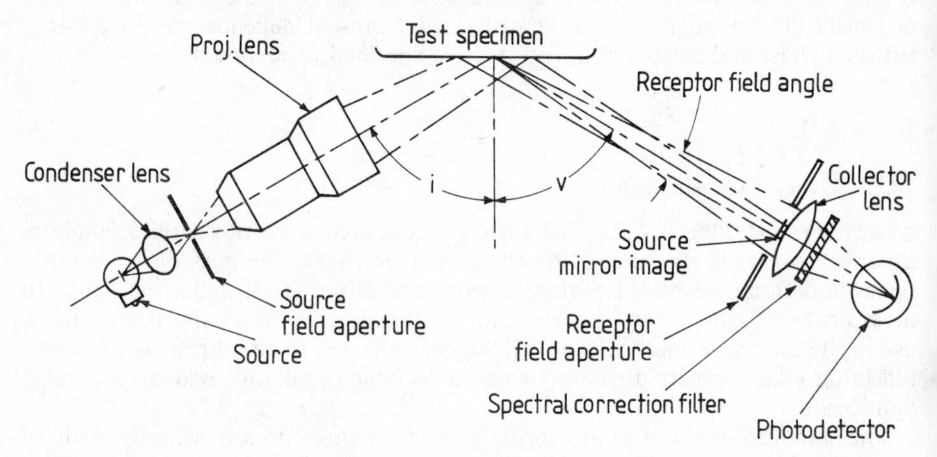

Fig. 14.3 Converging beam glossmeter showing apertures and source mirror-image position: ASTM D523

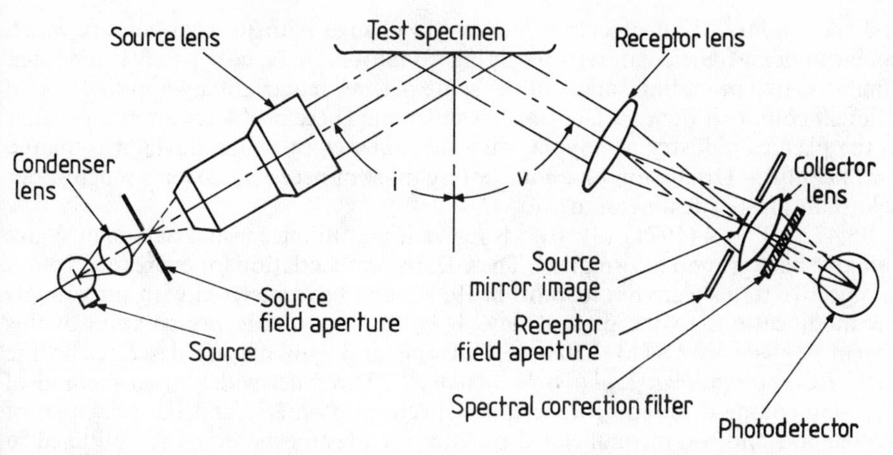

Fig. 14.4 Parallel-beam glossmeter showing apertures and source mirror-image position: ASTM D523

method for ceramics, ASTM C346–76 (1981).[39] A formula is given so that the appropriate specular reflection of a plane black surface may be calculated from the refractive index of the glass, the angle of incidence of the light and a tabulated scale factor related to the geometry. A note warns of the possibility of obtaining gloss values on clear films of more than 100 units with any one of the three geometries, because of reflection at both surfaces of the specimen. Three specimen mounting devices are described for ensuring the flatness so essential in gloss measurements.

Method 515B (1970) of BS 2782[40] is being re-written to conform to ASTM D523 (1985) and it is probably that the 45° geometry used formerly will be replaced by a 60° geometry, with alternatives of 20° and 85° as in the ASTM specification.

BS 3900: Part D5 (1980)[41] applies to paint films and uses the same geometries.

BS 3962: Part 1 (1980)[42] applies to wooden furniture but is relevant to plastics testing in that a polymeric finish may be applied to the wood. For this test an 85° geometry is used.

Most methods now refer to a primary standard as black glass of refractive index $n_d = 1{\cdot}567$ to which is assigned a gloss value of 100. When departures are allowed the change in gloss value associated with quoted change in refractive index is given. It is assumed, however, that in addition to a primary standard, a laboratory will hold a range of working standards which may be of ceramic tile, vitreous enamel, opaque glass or similar materials and all of which will be planar and calibrated against a primary standard.

14.6 COLOUR

Colour is not an absolute property but depends very much on the surface condition of the test piece and the viewing conditions. Even to specify 'daylight' or 'artificial light' is not enough, for the colour of the sun changes as it gets lower in the sky and the colours of all objects in our world change with it. Two colours which match under a particular set of lighting conditions may not match in another. Similarly, two mouldings made of the same plastics material may appear to be of different colours if their surface finish is different. These problems are not peculiar to the plastics industry as anyone who has taken cloth to the daylight to match it will confirm. Discussion of colour testing must of necessity cover a much wider field than plastics testing alone.

BS 4727: Part 4 (1971), (1980)[43] is relevant and Billmeyer and Saltzman[44] give valuable background information. The CIE recommendation for expressing colour in objective terms were put forward in 1931[45] and have survived with surprisingly few modifications to the present day. The most up-to-date presentation of this system is given in ASTM E308–85.[46] Terms and symbols are described in the CIE/IEC *International Lighting Vocabulary*.[47] The most widely used method of assessing colour is based on reflectance spectrophotometry and the principles of the method and the recommended practice for plastics materials are outlined in ISO 3557[48] (at present in the draft stage).

The assessment of small colour changes during weathering or of the clarity and colour of water, white solutions or materials, while coming under the general heading of colour, may present less of a problem than the full definition or specification of the colour being considered. It may be easier, therefore, to deal with these first of all. ISO 4582 (1980)[49] and ISO 877 (1976)[50] deal with the determination of the resistance to colour change of plastics exposed to daylight under glass and ISO 4892 (1981)[51] deals with artificial light sources. BS 4618 (1974)[52] uses the blue wool scale[53–55] to assess the degree of colour change which takes place when a given plastics test piece, protected by glass, is exposed to daylight or laboratory sources such as carbon arc or xenon lamp.

Method 530A[56] of BS 2782, (1976/1983) for the determination of yellowness index and Method 530B for the determination of the colour of near-white or near-colourless materials are both based on the CIE tri-stimulus values. Simpler methods include that of ISO 1600 (1975)[57] for the determination of light absorption of cellulose acetate before and after heating, where changes in optical density at wavelengths of 440 nm and 640 nm are reported and ISO 6271 (1981)[58] or BS 5339 (1976)[59] which use the platinum–cobalt scale to estimate the colour of clear liquids. In these methods the colour comparison is made with standard solutions by means of a simple colour comparator.

Returning now to the assessment of colour in general, a paper by Willmeyer[60] is still worth reading for background information on the physiology of perception of colour and in the same publication Vickerstaff and Walls[61] discuss matching and measurement of colour.

There is no doubt that the introduction of microcomputers has had a great influence in the field of colour measurement and prediction and this has been discussed in a most readable paper by Gall[62] and a later one by Allen.[63]

More recently, McLaren[64] has discussed the difference between acceptability and perceptibility in the context of colour difference equations and discusses improvements made between 1969 and 1983. Best,[65] in the same publication, discusses computer match prediction in the laboratory and factory.

The production of stable colour standards is important for all industries concerned with colour measurement and British Ceramic Research Co. Ltd are acknowledged leaders in this field. In a preliminary report in 1973 Malkin[66] compared several instruments using the original BCRA standard tiles. As stocks of these became exhausted a new series was developed as discussed by Clarke and Malkin[67] in 1981 and again by Malkin and Verrill[68] in 1983. The relationship of these new ceramic standards to others available elsewhere in the world has been discussed in detail recently by Malkin.[69]

Instruments have improved enormously in recent years and it is difficult to give up-to-date information. Billmeyer and Hemmedinger[70] give a good historical survey and a paper by Stanziola *et al.*[71] is also useful.

Billmeyer and Rich[72] describe various instruments suitable for plastics and Keane[73] points out the advantages of computer-assisted colorimetry.

Colour measurement and control are not easy and those thinking of entering these fields should seek advice from laboratories with experience, in addition to examining the most up-to-date instruments available.

REFERENCES

1. Meeten, G. H. (1986) *Optical Properties of Polymers*, Elsevier.
2. BS 4618: Part 5, Section 5.3 (1972) *Other Properties – Optical Properties.*
3. ISO 31/6 (1980) *Quantitites and Units of Light and Related Electromagnetic Radiations.*
4. Ross, G. and Birley, A. W. (1973) *Journal of Physics D, Applied Physics*, **6**, 795.
5. Venable, W., Jr (1973) *Coloring of Plastics 7, RETEC, Cherry Hill, NJ*, September, SPE Delaware Valley Section, p. 83.
6. Christie, J. S. and Hunter, R. S. (1975) *Coloring and Decorating of Plastics 9, RETEC, Cincinnati, Ohio, September, SPE Miami Valley Section*, p. 33.
7. Bugash, R. S. (1973) *Coloring of Plastics 7* (op. cit.), p. 53.
8. Nelkon, M. (1955) *Light and Sound*, 2nd edn, Heinemann.
9. James, D. I. (1974) *RAPRA Members Journal*, May, 134.
10. McCrone, W. C. (1963) *ASTM STP*, no. 348, p. 125.
11. ISO 489 (1983) *Plastics, Determination of the Refractive Index of Transparent Plastics.*
12. ASTM D542–50 (1977) *Standard Test Methods for Index of Refraction of Transparent Organic Plastics.*
13. DIN 53491 (1955) *Testing of Plastics – Determination of Refractive Index and Dispersion.*
14. Wiley, R. H. and Hobson, P. H. (1948) *Analytical Chemistry*, **20** (6), 520.
15. Brown, K. M. and McCrone, W. C. (1963) *The Microscope*, **13** (11), 311; Brown, K. M., McCrone, W. C., Kuhn, R. and Forlini, F. W. (1963) *The Microscope*, **14** (2), 39.
16. Billmeyer, F. W. (1974) *Journal of Applied Physics*, **18** (5), 431.
17. Ellis, R. H. (1957) *Review of Scientific Instruments*, **28** (7), 557.
18. Schael, G. W. (1964) *Journal of Applied Polymer Science*, **8** (6), 2717.
19. Hartshorne, N. H. and Stuart, A. (1970) *Crystals and the Polarizing Microscope*, 4th edn, Elsevier.
20. Stoves, J. L. (1957) *Fibre Microscopy: Its Technique and Application*, National Trade Press.
21. Webber, A. C. (1957) *Journal of the Optical Society of America*, **47** (9), 785.
22. Miles, J. A. C. and Thornton, A. E. (1962) *British Plastics*, **35** (1), 26.
23. Willmouth, F. M. (1976) *Plastics and Rubbers: Materials and Applications*, **1** (3/4), 101.
24. ASTM D1003–61 (1977) *Haze and Luminous Transmittance of Transparent Plastics.*
25. ASTM D1044–85 (1985) *Test Method for Resistance of Transparent Plastics to Surface Abrasion.*
26. Johnston, R. M., Keller, E. F., Edge, L. S. and Change, M. (1973) *Coloring of Plastics 7, RETEC, Cherry Hill, NJ, September, SPE Delaware Valley Section*, p. 39.

27. BS 4154 (1985) *Corrugated Plastics Translucent Sheets made from Thermosetting Polyester Resins (Glass Fibre Reinforced).*
28. BS 4203 (1980) *Extruded Rigid PVC Corrugated Sheeting.*
29. ASTM D1494–60 (1980) *Diffuse Light Transmission Factor of Reinforced Plastics Panels.*
30. Meyer, R. W. and Smith, D. (1958) *Reinforced Plastics*, **2** (5), 12.
31. Hunter, R. S. (1956) *Product Engineering*, **27** (2), 176.
32. Hammond, H. K. (1957) *American Paint Journal*, **41** (37), 95.
33. Dinsdale, A. and Malkin, F. (1955) *Transactions of the British Ceramics Society*, **54** (2), 94.
34. Hunter, R. S. (1952) *ASTM Bulletin*, **186**, 48.
35. Elm, A. C. (1961) *Official Digest of the Federation of Societies for Paint Technology*, **33** (433), 163.
36. Knittel, R. R. (1962) *Materials, Research and Standards*, **2** (3), 180.
37. ASTM D523 (1985) *Test Method for Specular Gloss.*
38. ASTM D2457–70 (1977) *Test Method for Specular Gloss of Plastics Films.*
39. ASTM C346–76 (1981) *Test Method for 45-deg. Specular Gloss of Ceramic Materials.*
40. BS 2782 Method 515B (1970) *Gloss (45°) of Sheet.*
41. BS 3900: Part D5 (1980) *Measurement of Specular Gloss of Non-metallic Paint Films at 20°, 60° and 85°.*
42. BS 3962: Part 1 (1980) *Methods of Test for Finishes for Wooden Furniture: Assessment of Low Angle Glare by Measurement of Specular Gloss at 85°.*
43. BS 4727: Part 4: (1971), (1980) *Terms Particular to Light and Colour. Group 01, Radiation and Photometry; Group 02, Vision and Colour Terminology; Group 03, Lighting Technology Terminology.*
44. Billmeyer, F. W. and Saltzman, M. (1967) *Principles of Colour Technology*, Interscience.
45. CIE (1931) *Colorimetry, Official Recommendations*, Publications no. 15.
46. ASTM E308–85 *Method for Computing the Colors of Objects by Using the CIE System.*
47. CIE/IEC (1970) *International Lighting Vocabulary*, 3rd edn.
48. ISO 3557 draft *Plastics – Recommended Practice for Spectrophotometry and Calculation of Colour in CIE Systems.*
49. ISO 4582 (1980) *Plastics – Determination of Changes in Colour and Variations in Properties after Exposure to Daylight under Glass, Natural Weathering or Artificial Light.*
50. ISO 877 (1976) *Plastics – Determination of Resistance to Change upon Exposure under Glass to Daylight.*
51. ISO 4892 (1981) *Plastics – Methods of Exposure to Laboratory Light Sources.*
52. BS 4618 Section 4.3 (1974) *Resistance to Colour Change Produced by Exposure to Light.*
53. BS 1006 (1978) *Methods of Test for Colour Fastness of Textiles and Leather.*
54. BS 1006 Section A02 (1978) *Grey Scale for Assessing Change in Colour.*
55. ISO 105 Section A02 (1984) *Textiles – Tests for Colour Fastness.*
56. BS 2782 Methods 530A and 530B (1976/1983) *Determination of Yellowness Index – Determination of the Colour of Near-white or Near-colourless Materials.*
57. ISO 1600 (1975) *Plastics – Cellulose acetate – Determination of Light Absorption before and after Heating.*
58. ISO 6271 (1981) *Clear Liquids – Estimation of Colour by the Platinum–Cobalt Scale.*
59. BS 5339 (1976) *Method of Measurement of Colour in Hazen Units (Platinum–Cobalt Scale) of Liquid Chemical Products.*
60. Willmeyer, E. N. (1953) *Journal of the Oil and Colour Chemists Association*, **36** (399), 491.
61. Vickerstoff, T. and Walls, I. S. M. (1953) *Journal of the Oil and Colour Chemists Association*, **36** (399), 507.

62. Gall, L., Computer colour matching, Colour 73, *Proceedings AIC Congress, York 2–6 July 1973*, Adam Hilger.
63. Allen, E., Advances in colorant formulation and shading, Color 77, *Proceedings AIC Congress, Troy, New York 10–15 July 1977*, Adam Hilger.
64. McLaren, K., The development of improved colour-difference equations by optimisation against acceptability data. *Golden Jubilee of Colour in the CIE, Soc. Dyers and Colourists, Bradford, 1981.*
65. Best, R. P., Computer match prediction in Laboratory and Factory. *Golden Jubilee of Colour in the CIE, Soc. Dyers and Colourists, Bradford, 1981.*
66. Malkin, F., *AIC Colour 73*, 351–352, Adam Hilger.
67. Clarke, F. J. J. and Malkin, F. (1981) *Journal of the Society of Dyers and Colourists*, **97**, 503.
68. Malkin, F. and Verrill, J. F. (1983) in *Proceedings CIE 20th Session, Amsterdam, Aug. 31–Sept. 8, 1983*, (Ed. J. Schanda), *OMIKK Budapest*, E37/1–2.
69. Malkin, F. (1987) Colour standards, in *Proceedings Symposium on Advances in Standards and Methodology in Spectrometry, Oxford, 14–17 Sept. 1986*, (Eds. C. Burgess and K. D. Mielenz), Elsevier.
70. Billmeyer, F. W. and Hemmendinger, H., Instrumentation for colour measurement and its performance, *Golden Jubilee of Colour in the CIE, Soc. Dyers and Colourists, Bradford, 1981.*
71. Stanziola, R., Momiroff, B. and Hemmendinger, H. (1979) *Color Research and Application*, **4**, 157.
72. Billmeyer, F. W. and Rich, D. C., Coloring of plastics 12, *SPE RETEC Conference, Oct. 1978, Sawmill Creek, Huron, OH, USA.*
73. Keane, J. T., Coloring and decorating of plastics 9, *SPE, RETEC Conference, Sept. 1975, Cincinnati, Ohio, USA.*

Chapter Fifteen

Thermal Properties

15.1 INTRODUCTION

This chapter deals with thermal conductivity, thermal diffusivity and specific heat. Other properties which are sometimes included under the umbrella term 'thermal properties' are dealt with in other parts of this volume.

15.1.1 Equation of conduction of heat

For the flow of heat in one direction the heat flux J_u is related to the temperature gradient $\partial\theta/\partial x$ by Fourier's law

$$J_u = -K\frac{\partial\theta}{\partial x} \qquad [15.1]$$

where K is the thermal conductivity. The minus sign indicates that the heat flows in the opposite direction to the temperature gradient. The form of Eqn [15.1] implies that heat conduction is a random process. If the energy were propagated without scattering then the heat flow would depend on the temperature difference between the end faces of the specimen instead of the temperature gradient.[1]

The general equation from which the time-dependent temperature distribution may be calculated is obtained from Eqn [15.1] and the equation of continuity

$$\frac{\partial J_u}{\partial x} = -\rho c\frac{\partial\theta}{\partial t} \qquad [15.2]$$

where ρ is the density, c is the specific heat, i.e. the heat capacity per unit mass, and t is the time. The equation of continuity is an expression of the conservation of energy. The heat flux can be eliminated between Eqns [15.1] and [15.2] to give

$$\frac{\partial^2\theta}{\partial x^2} + \frac{1}{K}\frac{\partial K}{\partial\theta}\left(\frac{\partial\theta}{\partial x}\right)^2 = \frac{1}{\alpha}\frac{\partial\theta}{\partial t} \qquad [15.3]$$

where $\alpha = K/\rho c$ is the thermal diffusivity. Equation [15.3] is the equation of conduction of heat, in the absence of heat generation and convection, for heat flow in one direction. If the conductivity is independent of temperature it reduces to

$$\frac{\partial^2 \theta}{\partial x^2} = \frac{1}{\alpha}\frac{\partial \theta}{\partial t} \qquad [15.4]$$

which is the equation usually referred to. It has been shown that for rubbers, and hence plastics except at melting transitions, the conductivity term in Eqn [15.3] is very small and Eqn [15.4] is adequate for most heat flow calculations.[2]

The parameter α was called the 'thermal diffusivity' by Kelvin and the 'thermometric conductivity' by Maxwell, but Kelvin's expression has been generally adopted. It measures the change in temperature which would be produced in unit volume of the substance by the quantity of heat which flows in unit time across unit area of a layer of the substance of unit thickness with unit temperature difference between its faces.[3] It is the parameter which determines the non-steady state temperature distribution in the absence of heat generation and convection and is therefore essential for transient heat transfer calculations.

In the above discussion the specific heat referred to is that at constant volume per unit mass, whereas in practice the sample is at constant pressure. However, a body containing temperature gradients usually also contains internal stresses and c_p is not quite correct either,[4] but is more appropriate than c_v. However, the difference between the two principal specific heats is small for polymers,[5] hence this is not a significant point. In the following sections c_v is used in theoretical arguments and c_p is used elsewhere, unless otherwise stated.

15.1.2 Units

A variety of units have been used for thermal properties which is a nuisance when different sets of results have to be compared. The two most common units for conductivity are the cal/cm °C and the BTU in/ft^2 h °F. There are two units of length in the imperial unit, because area is measured in square feet and thickness in inches, and this inconsistency is a potential pitfall for the unwary. A self-consistent conductivity unit VTU/ft h °F, is obtained if the temperature gradient is measured in °F/ft instead of °F/in, but this is not as common. For diffusivity the c.g.s. unit is the cm^2/s and the imperial unit is the ft^2/h. The SI unit for conductivity is the W/mK, and for diffusivity the unit is the m^2/s. In this chapter the SI unit for conductivity and a submultiple of the SI diffusivity unit, the mm^2/s are used. Conversion factors are given in Table 15.1.

15.1.3 Fundamental Significance of Diffusivity

Since $\alpha = K/\rho c$, diffusivity is often regarded as just a mathematical parameter rather than a material property. However, diffusivity is seen to be a fundamental material property if we think wholly in terms of energy.

The heat capacity at constant volume per unit volume is given by $c = (\partial u/\partial\theta)_v$, where u is the internal energy per unit volume. Obtaining from this an expression for the temperature gradient and substituting it into Eqn [15.1] gives

$$J_u = -\alpha\frac{\partial u}{\partial x} \qquad [15.5]$$

Table 15.1 Conversion factors for some thermal conductivity units

	W/mk	cal/cm s °C	W/cm °C	kcal/m h °C	BTU in/ft² h °F
W/mk	1	0·002 39	0·01	0·86	6·93
cal/cm s °C	419	1	4·19	360	2900
W/cm °C	100	0·239	1	86	693
kcal/m h °C	1·16	0·002 78	0·011 6	1	8·06
BTU in/ft² h °F	0·144	0·000 345	0·001 44	0·124	1

Thus thermal diffusivity is the parameter relating energy flux to energy gradient, whereas conductivity relates the energy flux to the temperature gradient. The origin of the units mm^2/s, which are perhaps meaningless in themselves, now becomes apparent.

15.1.4 Heat Transfer at Interface

Many heat transfer problems involve a resistance to heat transfer across the interface between the sample surfaces and the heating or cooling medium. Such a resistance may be caused by a stationary surface film (of gas or liquid) of small but indeterminate thickness through which the temperature gradually changes from that of the solid to that of the bulk fluid. Heat transfer across a solid/fluid interface depends on wetting, film thickness and temperature gradients in the film. If the fluid is moving it will also depend on whether the flow is laminar or turbulent, which in turn is determined by velocity and viscosity. Similarly heat transfer across the interface between two solids depends on adsorbed gases, pressure, surface finishes, hardness and trapped fluids.[6]

A surface heat transfer coefficient, h, can be defined as the quantity of heat flowing per unit time normal to the surface across unit area of the interface with unit temperature difference across the interface. When there is no resistance to heat flow across the interface h is infinite. The heat transfer coefficient can be compared with the conductivity: the conductivity relates the heat flux to the temperature gradient; the surface heat transfer coefficient relates the heat flux to a temperature difference across an unknown distance.

Some theoretical work has been done on this subject,[7] but since it is rarely possible to achieve in practice the boundary conditions assumed in the mathematical formulation, it is best to regard it as an empirical factor to be determined experimentally. Some typical values are given in Table 15.2. Cuthbert[8] has suggested that values greater than about 6000 W/m^2 K can be regarded as infinite. The spread in values is caused by different fluid velocities. Heat loss by natural convection also depends on whether the sample is vertical or horizontal.

15.2 THERMAL CONDUCTIVITY MEASUREMENT

In this section the methods of measuring conductivity are described. The methods

Table 15.2 Typical values of surface heat transfer coefficient, *h*

System	h ($W/m^2 K$)
Condensing steam	∞
Rubber/mould	∞
Heating or cooling with water	200–1500
Heating or cooling with air	4·5–90
Fluid bed	560

discussed have been chosen to illustrate the basic principles of conductivity measurement and the various ways of minimising the major sources of experimental error. The experimental methods can be divided into two groups: steady state and transient.

15.2.1 Steady State Methods

Under steady state conditions, i.e. when the temperature at any point does not change with time, the temperature distribution in the sample is governed by Laplace's equation. The sample geometries are chosen so that the temperature is a function of only one coordinate, and simple analytical solutions to Laplace's equation are used. There are two geometries which satisfy this condition: the parallel-faced slab with heat flow normal to the surfaces; or the hollow cylinder with radial heat flow. In the latter case although the heat flow is now in two directions the temperature is a function of only one coordinate, namely the radius, because of symmetry. The condition is also satisfied by the sphere but this is not relevant to the present discussion. The first case is illustrated in Fig. 15.1 which shows a parallel-faced section of thickness x and cross-sectional area A.

One face is maintained at a uniform temperature θ and the other at a uniform temperature $\theta - \Delta\theta$. The rate of heat flow q is in the x direction only and is normal to the two faces. The conductivity is obtained from Eqn [15.1]:

$$K = \frac{qx}{A\Delta\theta} \qquad [15.6]$$

The coaxial cylinder geometry is illustrated in Fig. 15.2. The cylinder is of length L, outside radius r_2 and inside radius r_1. The temperature difference between the inner and outer surfaces is $\Delta\theta$. Equation [15.1] can be integrated to give

$$K = \frac{q}{2\pi L\Delta\theta}\log_e\left(\frac{r_2}{r_1}\right) \qquad [15.7]$$

It is necessary to construct the apparatus so that the experimental conditions agree with the boundary conditions for simple analytical solutions. Failure to achieve this perfectly gives rise to three major sources of error.

1. The first major source of error is a result of a failure to obtain a normal or radial heat flow, depending on the geometry. Satisfying this boundary condition

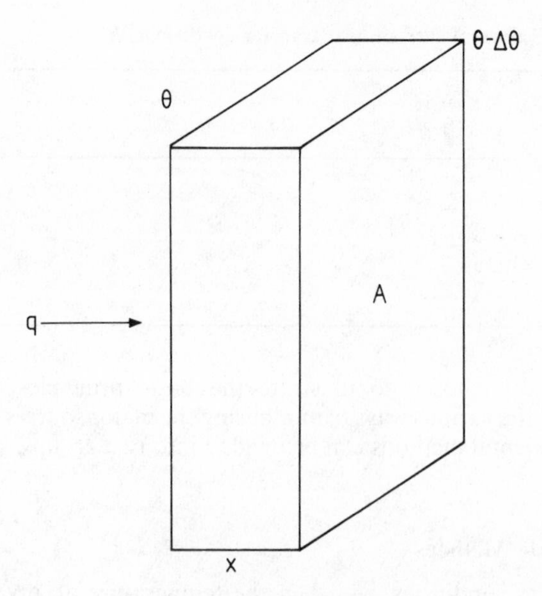

Fig. 15.1 Flat slab

is very difficult in practice; the problem has been tackled in three ways, the flux is constrained, or the unwanted heat flow is minimised and ignored, or estimated and allowed for in the calculations.

2. The second source of error is that the surfaces of the sample are not necessarily at the same temperature as the walls of the apparatus, i.e. there is a resistance to heat flow across the interface between the sample and the apparatus. In the majority of cases the temperatures of the cell walls are measured, and this gives a conductivity value which is too low.
3. The third source of error results from failing to satisfy the condition that the major surfaces must be isothermals. The size of this error depends on the degree of non-uniformity.

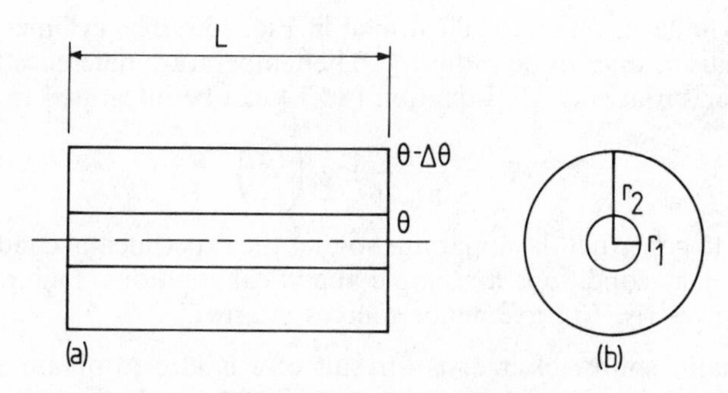

Fig. 15.2 Coaxial cylinder

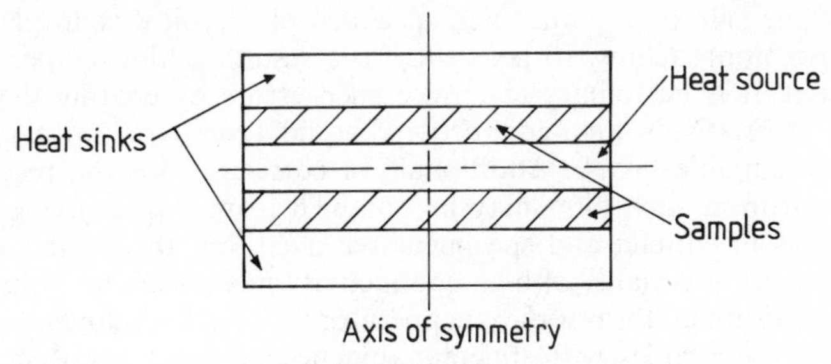

Fig. 15.3 Unguarded hot plate

The more important methods for measuring thermal conductivity are discussed below.

The Unguarded Hot Plate
The modern unguarded hot plate is based on the well known Lees' disc method[9] and the general arrangement is shown schematically in Fig. 15.3. Recommendations are given in BS 874[10] and ASTM C518 (1985).[11]

In this method two similar samples are placed on either side of an electrically powered heat source and sandwiched between two heat sinks. The heat sinks are controlled at a given temperature usually by circulating a liquid from a constant temperature bath, and a known power is supplied to the heat source. The apparatus is allowed to reach steady state and the temperature drop across the sample is measured, usually with thermocouples. The conductivity is calculated using Eqn [15.6].

The samples and the heat source are made as thin as possible and the apparatus is surrounded with a low-conductivity material such as vermiculite, but the side losses from the heater may still form a significant proportion of the total heat input. The heat loss may be evaluated by using a pair of specimens of low, known conductivity material, for example a cellular plastic calibrated by an absolute method. If the losses are known to, say 10 per cent and only represent 10 per cent of the heat input, the error due to this uncertainty is only 1 per cent. The side loss calibration must be carried out over the range of specimen thickness employed and over the appropriate temperature range, and it is to be noted that it is a function of both hot and cold face temperatures and of the ambient temperature. A further refinement is to use heat flow meters on one or both sides of the sample.

By using large-area thin specimens and very thin heater plates the side losses may be reduced to negligible proportions, but of course the preparation of such specimens becomes increasingly difficult the larger the area and the smaller the thickness.

The most convenient specimen size lies usually in the range from 50 to 200 mm across by 3 to 6 mm thick. The area of such specimens may easily be measured with the required precision, and the thickness similarly. The most important requirements for the specimens, however, are that the faces should be parallel and as flat as possible, since errors due to imperfect contact between specimens and heater and cold plates can be very large. As an example, an air film of thickness

0·03 mm on one side of a 3 mm thick specimen of a typical solid plastic, would cause an error approaching 10 per cent if the resulting film temperature drops were neglected. It is customary to reduce such errors by coating the specimens with a film of relatively high-conductivity liquid (compared with air) such as glycerol, or a suitable grease. Additionally a correction for the resulting small temperature drop in such films may be obtained from a measurement with hot and cold plates in contact and specimens removed, but the contacting medium present. Plates are invariably of high-conductivity metal such as copper, brass or aluminium to ensure uniform surface temperatures. A further refinement is to make measurements on samples with different thicknesses. Assuming that the surface heat transfer coefficient is constant it can then be eliminated from the calculations.

Temperatures (invariably of the plates rather than the specimens) are normally measured with fine wire thermocouples in conjunction with a DVM reading to 1 μV, and heat flows by measurement of heater input wattage, both measurements being capable of high precision. A stable power source is a necessity and batteries are still sometimes used, although the many admirable transistorised power supplies now available are more convenient, and do not suffer from the long-term drift associated with batteries. This is particularly important where measurements on low-conductivity materials are involved, since steady state conditions may not be reached for many hours after energising the heater.

The Guarded Hot Plate

An improvement on the unguarded apparatus is the guarded hot plate. This is the most accurate method available for solid materials (including foams), and is recommended by the standards organisations (ASTM C177,[23] BS 874[10]) for measurements on low-conductivity materials. The general arrangement is shown schematically in Fig. 15.4. The standards organisations recommend various dimension ratios for the guard width g, heater side length $2s$, and sample thickness l. If an accuracy of 1 per cent is to be achieved the limiting values of the ASTM specification should be used[12] and these are given in Table 15.3.

In this method the heat source is surrounded by a guard heater which has an independent power supply. The power to the guard is adjusted so that there is no temperature difference between the heat source and the guard. The heat from the heat source thus flows normally through the samples and the heat lost from the exposed edges comes from the guard.

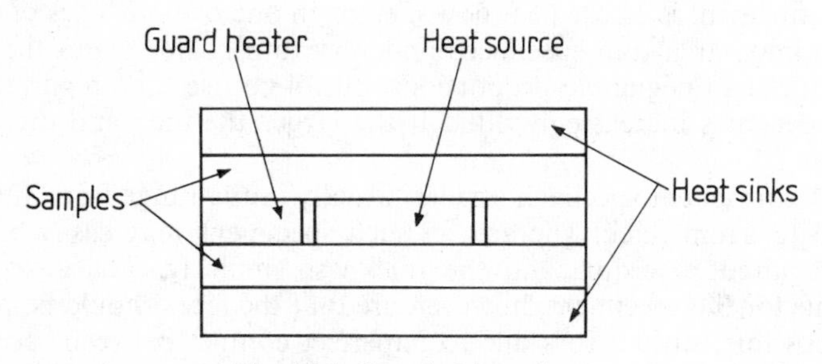

Fig. 15.4 Guarded hot plate

Table 15.3 Dimension ratios for guarded hot plate

Ratio	BS 874	ASTM C177
$\frac{2(g+s)}{l} \geqslant$	6	9
$\frac{g}{l} \geqslant$	1	1·5
$\frac{s}{l} \geqslant$	2	3

The heat sinks are controlled at a given temperature, usually by circulating a liquid from a constant temperature bath, and a known power is supplied to the heat source. The power supplied to the guard heater is adjusted to equalise the temperatures. The apparatus is allowed to reach steady state and the temperature drop across the samples is measured, usually with thermocouples. The conductivity is calculated using Eqn [15.6].

The attainment of complete equilibrium is even more important with low-conductivity materials than with solid plastics, since the long-time constant of cellular materials means that equilibrium times of the order of hours, or even a few tens of hours, are involved and the rate of change of temperature is slow and therefore difficult to detect. For this reason power supplies which are stable over long time periods are a necessity. Since temperature differences of the order of 0·05 °C between centre and guard heater plates have a significant effect on the measurement, the selection and construction of thermocouples is extremely important and it is necessary to determine the thermal e.m.f.s to within 1 μV.

As temperature differences across the specimens of as little as 10 °C or less may be used, cold plate temperatures must be maintained very constant and a thermostat bath temperature control of at least $\pm 0{\cdot}1$ °C is required.

The apparatus dimensions are usually 300 mm square and the sample thickness range is from 6 mm to 50 mm. There is a gap between the heat source and the guard heater of about 2 mm. The area used in Eqn [15.6] for calculating the conductivity is determined from the centre of this gap and is fractionally larger than the actual area of the heater.

In operation, unless the guards are controlled automatically, it is tedious to balance exactly guard and centre temperatures and it is usually more convenient to know the effect of a given amount of off-balance on the answer so that an appropriate correction can be applied. To do this, small adjustments to the guard energy are made around the balance point and the differences between guard and centre thermocouple readings plotted against apparent conductivity.

The centre energy must be maintained constant throughout and after each guard adjustment the apparatus must be left for equilibrium to be re-established. A typical plot is shown in Fig. 15.5 and the correct conductivity is given, of course, by the intercept at zero difference. Over small temperature differences the plot is linear.

The virtue of this refinement in technique is that, with experience, the slope of the plot indicates whether the apparatus is responding properly or not, and also

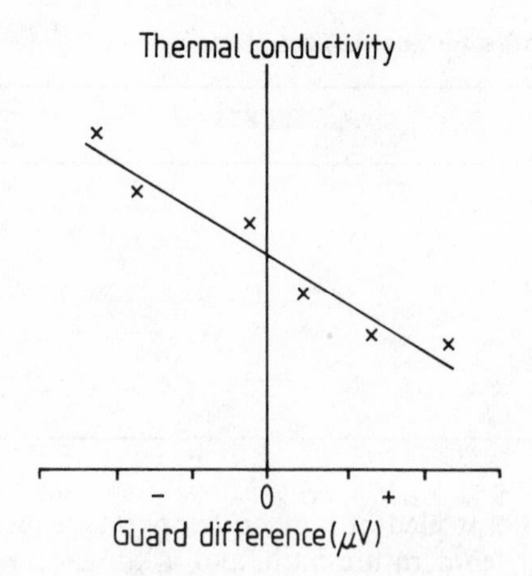

Fig. 15.5 Thermal conductivity plotted against guard difference

it is a measure of the efficiency of the plate, a high slope being undesirable since it indicates a high lateral conductivity between centre and guard portions of the hot plate.

Errors caused by a resistance to heat flow across the interface between the conductivity cell surfaces and the sample surfaces are more serious when measurements are being made on solid materials rather than on foams, because the resistance to heat flow across the interface is a larger fraction of the resistance of the sample. The errors can be minimised by wetting the surfaces with a heat transfer grease or oil. Measurements can be made on samples with different thicknesses as described in the previous section.

Coaxial Cylinder Methods

Methods based on the coaxial cylinder geometry have the advantage that the exposed areas, from which heat can be lost, are much smaller compared with the area of the heat source than is the case with the hot plate methods. Thus the effects of lateral heat flow can be minimised without introducing guard rings. The essential features of a cylindrical conductivity cell are shown schematically in Fig. 15.6.

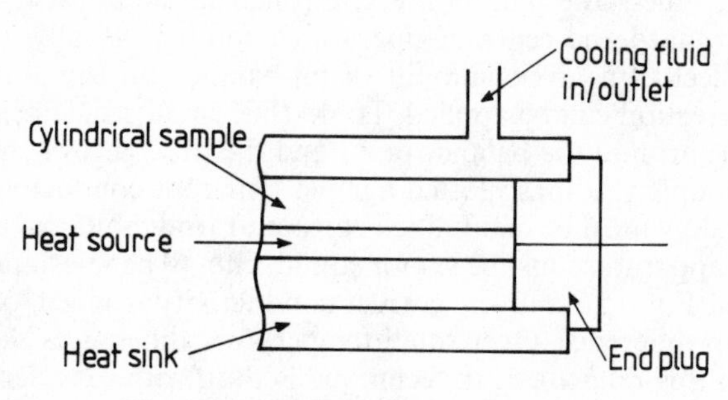

Fig. 15.6 Coaxial cylinder apparatus

The apparatus comprises a rod-shaped heat source surrounded by a tubular sample inside a cylindrical heat sink. The heat source and the heat sink have the same axis. The temperature of the heat sink is controlled by circulating either water or oil from a constant temperature bath. A known power is supplied to the heat source, and the temperature difference between the heat source and heat sink is measured after steady state conditions have been reached. The conductivity is obtained from Eqn [15.7].

The apparatus is constructed with a large length-to-diameter ratio so that the heat lost from the ends is small compared with the heat transferred radially through the sample. This was checked for solid samples by Kline[13] who made cells with length to diameter ratios of 8:1, 12:1 and 24:1. The longer pair gave identical results while results from the shorter one were 10 per cent lower. However, the greater the relative heat loss the greater should be the calculated conductivity. This shows that the major source of error is not heat loss from the ends but resistance to heat flow across the surfaces, because it is impossible to ensure that the sample is a perfect fit in the apparatus.

Quasi-stationary Method

Eiermann *et al.*[14] developed a quasi-stationary method of measuring the thermal conductivity of solid polymers. The apparatus consisted of two heat sources and two thin samples on either side of a copper heat sink the temperature of which was monitored and which was surrounded by an adiabatic shield to minimise the heat loss. The instantaneous heat flow rate was determined from the rate of temperature rise of the heat sink

$$q_1 + q_2 = Mc\frac{\mathrm{d}\theta}{\mathrm{d}t} \qquad [15.8]$$

where q_1 and q_2 are the heat flow rates across the two samples, M is the mass of the heat sink, c is the specific heat of the heat sink, and $\mathrm{d}\theta/\mathrm{d}t$ is its rate of temperature rise. Since $q = -KA(\Delta\theta/x)$ this gives for the conductivity

$$K = -\frac{Mcx}{A(\Delta\theta_1 + \Delta\theta_2)}\frac{\mathrm{d}\theta}{\mathrm{d}t} \qquad [15.9]$$

The temperature range covered by the method was from $-190\,^\circ\mathrm{C}$ to $+90\,^\circ\mathrm{C}$.

The heat capacity of the samples was involved in the analysis of the results because the experiment was dynamic. This was eliminated by repeating each experiment with a heat sink of different thermal capacity. Thermal resistance was eliminated by making measurements in a helium atmosphere and then in a nitrogen atmosphere. Thus, to obtain conductivity values for one material four separate measurements were required. Apparatus of this type has recently been used by Ott.[15]

15.2.2 Transient Methods

The time-dependent temperature distribution in a transient experiment is governed by Eqn [15.4] and usually the related parameter, thermal diffusivity, is obtained. However, under certain circumstances the solution of the heat equation contains the thermal conductivity as well as the diffusivity, and by choosing a suitable method the diffusivity can be eliminated from the answer. The more important methods are the line and plane source heater methods.

Line Source Method

The most common transient techniques which gives the thermal conductivity directly is the continuous line source method. For a line source heater of infinite length in an infinite mass of sample the temperature at a distance r from the line source is given by[16]

$$\theta = \frac{q}{4\pi K}\int_a^{\infty}\frac{e^{-u}}{u}\,du \qquad [15.10]$$

where $a = r^2/4\alpha t$, and q is the rate of heat generation per unit length. For small values of a the integral reduces to $\log_e(4\alpha t/r^2) - \text{const}$. Thus if the diffusivity and conductivity are constant over the temperature range $\Delta\theta$ then

$$\theta_2 - \theta_1 = \Delta\theta = \frac{q}{4\pi K}\log_e\frac{t_2}{t_1} \qquad [15.11]$$

where θ_1 is the temperature at r at time t_1 and θ_2 is the temperature at time t_2.

Rearrangement gives

$$K = \frac{q}{4\pi\Delta\theta}\log_e\frac{t_2}{t_1} \qquad [15.12]$$

Hence the conductivity can be obtained directly by measuring q and the change of temperature with time.

It is necessary to make small corrections for finite heater wire diameter, finite wire length, and finite sample mass. These have been discussed in detail for liquids by Horrocks and McLaughlin,[17] who found that the corrections were only significant for short heating times. This technique has been used with polymers,[18,19] and has also been developed in the form of a conductivity probe[20–22] for making single-point measurements on, for example, foams.

Plane Source Method

An alternative to a line source is to use a plane source of constant flux, J. If the sample is assumed to be a semi-infinite solid ($0 \leqslant x < \infty$) then the temperature at a point x and time t is given by[23]

$$\theta = \frac{2J}{K}\sqrt{(\alpha t)}\,\operatorname{ierfc}\frac{x}{2\sqrt{(\alpha t)}} \qquad [15.13]$$

At the point $x = 0$

$$\theta = \frac{2J}{K}\left(\frac{\alpha t}{\pi}\right)^{1/2} \qquad [15.14]$$

The temperature at the surface, $x = 0$, and at a point in the sample, $x = 1$, say, are measured as functions of time, and from this both K and α can be determined. The method was developed by Harmathy[24] for measurements on building materials, and in his paper he also discussed the limitations on the technique imposed by the finite size of the samples. The method has been used for polymers by Steere.[25,26] However, to obtain a sufficiently thick sample he stacked thin films together but ignored the resistance to heat flow at the interfaces. A further difficulty with Steere's technique is that thin films are invariably oriented.

15.3 THERMAL DIFFUSIVITY MEASUREMENT

To measure diffusivity it is only necessary to know the change in temperature with time at three colinear points in the direction of the heat flow. For conductivity, although the temperature gradient can easily be obtained, the heat flux cannot be measured at a point; the total power to the heat source is known but it is difficult to control or calculate the movement of all the energy. Although diffusivity is easier to measure, and although it is the essential parameter for transient heat flow calculations, it has received less attention than conductivity. The scatter in the diffusivity data in the literature is not as great as for conductivity (possibly because not as many people have tried to measure it!).

15.3.1 Basic Principles

Boundary Conditions

For certain boundary and initial conditions analytical solutions to Eqn [15.4] can be obtained. Apart from a continuous heating method,[2] which is based on a numerical calculation, diffusivity measurement methods are all based on such solutions. The experimental conditions are matched to these mathematical conditions as closely as possible and the appropriate solution is used to give a value for the diffusivity. The experiment can be repeated at different temperatures in order to obtain the temperature-dependence of the diffusivity. The experimental procedure has been criticised[27] because if the diffusivity changes with temperature then almost invariably the conductivity is also temperature-dependent and Eqn [15.3], which would not have given an analytical solution, should have been used instead of Eqn [15.4]. However, Hands and Horsfall[28] have shown that, except near melting transitions, thermocouples sensitive to 0·002 °C would be needed to detect the effect of the conductivity term in Eqn [15.3]. Hence generally the simpler equation is adequate for diffusivity measurement and for heat flow calculations.

Diffusivity measurement methods based on analytical solutions to Eqn [15.4] have all had the same initial condition that the whole sample is at a constant uniform temperature. But three different types of boundary conditions have been employed: first the sample surface is subjected to a step change in temperature; second the surface is subjected to a linear rate of temperature rise; and third the surface is subjected to a periodic temperature fluctuation.

The solutions to Eqn [15.4] for these three types of boundary condition are usually presented in terms of dimensionless groups and are applicable to infinite solids. Before proceeding to discuss the experimental methods it is necessary to consider these points in more detail and also the application of the solutions to finite solids. The word 'solid' here is a mathematical term and does not refer to the phase of the sample.

Dimensionless Groups

The solutions to Eqn [15.4] may be presented graphically[29,30] and in order to give the greatest amount of information with the minimum number of curves the graphs are presented in terms of four dimensionless groups.

In problems involving a step change in surface temperature the unaccomplished temperature change of a point in the sample is related to the maximum possible temperature change. Hence the dimensionless temperature is given by

$$Y=\frac{\theta_1-\theta}{\theta_1-\theta_0} \qquad [15.15]$$

where θ_0 is the initial uniform temperature, θ_1 is the temperature of the heating or cooling medium, and θ is the temperature at any point in the sample for $t>0$. All values of Y lie between 0 and 1. The extreme values indicate the heating or cooling temperature $\theta=\theta_1$, and the initial temperature $\theta=\theta_0$, respectively.

Some authors use an alternative dimensionless temperature

$$\frac{v}{V}=1-Y=\frac{\theta-\theta_0}{\theta_1-\theta_0} \qquad [15.16]$$

and this is the ratio of the actual temperature change to the maximum possible temperature change.

The times required for similar temperature changes in similar bodies are proportional to the square of the dimensions and inversely proportional to the thermal diffusivity. It is therefore convenient to use a dimensionless group given by

$$\phi=\frac{\alpha t}{a^2} \qquad [15.17]$$

where a is the radius of a sphere of cylinder, of the semi-thickness of a slab.

At any time, the temperature at a point in the sample is related to the ratio of the distance of the point from the centre to the radius or semi-thickness. Hence the dimensionless length is given by

$$\xi=\frac{r}{a} \qquad [15.18]$$

where r is the distance measured from the centre along a radius of a sphere of cylinder, or the distance measured from the centre plane of a flat slab normal to the surfaces.

The fourth dimensionless group must be introduced for problems involving a resistance to heat transfer across the interface. The heat transfer coefficient can be combined with the conductivity and the radius or semi-thickness to give a dimensionless resistance ratio.

$$m = \frac{K}{ah} \qquad [15.19]$$

When h is infinite m is zero and there is no resistance to heat flow across the interface.

It should be pointed out that although the reciprocal of m looks like a Nusselt number it is not one. For a Nusselt number the conductivity would be that of the heating or cooling fluid and not, as in this case, the conductivity of the sample.

Infinite and Non-infinite Solids

The analytical solutions to the heat equation are for one dimensional problems; this implies that the heat flow is in one direction only or that the equation reduces to one dimension because of symmetry, i.e. the temperature is a function of one length variable only. This condition is satisfied for three sample geometries: the infinite flat slab, the infinite circular cylinder and the sphere. An infinite slab is a parallel-faced slab of finite thickness and infinite length and breadth; an infinite cylinder is a cylinder of finite diameter and infinite length.

Real samples are obviously not infinite and the practical definition of an infinite solid depends on how much error can be tolerated. For example if one lateral dimension of the slab is reduced from infinity to five times the thickness, or the length to diameter ratio of a cylinder is 5:1, then a 1 per cent error is introduced if lateral heat flow is ignored.

The solutions to Eqn [15.4] for infinite solids are of the form

$$Y = f(\phi, \xi, m) \qquad [15.20]$$

and are in the form of infinite series. For samples in the form of cubes, rectangular parallelepipeds (bricks) and short right circular cylinders, where lateral heat flow cannot be ignored, the infinite solid solutions can still be used.[31] For a brick, by considering heat transfer through each pair of faces in turn (imagining the lateral dimensions to be infinite) three values of the dimensionless temperature are obtained. The final solution is the product of these three, $Y = Y_x Y_y Y_z$. Similarly the solution for a short cylinder is given by the product of the solutions for an infinite slab and an infinite cylinder, $Y = Y_z Y_r$.

This approach is based on separation of variables and it can be shown by substitution that a product of solutions is itself a solution. Separation of variables is only valid for certain boundary conditions and imposes restrictions on geometry which are, however, satisfied by the examples under consideration.

Sources of Error

In the experimental techniques, which will be described below, there are a number of recognised sources of error. These have not always been avoided and although the measurement is basically straightforward the results in the literature are not always reliable. The significance of the errors will be discussed in connection with the methods of measurement and only a summary is given here:

1. Assuming an infinite surface heat transfer coefficient between the sample and the heating or cooling medium.

2. Ignoring lateral heat flow when it is significant.
3. Conduction along thermocouple leads.
4. Assuming that diffusivity is temperature independent.
5. Ignoring the thermal expansion of the samples.

15.3.2 Quenching Methods

Thermal diffusivity has usually been measured using a quenching method, i.e. the solid sample at a uniform temperature is immersed in a temperature-controlled bath at a different temperature. The rate of change of temperature at the centre is then monitored with an embedded thermocouple. The sample dimensions are usually chosen so that lateral heat flow can be ignored and regular sample geometries, i.e. 'infinite' flat slabs, 'infinite' cylinders, or spheres, are used.

Quenching methods are based on the assumption that the thermal diffusivity is constant over the experimental temperature range. To obtain the temperature-dependence of the diffusivity, the temperature range of interest is covered in a series of steps which are small enough for this assumption to be valid. The size of the step temperature change is a compromise between this requirement and the need to make accurate temperature difference measurements. The value which is usually chosen is about 5 °C. In some experiments this point has been ignored and the measurement made with a much larger step. For example the sample may be conditioned at room temperature and then immersed in boiling water. An average value for the diffusivity is obtained and the error depends on the temperature dependence of the diffusivity and hence changes from material to material.

Two methods have been used to obtain a value for the diffusivity from the experimental results. As was shown above, for the geometries in question the analytical solutions to the heat equation, which are in the form of infinite series, can be conveniently represented graphically. The first such collection of graphs was published by Williamson and Adams.[39] By reference to such graphs a value for ϕ and hence α can be obtained.

The second method is based on the rapid convergence of the series solutions. For example after about 10 per cent of the time for the centre temperature to reach 99 per cent of the surface temperature has elapsed, the second term in the series is about $\frac{1}{2}$ per cent of the first term. Thus after a certain time, only the first time is relevant. Now Y is always an exponential function of ϕ; for example consider an infinite slab of thickness $2a$, the first term of the series for the dimensionless temperature at the centre is given by

$$Y = \frac{4}{\pi}\exp\left(-\frac{\pi^2}{4}\phi\right) \qquad [15.21]$$

For long times this is the only significant term and the diffusivity can be obtained from the slope of a graph of $\log_e Y$ against t since $\phi = \alpha t/a^2$.

15.3.3 Linear Heating Method

The second type of boundary condition to be considered is that in which the sample surface is subjected to a linear rate of temperature rise. A method based on this has been developed by Shoulberg[32] for diffusivity measurements on polymer melts. He used two discs of his material with a thermocouple sandwiched between them; the diameter-to-thickness ratio was such that his sample could be regarded as an infinite flat slab. The sample completely filled a cavity in an aluminium block and was melted in the apparatus. The aluminium block was heated electrically and the power was adjusted to give an approximately linear rate of temperature rise. Under his experimental conditions this lasted for about 30 °C.

For this boundary condition the solution to the heat equation for the temperature at the centre is[33]

$$\theta = kt - \frac{ka^2}{2\alpha} + \frac{16ka^2}{\alpha\pi^3}\sum_{n=0}^{\infty}\frac{(-1)^n}{(2n+1)^3}\exp\left(\frac{-(2n+1)^2\pi^2\phi}{4}\right) \qquad [15.22]$$

where k is the linear rate of temperature rise. After a long time (in Shoulberg's case about 12 min) the summation term is negligible and the temperature difference between the surface and the centre becomes constant. The diffusivity is then given by

$$\alpha = \frac{ka^2}{2\Delta\theta} \qquad [15.23]$$

where $\Delta\theta$ is the temperature difference between the surface and the centre of the sample, and a is the semi-thickness of the sample, i.e. the thickness of one of the discs. The temperature range of interest was covered in a series of 30 °C steps, and using Eqn [15.24] he obtained an average value for the diffusivity for each step.

15.3.4 Periodic Heating Method

A method for measuring the diffusivity of solid polymers based on this type of boundary condition has been developed by Berlot.[34,35] A disc sample of thickness $2a$ is held at a uniform temperature and then a sinusoidal temperature fluctuation of angular frequency ω is imposed on the outer surfaces. The amplitude ratio and phase of the temperature at the centre are monitored with a thermocouple. Under these conditions the amplitude ratio, A, and phase, ϕ, are given by[36]

$$A = \left(\frac{2}{\cosh 2ka + \cos 2ka}\right)^{1/2} \qquad [15.24]$$

$$\phi = \arg\left(\frac{1}{\cosh ka(1+\mathrm{i})}\right) \qquad [15.25]$$

$$k = \left(\frac{\omega}{2\alpha}\right)^{1/2} \qquad [15.26]$$

k can be obtained from Eqn [15.24] or Eqn [15.25] and hence α from Eqn [15.26].

15.3.5 Continuous Heating Method

If the temperature-dependence of diffusivity is taken into account then the temperature range from ambient up to, say, 250 °C can be covered in one experiment. With a quenching method a true step change in temperature is difficult to achieve because liquids capable of withstanding high temperatures tend to have high viscosities and this results in large temperature gradients close to the sample surface. Another objection to a quenching method is that the predominance of the quenching temperature throughout the experiment coupled with the extremely large initial temperature difference could give rise to computational difficulties. These problems, inherent in the quenching method, do not occur in a continuous heating method.

A method based on this principle was developed by Hands and Horsfall[2] and has been further developed by Smith.[37] The apparatus is shown schematically in Fig. 15.7. Two disc-shaped samples are placed together and a thermocouple is sandwiched between them for monitoring the centre temperature. The outside temperatures are measured with two further thermocouples which are in intimate contact with the surfaces of the sample. The three thermocouple junctions lie on the axis of symmetry. The sample sandwich is contained in a brass ring and two brass end plates which contain electrical heaters.

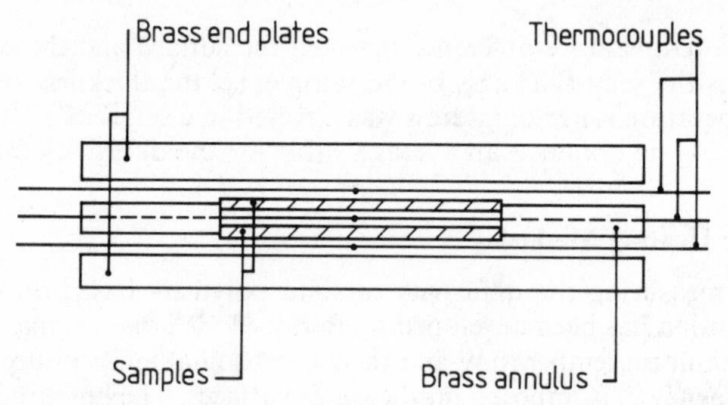

Fig. 15.7 Schematic diagram of apparatus for continuous heating method

Enclosing the samples allowed measurements to be made on molten samples and eliminated errors caused by ignoring thermal expansion. In quenching methods, for example, the appropriate sample dimension is measured at room temperature and this is the value which is used in the calculations. Since diffusivity depends on the square of the thickness any error caused by ignoring expansion is automatically doubled. This error can be a few per cent depending on the temperature range covered.

In the original apparatus the end plates contained expansion holes and the excess polymer extruded out during the course of a measurement. In Smith's modified apparatus the samples are contained by O-rings and the end plates are allowed to move against springs; the change in thickness is measured and used in the calculation.

In use the heaters are energised and the thermocouple outputs monitored at 20 s time intervals. The temperature range up to 300 °C is covered in about 20 min. The heat equation is solved numerically[38] to give thermal diffusivity as a function of temperature.

15.3.6 Pulse heating method

If a disc-shaped sample is irradiated on one surface for a short time using a flash tube or a pulsed laser, then the temperature–time curve for the back face depends on the thermal diffusivity of the sample and the heat losses. If the pulse time τ is very much shorter than the time for the pulse to pass through the sample, then the heat losses can be ignored. Cape and Lehman[39] have given a criterion for this condition

$$a^2/\pi^2\alpha' \geqslant 10\,\tau \qquad [15.27]$$

where a is the thickness of the disc. Under these conditions the ratio of the temperature to the maximum technique for the back face is given by[40]

$$\frac{\theta}{\theta_{max}} = 1 + 2\sum_{n=1}^{\infty}(-1)^n \exp\left(-\frac{n^2\pi\alpha t}{a^2}\right) \qquad [15.28]$$

The temperature as a function of time on the back face is usually measured with a radiation pyrometer.[40] Analysis of the results has been discussed by Taylor[41] and Stuckes.[42]

15.4 SPECIFIC HEAT

Specific heat, c, is the heat capacity, C, per unit mass or per unit volume; usually the term refers to the mass specific heat. Heat capacity is defined by

$$C = \frac{dQ}{dT} \qquad [15.29]$$

where Q is the heat energy absorbed and T is the temperature. The heat capacities of greatest interest are those at constant pressure C_p, and constant volume C_v. Substituting into Eqn [15.28] for the heat energy absorbed, from the first law of thermodynamics, assuming that the work done is of a mechanical nature only, the heat capacities are given by

$$C_p = \left(\frac{\partial U}{\partial T}\right)_p + P\left(\frac{\partial V}{\partial T}\right)_p \qquad [15.30]$$

and

$$C_v = \left(\frac{\partial U}{\partial T}\right)_v \qquad [15.31]$$

where U is the internal energy of the system, P is the pressure, and V is the volume.

The heat capacity at constant volume is of more interest from a theoretical point of view because it is directly related to the internal energy of the system. However, it is almost impossible to measure C_v directly. To obtain a value for C_v it is necessary to measure C_p and to calculate C_v from the formula[43]

$$C_p - C_v = TVB\beta^2 \qquad [15.32]$$

where B is the bulk modulus, and β is the volume expansion coefficient. In terms of the principal specific heats this is

$$C_p - C_v = \frac{TB\beta^2}{\rho} \qquad [15.33]$$

Putting in some typical values, the difference between the principal specific heats for polymers at room temperature is of the order of 0·004 J/g K, which is usually small enough to be ignored.[5] Typical specific heat values for solids, including plastics, range from 0·4 J/g K to 4 J/g K. Data is given in the review by Wunderlich.[44]

15.4.1 Adiabatic Calorimeters

One of the most precise ways of measuring specific heat is by means of an adiabatic calorimeter; it is also the most relevant and direct method since a small measured amount of heat is applied to the specimen and its resultant temperature rise recorded. It is adiabatic since the temperature of the system is allowed to rise, but no heat exchange with the surroundings is permitted and this is normally contrived by surrounding the calorimeter proper by a jacket maintained at the same temperature to within very close limits (of the order of 0·01 °C or less). All the heat applied to the specimen, less that used in raising the temperature of the calorimeter itself, is therefore employed in raising the specimen temperature; the specific heat is simply calculated. The thermal capacity of the calorimeter is determined in a separate experiment.

A number of workers have described adiabetic calorimeters and the constructional and experimental difficulties.[45–58]

15.4.2 Drop calorimeters

Drop calorimeters are widely used on account of their simplicity. A specimen, often contained in a metal capsule, is heated to some appropriate constant temperature in an oven or furnace and allowed to drop into liquid in a stirred calorimeter. The temperature of the calorimeter is plotted against time and from the curve the temperature rise, allowing for heat losses or gains to or from the environment, is obtained. The specific heat is calculated from this rise after applying corrections for the water equivalent of the calorimeter, etc., determined separately.

ASTM C351 (1982)[59] *Mean Specific Heat of Thermal Insulation* is typical, although not intended specifically for plastics.

Sometimes the drop calorimeter and adiabatic methods are combined, the heated capsule being dropped into an adiabatic container. Griskey and Hubbell[61] describe such a method as applied to measurements on methacrylic polymers in the range from 120 to 300 °C.

15.4.3 Scanning calorimeters

The usual techniques are differential thermal analysis (DTA)[61] and differential scanning calorimetry (DSC).[62–64] In these methods it is assumed that the heat loss from the calorimeter is a function of temperature only. By comparing the rate of heat input and temperature rise for a polymer sample with that of a standard, usually synthetic sapphire, the specific heat of the polymer can be calculated.

DTA measures the difference in temperature between the sample and a standard for the same rate of heat input. DSC compares the rate of heat inputs for the same rate of temperature rise. The latter is easier to analyse as it gives a direct measure of the rate of heat input. The method is based on the assumption that the samples are so small that thermal equilibrium is obtained almost immediately. For polymers this is not correct and errors from this source are discussed by Strella and Erhardt.[66]

Richardson[66–68] and Laye[69] have discussed methods of calibrating the DSC to improve the accuracy of the results.

REFERENCES

1. Kittel, C. (1966) *Introduction to Solid State Physics*, 3rd edn, Wiley, p. 168.
2. Hands, D. and Horsfall, F. (1977) *Rubber Chemistry and Technology*, **50**, 253.
3. Carslaw, H. S. and Jaeger, J. C. (1959) *Conduction of Heat in Solids*, 2nd edn, Oxford University Press, p. 9.
4. Parrott, J. E. and Stukes, A. D. (1975) *Thermal Conductivity of Solids*, Pion, p. 5.
5. Choy, C. L. (1975) *Journal of Polymer Physics*, **13**, 1263.
6. Howard, J. R. (1975) *Engineering*, **215**, 220.
7. McAdams, W. H. (1957) *Heat Transmission*, 3rd edn, McGraw-Hill.
8. Cuthbert, C. (1954) *Transactions of the Institution of the Rubber Industry*, **30**, 16.
9. Lees, C. H. (1898) *Philosophical Transactions of the Royal Society*, **A191**, 399.
10. BS 874 (1973) (1980) *Methods for Determining Thermal Insulating Properties.*
11. ASTM C518 (1985) *Steady-State Thermal Transmission Properties by Means of the Heat Flow Meter.*
12. Parrott, J. E. and Stukes, A. D. (1975) op. cit., p. 19.
13. Kline, D. E. (1961) *Journal of Polymer Science*, **50**, 441.
14. Eiermann, K., Hellwege, K. H. and Knappe, W. (1961) *Kolloidzeitschrift*, **174**, 134.
15. Ott, H. J. (1981) *Plastics and Rubber Processing Applications*, **1**, 9.
16. Carslaw, H. S. and Jaeger, J. C. (1959) op. cit., p. 261.
17. Horrocks, J. K. and McLaughlin, E. (1963) *Proceedings of the Royal Society*, **A273**, 259.
18. D'Eustachio, D. and Schreiner, R. E. (1952) *Journal of the American Society of Heating and Ventilating Engineers*, **58**, 331.
19. Vos, B. H. (1956) *Applied Scientific Research*, **A5**, 425.
20. Mann, G. and Forsyth, F. G. E. (1956) *Modern Refrigeration*, **188**, June.
21. Hooper, F. C. and Lepper, F. R. (1950) *Heating Piping and Air Conditioning*, **20**, August, 129.
22. Hooper, F. C. and Chang, S. C. (1952) *Heating Piping and Air Conditioning*, **22**, October, 125.
23. Carslaw, H. S. and Jaeger, J. C. (1959) op. cit., p. 75.
24. Harmathy, T. Z. (1964) *Journal of Applied Physics*, **35**, 1190.
25. Steere, R. C. (1966) *Journal of Applied Physics*, **37**, 3338.
26. Steere, R. C. (1966) *Journal of Applied Polymer Science*, **10**, 1673.
27. Martin, B. (1970) *Polymer*, **11**, 287.
28. Hands, D. and Horsfall, F. (1971) *Polymer*, **12**, 145.
29. Williamson, E. D. and Adams, L. H. (1919) *Physical Review*, **14**, 99.
30. Hands, D. (1971) *RAPRA Technical Review*, 60.
31. Carslaw, H. S. and Jaeger, J. C. (1959) op. cit., p. 33.
32. Shoulberg, R. H. (1963) *Journal of Applied Polymer Science*, **7**, 1597.

33. Carslaw, H. S. and Jaeger, J. C. (1959) op. cit., p. 104.
34. Berlot, R. (1966) *Plastiques Modernes et Elastomères*, **18**, 231.
35. Berlot, R. (1969) *Plastiques Modernes et Elastomères*, **19**, 117.
36. Carslaw, H. S. and Jaeger, J. C. (1959) op. cit., p. 105.
37. Smith, D. I. (1987) Ph.D. Thesis, University of Bradford.
38. Horsfall, F. (1976) M. Sc. Thesis, University of Bradford.
39. Cape, J. A and Lehman, G. W. (1963) *Journal of Applied Physics*, **34**, 1909.
40. Parker, W. J., Jenkins, R. J., Butler, C. P. and Abbott, G. L. (1961) *Journal of Applied Physics*, **45**, 2321.
41. Taylor, R. (1965) *British Journal of Applied Physics*, **16**, 509.
42. Parrott, J. E. and Stuckes, A. D. (1975) op. cit., p. 37.
43. Zemansky, M. W. (1957) *Heat and Thermodynamics*, 4th edn, McGraw-Hill, (1970), p. 251.
44. Wunderlich, B. and Baur, H. (1970) *Advances in Polymer Science*, **7**, 151.
45. Southard, J. C. and Brickwedde, F. G. (1933) *Journal of the American Chemical Society*, **55**, 4378.
46. Bekkedahl, N. and Matheson, H. (1935) *Journal of Research of the National Bureau of Standards*, **15**, 503.
47. Scott, R. B., Myers, C. H., Rands, R. D. Jr, Brickwedde, F. G. and Bekkedahl, N. (1945) *Journal of Research of the National Bureau of Standards*, **35**, 39.
48. Williams, J. W. and Daniels, F. (1924) *Journal of the American Chemical Society*, **46**, 903.
49. Dole, M., Hettinger, W. P. Jr, Larson, N. R., Wethington, J. A. and Worthington, A. E. (1951) *Review of Scientific Instruments*, **22**, 812.
50. Worthington, A. E., Marx, P. C. and Dole, M. (1955) *Review of Scientific Instruments*, **26**, 698.
51. West, E. D. and Ginnings, D. C. (1958) *Journal of Research of the National Bureau of Standards*, **60**, 309.
52. Richardson, M. J. (1965) *Transactions of the Faraday Society*, **61**, 1876.
53. Aukward, J. A., Warfield, R. W., Petree, M. C. and Donovan, P. (1959) *Review of Scientific Instruments*, **30**, 597.
54. Tautz, H., Gluck, M., Hartmann, G. and Leuteritz, R. (1963) *Plastik und Kautschuk*, **10**, 648.
55. Hellwege, K. H., Knappe, W. and Wetzel, W. (1962) *Kolloidzeitschrift*, **180**, 126.
56. Wunderlich, B. and Dole, M. (1957) *Journal of Polymer Science*, **24**, 201.
57. Bowring, R. W., Garton, D. A. and Norris, H. F. (1960) *UKAEA Report AEEW–R. 38*, HMSO.
58. Dainton, F. S., Evans, D. M., Hoare, F. E. and Melia, T. P. (1962) *Polymer*, **3**, 263.
59. ASTM C351 (1982) *Mean Specific Heat of Thermal Insulation.*
60. Griskey, R. G. and Hubbell, D. O. (1968) *Journal of Applied Polymer Science*, **12**, 853.
61. Slade, P. E. Jr and Jenkins, Ll. T. (Eds). (1966) *Techniques and Methods of Polymer Evaluation*, vol. 1, *Thermal Analysis*, Arnold.
62. Watson, E. S., O'Neill, M. J., Justin, J. and Brenner, N. (1964) *Analytical Chemistry*, **36**, 1233.
63. O'Neill, M. J. (1964) *Analytical Chemistry*, **36**, 1238.
64. O'Neill, M. J. (1966) *Analytical Chemistry*, **38**, 1331.
65. Strella, S. and Erhardt, P. F. (1969) *Journal of Applied Polymer Science*, **13**, 1373.
66. Richardson, M. J. and Burrington, P. (1974) *Journal of Thermal Analysis*, **6**, 345.
67. Richardson, M. J. (1976) *Plastics and Rubber: Materials and Applications*, **1** (3–4) 162.
68. Richardson, M. J. (1984) *Polymer Testing*, **4** (2–4).
69. Laye, P. G. (1980) *Analytical Proceedings*, **17**(6), 226.

Chapter Sixteen

Effect of Temperature

16.1 INTRODUCTION

Plastics materials as a whole, and thermoplastics in particular, are sensitive to temperature; whereas those that are ductile or flexible may demonstrate shortcomings in becoming brittle or rigid at some low temperature, far more serious as a rule is the softening and even degradation that may occur when the temperature is raised to only moderate levels – that is, 'moderate' in the eyes of the designer, engineer or artisan more familiar with traditional materials such as metal, wood and stone.

Much effort has been expended in designing tests to measure the upper and lower temperature limits of performance of plastics, for most of them do not demonstrate sharp melting points or first-order transitions. On the contrary, they change their physical condition over a range of temperatures, the extent and position of which is dependent on a host of variables such as physical form, processing cycle, moisture content, time-scale of measurement and physical parameter of interest.

For those few plastics which show well defined melting points suitable methods of tests are described in this chapter.

Standardised procedures for assessing the softening characteristics of plastics are invariably *ad hoc* in nature, selecting some mechanical property – usually bending stiffness or hardness – which is characterised indirectly by a measurement of deflection, the end-point being taken when some arbitrarily chosen value of deflection is reached. Rate of heating and level of stress are equally arbitrarily chosen. These tests are described in this chapter, but it will be realised at the outset that such tests yield figures which have no absolute significance; they are only intended as quality control methods and the data they yield should not be regarded as anything else. In the ultimate, the only satisfactory way of determining the upper or lower limits of usage temperatures is to identify the property or properties of significance for the intended use (they may be electrical rather than mechanical) and then to make appropriate fundamental measurements to follow the deterioration of value with change in temperature. Some hints on this subject are given in the following sections.

Prior to all this, however, in considering reversible effects of temperature, the methods for measuring thermal expansion and the related shrinkage will be described together with a brief description of methods of finding the glass transition temperature.

Most plastics materials, being based on a carbon–carbon chain backbone or containing a high content of this bond and of carbon–hydrogen bonding, have very limited resistance to actual degradation by heat. Depending on the temperature, the nature of the polymer and the presence or absence of oxygen, so initially breakdown or crosslinking (of thermoplastics) may result; ultimately, and invariably, total breakdown occurs into a variety of products ranging from carbon through to virtually pure monomer – again depending on the conditions, but particularly on the nature of the polymer. The last sections of this chapter are therefore devoted to methods of test and investigational methods for the irreversible effects of temperature, but do not include fire tests which are dealt with in Ch. 18.

16.2 THERMAL EXPANSION

The coefficients of linear and volume expansion are defined, respectively, as

$$\alpha = \frac{1}{l}\frac{(\partial l)}{(\partial T)_p} \quad \text{and} \quad \beta = \frac{1}{V}\frac{(\partial V)}{(\partial T)_p} \qquad [16.1,2]$$

where l is length, V is volume and T is temperature. For isotropic and homogeneous materials the two are related by $\beta = 3\alpha$. However, plastics are not often isotropic and homogeneous and if accurate values are needed it is necessary to measure the linear expansion in orthogonal directions. A useful review of methods of measuring the thermal expansion of polymers has been given by Griffiths.[1]

ASTM D969 (1979)[2] gives a method for determining the coefficient of linear thermal expansion using a quartz dilatometer and a dial gauge. Linear expansion can also be measured with the commercially available thermomechanical analysers (TMA). These instruments usually cover a wider temperature range and use a linear displacement transducer for measuring the displacement.

Volume expansion is usually measured with a liquid-in-glass dilatometer [ASTM D864 (1978)[3]]. A detailed account of this technique has been given by Bekkedahl.[4] The sample is placed in the dilatometer and covered with a known mass of liquid which must not react with the sample. The liquid which is most commonly used is mercury. The dilatometer is suspended in a temperature-controlled bath and allowed to reach equilibrium. The position of the liquid meniscus in the capillary tube is determined with a cathetometer. The temperature is changed and after thermal equilibrium has been achieved the change in the liquid height is measured. The temperature range of interest is covered in a series of small steps. After correction for the expansion of the containing liquid and the dilatometer, the volume expansion of the sample can be calculated for each step. Great care has to be taken in calibrating and operating the apparatus, although the dilatometer method is essentially simple and can yield very accurate results.

16.3 SHRINKAGE

Closely related to the coefficient of linear thermal expansion is the property known as 'shrinkage'; this general-sounding term is normally taken to apply to the difference between the linear dimensions of a moulding and the corresponding dimensions of the mould cavity from which the moulding was produced. This property is of importance to the processor, as has already been mentioned in Ch. 5. In addition there is the shrinkage which takes place when a resin polymerises, as, for example, when using an epoxy resin in a casting application, and shrinkage due to heat is also an important property for some plastics film products.

ISO 2577 (1984)[5] is the international standard for the determination of shrinkage of compression-moulded test specimens in the form of bars. In this method, test pieces in the form of bars, length greater than 80 mm, width from 10 to 15 mm and thickness from 4 to 10 mm are moulded in a positive or semi-positive compression mould. After slow cooling to 23 °C ± 2 °C in an atmosphere 50 ± 5 per cent r.h. for not less than 16 h and not more than 72 h, their length is measured to the nearest 0·02 mm (L_1) and compared with the length of the mould cavity measured to the nearest 0·01 mm (L_0). The moulding shrinkage (MS) is then calculated as a percentage by the relationship:

$$MS = \frac{L_0 - L_1}{L_0} \times 100 \qquad [16.3]$$

It is permissible to use engraved lines on the mould for the measurements and the standard requires that the test pieces are examined for warpage after moulding by testing them on a flat surface (e.g. glass or engineers' surface plate). Test pieces showing significant warpage are rejected.

The same standard also allows the determination of post-moulding shrinkage. Here the moulded and measured test pieces are conditioned at 80 °C ± 2 °C for urea formaldehyde moulding materials and 110 °C ± 3 °C for all other thermosetting moulding materials. For the normal determination the heating period is 168 ± 2 h, but for a rapid determination 48 ± 1 h is used. After the heating period the test pieces are allowed to cool and remeasured. The post-shrinkage (PS) is calculated as a percentage by subtracting the length after heating from the length before heating, dividing by the length before heating and multiplying by 100.

BS 2782 Method 640A (1979)[6] follows ISO 2577.

ASTM D955 (1979)[7] covers the subject somewhat more comprehensively in making specific reference to compression, transfer and injection moulding; for the last two a longitudinal bar is used. The mode of expression of results is in terms of change of length per unit length.

ASTM D1299 (1979)[8] describes a method for finding the shrinkage of moulded and laminated thermosetting plastics at elevated temperature. Discs 100 mm in diameter or 100 mm squares, both 3·2 m thick, are used as test pieces. A temperature of 70 °C, 90 °C, 105 °C, 130 °C, 180 °C or 230 °C is used, as appropriate, and the shrinkage is expressed as change in dimension per unit dimension.

DIN 53464 (1962),[9] which makes specific reference to phenolic, urea formaldehyde and melamine formaldehyde moulding materials, lays down precise moulding conditions for those materials and uses test bars 120 ± 2 mm long by 15 ± 0·5 mm wide by 10 ± 0·5 mm thick.

ASTM D701 (1981),[10] Section 7.1.8 gives a method for measuring shrinkage due to exposure for 30 min at 140 °C. Sheets 305 mm (12 in) square are used, and the separation of gauge marks placed thereon is measured to the nearest 0·25 mm (0·01 in). This test is specified for cast acrylic sheet material only but the standard states that it may be used for other forms of cast methacrylate products with suitable modification of technique.

There are several other standardised methods of tests for shrinkage of thermoplastics which are specific for certain types of product. BS 2782 Method 641A (1985),[11] details the determination of the dimensional stability at 100 °C of flexible PVC sheet. A strip of the sheet 250 mm long by 6·4 mm wide is cut and its length measured. The length is again measured after immersion for 15 ± 1 min in a water bath at 100 °C ± 1 °C. The result is expressed as a percentage change of length on the original length of test piece. ASTM D2732 (1983)[12] details a broadly similar test but uses test pieces 100 mm square. BS 3794: Part 2, Method 9[13] is a test method for finding the dimensional change at elevated temperature of decorative laminated sheet. It is designed to give an indication of the lateral dimensional changes of specimens taken from the sheet under test over an extreme range of relative humidities at elevated temperatures.

Test pieces 140 mm long and 12·7 mm wide are measured for length to the nearest 0·02 mm and heated at 70 °C ± 2 °C for 24 h, then cooled in a desiccator and the length is re-measured. Other measured test pieces are conditioned at 40 °C ± 2 °C and 90–95 per cent r.h. for 96 ± 4 h and the length is measured after removal of surface water. The percentage decrease in length (shrinkage s) is calculated after the oven treatment and the percentage increase in length (d) is calculated after the humidity treatment. The dimensional change of the sample under test is then expressed as the sum of s and d.

BS 2782 Method 643A[14] and ASTM D2838[15] describe test methods for determining the shrink properties of heat-shrinkable plastics films. Method 643A is a simple 20 ± 1 s immersion in a heating bath maintained at a specified temperature with a change in length determined and expressed as a percentage of the original length of the test piece. D2838 details the determination of shrink tension and orientation release stress of plastics film and thin sheeting. This test recommends the use of a commercially available apparatus and it is claimed to be useful to predict the appearance and performance of the film in shrink packaging applications. It does not determine the shrinkage on heating of film intended for shrink-wrapping applications, as does Method 643A of BS 2782. Method 634A is quite adequate for normal quality control purposes, while ASTM D2838 is more appropriate for research and development.

Finally, two published standards for determining the shrinkage of casting resins during cure, ISO 3521 (1976)[16] and ASTM D2566 (1979),[17] give two quite different methods: ISO 3521 determines the volume change by comparing the density of the cured resin with that of the uncured resin, while ASTM D2566 uses the more conventional method of comparing the length of a cured casting with the length of the mould used to make the casting.

BS 2782 Method 644A (1979)[18] is identical to ISO 3521.

16.4 TRANSITION

Unlike crystalline materials, amorphous materials do not melt when they are heated, but go through a glass transition. In the glass transition temperature range many physical properties of amorphous materials undergo a more-or-less drastic change. The most obvious change is that from a viscous or flexible material above T_g to a hard, sometimes brittle, solid below. On a molecular level, as the temperature is raised to the glass temperature and above, segments of the molecules are able to rotate. The actual value of T_g for a polymer is clearly method- and rate-dependent; T_g is that temperature at which the molecular relaxation times are of the same order as the experimental times.

Brysdon,[19] for example, quotes values for a polyoxacylobutane polymer which ranged from 7 °C to 32 °C for dilatometric and electrical methods where the frequencies were 10^{-2} Hz and 100 Hz, respectively. For a detailed discussion of the glass transition the reader is referred to the review articles by Shen and Eisenberg.[20,21]

A crystalline solid melts when it is heated and this change of phase from solid to liquid is known as a first-order transition, because there is a discontinuity in the first-order derivatives of Gibbs function, such as the volume.[22] In a second-order phase transition there is a discontinuity at the transition temperature in the second-order derivatives of Gibbs function, such as the specific heat.[23] For a true second-order transition in the Ehrenfest sense the phase change must take place at constant temperature and pressure with no change of entropy or volume. A glass transition, therefore, is not a second-order transition, although it approximates to one.[24] Gibbs and DiMarzio[25,26] developed a theory of the glass transition by assuming that there is an underlying second-order transition,[27] and that the value of the glass transition temperature obtained from an experiment done at an infinitely slow rate would be the same as the thermodynamic second-order transition temperature. This theory is not without its critics.[28]

From the structural point of view the molecular arrangements in an amorphous material (or glass) below T_g possess the permanence of crystals and the randomness of liquids. As a crystalline arrangement of the molecules has a lower energy, for a glass to form on cooling from the melt there must be some barrier to crystallisation. This may be caused by steric hindrances in the molecule such as bulky side groups, or, if the material is cooled quickly enough, by insufficient thermal energy, or by a combination of both. T_g is always less than T_m (melting temperature) since a crystal has a lower energy than an amorphous structure.

The glass transition does not occur at a single temperature but takes place over a short temperature range, because real polymeric materials have a molecular weight distribution. The single point temperatures which are quoted are obtained by interpolation. T_g can be measured in a number of ways because most material properties change on going through it. It is not sufficient just to quote a value for T_g; it is important to give the method and the rate of measurement.

A number of methods for measuring T_g have been reported in the literature. T_g can be determined by measuring the expansion, either linear or volume, since the expansion coefficient above T_g is greater than that below.

The change in the dynamic mechanical properties, as discussed in Ch. 9, can also be used, since the dissipation factor plotted against temperature shows a pronounced peak. A similar effect is observed with the electric loss factor as described in Ch. 13.

Optical properties also change when the sample is heated or cooled through the glass transition range. A value for T_g can be obtained by measuring refractive index[29] or birefringence.[30]

Thermodynamic properties such as the specific heat can also be used because there is a step change at the transition temperature. A differential scanning calorimeter or a differential thermal analyser allows the temperature range of interest to be scanned quite quickly, making this a very convenient method to use. The value of T_g depends on the scan rate and it may be necessary to make a number of measurements at different scan rates and extrapolate to zero rate. ASTM D3418 (1982)[31] is the only published standard method for measuring T_g and is based on thermal analysis.

16.5 MELTING POINT

16.5.1 General Considerations

Under this heading only those tests will be considered which are suitable for materials with a fairly sharp change of state (solid to liquid), i.e. those which behave somewhat like an ordinary low-molecular-weight organic compound. In the characterisation of these, most of the standard methods depend on a visual observation of the transition. These melting point tests are generally unsuitable for the majority of plastics, or more correctly thermoplastics, which soften slowly over a temperature range which may be quite extensive and frequently never reach a free-flowing condition before thermal decomposition sets in. With such an indefinite change of state there is no melting point in the usual sense of the term and it is necessary, if it is required to put a temperature figure to such an indefinite property, to follow the change in some parameter – usually mechanical – with rise of temperature and ascribe a softening point to that temperature at which the property being observed reaches some prescribed value. This type of test is discussed more fully in §16.6 below.

There are several tests available for melting point determination and attention is given below to those which have been standardised. However, mention is made of the simple Durrans method,[32] where 3 g of resin is gently melted in the bottom of a test tube, and the bulb of a thermometer is inserted in the resin which is then cooled to solidify. A weight of 50 g of mercury is then poured on top of the resin and the whole assembly is heated so that the resin temperature rises at a rate of 2 °C/min until the resin melts and floats on the mercury. The temperature at which this first occurs is recorded as the melting point of the resin. *Care is extremely necessary in performing this test and it should only be done by qualified personnel with the proper precautions to avoid exposure to mercury vapour which is extremely poisonous.*

16.5.2 'Ring and Ball' Method

This appears as Method 103A of BS 2782 (1970)[33] and is generally used for synthetic resins of the novolak type (Fig. 16.1).

The sample under examination is reduced to small pieces (not powder, as bubbles may result on melting) and 1·6 g is weighed into a ring (as shown) which is placed on a metal plate, of suitable finish to avoid sticking. The assembly is then placed in an oven at a temperature 10 °C–20 °C above the expected melting point until

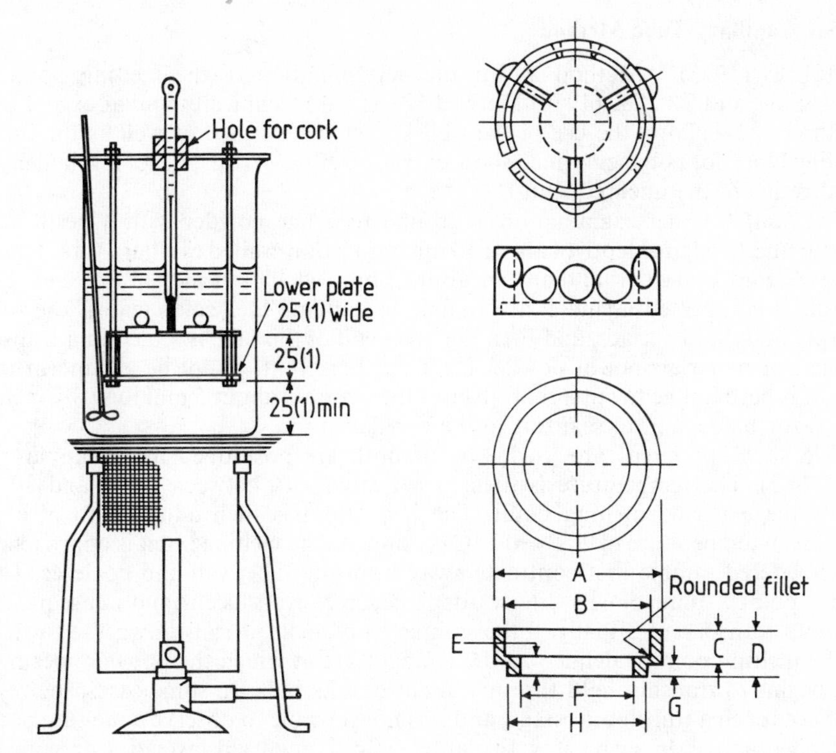

Fig. 16.1 'Ring and ball' softening point apparatus: BS 2782 (dimensions in millimetres (inches))

Ring dimensions:

A	23 ± 0·1	(0·906 ± 0·004)
B	19·9 ± 0·1	(0·781 ± 0·004)
C	4·4 ± 0·1	(0·172 ± 0·004)
D	6·4 ± 0·1	(0·250 ± 0·004)
E	2·8 ± 0·1	(0·109 ± 0·004)
F	15·9 ± 0·1	(0·625 ± 0·004)
G	2·0	(0·08)
H	19·0 ± 0·1	(0·750 ± 0·004)

the resin has melted (time should not exceed 15 min) removed, placed on a flat metal surface, and any excess resin over the top of the ring is immediately removed by scraping with a hot knife. Having been allowed to cool for at least 15 min, the ring filled with resin is removed from the plate, if necessary using gently tapping.

A steel ball, 9·53 mm ($\frac{3}{8}$ in) diameter and mass 3·45–3·55 g, is placed centrally on the cast resin, using a centring guide (as illustrated) which rests on the ring.

Two such specimens are prepared and placed in the apparatus illustrated, using as heat transfer medium a liquid without effect on the resin (glycerol will often be suitable). The heating bath is raised in temperature by 5 °C ± 0·5 °C per min, with constant mechanical stirring, and the melting point is recorded as that temperature at which the ball or surrounding resin first touches the lower plate. The agreement between the duplicate results must be within 1 °C.

16.5.3 Capillary Tube Method

ISO 1218 (1975)[34] Method A, for the determination of the melting point of polyamide, and BS 2782 (1976) Method 123A[35] are technically similar except that Method 123A allows the use of the well known Thiele tube as well as the metal heating block for polymers which show only a sinter point (e.g. phenol formaldehyde solid resins in the uncured state).

The sample under examination is ground to a fine powder with a pestle and mortar and filled to a depth of about 10 mm into a thin-walled capillary tube, length about 60 mm, and internal diameter about 1 mm, which is sealed at one end. The sample is compacted as much as possible by tapping the sealed end of the tube sharply on a hard surface, and then the open end of the tube is sealed in a Bunsen flame. For materials not in powder form, or those that cannot be ground in this way, it is permissible to cut a thin silver of the material about 5 mm long, by means of a razor blade, and to seal this in the capillary tube.

Two such specimens are used; one or both are positioned in the apparatus (Fig. 16.2). The temperature is then raised rapidly to between 10 °C and 20 °C below the expected melting point. The heat input is then adjusted to give an increase in temperature of $2\,°C \pm 0{\cdot}5\,°C$ per min. As the melting point is approached, the powdered sample first contracts away from the tube wall and coalesces (the 'sinter point'). Immediately afterwards it becomes translucent and subsequently becomes transparent. Finally, if the sample truly melts, it runs down the tube.

The melting point is defined as the temperature at which the powder specimen first begins to transmit light through about half its bulk. In some cases, the resin does not reach a translucent stage, and it is not possible to observe a melting point as defined above. In such cases, the sinter point is reported instead. For samples which are not in powder form the melting point is defined as the temperature at which the sharp edge of the specimen disappears.

The determination is carried out in duplicate and if the difference between the two results exceeds 3 °C two additional test portions have to be tested.

DIN 53736 (1973)[36] is similar to ISO 1218 Method A.

16.5.4 Hot Plate Method

ISO 1218 (1975)[34] Method B and BS 2872 (1976) Method 123B[37] are technically similar but ISO 1218 gives a more precise description of the hot plate apparatus. The method is described as specifically for polyamides, but should be suitable for any material with similar melting characteristics.

The test sample is a single granule of carefully defined size (related to a certain sieve size) which is placed on a microscope slide resting across the centre of a hot plate. A few drops of silicone fluid are also placed on the slide, adjacent to the granule, and the spacing is such that a cover glass placed on the latter is slightly tilted towards the fluid which makes contact with only a part of the underside of the glass (Fig. 16.3).

The temperature of the hot plate, which has a thermometer inserted in a cavity immediately below its centre, is raised at a rate of $2\,°C \pm 1\,°C$ per min and the boundary line of the silicone fluid is carefully observed. The melting point is taken

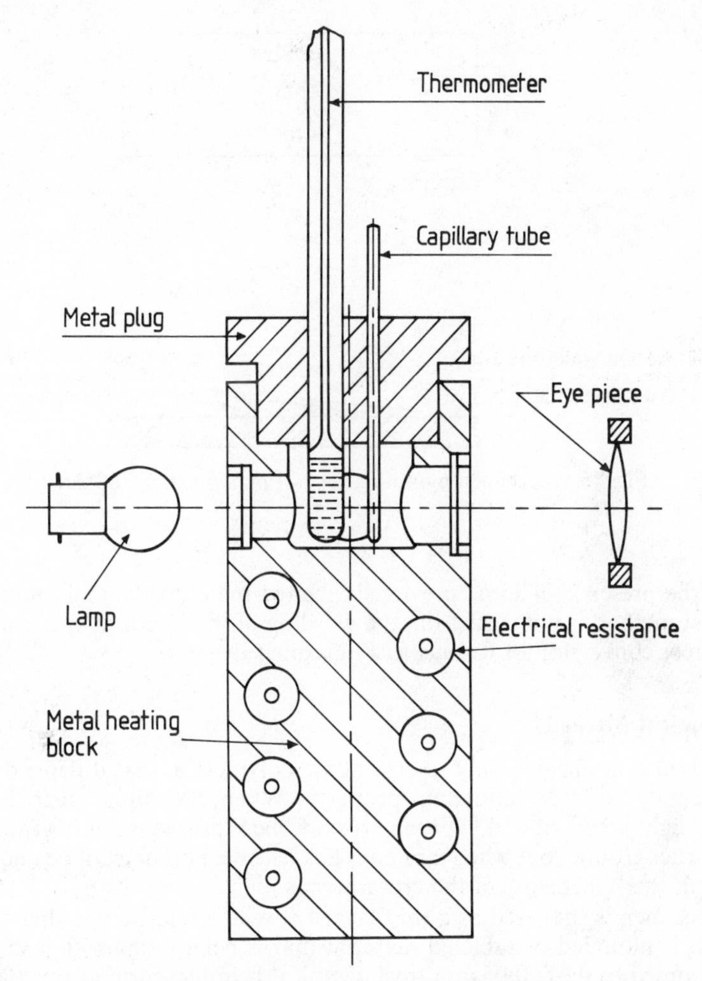

Fig. 16.2 Melting apparatus with capillary tube

as that temperature at which the boundary line moves rapidly across the cover glass, i.e. when the granule no longer supports the glass. Two such tests are carried out.

This is the basis of the methods described in ASTM D789 (1981),[38] a specification for nylon materials, and in ASTM D2133 (1981)[39] for acetal resin materials, where the use of a commercial apparatus, the Fisher–Johns melting point apparatus, is specified. Using the same apparatus, appearance changes are used to measure the melting point of fluorinated ethylene propylene copolymer (ASTM D2116 (1983)).[40] When measuring the melting characteristics of polytetrafluoroethylene resins the method is not so successful. ASTM D1457 (1983)[41] specifies the use of thermal analysis to determine if the resin has been melted previously or that it contains some previously melted resin. It does this by examining the melting peak temperature on successive runs. The presence of any peak temperature which is less than 5 °C above the peak temperature obtained during the first melting may

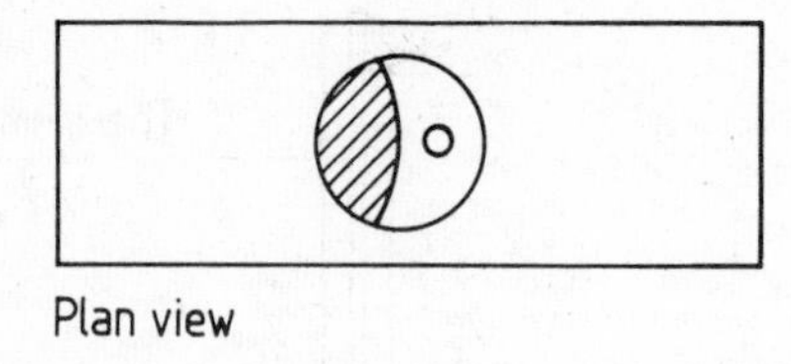

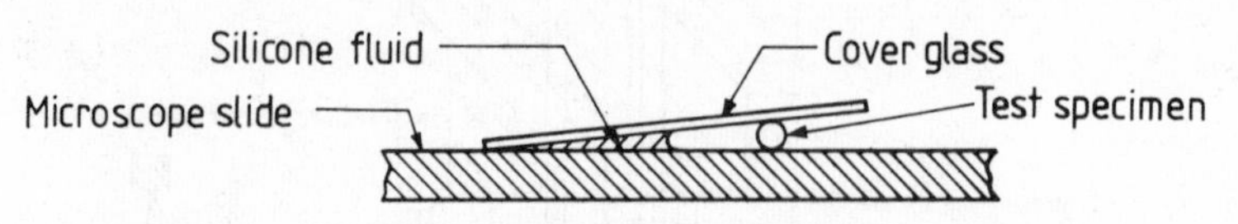

Fig. 16.3 Specimen mounting for hot plate method: BS 2782

indicate the presence of some previously melted material. The full interpretation of this test method is not clear from the wording of the standard and is not helped by incorrect conversion of Celsius to Fahrenheit degrees.

16.5.5 Optical Method

This method is applicable only to crystalline polymers in that it depends on their birefringent, or double refracting, property. When crystalline, such a material transmits light when viewed between crossed Nicol prisms, i.e. it is visible against the dark background, but when molten it has become non-crystalline and becomes invisible in the 'blackness' of the cross prisms.

The specimen is prepared as a thin layer between a microscope slide and cover glass. If it is moulded or tableted material that is being examined, it may simply be microtomed to 0·01–0·05 mm thickness; if it is in film form of this thickness it may be used direct. To avoid light scattering and ensure good thermal contact it is advisable to immerse the specimen in a liquid, preferably of similar refractive index, which of course must not affect the material. Powdered material, or thicker film or sheet, is placed between the microscope slide and cover glass, which together are put on a hot plate and heated to a temperature about 25 °C above the melting point. This cover glass is pressed down to reduce the specimen to a thickness of 0·01–0·05 mm and, having been kept at elevated temperature for about 10 min, the assembly is then allowed to cool slowly, to induce optimum crystallinity. If the material is liable to thermal decomposition under these conditions the temperature may have to be lowered and the cooling started immediately.

A microscope is used, fitted with analyser and polariser and capable of giving × 50 and × 100 magnification. It is also fitted with a hot stage, which is mounted just above the microscope stage and consists of an insulated metal block with a central hole for the passage of light, and with a recess close to the hole for the insertion of a thermometer or thermocouple. The hot stage is electrically heated at controlled rates and should be designed so that it can be enclosed in an atmosphere of nitrogen if necessary.

At the start of the test, the polariser and analyser are adjusted for complete extinction of light and the specimen assembly is placed on the hot stage. The polariser and analyser are then rotated again, if necessary, to obtain maximum brightness of the specimen and thus overcome the effects of strain in the latter. The hot stage is then heated at a (linear) controlled rate, usually 1 °C ± 0·25 °C per min and the melting point is taken as that temperature at which the field becomes completely dark. For high melting point materials, the test may be conveniently started 50 °C below the expected softening point.

A test of this type is given as ISO 3146 (1985)[42] and is essentially similar to that described in ASTM D2117 (1982).[43] BS 2782 (1976) Method 123C[44] is technically similar to ISO 3146 but differs in respect of rate of heating and gives more precise instructions on the preparation of the test sample.

16.6 SOFTENING POINT

16.6.1 General Considerations

The results of the *ad hoc* tests, which are referred to in §16.1 and which have found wide favour as standardised methods, flourish under a number of informal names, particularly: 'softening point', 'plastic yield', 'heat distortion point', 'deformation under load' and 'temperature of deflection under load'. They measure the temperature response of the material under test to a variety of stress conditions, especially in bending or under a penetrative load. There are two methods of obtaining an end-point; the most popular is to select some arbitrary stress condition and raise the temperature until the strain has reached some arbitrary level. This is essentially a measure of the temperature at which the modulus deteriorates to some prescribed level (but without reference to the value of the modulus at normal ambient conditions). The other principal variant is to select the test temperature and measure the results in terms of the degree of yield resulting, at this temperature, from standardised conditions of stress.

Crofts and Brown[45] have comprehensively reviewed methods for measuring softening point and provide a useful bibliography.

The size of the test piece required for the various types of test are quite different; for instance, a penetration-type test requires only a small area and thickness whereas a cantilever bending method may need a test piece of substantial cross-section. Since the temperature of the surrounding heat exchange medium (water bath, air or whatever) must reflect accurately the temperature of at least the stressed area of the specimen, the rates of temperature rise of the heat exchange media may have to be quite different for the various methods.

While acknowledging the value of softening point determinations and the like as quality control tests, it is necessary to stress just how *ad hoc* and arbitrary they are and how incapable of comparison their various results.[46–49] Table 16.1 serves to emphasise this point: not only will the wide differences between results in each column be noted, but also the inconsistent differences and, in some cases, the changed order of ranking.

Table 16.1 Softening points (°C) determined by various standard methods

Material	Cantilever BS 2782 102C	DTI (ASTM D648) 455 kPa	DTI (ASTM D648) 1826 kPa	Vicat BS 2782 120A
Polystyrene	95	90	97	98
Toughened polystyrene	84	72	85	86
ABS	94	84	96	95
Poly(methyl methacrylate)	95	80	97	90
Cellulose acetate	76	64	77	72
Rigid PVC	78	70	82	82
Polyethylene (LD)	too flexible	too flexible	45	85
(MD)	90	35	69	105
(HD)	115	45	75	125
Polypropylene	145	60	140	150
Nylon 66	180	75	183	185
Acetal	170	120	165	175

More recently, however, Abolins[50] has demonstrated that a linear relationship exists between the Vicat softening point and the temperature of deflection under load as measured by ASTM D648 (1982),[51] particularly for annealed test pieces.

16.6.2 Cantilever Techniques

BS 2782 (1978) Method 115A[52] determines the 'plastic yield' and requires two test pieces each 200 ± 1 mm long by a square cross-section of side 15 mm + 0, − 0·3 mm. A notch is moulded or machined into one long face of each test piece, the root of the notch being 5 mm from one end. The notched surface is one of those perpendicular to the direction of moulding pressure.

The test is carried out in an oven capable of being maintained throughout the test within the specified tolerances of the specified test temperature p; this latter is normally quoted in the relevant product specification but if not it is to be one of 40 °C, 55 °C, 70 °C, 85 °C, 105 °C, 175 °C or 200 °C, the temperature being selected so that the result ('plastic yield') does not exceed 6 mm. The test set-up is shown in Fig. 16.4.

The clamped test piece, without weight W but with stirrup plus attachment (combined weight not exceeding 20 g), is placed in the oven at the test temperature; 15 min afterwards the height of the stirrup is taken, with respect to some suitable datum point, to the nearest 0·1 mm and then the weight W is attached to the stirrup (W + stirrup + attachment = 450 g).

After a further 6 h $\pm$ 10 min the height of the stirrup with respect to the same datum is again taken. The difference between the two height values (mm) is the plastic yield; the results for the two test pieces are averaged.

Finally, in cantilever methods, there is Method 102C of BS 2782 (1970).[53]

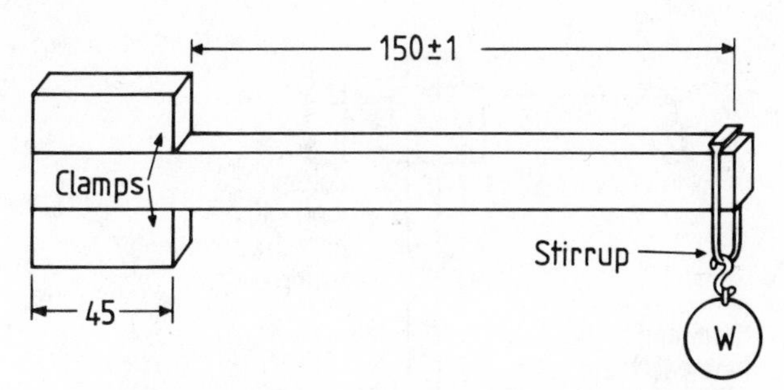

Fig. 16.4 Plastic yield test: BS 2782 (dimensions in millimetres)

Although it was intended to withdraw this method from later editions of BS 2782 it appears in the 1986 reprint of BS 2782 (1970) because it still features in the *UK Building Regulations.* It enjoyed many years of popularity and was often used to define the softening point of thermoplastics in the trade literature and so is still worth describing even though it has been rendered obsolete by the Vicat test (§16.6.6). The test piece is shown in Fig. 16.5.

The test piece is mounted horizontally in a clamp (Fig.16.6). Initially the 20 g weight is supported so that there is no load on the test piece. The assembly is immersed in a bath of appropriate liquid, at a temperature 25–30 °C below the expected softening point. (The 'appropriate liquid' must be without effect on the material under test: liquid paraffin is suitable for cellulose acetate and glycerol for polystyrene, toughened polystyrene and rigid PVC.) The support is then removed from under the load and the temperature of the bath is raised at 1·0 °C ± 0·2 °C per min with adequate stirring.

The temperature of the bath is noted at which the under edge of the free end of the test piece coincides with a 30° graduation marked on the quadrant plate.

The mean of the temperatures obtained from two specimens is taken as the softening point.

16.6.3 Three-point Bending Techniques

ISO 75 (1974)[54] describes the determination of temperature of deflection under

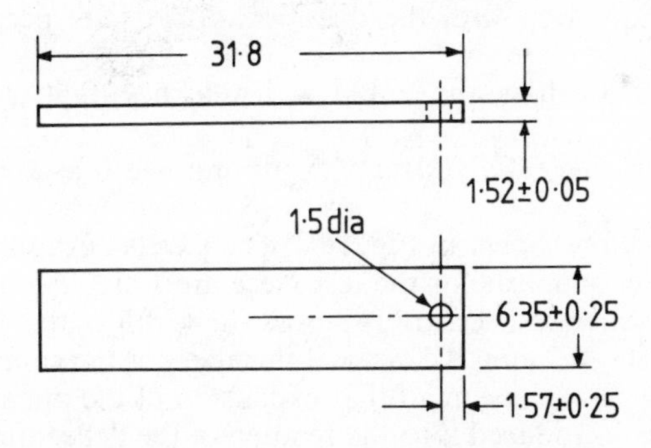

Fig. 16.5 Softening point test piece: BS 2782 Method 102C (dimensions in millimetres)

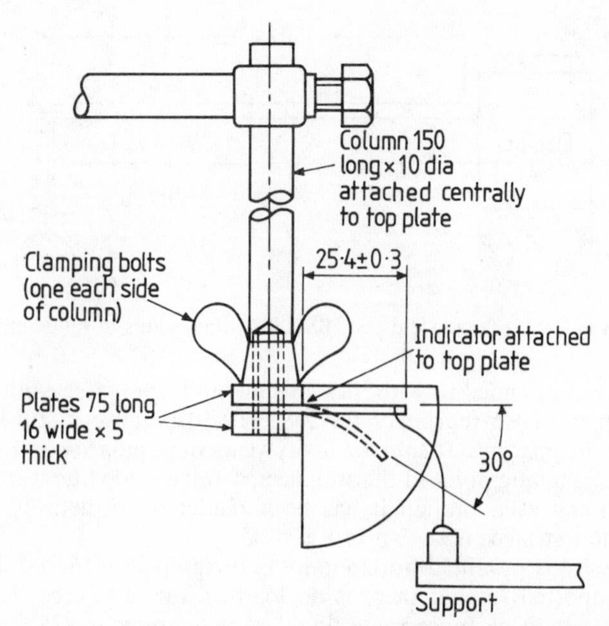

Fig. 16.6 Test assembly for softening point test: BS 2782 Method 102C (dimensions in millimetres)

load of plastics and ebonite. This test method was directly derived from ASTM D648.[51] BS 2782 (1976) Methods 121A and 121B[55] are in general agreement with ISO 75. DIN 53461 (1984)[56] is similarly related to the ISO method. As the various national and international standards are so closely related, only the BS method will be described.

To accommodate materials of different initial (room temperature) stiffness, two variants of the method are available, one applying a maximum surface stress of 1·8 MPa (Method 121A), and the other one of 0·45 MPa (Method 121B). Test pieces are as follows:

Moulding material and extrusion compounds: rectangular bar of minimum length 110 mm, width 9·8–12·8 mm and thickness 3·0–4·2 mm moulded as appropriate to the material in question with the direction of pressure perpendicular to the largest faces of the test piece.

Sheet: length and width as above, but with thickness that of the sheet (which is to be 3–13 mm).

Casting and laminating resin systems: length and width as above but thickness 3–7 mm.

The apparatus used is shown in Fig. 16.7. The parallel cylindrical metal blocks which form the outer supports for the test piece are 100 ± 2 mm apart; these and the loading block, which rests centrally across the width of the test piece, all have contacting radii of 3 ± 0·2 mm. All vertical members of the supports should have the same coefficient of expansion so that expansion of the apparatus cancels out and no error is thus introduced into the reading of the deflection of the test piece through uncompensated expansion in the apparatus; even so, it is as well to check

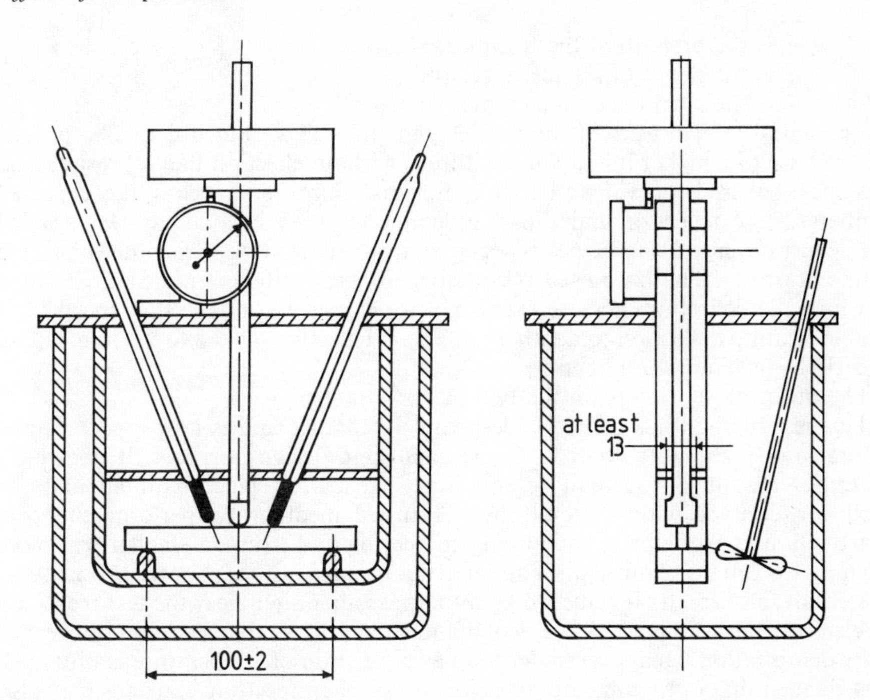

Fig. 16.7 Apparatus for determination of temperature of deflection under load: BS 2782 (dimensions in millimetres)

this by carrying out a 'control' run beforehand using a test piece of very low expansion coefficient, for example Invar alloy or borosilicate glass.

The heating bath into which the above assembly is immersed is filled with a liquid which has been shown to have no effect on the material under test (liquid paraffin, transformer oil, glycerol and silicone fluid are suggested for trial). The bath must be capable of being heated so that the temperature can be raised at a rate of 2 °C per min, with the temperature at any time being not more than 1 °C from the value corresponding to this rate.

To undertake the test, the cross-sectional dimensions of the test piece are measured to the nearest 0·02 mm and then it is placed in the apparatus so that the directions of application of load is parallel to the plane of the largest faces. In this position, the vertical height is designated depth d and horizontal width from one large vertical face to the other is designated breadth b. The test piece is loaded under a bending stress of 1·8 MPa or 0·45 MPa (as appropriate) by applying a force as follows (making due allowance for any force exerted by the spring of the dial gauge):

$$F = \frac{2abd^2}{3l} \qquad [16.4]$$

where F is the force (N)
a is the bending stress (1·8 MPa for Method A, 0·45 MPa for Method B)

b is the breadth of the test piece (mm)
d is the depth of the test piece (mm)
l is the span between supports (mm).

The initial temperature of the bath is generally 23 °C, but if it can be proved that the use of a higher initial temperature is without effect on the test result, then this may be used, provided that it is not less than 30 °C below the expected temperature of deflection under load. Five minutes after the load has been applied the deflection-measuring device is set to zero (or its reading is recorded) and the temperature of the bath is raised as described above. The temperature (°C) is noted at which the deflection increase is reached which corresponds to that specified in a table relating deflection to depth d (0·33 mm deflection for 9·8 to 9·9 mm, falling linearly to 0·25 mm for 12·8 mm).

The mean of the results on the two pieces is taken.

This test method has gained widespread acceptance and is now used in many laboratories for quality control, development and design purposes. It does have its critics mainly on account of the large size of test piece required and the need for contact with a liquid heat transfer medium to perform the test. Martinelli and Hodgkin[57] have modified the method to use a smaller test piece (25 mm × 6·3 mm × 3 mm) and an electric furnace as the source of heating. Presentation of results is achieved by a thermocouple sited near the test piece and an electrobalance connected to a two-channel recorder so that a continuous record of its deformation behaviour under load as a function of temperature is obtained. It is claimed that optimum cure times for certain thermosetting epoxides may also be determined by the modified method.

The extension of the technique is taken one stage further by BS 2782 Method 121C (1976)[58] for rigid thermosetting resin bonded laminated sheet, by applying a stress equivalent to 10 per cent of the measured ultimate crossbreaking strength. In this method the span is variable so that it is adjusted to 30 ± 2 times the test piece thickness but the technique of obtaining the deformation point is precisely as before. This test is somewhat inelegant and the choice of 10 per cent of the normal temperature crossbreaking strength rather than some fraction of the bending modulus is surprising.

16.6.4 Four-point Bending Techniques

The Martens test is described in DIN E53458 (1982)[59] and is an ingenious way of converting four-point bending into magnified vertical movement of the end of an arm (Fig. 16.8).

The apparatus is carefully specified and a value of weight G is given for each of the three different sizes of test piece permitted. The rate of heating is 50 °C ± 1 °C/h and the Martens temperature is taken when the deflection at the end of the lever reaches 6 mm. The values of G are selected to give the same maximum surface stress to each of the three alternative forms of test piece, but the maximum strain is necessarily different. The ratio of the two (stress–strain) is stiffness, which is a function of temperature and therefore test results differ according to which of the standard test pieces has been used.

Gohl[60] discusses the limitations of the Martens test and concludes that it is unsuitable for materials of modulus less than about 5000 MPa.

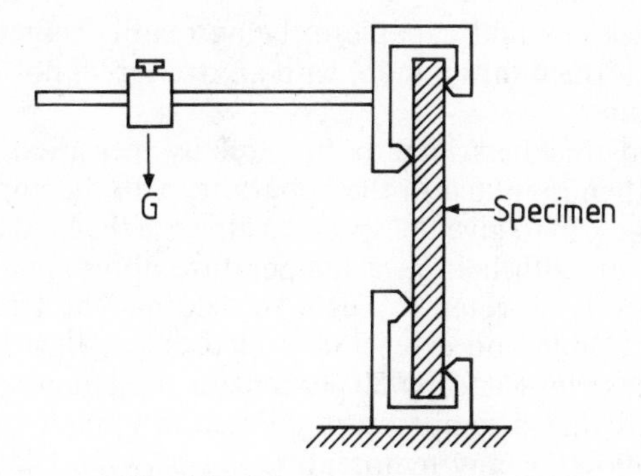

Fig. 16.8 Martens test: DIN 53458

16.6.5 Tensile Techniques

For sheets and films of thickness from 0·025 mm to 1·5 mm and of modulus greater than 69 MPa at 23 °C, a test for tensile heat distortion temperature is described in ASTM D1637 (1983).[61] It measures the temperature at which thermoplastic sheeting begins to deform appreciably (either to stretch or shrink) under a small tensile force.

Two forms of apparatus are shown in Fig. 16.9. The oven or heating bath must be capable of being raised in temperature at a constant rate of 2 °C ± 0·2 °C/min. The weight(s) required must be appropriate to apply a stress of 0·345 MPa to the test piece.

The test pieces are strips from 6 to 25 mm wide and from 50 to 180 mm long. The initial distance between the grips is between 25 and 125 mm. Materials of less

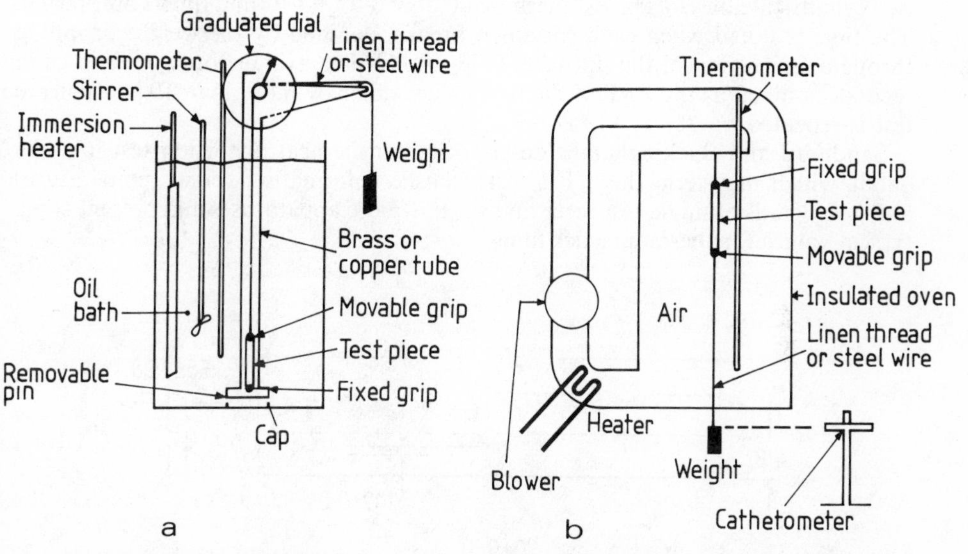

Fig. 16.9 Tensile heat distortion apparatus: ASTM D1637: (a) heat bath method; (b) air oven method

than 0·075 mm in thickness and capable of being readily cemented together (or otherwise joined) are formed into a loop, with an overlap of not more than 6 mm, and tested in this manner.

After the width and thickness have been carefully measured, the test piece is pre-conditioned and then mounted in the apparatus with the appropriate weights attached to the moving grip to give the specified stress of 0·345 MPa. The assembly is placed in the oven or both held at a temperature approximately 20 °C below that at which the piece is expected to begin to deform. The temperature is then raised by 2 °C ± 0·2 °C/min and extension is recorded against temperature until either the test piece has elongated by 50 per cent or the temperature has reached 250 °C. It is suggested that a temperature correction (which should not exceed 3 °C) be applied to allow for any temperature lag between heating medium and test piece, which correction can be determined once and for all for each apparatus by attaching a thermocouple to a test piece and comparing the heating curve for the thermocouple with that from the thermometer in the oven or bath.

The tensile heat distortion temperature is determined from a plot of percentage change in length against temperature and is the temperature at which a 2 per cent shrinkage or a 2 per cent extension occurs. Bourgault[62] has shown that with properly designed apparatus for this test it is a useful tool in the investigation of the properties of heat shrinkable and crosslinked polymers.

In the product specification ASTM D1430 (1981)[63] *polychlorotrifluoroethylene (PCTFE) Plastics*, a test is included for 'zero strength time' (ZST) which it is stated is useful for control of molecular weight of this type of polymer. The test piece is a notched strip (Fig. 16.10) cut or punched out of sheet 1·58 ± 0·08 mm thick pre-formed and moulded from the powder in specified form.

One end of the test piece is hung from a specimen holder and a weight of 7·5 ± 0·1 g is attached to the other. Two such weighted test pieces are inserted one each into two cylindrical holes in a brass 'thermostat' at 250 °C ± 1 °C and timers are started. The time is noted when each specimen breaks, denoted by the weight dropping through the bottom of the 'furnace' (Fig. 16.11). The average of two readings in seconds is taken as the ZST; if the two values differ by more than 10 per cent the test is repeated.

Sandiford and Buckingham[64] describe the tensile heat distortion test in some detail, which they term the 'TDT test' (tensile deformation versus temperature) and give details of single-test piece and six-test-piece apparatuses and present some typical values for thermoplastics films.

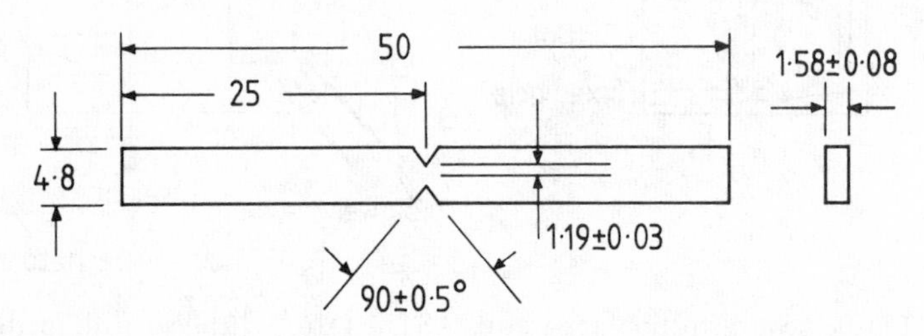

Fig. 16.10 ZST test piece: ASTM D1430 (dimensions in millimetres)

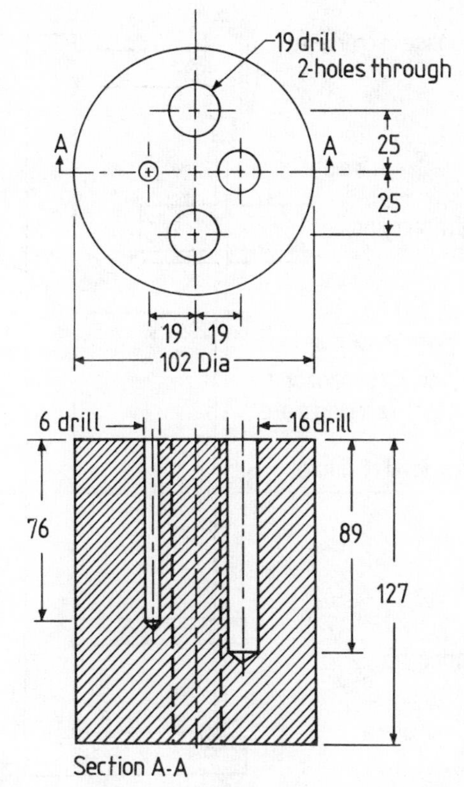

Fig. 16.11 ZST thermostat: ASTM D1430 (dimensions in millimetres)

16.6.6 Penetrometer Techniques

The most important penetrometer method and one which has gained almost universal acceptance is the *Vicat softening point* test. A test piece at least 10 mm square and not less than 2·5 mm thick is subjected to a round indenting tip of cross-section area $1{\cdot}00 \pm 0{\cdot}015\ mm^2$. The lower surface of the indenting tip is free of burrs and is plane and perpendicular to the axis of the rod to which it is attached (Fig. 16.12). A specified load is applied to the indentor tip and the degree of movement of the tip is indicated by a dial gauge. The test piece and tip are immersed in an oil bath and the temperature is raised at a specified rate. The softening point is defined as the temperature at which the indentor tip penetrates the test piece to a depth of 1 mm.

ISO 306 (1974),[65] BS 2782 (1976) Methods 120A to 120E.[66] ASTM D1525 (1982)[67] and DIN 53460 (1976)[68] are all similar in describing the Vicat softening point test for thermoplastics. All allow the use of a force of 9·871 N on the indentor (this is rounded to 10 N in BS 2782, while ASTM D1525 specifies the force in terms of a total mass of 1000 g applied to the indentor) but BS 2782, ISO 306 and DIN 53460 offer alternative forces of 49 N. The standard rate of temperature rise

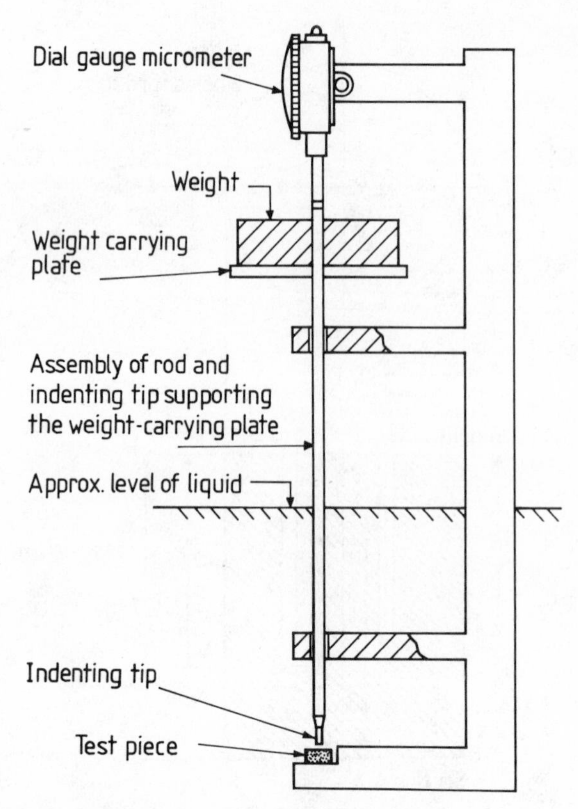

Fig. 16.12 Vicat softening point apparatus: BS 2782. It is recommended that the frame and indenting rods are constructed from low-expansion alloy

during the test is $50 \pm 5\,^{\circ}\mathrm{C/h}$, but for some materials at a rate of $120\,^{\circ}\mathrm{C} \pm 5\,^{\circ}\mathrm{C/h}$ is permissible.

Test pieces are between 2·5 and 6·5 mm thick (BS 2782) or between 3 and 6 mm thick (ISO 306 and ASTM D1525) but for samples, say, of sheet, of thickness less than 3 mm the plying up of samples to give a composite thickness of from 2·5 mm to 3 mm is allowed, provided that the stack is composed of not more than three pieces. For samples of thickness greater than 6 mm (6·5 mm for BS 2782), the test piece is reduced to a thickness of 3 mm by machining one surface only prior to testing. Two test pieces of each sample are tested and if the results show a difference greater than 2 °C the sample must be re-tested.

BS 2782 Method 120C describes the so-called 1/10 Vicat softening point. Here a force of 10 N is applied to the indentor and the softening point is defined as the temperature at which there is a penetration of 0·1 mm when the rate of heating is 50 °C/h. The respective uses of the normal and 1/10 Vicat softening point tests are discussed by Stephenson and Willbourn.[47]

With wide adoption of the Vicat method it is a pity that the various national and international standards differ in minor details although the difference in results obtained is probably insignificant. Crofts[69] has reported the effect on Vicat softening point of the rate of heating and of the force applied to the indentor. By extrapolating curves to the hypothetical softening point at zero force and zero rate of heating be obtained good agreement with the glass transition temperature of certain thermoplastics.

Several workers have been prompted to refine the Vicat softening point apparatus. Automatic indication of the end-points of Vicat softening point (and deflection temperature under load) tests is achieved by Graves and Loveless[70] using a lever actuating a mercury switch to cause a panel light to glow; they also describe VSP tests at rates of temperature increase of both 50 °C/h and 120 °C/h (i.e. 2 °C/min), quoting results for seven thermoplastics to justify the use of the faster rate.

Ehlers and Powers[71] use a transducer attached to the top of the penetrometer shaft of the Vicat apparatus to record depth of penetration against temperature fed to the recorder from a thermocouple.

BS 2782 (1976) Method 122A[72] describes a penetrometer method entitled *Determination of Deformation under Heat of Flexible Polyvinyl Extrusion Compound.* A test piece 13 mm in diameter and 1·27 ± 0·07 mm thick is accurately measured for thickness under specified conditions and placed on a flat horizontal base. A force of 9·8 N is applied to the test piece with a vertical plunger, 3 mm in diameter and flat, and the whole assembly is placed in an oven for 24 h at 70 °C ± 1 °C. After this the assembly is removed from the oven and allowed to cool for 1 h at room temperature. Finally the thickness of the test piece in the deformed area is re-measured between 3 and 5 min after removal of the load.

The deformation under heat of the test piece is taken from the difference between the initial and final thicknesses calculated as a percentage of the initial thickness, and the result is expressed as the mean of two readings.

Another test which determines deformation under load at both room and elevated temperatures is described in ASTM D621–64 (1976).[73] This contains two test methods: Method A is for rigid plastics and is designed to assess the ability of such materials to withstand compressive load, e.g. when held by bolts, without yielding and loosening the assembly; Method B is intended to determine the ability of non-rigid plastics to return to their original shape after having been deformed.

Method A

The essentials of the apparatus are that it shall be capable of exerting constant loads of 113 kg, 227 kg and 454 kg, all ± 10 per cent, and that it is equipped with a dial gauge measuring the relative movement of the faces of the anvils to 0·02 mm or less.

It will be realised that many compression testing mechanisms (Ch. 8), if fitted with temperature-controlled ovens, are adaptable to meet these specification requirements.

Test pieces are 12·7 mm cubes, either solid or composite, machined down or built up as necessary. Surfaces must be plane and parallel.

After pre-conditioning, if necessary, the test piece is placed between the anvils and the load is applied without shock. After 10 s the dial reading is taken and again after 24 h at the test temperature of 23 °C, 50 °C or 70 °C, all ± 1 °C. The test piece is removed, its thickness is measured and the original height, B, is calculated from this plus the change in dial reading, A:

$$\text{Deformation, per cent} = A/B \times 100$$

Method B

This is basically the same as Method A but the stress is 0·69 MPa ± 1 per cent and the period of test only 3 h.

The test pieces are 28·67 mm diameter by 12·7 mm thick. After pre-conditioning, the procedure of Method A is followed with the exception given above and a difference in the dial reading sequence. The zero point is obtained at the beginning of the test by determining the dial reading with the anvils together under full load and the thickness of test piece (H_0) is measured by micrometer, with low-pressure ratchet attachment, to the nearest 0·0002 in (0·005 mm).

After the thickness of the test piece has been determined at the end of 3 h (H_1), it is removed from the apparatus and left in the test chambers for 1 h. Finally the test piece is removed from the chamber, kept at room temperature for ½ h and then its thickness is re-determined by micrometer (H_2).

$$\text{Deformation per cent} = [(H_0 - H_1)/H_0] \times 100$$

$$\text{Recovery, per cent} = [(H_2 - H_1)/(H_0 - H_1)] \times 100$$

The reader will have noticed the curious mixture of units given in this standard. Although ASTM D621 was re-approved in 1976 it needs editing to bring it into line with the current practice of using SI units.

16.7 LOW-TEMPERATURE BRITTLENESS AND FLEXIBILITY TESTS

16.7.1 General Considerations

Most applications of rigid plastics make use of their stiffness, even if this is low by comparison with metals, wood and stone. It is therefore logical that in the majority of quality control tests to check the thermal properties of such rigid plastics it is loss of stiffness that is used as the parameter of measurement. Likewise, as many of the applications of flexible plastics utilise their characteristic 'give', the corresponding quality control for these follows their loss of flexibility with decrease in temperature. Amongst the standardised tests, however, will be found those which are slow in action, for example plotting torsional stiffness against temperature, and those which are fast-acting, such as those which assess the temperature at which impact shock causes shattering. Early background to both types of test is described by Clash and Berg.[74]

16.7.2 Torsional Stiffness Methods

Welding[75] critically reviewed torsion tests before some of them had been standardised and concluded that they could be simple and accurate. The Clash and Berg type of test has been standardised by ISO, BS, ASTM and DIN and is commonly used for testing plastics. The apparatus is characterised by the torque to the test piece being applied by a pulley, cords and two weights. The Gehman test uses an apparatus where the torque is applied by calibrated torsion wire. Williamson,[76] in an early paper found the Clash and Berg type of apparatus too insensitive at low values of modulus, and described a simple torsion apparatus based on calibrated torsion wires; an automated version of this type of equipment has been described by Wheeler.[77] The Gehman test is used extensively for testing rubbers and has been standardised by ISO, BS and ASTM (see Brown[78] p. 267).

ISO 458: Part 1 (1985)[79] BS 2782 (1976) Method 150A[80] ASTM D1043 (1984)[81] and DIN 53447 (1981)[82] are similar; all are for the determination of stiffness in torsion as a function of temperature. The apparatus is shown diagrammatically in Fig. 16.13. The weights and pulley of the apparatus are such that a torque in the range from 0·01 Nm to 0·12 Nm can be applied, through clamps, to the test piece which is immersed, in a Dewar flask, in a suitable heat transfer medium. Good stirring is essential during heating and cooling and commercially available apparatus usually incorporates an efficient electrically driven stirrer in the Dewar flask. For cooling, mechanical refrigeration or dry ice is recommended and an immersion heater in the flask is commonly used for heating.

Test pieces are rectangular strips and various sizes are allowed as given in Table 16.2. The ASTM method allows the use of different size test pieces where the span (i.e. distance between the grips of the machine) is variable. Spans of from

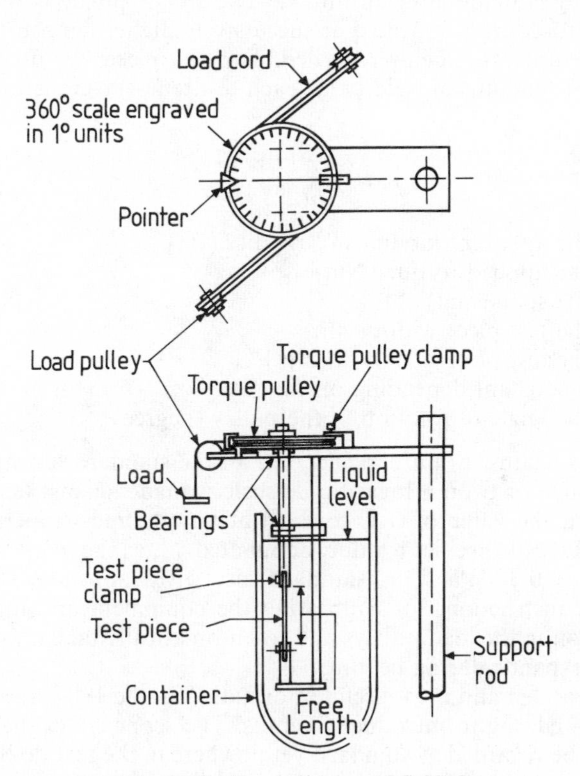

Fig. 16.13 Torsion tester: BS 2782

Table 16.2 Test piece dimensions for stiffness in torsion as a function of temperature

Standard method	Length (mm)	Width (mm)	Thickness (mm)
ISO R458	70 ± 10	6·0 – 6·3	1–5
BS 2782 Method 150A	60 – 80	6·0 – 6·3	1–5
ASTM D1043	63·5 ± 0·025	6·35 ± 0·025	1–5
DIN 53447	60 – 65	6·0 – 6·3	1–5

38 to 100 mm may be used provided that the span-to-width ratio of the test piece is between 6 and 8.

In operation the width and thickness of the test piece are measured to an accuracy of ± 0·02 mm and ± 0·01 mm, respectively, and then mounted in the grips of the apparatus with the torque pulley clamped at the zero position. The heat transfer medium is brought to the required temperature (which is somewhat below the lowest temperature of the range to be studied) and the test piece is immersed. After 3 min, + 15, − 0 s the pulley clamp is released and the reading of angle of movement (θ) of the pulley is taken after an interval of 5 s. The pulley is then returned to zero and the procedure is repeated at successively higher temperatures, the value of θ for each temperature being recorded. Two test pieces are used for each test.

The apparent modulus of rigidity for each test temperature is calculated by the relationship

$$G = \frac{917 \times Tl}{b(d^3) \times \mu\theta} \qquad [16.5]$$

where G is the apparent modulus of rigidity (Pa)
T is the applied torque (Nm)
l is the span (mm)
b is the test piece width (mm)
d is the test piece thickness (mm)
μ is a constant depending on the ratio b/d
θ is the angle of rotation of the pulley (degrees).

Tables giving μ against b/d are included in all the standard test methods. A plot of the mean value of G on a logarithmic scale is made against temperature on a linear scale and the value of G is read off at any desired temperature from the curve. ASTM D1043 specifies a value, designated T_F, as the temperature at which G is equal to 310·3 MPa. This standard also gives appendices which include comprehensive instructions for calibrating the equipment and for modifying to eliminate friction in the load pulleys and to automatically take up slack in the test piece when it expands during heating.

Despite this added and most useful information in the 1984 edition this ASTM standard needs editing to introduce SI units. The scope states that the values in SI units are to be regarded as standard yet nowhere in the text do SI units appear. One is still required to express the results in kgf/cm^3, (or lb/in^2).

A modification of the test is used to specify the cold flex temperature of flexible polyvinyl chloride compound in BS 2782 (1985) Method 150B.[83] Here a test piece of closely specified dimensions (65 mm long, from 6·2 to 6·4 mm wide and $1{\cdot}30 \pm 0{\cdot}08$ mm thick) is tested in the Clash and Berg apparatus at three temperatures chosen so that an angular deflection of 200° occurs between the highest and lowest temperatures. A graph of angular deflection versus temperature is plotted and the temperature at which a deflection of 200° occurs is read off from the graph. The mean of two test results is reported as the cold flex temperature of the sample under test.

16.7.3 Extensibility Method

Resistance to low temperature of flexible PVC sheet is measured by the extensibility in tension at a specified temperature according to Method 150C of BS 2782 (1983).[84]

A constant-rate-of-loading instrument is required, capable of increasing stress on a rectangular test piece (180 mm long $\times$ $6{\cdot}3 \pm 0{\cdot}1$ mm $\times$ sheet thickness) uniformly from zero at a rate of $20{\cdot}7 \pm 0{\cdot}4$ MPa per min. The free (unclamped) length of test piece is 125 mm and an extensometer device is needed to measure extension to the nearest 1 per cent. A cooling bath at $-5 \pm 0{\cdot}5$ °C, comprising one volume of industrial methylated spirit and three volumes of water, to which is added solid carbon dioxide as necessary, is provided for immersing that part of the apparatus which contains the clamped test piece.

The load required to give a stress of 10·3 MPa on the test piece is calculated from the width and thickness obtained gravimetrically (Ch. 7). With the grips set $127 \pm 1{\cdot}3$ mm apart, the test piece is mounted symmetrically in them, first in the upper one and then, with a 20 g weight attached to its lower end to keep it taut, in the lower one. The mounted test piece is then completely immersed in the bath, for from 30 to 60 s according to thickness, and after this the load is applied uniformly so that 10·3 MPa is reached in 30 s. When this stress is reached the extension of the test piece is read (all the extended specimens to be in the bath) to the nearest 1·3 mm.

Three such determinations are carried out and their results are averaged.

16.7.4 Flex Cracking Methods

The 'cold bend temperature' of flexible PVC extrusion compounds, described in Method 151A of BS 2782 (1984)[85] measures the lowest temperature, measured in multiples of 5 °C, at which none of a set of three test pieces fracture or crack when wound on to a standard mandrel.

At least six test pieces, each 100 mm long and 4·8 mm wide, are cut from specially prepared sheet of thickness $1{\cdot}27 \pm 0{\cdot}08$ mm, they are tested in an apparatus of the type shown in Fig. 16.14 in which all the essential features are given.

Three test pieces are placed in the guides and one end of each is secured by a clamp to the mandrel. Test pieces and winder are than immersed in a cooling bath of industrial methylated spirit maintained at the required temperature by solid carbon dioxide. After 10 min at the temperature, and while still immersed, the test pieces are wound tightly around the mandrel for three complete helical turns at the rate of one revolution per second. They are then withdrawn from the bath and examined for signs of failure – fracture or surface cracking. The test is repeated,

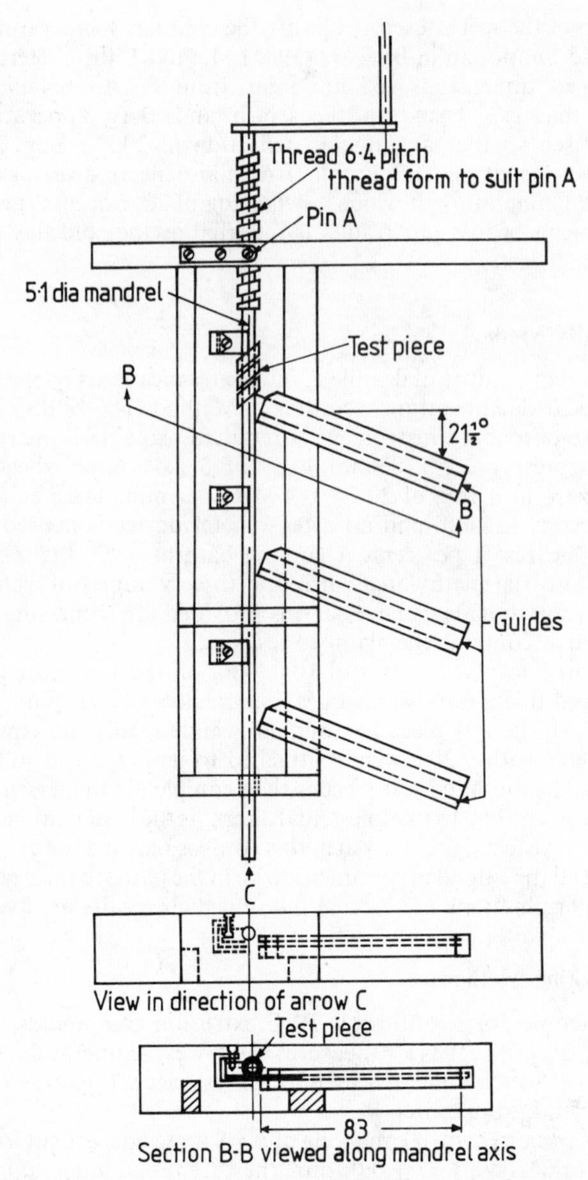

Fig. 16.14 Cold blend test: BS 2782 (dimensions in millimetres)

on fresh sets of test pieces, at various temperatures until two temperatures, differing by 5 °C, are found such that at the higher no test piece fails and at the lower one or more fail.

The cold bend temperature is the higher of these two temperatures.

In the course of an investigation into low-temperature failures of handbag materials, Wormald[86] developed a flexing tester operated at −5 °C in which the test result was the number of flexings required to cause failure of a standard test piece when repeatedly flexed under controlled conditions, particularly incorporating in each cycle a rapid fold followed by an appreciable waiting period to simulate service conditions.

16.7.5 Impact Methods

ISO 974 (1980)[87] and ASTM D746 (1979)[88] both describe the determination of brittleness temperature by impact of plastics. They determine the temperature at which 50 per cent of the test pieces would fail under the conditions of the test (but see below).

Test pieces are die punched either as 6·35 ± 0·51 mm wide pieces or as shown in Fig. 16.15. In either case they are 1·91 ± 0·25 mm thick, of suitable length to be clamped in the apparatus as shown in Fig. 16.16. The alternative test piece is clamped so that the entire tab is inside the jaws for a minimum distance of 3·18 mm. It should be noted that ISO 974 specifies a different mode of test piece support there being a large radius on the under edge of the clamping device (Fig. 16.17).

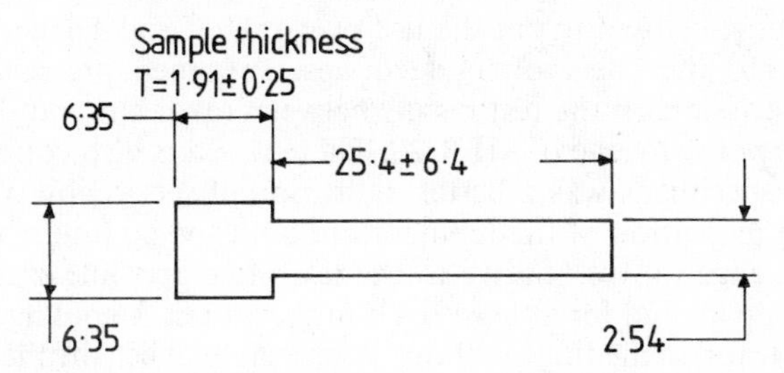

Fig. 16.15 Alternative test piece for brittleness temperature test: ASTM D746 (dimensions in millimetres)

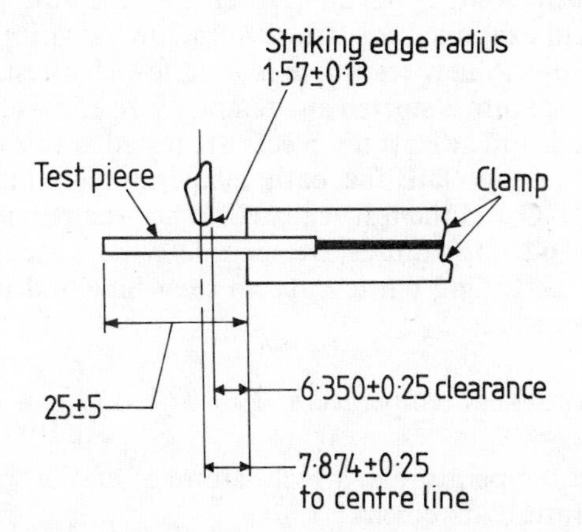

Fig. 16.16 Clamp and test piece for brittleness temperature test: ASTM D746 (dimensions in millimetres)

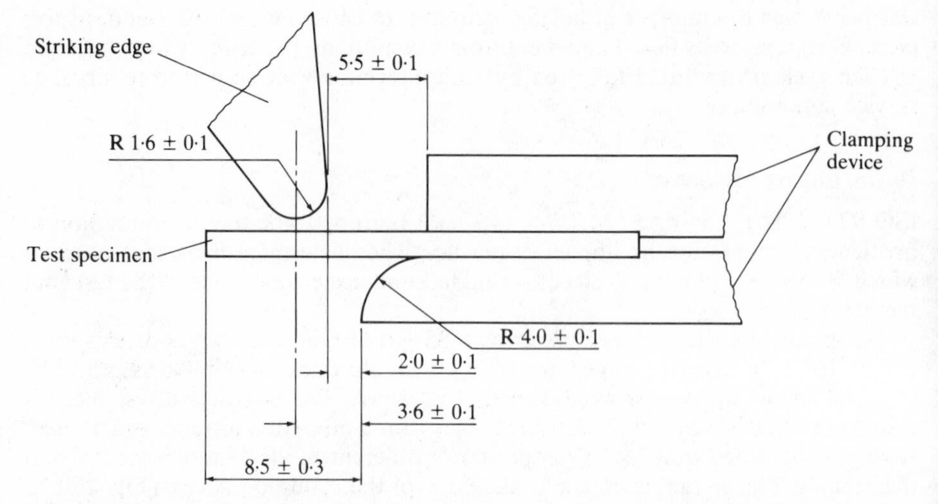

Fig. 16.17 Dimensional details of striking edge and clamping device ISO 974 (dimensions in mm)

Sharp dies must be used to cut the test pieces. Hoff and Turner[89] conducted an investigation into the effect of mode of test specimen preparation and found considerable variation on the test results between razor cut and die cut test pieces on low-density polyethylene of MFR 20. These workers also concluded that a test using notched specimens was a better indication of serviceability because it gave a more precise indication of the temperature limit for no failures.

The striking edge moves relative to the test piece at a linear speed of from 1·8 to 2·1 m/s at impact and for at least 6·4 mm thereafter. A cooling bath is required, of liquid heat transfer medium without appreciable effect on the test pieces, and held within $\pm 0{\cdot}5\,^\circ$C of the desired value.

The test piece is mounted in the apparatus and immersed in the bath at the required temperature for $3 \pm 0{\cdot}5$ min. After this the test piece is impacted as described above and examined for failure – fracture or visible crack (as shown by bending through 90°). A new test piece is used for each test.

The complete procedure is started at a temperature at which a 50 per cent change of failure is expected, and at least ten pieces are tested at this temperature. If all the test pieces fail or do not fail, the bath temperature is increased or decreased (respectively) by 10 °C and the test repeated. When temperatures have been found where all failures and non-failures are recorded, the temperature is changed by steps of 2 °C or 5 °C, testing ten test pieces each time and including the 'all fail' and 'no fail' temperatures.

$$\text{Brittleness temperature, } T_{\mathrm{B}} = T_{\mathrm{h}} + \Delta T\left[\frac{S}{100} - \frac{1}{2}\right] \qquad [16.6]$$

where T_{h} highest temperature at which failure of all the specimens occurs
ΔT temperature increment (°C)
S sum of the percentage breaks at each temperature (from a temperature corresponding to 'no fail' down to and including T_{h}).

Alternatively, a graphical method may be used for determining T_B or, for quality control purposes, the test may be carried out at one standard temperature and the material deemed to pass if not more than five test pieces fail out of ten.

Although the title of ASTM D746 refers to 'plastics and elastomers', the brittleness at low temperature for rubber is defined differently in the test methods for rubber, Part A25 of BS 903 (1968),[90] yet the apparatus is essentially the same (see Brown,[78] p. 270).

The development and study of the ASTM D746 method is described by Webber.[91] A number of authors[92–94] recommend notching the test pieces to obtain more reproducible results; while this departs from the ASTM procedure, it has been recognised as a useful supplementary technique by ISO 974 and is permitted by this international standard test. ISO 974 also includes an Appendix which specifies three procedures for preparing test pieces from polyolefin samples.

A modification to this test method may be found in BS 3434: Part 8[95] Method 10C which permits the use of various thicknesses of test piece. The distance from the rear striking edge of the striker to the face of the under clamp is varied according to the thickness of the test piece ranging from 4·8 to 6·4 mm according to thicknesses of from 0·1 to 2·15 mm, respectively.

Two brittleness impact tests at low temperature for plastics film and coated fabrics have been standardised. BS 2782 (1983) Method 150D,[96] *Cold Crack Temperature of Film and Thin Sheeting*, and ASTM D1790 (1983)[97] are basically similar to a test described by Williams.[98] In the BS test a test piece 32 mm long and 6·5 mm wide is folded into a loop, the free end being clamped, and the looped test piece is immersed in a bath of coolant. The test piece is struck with a flat-ended striker (25 mm in diameter) at a velocity of 2 m/s and a nominal energy of 2·6 kJ. Tests are carried out at increments of temperature of 2 °C on at least twelve test pieces and the cold crack temperature is calculated as the temperature at which 50 per cent of the test pieces would have cracked. The BS test in a less sophisticated form also appears in BS 3425: Part 8 (1983) Method 10A.[99]

16.8 HEAT AGEING

16.8.1 General Considerations

So far in this chapter we have been dealing mainly with *reversible* processes, measurements designed to study or check the influence of change of temperature on plastics materials wherein any change of property so induced may be cancelled by return to the original datum temperature: i.e. no 'permanent' damage has been done. We now consider the test for measuring the *irreversible* effects of change of temperature – thermal stability and degradation; as a consequence, we are only concerned with effect of heat, since lowering the temperature reduces the energy in a system and of itself causes no permanent change, though, depending on circumstances, the process of returning to the original state may be slow. Elevation of temperature causes loss of volatile constituents, such as an evaporation of plasticiser if present, as a first irreversible step; however, considering the polymer alone, since most polymers are based on a wholly or partially carbon backbone they are susceptible to bond scission at relatively modest temperatures – perhaps as low as 100 °C depending on polymer type, environment and duration.

The mode of polymer degradation varies widely with polymer type; for instance, polyethylene in oxygen initially crosslinks, PVC degrades by losing hydrogen chloride and poly(methyl methacrylate) depolymerises almost totally into monomer. It is not therefore surprising that standardised tests for thermal stability tend to differ from polymer to polymer: clearly, with an unknown material it is not sufficient, say, only to measure weight loss due to heating – there may be in fact a gain.

A general standard on the subject is ISO 2578[100] *Plastics: Determination of Time–Temperature Limits after Exposure to Prolonged Action of Heat*. This defines a time–temperature limit as the highest temperature to which the particular plastics material can be subjected for a determined time before the numerical value of a chosen characteristic reaches a pre-determined critical (threshold) value. For general purposes a period of 20 000 h is recommended but the choice of the characteristic and the threshold value of that characteristic is left to the user, or is obtained from the specification of the material under test. This standard is, at best, a guide only; selection of the temperature for exposure involves predicting, or knowing beforehand, the approximate temperature range in which the time–temperature limit of the material to be tested is located, otherwise exploratory tests must be made. It would have been helpful if a list of the common plastics materials had been included, with a guide to the temperature limit at 20 000 h. IEC publication 216,[101] *Guide for the Preparation of Test Procedures for Evaluating the Thermal Endurance of Electrical Insulating Materials*, was taken into account during the preparation of ISO 2578. ASTM D3045 (1984)[102] and DIN E53446[103] are similar in content to the international standard. IEC publication 216 was also used as the basis for BS 4618 (1974) Section 4.6[104] on the presentation of plastics design data in thermal endurance tests. Underwriters Laboratories Inc., Standard No. UL 746 (1975)[105] on long-term property evaluation is similar to ISO 2578 but is expanded to give recommended test methods and it also gives a list of thermosetting materials and their approximate thermal indices, i.e. the maximum temperature for which they have indefinite service life.

Caution needs to be used in interpreting the results of these tests in terms of actual service life at elevated temperatures. Mair and Wiebusch[106] have shown that it is unwise to use time and temperature limits to predict permanent service temperature as this requires the taking into account of all properties relevant to the application, such as mechanical or electrical loading throughout the service life of the material. Parthum[107] has investigated the ageing behaviour of various moulding grades of phenol-formaldehyde-based materials under prolonged heating to determine the limiting temperature at 25 000 h (3 years). The results obtained were not in agreement with practical experience in the use of the various materials. Meacham[108] has made a plea for a 'standard' heat resistance test and has proposed the use of ASTM D794 (1982)[109] for this purpose. He also proposes the criterion of heat resistance to be retention of flexural strength. Unfortunately, ASTM D794 is rather vague, does not specify the oven to be used and, for continuous heat tests, specifies the temperature–time limit be determined without giving precise instructions as to how to do so. In view of this it would seem more appropriate to use ISO 2578 as the basis for Meacham's proposal.

16.8.2 Standard Test Methods

The procedures mentioned in §16.6 come into this category in certain instances, in that some of them may be carried out at temperatures which cause irreversible changes. ASTM D700 (1981)[110] for phenolic moulding compounds refers to ASTM D794 under the requirements for heat resistance and requires the determination of flexural strength after 7 days' ageing at the specified temperature followed by reconditioning to normal temperature. If the flexural strength so determined is less than 75 per cent of the initial flexural strength the material is deemed to have failed.

Similarly Methods 131C and 131D of BS 2782 (1983)[111] determine the crushing strength after heating (heat resistance) of, respectively, thermosetting moulding material and thermosetting laminated sheet or mouldings. The test is basically a compression strength measurement (Ch. 8) after the test pieces have been cooled following heating in an air oven for 17 ± 1 h at $135\,°C \pm 5\,°C$, followed by 6 h ± 15 min at $170\,°C \pm 5\,°C$ and total immersion in fusible metal at $400\,°C \pm 10\,°C$ for 30 ± 2 min. The test pieces are either cylinders (Method 131C) of length $10 \pm 0{\cdot}2$ mm and similar diameter or cubes (Method 131D) of side of the same dimension; the latter may be a composite specimen made by building up thin sheet. The compression load on the (heat treated) test pieces is increased steadily so that failure occurs within 30 m ± 15 s.

Two methods are available for looking at the effect of ageing at high temperature of polyethylene. BS 2782 Method 108A[112] requires that a sample of material be milled on an open two roll mill at $160\,°C \pm 5\,°C$ for 3 h. Method 108B[113] uses a specimen approximately $76 \times 76 \times 1{\cdot}3$ mm which is heated for 48 ± 2 h at $140\,°C \pm 1\,°C$ in an oven with fan circulation. The test piece rests on a sheet of polytetrafluoroethylene or a sheet of metal coated with that polymer. These methods of ageing are particularly used for examination of polyethylene used for cable insulation or sheet and is usually followed by determination of power factor at 1–20 MHz (Ch. 13) or, for black compounds, by determination of melt flow index, (Ch. 5).

Methods Based on Weight Loss

There are a considerable number of 'loss in weight on heating' tests which have been standardised. Most, however, are not ageing tests, but rather control tests of solids content of resin solutions and monomer or resin content of lamination resins; as such they are not considered to fall within the scope of this book (being more analytical in character). Nevertheless this chapter would not be complete without mention of certain of the tests.

ISO 1137 (1975),[114] for the determination of behaviour in a ventilated tubular oven, allows for the loss in mass (and change in mechanical properties) after heating for a specified period of time and temperature in a tubular oven with an air velocity of 100 m/min. ASTM D1870 (1983)[115] is similar but allows a greater freedom of choice of temperatures and air velocities. Both of these methods are useful where it is desired to determine the long-term effect of heat on plastics under carefully controlled and contamination-free conditions.

ISO 176 (1976)[116] *Determination of Loss of Plasticisers – Activated Carbon Method*, Methods A and B has been issued as a dual numbered standard by BSI in BS 2782 Methods 465A and 465B (1979).[117] It is a test designed to find the loss of plasticiser when a specified test piece is heated in the presence of activated carbon. Both methods specify activated carbon with a grain size of about 4 to 6 mm, free from powder. In the A method the test piece is in direct contact with the activated carbon. For the B method cylindrical cages constructed from bronze gauze having an aperture of about 600 μm, 60 mm in diameter and 6 mm high are used to keep the test pieces out of contact with the activated carbon. The test is carried out by loading the test pieces into metal cans (about 100 mm diameter and 120 mm high) fitted with non-airtight covers. These containers are heated in a thermostatically controlled bath or oven to within ±1 °C of the prescribed temperature.

For Method A a layer of about 120 cm^3 of activated carbon is spread evenly over the bottom of a container and a test piece is placed on top. A further 120 cm^3 of carbon is placed on top of this, followed by two further test pieces each covered by 120 cm^3 of carbon. After the lid of the container is secured, the whole is heated for 24 h at 70 °C ± 1 °C. For Method B the procedure is the same except that each test piece is put into a gauze cage to avoid direct contact with the carbon and the heating is carried out for 24 h at 100 °C ± 1 °C. By weighing each test piece to within 0·001 g before and after heating, the change in mass is calculated as a percentage of the original mass and the result is expressed as the arithmetic mean of the values obtained from the three test pieces.

ASTM D1203 (1981)[118] is similar to ISO 176.

ISO 3671 (1976)[119] of aminoplastics moulding materials, determines the loss in mass (assumed to be predominantly water) after heating for 3 h at 55 °C ± 1 °C a 5 g sample of the moulding powder. In addition, there are many mass loss tests given in the various ASTM materials specifications.

In ISO 177 (1976),[120] the mass loss of plasticised material at 70 °C is determined when a sample is placed between specified 'absorbing discs', to simulate certain end-uses.

Methods Specifically for Vinyl Chloride Polymers

ISO R182 (1970)[121] Method A, BS 2782 (1976) Method 130A[122] ASTM 4204 (1982)[123] and DIN 53381: Part 1 (1983),[124] all assess the time taken for the vinyl chloride polymer or copolymer to evolve sufficient hydrogen chloride to change the colour of a piece of Congo Red paper placed with its lower edge 25 mm above the top of the polymeric sample, the upper end of the paper being held in position by a plug of glass wool inserted in the mouth of the tube. The assembly is placed in a constantly stirred oil bath at 180 °C ± 1 °C to the level of the upper surface of the material in the tube. The heat stability (mean of two determinations) is taken as the time in minutes between insertion of the tube in the hot oil bath and the first signs of a change of the indicator paper from red to blue.

This test method has enjoyed some popularity in the past and is still widely used as a quality control test but it does have one disadvantage. Certain stabiliser systems (notably those based on thiotin compounds) cause bleaching of the Congo Red paper which makes the end-point difficult to determine. In such cases a fresh piece of indicator paper may have to be introduced half-way through the test. All these standards take account of this problem by suggesting that if the colour change is indistinct then two times are recorded corresponding to the first sign of colour change (from red to violet) and to the permanent change (from violet to blue).

PVC compositions discolour markedly when degradation sets in, as a result, it is believed, of the creation of conjugated double bond systems by loss of hydrogen chloride. Oven ageing tests which depend on visual assessment of such degradation are popular and are reckoned to give more practical information on the thermal degradation of PVC than the Congo Red test. ISO 305 (1976),[125] ASTM D2115–67 (1980)[126] and DIN 53381: Part 2 (1983),[127] are three such standard tests. There is nothing particularly profound in these methods and the reader is referred to the original standards for details. It should be noted that ISO 305 specifies non-standard size test tubes in the test method and is the subject of criticism becauseof this.

Because of the limitations of the Congo Red test and the subjective judgement of colour change to indicate the level of degradation of PVC in the above tests, other ways of more precisely measuring the evolution of hydrogen chloride have been devised. For some years standard tests existed in which the PVC material under test was heated in a stream of nitrogen (or other gas) at temperatures up to 200 °C and the hydrogen chloride absorbed in a suitable medium and the acid determined by a standard chemical method (e.g. addition of silver nitrate followed by titration of the excess silver nitrate with ammonium thiocyanate).

This type of test is time-consuming particularly if a plot of amount of acid evolved against time is required and the tests are now obsolete. Such a test is described in the [now discontinued] (1983) ASTM D793 (1976).[128]

It is now the more common practice to use pH change to determine the time of heating of a PVC material to cause the first evolution of hydrochloric acid gas as evidence of decomposition. ISO R182 (1970)[121] Method B, and BS 2782 (1976) Method 130B[129] are all similar and based on this method of measurement. A weighed portion (1 g) of the material is placed in a test tube (150 mm long and 17 mm diameter) fitted with glass inlet and outlet tubes. The outlet tube is taken to a pH-measuring cell fitted with calomel and glass electrodes and filled with 60 ml of 0·1 N potassium chloride solution of pH 6·0. The inlet tube is connected to a gas supply (either air or nitrogen), which must be dry and free of carbon dioxide, at a rate of 6·0 l/h. To carry out the test, the gas flow is started and after 1 min the test tube is immersed in an oil bath at 180 °C ± 1 °C and the time is noted. The test is continued until the pH of the potassium chloride solution in the pH cell reaches 3·9 ± 0·1, when the time is again noted. The thermal stability is expressed as the time, in minutes, from insertion of the tube containing the test portion in the oil bath until the pH has reached 3·9. Two test portions are used and the arithmetic mean of the two results is recorded as the thermal stability. When the two values are more than ± 10 per cent apart from their average the test is repeated.

The pH method has the advantage over the Congo Red method that different gases can be used to simulate oxidative (air) and neutral (nitrogen) conditions of use.

Other Standard Tests for Plastics

In ASTM D1457[41], *PTFE Moulding and Extrusion Materials*, there is a thermal instability test, the index of which is obtained by multiplying by 100 the difference in specific gravity between test pieces before and after sintering under specified conditions. These sintering conditions vary according to the type of material under test and are closely laid down in Table 3 of the standard. They include the initial temperature, rate of heating, hold time, cooling rate, final or second hold temperature, second hold time and the period to room temperature.

The principle of the test is change in molecular weight as indicated by change in specific gravity. The text of the method leaves much to be desired as two types of specific gravity are used; *standard* specific gravity and *extended* specific gravity. The difference between these two properties in relation to the test is not clear as one is measured by the displacement method and the other by density gradient column.

In the dimensional stability test for extruded acrylic plastic sheet in ASTM D1547 (1981),[130] the presence of bubbles is studied after specimens 6 in (152 mm) square have been heated at either 143 °C ± 6 °C or 154 °C ± 6 °C for 15 or 30 min according to, respectively, grade and thickness of material. The specimens, after pre-conditioning at 50 °C ± 6 °C, are tested on a pre-heated glass plate sprinkled with talcum powder.

For testing the thermal oxidative stability of propylene plastics ASTM D2445 (1981),[131] a quantity of pellets or granules is placed in a U-tube which is partially immersed in an oil bath at 150 °C ± 0·5 °C. Oxygen is metered in at one end of the tube at a rate of 10 cm^3/min, and the pellets or granules are inspected at regular intervals for signs of failure as manifested by the appearance of surface crazing. Embrittlement is confirmed by removing a few of the granules and crushing them; if they are readily reduced to powder, embrittlement is taken as complete.

Finally, ISO 1599 (1975),[132] for the determination of viscosity loss on moulding of cellulose acetate, requires the test sample to be compression-moulded under a specified moulding cycle. The moulding is then ground up, the granulate is dried and its viscosity in dilute solution is determined according to ISO 1157 (1975).[133] The percentage viscosity loss is calculated from the values of viscosity before and after moulding.

16.9 PHYSICAL TESTS AT NON-AMBIENT TEMPERATURES

16.9.1 General Considerations

If more meaningful data on temperature effects are required, more useful for design purposes than the arbitrary values obtained from the *ad hoc* thermal yield, brittleness and stiffening tests described in §16.6 and §16.7, it is essential to perform relevant 'absolute' tests at the temperature or temperatures of interest. By and large, their execution follows the description of the tests at standard temperatures, the variation being in creating an atmosphere of the appropriate temperature around the specimens – and perhaps the whole apparatus – with the minimum of variation from point to point and at any one point. There may, of course, be practical difficulties in enclosing an apparatus, or even the essential part of it, in a temperature-controlled atmosphere, or, if this is successfully achieved, of actually operating it in such an enclosure. Carrying out impact tests with a pendulum machine is a particular case in point but, depending on the case in question, simple solutions are often available. Thus, impact testing of this type is a fast operation, and it has been found[134] that appropriate pre-cooling of the test pieces, followed by testing immediately after removal from the cooling cabinet, is a satisfactory alternative. There may, however, be unexpected pitfalls with otherwise simply accommodated tests, e.g. ice formation on the guides of compression chucks causing apparently high results, or on the surface of test pieces, leading to low results in electrical resistivity measurements.

ISO 3205 (1976)[135] and ASTM D618 (1981)[136] give lists of preferred test temperatures from −269 °C to 1000 °C, but there are differences between these two standards (see Ch. 4).

16.9.2 Standard Test Methods

ISO 2578 (1974),[100] BS 4618 (1974) Section 4.6,[104] ASTM D3045 (1984),[102] DIN 53446 (1982)[103] and UL 746B (1984)[105] have already been described in §16.8.

The product specification BS 3953[137] for synthetic resin bonded woven glass fabric laminated sheet requires the flexural stress at rupture of prepared laminates to be measured by Method 335A of BS 2782[138] at 150 °C after 1 h at 150 °C and after 100 h at 200 °C. The related, but earlier, standard BS 4045[139] replaces the high-temperature test by one at room temperature after exposure for long periods at elevated temperature, and terms it a 'thermal endurance' test. The ability to resist deterioration due to increased temperatures, as measured by these tests, is of course quite different as is the design characteristic where one is looking for ability of the product to be used under load in high-temperature conditions.

16.9.3 Test Equipment

Clearly the design of test equipment and test chambers for use at very high and very low temperatures is very important if satisfactory results are to be obtained. Account must be taken of problems which are likely to arise with the use of testing machines which operate satisfactorily at ordinary temperatures (see §16.9.1 above). Thermal insulation of test chambers is all-important if adequate temperature control is to be achieved. Many suppliers of test machines, market ovens and chambers suitable for total enclosure of the machine or, more often, of the area immediately surrounding the test piece and some of these are capable of working from well below ambient temperatures up to a few hundred degrees centigrade.

For very low-temperature testing recourse usually has to be made to the use of liquefied gases; the relative merits and disadvantages and dangers of the various gases have been summarised by Mathes.[140] Liquid nitrogen vapour is often used and, properly handled, is reasonably safe. Following the ideas of Wessel and Olleman,[141,142] Ives and Mead[143] describe a cryostat for low-temperature mechanical testing using the vapour from liquid nitrogen which gives good temperature control up to near room temperature.

The liquid nitrogen moves from its storage flask to the cold chamber of the apparatus through a dip tube which passes from the bottom of the vessel up to a well insulated pipe (as short as possible) and thence to the chamber; no valves or other restrictions are included in this line. The liquid nitrogen container is otherwise sealed, except for one vent pipe connected to the vapour above the liquefied gas; the increase in pressure of this vapour, caused by continuous slow evaporation, forces liquid or cold vapour into the working chamber (Fig. 16.18).

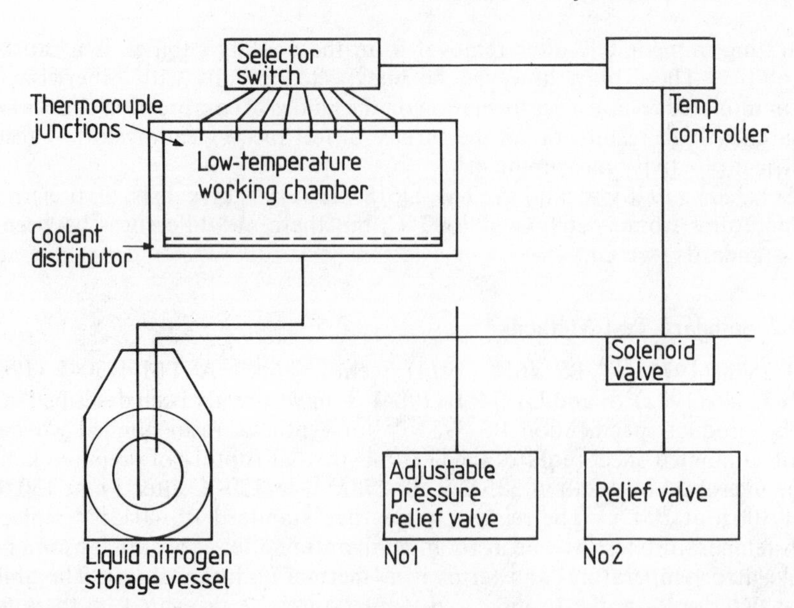

Fig. 16.18 Schematic diagram of low-temperature apparatus (according to Ives and Mead[141])

Temperature is controlled by releasing or reducing the excess pressure in the storage flask and to this end the vent pipe is connected directly to an adjustable pressure relief valve (No. 1) and by way of a solenoid valve to relief valve No. 2. In operation, the settings of the valves are adjusted so that valve No. 1 permits the driving force in the flask to be high enough to ensure an adequate supply of coolant to the chamber, but the pressure is such that the supply is not excessive and overshooting is avoided. The second pressure relief valve is brought into action by the opening of the solenoid valve, which in turn is actuated by a temperature controller with a sensing element in the cold chamber. This second valve is adjusted to control the pressure at a value slightly below that necessary to ensure a supply of coolant to the working chamber, and thus an enhanced demand for coolant, initiated by a rise in chamber temperature, can be met almost instantaneously.

The subject of testing at low temperatures has been reviewed by Lieb and Mowers.[144]

Comments on the types of chamber commercially available are given by Brown.[145] General directions for achieving elevated or subnormal temperatures are given in ISO 3383 (1985)[146] which is a useful document but is of an elementary nature and its shortcomings are highlighted by Brown,[78] p. 70.

When testing at non-ambient temperature it is necessary to condition the test piece for a time long enough for it to reach the test temperature, although excessive conditioning will cause heat ageing. Attention is drawn to the conditioning times given in the Appendix to Ch. 4.

REFERENCES

1. Griffiths, M. D. (1974) *RAPRA Members Journal*, July, 187.
2. ASTM D696 (1979) *Coefficient of Linear Expansion of Plastics.*

3. ASTM D864 (1978) *Coefficient of Cubical Thermal Expansion of Plastics.*
4. Bekkedahl, N. (1949) *Journal of Research of the National Bureau of Standards*, **42** (Rp. 2016), 145.
5. ISO 2577 (1984) *Plastics – Determination of Shrinkage of Compression Moulded Test Specimens in the form of Bars.*
6. BS 2782 Method 640A (1979) *Mould Shrinkage of Thermosetting Moulding Materials.*
7. ASTM D955 (1979) *Measuring Shrinkage from Mold Dimensions of Moulded Plastics.*
8. ASTM D1299 (1979) *Shrinkage of Molded and Laminated Thermosetting Plastics at Elevated Temperatures.*
9. DIN 53464 (1962) *Testing Plastics, Determination of Shrinkage Properties of Molded Thermosetting Materials.*
10. ASTM D702 (1981) *Cast Methyacrylate Plastics Sheets, Rods, Tubes and Shapes.*
11. BS 2782 Method 641A (1985) *Dimensional Stability at 100 °C of Flexible PVC Sheet.*
12. ASTM D2732 (1983) *Unrestrained Linear Thermal Shrinkage of Plastic Film and Sheeting.*
13. BS 3794: Part 2, *Decorative Laminated Sheets*, Method 9 (1982) *Dimensional Changes at Elevated Temperatures.*
14. BS 2782 Method 643A (1976) *Shrinkage on Heating of Film Intended for Shrink Wrapping Applications.*
15. ASTM D2838 (1983) *Shrink Tension and Orientation Release Stress of Plastic Film and Thin Sheeting.*
16. ISO 3521 (1976) *Plastics – Polyester and Epoxy Coating Resins – Determination of Total Volume Shrinkage.*
17. ASTM D2566 (1979) *Standard Test Method for Linear Shrinkage of Cured Thermosetting Casting Resins During Cure.*
18. BS 2782 Method 644A (1979) *Total Volume Shrinkage of Polyester and Epoxy Coating Resins.*
19. Brydson, J. A. (1975) *Plastics Materials*, 3rd edn, Butterworth.
20. Shen, M. C. and Eisenberg, A. (1970) *Rubber Chemistry and Technology*, **43**, 95.
21. Eisenberg, A. and Shen, M. C. (1970) *Rubber Chemistry and Technology*, **43**, 156.
22. Zemansky, M. W. (1957) *Heat and Dynamics*, 4th edn, McGraw-Hill, p. 318.
23. Zemansky, M. W. (1957) op. cit., p. 332.
24. Gordon, M. (1957) in *Physics of Plastics*, (Ed. P. D. Ritchie), Illife.
25. Gibbs, D. H. and DiMarzio, E. A. (1958) *Journal of Chemical Physics*, **28**, 373.
26. DiMarzio, E. A. and Gibbs, J. H. ibid., 807.
27. Staverman, A. J. (1968) *Rubber Chemistry and Technology*, **41**, 544,
28. Rehage, G. and Borchard, W. (1973) in *The Physics of Glassy Polymers* (Ed. R. N. Haward), Applied Science.
29. Wiley, R. H. (1947) *Journal of Polymer Science*, **2**, 10.
30. Sheldon, R. P. (1962) *Journal of Applied Science*, **6**, 343.
31. ASTM D3418 (1982) *Temperature Transition of Polymers by Thermal Analysis.*
32. Durrans, T. H. (1929) *Journal of The Oil and Colour Chemists Association*, **12**, 108.
33. BS 2782 Method 103A (1970) *Softening Point of Synthetic Resin (Ring and Ball Method).*
34. ISO 1218 (1975) *Plastics – Polyamide – Determination of Melting Point.*
35. BS 2782 Method 123A (1976) *Determination of Melting Point of Synthetic Resins.*
36. DIN 53736 (1973) *Testing of Plastics; Determination of the Melt Temperature of Semi Crystalline Plastics.*
37. BS 2782 Method 123B (1976) *Determination of the Melting Point of Polyamides.*
38. ASTM D789 (1981) *Nylon Injection Molding and Extrusion Materials.*
39. ASTM D2133 (1981) *Acetal Resin Injection Molding and Extrusion Materials.*
40. ASTM D2116 (1983) *FEP – Fluorocarbon Molding and Extrusion Materials.*
41. ASTM D1457 (1983) *PTFE Molding and Extrusion Materials.*
42. ISO 3146 (1985) *Plastics – Determination of the Melting Point on Semi-crystalline*

Polymers – Optical Method.
43. ASTM D2117 (1982) *Melting Point of Semi-crystalline Polymers by the Hot Stage Microscopy Method.*
44. BS 2782 Method 13C (1976) *Melting Point of Crystalline Polymers with Polarised Light.*
45. Crofts, D. and Brown, R. P. (1971) *RAPRA Technical Review*, no. 61, August.
46. Tordella, J. P., Webber, A. C. and Cooper, E. R. (1953) *ASTM STP*, no. 14.
47. Stephenson, C. E. and Willbourn, A. H. (1959) *ASTM STP*, no. 247, p. 169.
48. Sherr, A. E. (1965) *SPE Journal*, **21** (1), 67.
49. Horsely, R. A. (1965) *Transactions of the Plastics Institute, London*, **33** (106), 119.
50. Abolins, V. (1975) *Materials Engineering*, **81** (4), 52.
51. ASTM D648 (1982) *Test for Deflection Temperature of Plastics under Flexural Load.*
52. BS 2782 Method 115A (1978) *Plastic Yield.*
53. BS 2782 (1970) Method 102C *Softening Point of Thermoplastic Moulding Material (Bending Test).*
54. ISO 75 (1974) *Plastics and Ebonite – Determination of Temperature of Deflection under Load.*
55. BS 2782 Methods 121A and 121B (1976) *Determination of Temperature of Deflection under a Specified Bending Stress of Plastics and Ebonite.*
56. DIN 53461 (1984) *Testing of Plastics – Determination of Temperature of Deflection under Load According to ISO 75.*
57. Martinelli, F. J. and Hodgkin, J. H. (1973) *Journal of Applied Polymer Science*, **17**, 1443.
58. BS 2782 Method 121C (1976) *Determination of Temperature of Deflection under a Specified Bending Stress of Rigid Thermosetting Resin Bonded Laminated Sheet.*
59. DIN E53458 *Testing of Plastics; Determination of Temperature of Deflection under Load (Martens Method).*
60. Gohl, W. (1959) *Kunststoffe*, **49** (5), 228.
61. ASTM D1637 (1983) *Tensile Heat Distortion Temperature of Plastics Sheeting.*
62. Bourgault, C. V. (1974) *Plastics Design and Processing*, **14** (7), 17.
63. ASTM D1430 (1981) *Polychlorotrifluoroethylene (PCTFE) Plastics.*
64. Sandiford, D. J. H. and Buckingham, K. A. (1961) *British Plastics*, **34** (11), 594.
65. ISO 306 (1974) *Determination of the Vicat Softening Temperature of Thermoplastics.*
66. BS 2782 Methods 120A to 120E (1976) *Determination of Vicat Softening Temperature on Thermoplastics.*
67. ASTM D1525 (1982) *Vicat Softening Temperature of Plastics.*
68. DIN 53460 (1976) *Determination of Vicat Softening Temperature of Thermoplastics.*
69. Crofts, D. (1972) *RAPRA Research Report*, no. 200.
70. Graves, F. L. and Loveless, H. S. (1963) *Materials Research and Standards*, **3** (1), 33.
71. Ehlers, G. F. L. and Powers, W. H. (1964) *Materials Research and Standards*, **4** (6), 298.
72. BS 2782 Method 122A (1976) *Determination of deformation under Heat of Flexible Polyvinyl Chloride Extrusion Compound.*
73. ASTM D621–64 (1976) *Deformation of Plastics under Load.*
74. Clash, R. F. and Berg, R. M. (1944) *ASTM Symposium on Plastics*, p. 54.
75. Welding, G. N. (1955) *Plastics, London*, **20** (214), 158.
76. Williamson, I. (1950) *British Plastics*, **23** (256), 87.
77. Wheeler, A. (1969) *Plastics and Polymers*, **37** (131), 469.
78. Brown, R. P. (1986) *Physical Testing of Rubber*, 2nd edn, Elsevier Applied Science.
79. ISO 458: Part 1 (1985) *Determination of Stiffness in Torsion as a Function of Temperature.*
80. BS 2782 Method 150A (1976) *Stiffness in Torsion as a Function of Temperature.*
81. ASTM D1043 (1984) *Stiffness Properties of Plastics as a Function of Temperature by Means of a Torsion Test.*
82. DIN 53447 (1981) *Determination of Stiffness in Torsion (Clash–Berg Method).*
83. BS 2782 Method 150B (1983) *Cold Flex Temperature of Flexible Polyvinyl Chloride*

Compound.
84. BS 2782 Method 150C (1983) *Determination of Low Temperature Extensibility of Flexible Polyvinyl Chloride Sheet.*
85. BS 2782 Method 151A (1984) *Determination of Cold Blend Temperature of Flexible Polyvinyl Chloride Extrusion Compound.*
86. Wormald, D. (1958) *British Plastics*, **31** (9), 392.
87. ISO 974 (1980) *Method of Determining Brittleness Temperature by Impact.*
88. ASTM D746 (1979) *Brittleness Temperature of Plastics and Elastomers by Impact.*
89. Hoff, A. E. W. and Turner, S. (1957) *ASTM Bulletin*, 225, 58.
90. BS 903: Part A25 (1968) *Brittleness Temperature by Impact.*
91. Webber, A. C. (1958) *ASTM Bulletin*, 227, 40.
92. Bestelinki, P. N. and Turner, S. (1958) *ASTM Bulletin*, 231, 68.
93. Birks, A. M. and Rudin, A. (1960) *ASTM Bulletin*, 264, 49.
94. Turner, S. (1958) *British Plastics*, **31** (12), 526.
95. BS 3424: Part 8, Method 10C (1983) *Low Temperature Impact Test for Coated Fabrics.*
96. BS 2782 Method 150D (1983) *Cold Crack Temperature of Film and Thin Sheeting.*
97. ASTM D1790 (1983) *Brittleness Temperature of Plastic Sheeting by Impact.*
98. Williams, H. O. (1958) *British Plastics*, **31** (3), 107.
99. BS 3424: Part 8, Method 10A (1983) *Determination of Cold Crack Temperature of Coated Fabrics.*
100. ISO 2578 (1974) *Plastics: Determination of Time–Temperature Limits after Exposure to Prolonged Action of Heat.*
101. IEC Publication 216 (1966) *Guide for the Preparation of Test Procedures for Evaluating the Thermal Endurance of Electrical Insulating Materials.*
102. ASTM D3045–74 (1984) *Heat Ageing of Plastics without Load.*
103. DIN E53446 (1982) *Testing Plastics, Determination of Temperature–Time Limits.*
104. BS 4618 Section 4.6 (1974) *Recommendations for the Presentation of Plastics Design Data; Thermal Endurance Test.*
105. UL 746B (1984) *Polymeric Materials – Long Term Property Evaluation* (Underwriters Laboratories Inc.).
106. Mair, H. J. and Wiebusch, K. (1977) *Kunststoffe*, **67** (9), 530.
107. Parthum, W. (1975) *Plastik und Kautschuck*, **22**(4), 334.
108. Meacham, S. E. (1978) *Plastics Technology*, **24** (1), 83.
109. ASTM D794 (1982) *Determining Permanent Effect of Heat on Plastics.*
110. ASTM D700 (1981) *Phenolic Molding Compounds.*
111. BS 2782 Methods 131C and 131D (1983) *Crushing Strength after Heating (Heat Resistance) of Thermosetting Material, Laminates and Mouldings.*
112. BS 2782 Method 108A (1970–1986) *Ageing of Polythene by Hot Milling.*
113. BS 2782 Method 108B (1970–1986) *Ageing of Polythene by Hot Air Oven.*
114. ISO 1137 (1975) *Plastics – Determination of Behaviour in a Ventilated Tubular Oven.*
115. ASTM D1870 (1983) *Elevated Temperature Ageing using a Tubular Oven.*
116. ISO 176 (1976) *Determination of Loss of Plasticisers – Activated Carbon Method.*
117. BS 2782 Methods 456A and 456B (1979) *Determination of Loss of Plasticisers (Activated Carbon Method).*
118. ASTM D1203 (1981) *Loss of Plasticisers from Plastics (Activated Carbon Methods).*
119. ISO 3671 *Determination of Volatile Matter of Aminoplastics Moulding Materials.*
120. ISO 177 (1976) *Plastics – Determination of Migration of Plasticisers.*
121. ISO R182 (1970) *Determination of the Thermal Stability of Polyvinyl Chloride and Related Co-polymers and their Compounds by Splitting off of Hydrogen Chloride.*
122. BS 2782 Method 130A (1976) *Thermal Stability of PVC; Dehydrochlorination Methods.*
123. ASTM D4204 (1982) *Thermal Stability of Poly(Vinyl Chloride) (PVC)* Resin.
124. DIN 53381: Part 1 (1983) *Determination of Thermal Stability of PVC; Dehydrochlorination*

Methods.

125. ISO 305 (1976) *Determination of Thermal Stability of Polyvinyl Chloride, Related Chlorine Containing Polymers, and their Compound-Discolouration Method.*
126. ASTM D2115–67 (1980) *Oven Heat Stability of Poly(Vinyl Chloride) Compositions.*
127. DIN 53381: Part 2 (1983) *Determination of Thermal Stability of PVC-Discolouration Methods.*
128. ASTM D793 (1976) *Short Term Stability at Elevated Temperatures of Plastics Containing Chlorine.*
129. BS 2782 Method 130B (1976) *Thermal Stability of Polyvinyl Chloride Compounds (pH Method).*
130. ASTM D1547 (1981) *Extruded Acrylic Plastics Sheet.*
131. ASTM D2445 (1981) *Thermal Oxidative Stability of Propylene Plastics.*
132. ISO 1599 (1975) *Cellulose Acetate – Determination of Viscosity Loss on Moulding.*
133. ISO 1157 (1975) *Cellulose Acetate in Dilute Solutions – Determination of Viscosity Number and Viscosity Ratio.*
134. Ives, G. C. and Mead, J. A. (1959) *SCI Monograph*, no. 5, Society of Chemical Industry, p. 80.
135. ISO 3205 (1976) *Preferred Test Temperatures.*
136. ASTM D618 (1981) *Conditioning Plastics and Electrical Insulating Materials for Testing.*
137. BS 3953 (1976) *Synthetic Resin Bonded Woven Glass Fabric Laminated Sheet.*
138. BS 2782 Method 335A (1978) *Determination of Flexural Properties of Rigid Plastics.*
139. BS 4045 (1966) *Epoxy Resin Pre-impregnated Glass Fibre Fabrics.*
140. Mathes, K. N. (1964) *SPE Journal*, **20** (7), 634.
141. Wessel, E. T. and Olleman, R. D. (1953) *ASTM Bulletin*, 187, 56.
142. Wessel, E. T. (1956) *ASTM Bulletin*, 211, 40.
143. Ives, G. C. and Mead, J. A. (1961) *Materials Research and Standards*, **1** (3), 194.
144. Lieb, J. H. and Mowers, R. E. (1966) in *Testing of Polymers*, vol. 2 (Ed. J. V. Schmitz), Interscience, Ch. 3.
145. Brown, R. P. (Ed.) (1979) *Rapra Guide to Rubber and Plastics Testing.*
146. ISO 3383 (1985) *General Directions for Achieving Elevated or Sub-normal Temperatures.*

Chapter Seventeen

Environmental Resistance

17.1 INTRODUCTION

The subjects considered in this chapter have one factor in common – they are test methods designed to evaluate the resistance of plastics materials to change as a result of exposure to some environment other than high or low temperature. Many are accelerated tests and invariably these have at least one common shortcoming which is the very result of their aim to provide a forecast of durability within a finite period of time: unless they are 'accelerated' they offer nothing over a field trial, but to achieve this acceleration the stringency of test conditions may be so enhanced as to scale an energy barrier and create an effect which would never occur in normal usage. Even if this does not happen, the equation of the results of accelerated ageing tests into likely service life is difficult if not impossible because the accelerated conditions are usually so idealised that they may not truly simulate the sequence of events if the field. This, and the 'energy barrier' consideration, may lead to ranking orders of merit from the accelerated ageing tests which are not fulfilled in practice.

Temperature plays a part in all environmental tests. The effects of elevated temperature alone have already been dealt with in Ch. 16, except for that of fire which is considered to be more of an environmental effect than simply one of exposure of plastics to very high temperatures and which is described in Ch. 18.

It cannot be claimed that all the degradative influences which plastics materials are likely to encounter are covered in this chapter, but it is the intention to consider the more important ones.

17.2 MOIST HEAT AND STEAM TESTS

There would appear to be no great concern over the long-term effects of moist heat on plastics, although conditioning of certain materials at specified relative humidity is very important if consistent and meaningful results are to be obtained from certain test procedures (see Ch. 4).

There is, however, an international standard, ISO 4611 (1980),[1] which provides for damp heat tests at 40 °C and 93 per cent r.h., both static and cycling, with or without water or salt mist spray.

Perhaps the most comprehensive standardised test method which is relevant to this subject is ASTM D756–78 (1983),[2] *Practice for Determination of Weight and Shape Changes of Plastics under Accelerated Service Conditions.* In the words of the defined scope of this standard: 'This practice covers the determination of the weight and shape changes occurring in plastics under various conditions of use, not where exposure to direct sunlight, weathering, corrosive atmospheres or heat alone is involved, but where changes in atmospheric temperature and humidity are encountered. This embraces the interior of buildings, and the interior of transport facilities such as motor vehicles, airplane cargo spaces or wing interiors, holds of ships and railroad cars.'

Notwithstanding this explanation, Procedure B of ASTM D756 involves dry heat ageing only (72 h at 60 °C ± 1 °C). Others specified are:

Procedure A 24 h at 60 °C ± 1 °C at 88–89 per cent r.h. followed, within 2 h, by 24 h at 60 °C ± 1 °C in an oven.

Procedure C 24 h at 70 °C ± 1 °C at 70–75 per cent r.h. followed, within 2 h, by 24 h at 70 °C ± 1 °C in an oven.

Procedure D 24 h at 80 °C ± 1 °C over distilled water in an oven followed, within 2 h, by 24 h at 80 °C ± 1 °C in an oven.

Procedure E 24 h at 80 °C ± 1 °C at 70–75 per cent r.h. followed, within 30 min, by 24 h at − 40 °C ± 2 °C or − 57 °C ± 2 °C, then, within 2 h, by 24 h at 80 °C ± 1 °C in an oven and, finally, within 30 min by 24 h again at − 40 °C ± 2 °C or − 57 °C ± 2 °C.

Procedure F 24 h at 38 °C ± 1 °C at 100 per cent r.h. followed, within 2 h, by 24 h at 60 °C in an oven.

Procedure G As Procedure F except that the temperature is 49 °C ± 1 °C in both humid and dry heat periods.

Desiccator conditioning is used between each part of the various cycles.

Specimens are weighed, their dimensions measured and visual changes noted after each of the distinct parts of the various cycles.

Also of interest, because of the use of plastics in electrical insulation applications, is the series of BS 2011 dealing with components for telecommunication and allied electronic equipment, particularly Part 2: 1 Db (1981),[3] *Damp Heat, Cyclic*, which specifies a 24 h cycle of the following:

1. 25 °C ± 10 °C ('laboratory temperature') to 55 °C ± 2 °C in $1\frac{1}{2}$–$2\frac{1}{2}$ h during which period the relative humidity must be between 80 and 100 per cent and condensation must occur on the components.
2. 55 °C ± 2 °C for 16 h, during which a periodical excursion of 2 °C–3 °C variation in temperature must occur at least four times per hour. The relative humidity during this part of the cycle is 96–100 per cent and condensation will occur.
3. Cool to 'room temperature', the relative humidity remaining at 80–100 per cent to complete the 24 h cycle. During this period droplets of water must not appear on the components.

Similarly interesting, from a related industry is Part F2 (1983) *Determination of Resistance of Humidity* (*cyclic condensation*) of BS 3900,[4] *Methods of Test for Paints* (originally derived from Ministry of Defence Specification DEF 1053). Here the cycle involves the temperature cycling continuously from 42 °C to 48 °C and back to 42 °C in 60 ± 5 min, the heating and cooling periods being approximately equal. Heating is effected through an open water bath so that 100 per cent r.h. is achieved and copious condensation occurs on the test panels.

17.3 EFFECTS OF LIQUIDS AND CHEMICALS

There are a number of published test methods for determining the matter extractable from a plastics material by a specific chemical reagent (for example, ISO 59 (1976) *Plastics – Phenolic Mouldings – Determination of Acetone Soluble Matter*). These are considered to be chemical tests and will not be discussed as being outside the scope of this book. By far the greatest number of published test methods involve the use of distilled water as the test medium and even for chemicals these are mostly offered as a solution in water as the test medium.

17.3.1 Water Absorption and Water-soluble Matter

The resistance of plastics materials to water varies significantly according to their chemical nature and, if they are filled or reinforced, the nature of additives or reinforcements present. For example the sensitivity to moisture is quite different for the relatively hydrophilic secondary cellulose acetate and the decidedly hydrophobic polystyrene. In practical terms of influence on mechanical properties, reference has already been made (Ch. 4) to the twenty-fold increase in impact strength on nylon 6 between dryness and moisture saturation.

A particularly interesting effect is described by Uijlenburg,[5] who studied the water absorption of unplasticised PVC at various temperatures up to 140 °F and then measured certain mechanical properties, specifically related to the material in pipe form. It was found that at 140 °F effects were produced by water absorption which were not encountered at more normal temperatures and therefore a proposal for a speed measurement of the long-term bursting pressure of unplasticised PVC pipe, by testing at 140 °F, was likely to yield very misleading results as to performance at more normal ambient temperatures.

Braden[6] has analysed the processes involved in the absorption of water by plastics materials, which he concluded depend on only two parameters: the diffusion coefficient and the equilibrium uptake. The former governs the kinetics of water absorption and is highly temperature-dependent. Equilibrium uptake is virtually unaffécted by temperature. Blank[7,8] who examined five equations proposed for relating water absorption to time, concluded that none was universally applicable and deduced an equation for obtaining equilibrium absorption (Q) from values of absorption (q_1, q_2 and q_3) measured at equally spaced short intervals:

$$Q = \left[\frac{q_2^4 - q_1 2 q_3^2}{2q_2^2 - (q_1^2 + q_3^2)} \right]^{1/2} \qquad [17.1]$$

The effect of fillers is well illustrated by the differing requirements for maximum water absorption specified in BS 771 (1980),[9] *Phenolic Moulding Materials.*

The standard water absorption test is designed as no more than a quality-control test, for use purely as a guide to water sensitivity, by measuring water take-up under a specified but arbitrary set of experimental conditions with no pretence to absolute significance or even to an assessment of the effect of moisture attack on, say, electrical or mechanical properties (for this see, for example, Hauck[10]). Even so, great care is needed to obtain reproducible and comparable test data from such water absorption tests. Control of temperature is obviously of paramount importance in a diffusion or rate process. The presence of grease, even finger-marks, on the surface may influence results. The physical state of the actual surface must be specified, for instance whether 'as moulded' (with a resin skin) or as left after machining. For hydrophobic materials in particular, the precise mode of removal from the water immersion bath and of drying, and the time taken to weigh, may all significantly influence the test results. The head of water could have an effect and, since all plastics materials are relatively resistant to water, water absorption tests generally must take heed of surface-to-volume ratio of test pieces.

In practically all standard tests, since they are essentially short-term, equilibrium is not obtained throughout the test piece thickness (this could take months, or even years) and therefore the effect measured is basically a surface one. Thus, the total surface area of the test piece is of paramount importance in a test measuring water absorption by simply weight increase, as in the ratio of surface-to-volume in tests measuring water absorption on a weight-to-weight basis. Finally, a constituent may well be present which is significantly soluble in water, hence it will be leached out in the test, or the material may even hydrolyse. A quantity of the test piece lost by solution in the water will obviously affect the results of a test which is designed to assess attack of water by weighing water uptake; the loss of soluble matter may cause a much lower net gain in weight, may cancel out the gain or even exceed it. If water solubility is anticipated it should be measured in the test.

The work of Braden[6] demonstrates that two materials of the same equilibrium uptake of water could, if their diffusion coefficients were different, give quite dissimilar water absorption test results after some arbitrarily selected short test period of, say, 24 h.

Standard Test Methods

The earlier ISO standards for water absorption which separated tests at 23 °C and in boiling water have now been combined into one standard ISO 62.[11] This has four methods each allowing a variety of test pieces depending upon the form of the material being tested. The standard circular (moulded) test piece is 50 mm in diameter and 3 mm thick but in certain cases a square test piece (50 mm side and 4 mm thick) may be used by agreement with the parties concerned. Methods 1 and 2 specify immersion in distilled water at 23 °C and Methods 3 and 4 require boiling in distilled water for 30 min. Methods 1 and 2 are for finding the water absorption and Methods 3 and 4 take into account any significant water soluble matter of the material under test by heating the test pieces after immersion for 24 h at 50 °C.

The results are expressed by:

$$\text{water absorption (Method 1 and 2)} = m_2 - m_1 \text{ (mg)} \qquad [17.2a]$$

$$\text{water absorption (Method 3 and 4)} = m_2 - m_3 \text{ (mg)} \qquad [17.2b]$$

where m_1 is the original mass of the test piece
m_2 is the mass after immersion in water
m_3 is the mass after post-immersion drying.

BS 2782 Methods 430A to 430D[12] are identical to ISO 62 Methods 1 to 4, respectively, and these incorporate nine of the ten methods in the, now obsolete, original 1970 version of BS 2782. The one exception is the determination of water-soluble matter of flexible PVC compound (old method 502C). This has not been included in the 1986 reprint of BS 2782 (1970) nor issued as a separate method in the later edition, yet it is still referred to in BS 2572 (*Flexible PVC Compounds*).[13] Using the same notation as in Eqns [17.2a] and [17.2b], water-soluble matter (mg) is given by:

$$m_1 - m_3$$

Both the ISO and BS tests permit the expression of water absorption results as a percentage of the original mass of the test piece as well as the (preferred) change of mass reported in mg.

ASTM D570 (1981),[14] *Water Absorption of Plastics*, describes seven procedures which differ in their severity of duration and temperature: different test pieces are used according to the physical form of the sample under test. The conditions of test are rather different from the ISO and BS tests. This standard also restricts the reporting of results to a percentage change in mass to the nearest 0·01 per cent.

Test temperatures and period of immersion are summarised in Table 17.1, but it is necessary to refer to the individual tests for details of the methods.

The Germans have followed the British and ISO example and have incorporated all of the water absorption methods into the one standard DIN 53495.[15] This includes the ISO methods and also has an additional method for finding the absorption after conditioning in an atmosphere of 93 ± 2 per cent r.h. at $23\,°C \pm 2\,°C$ for 24 h. There are two other differences between this DIN standard and the ISO methods: the option of not pre-drying the test piece before test and exclusion of test pieces cut from forms other than sheet; for such tests reference has to be made to the relevant product standards.

This section would not be complete without reference to BS 2782 (1970) (as reprinted 1986) Method 504A,[16] *Water Vapour Absorption of Cellulose Acetate Moulding Material.* This uses rectangular test pieces of specified thickness ($1{\cdot}52 \pm 0{\cdot}5$ mm), exposed for 48 h at $23\,°C \pm 1\,°C$ and 50 ± 5 per cent r.h. After this treatment the test pieces are removed and supported over distilled water in a closed vessel at $23 \pm 1\,°C$ for 72 ± 2 h. The water vapour absorption is then calculated as a percentage by:

$$\frac{W_2 - W_1 \times 100}{W_1}$$

where W_1 is the mass after conditioning at 50 per cent r.h. and W_2 is the mass after the conditioning over distilled water.

17.3.2 Effect of Chemicals

Water absorption and water solubility are in reality but one aspect of the general

Table 17.1 Standard water absorption tests

Standard	Immersion time	Temperature (°C)	Post-immersion heating Time	Post-immersion heating Temperature (°C)
ISO 62 Method 1 (BS 2782 430A)	24 ± 1 h	$23 \pm 0{\cdot}5$	—	—
ISO 62 Method 2 (BS 2782 430B)	24 ± 1 h	$23 \pm 0{\cdot}5$	24 ± 1 h	50 ± 2
ISO 62 Method 3 (BS 2782 430C)	30 ± 1 min	100	—	—
ISO 62 Method 4 (BS 2782 430D)	30 ± 1 min	100	24 ± 1 h	50 ± 2
BS 2782 (obsolete 2970 edition) Method 502C	48 ± 1	50 ± 1	Dry to constant mass over calcium chloride	

ASTM D570

Procedure

1. 24–0, $+\frac{1}{2}$ h in distilled water at 23 °C ± 1 °C on edge.
2. As (1), but for 120 ± 4 min.
3. As (2), followed by (1), total immersion period being 24 h.
4. After (1), re-immerse, weigh at end of first week and thereafter at 2-weekly intervals until equilibrium is substantially obtained (as defined).
5. 120 ± 4 min, in boiling distilled water, on edge; cool in distilled water at room temperature for 15 ± 1 min.
6. As 5 but for 30 ± 1 min.
7. As 5, but in distilled water at 50 °C ± 1 °C for 48 ± 2 h.

subject of 'chemical resistance' and probably the only excuse for treating the effect of water as a separate entity – and before all others – is the profusion of the medium. By their very nature the basic polymers as such would not be expected to be subject to chemical attack except by specific compounds or classes of compounds. The most aggressive media will be the organic solvents since the polymers themselves are mainly organic and, subject to the influence of molecular weight and crystallinity, non-polar polymers are generally susceptible to the attack of non-polar solvents and polar polymers to polar solvents.

Organic plastics are generally resistant to the attack of aqueous media, for the reasons briefly outlined in §17.3.1 above, but if the polymer chain is prone to oxidative attack, concentrated nitric acid for instance may have a pronounced degradative effect. Much may be gained from the *ad hoc* approach of immersion of small strip specimens in the medium, preferably at various temperatures, and observation of the effects – discoloration, crazing, cracking, complete degradation, etc. Partial immersion has much to recommend it, as the attack at an air–liquid interface, with oxygen accessibility, may be much more severe. This is the principle employed in the draft CEN standard (also a draft BS) for testing domestic baths made from acrylic sheet.[17] Here water at 70 °C and 12 °C is alternately run into and drained from the bath to a depth of 300 mm at 10 min intervals for 100 cycles. After this treatment the bath must show no signs of surface crazing or other deterioration when tested with an eosin dye solution.

Monitoring the attack, by measurement of the change of some important physical property such as hardness and transparency, is a valuable refinement: Jessup[18] used flexural strength, hardness and electrical properties when examining the effects of dilute sulphuric acid and certain organic solvents on a range of thermosetting plastics.

Standard Test Methods

ISO 175 (1981)[19] proposes the use of thirty-eight test fluids based on laboratory chemicals with an additional ten miscellaneous products. The standard specifies preferred test durations of 24 h, 1 week and 16 weeks but allows test periods up to 5 years and temperatures up to 150 °C although the preferred test temperatures are 23 °C and 70 °C.

Details are given for the determination of change in dimensions, mass and appearance but the choice of other properties for determination of changes in a material after immersion in a chemical is left to the discretion of the user. This standard would undoubtedly be more useful if more precise guidance was given about the physical properties to use, as in BS 4618, Section 4.1 (1972),[20] *Recommendation for the Presentation of Data on the Chemical Resistance of Plastics to Liquids*. This lays down the test temperature, recommends the exposure period and offers a list of reagents based on those of ISO 175. The criteria for assessing the effect of the chemical liquid are change of dimensions, change in appearance and change of mechanical properties; a list of recommended tests is given.

The original 1970 edition of BS 2782 contained a method (505A) for determining the effect of sulphuric acid on rigid PVC compounds. This method had the merit of being quantitative because it was based on the comparison of shear strength before and after immersion in concentrated sulphuric acid at 95 °C–100 °C for 24 h. Over the years this test has not found favour and has now been deleted from the later version of BS 2782. The uPVC pipe specifications BS 3505[21] and BS 3506[22] have a test for resistance to sulphuric acid which requires that test pieces are cut from the pipe having a total surface area of $45 \pm 3\ cm^2$. These are weighed, immersed in $93 \pm 0{\cdot}5$ per cent (m/m) sulphuric acid for 14 days at $55\ °C \pm 2\ °C$ and then washed in running water and re-weighed. The weight change must be within the limits of $+0{\cdot}32$ g and $-0{\cdot}013$ g. These tests for resistance to sulphuric acid were originally devised to ensure that the rigid PVC is truly unplasticised, as inclusion of quite small quantities of ester plasticisers in the material results in a marked loss of chemical resistance and mechanical properties.

ASTM D543 (1982)[23] includes some useful early references to chemical resistance of plastics and lays down general procedures for the examination of fifty specified reagents and solvents by following weight and dimensional changes and alteration in mechanical properties.

ASTM D2299 (1982)[24] describes how to carry out tests for determining relative stain-resistance of plastic, not so much by chemicals as by the hazards of everyday life such as coffee, tea, lipstick, shoe polish, and so on.

Some plastics, notably PVC, which contain lead and cadmium compounds can show staining if they come into contact with hydrogen sulphide. ASTM D1712[25] deals with this specific hazard by immersing test pieces in a saturated aqueous solution of hydrogen sulphide for 15 min at 23 °C. Visual comparison with similar test pieces which have not been exposed to hydrogen sulphide decides the degree of staining. The British Standards for PVC unsupported calendered sheeting BS 1763[26] and BS 2739[27] deal with this matter by immersing test pieces in a solution of sodium sulphide acidified with acetic acid. This is a rather safer method as emission of hydrogen sulphide, a very poisonous gas, is minimised.

A variation on the sulphide staining test is given in ASTM D2151,[28] *Staining of Poly (Vinyl Chloride) Compositions by Rubber Compounding Ingredients.* Here test samples of a vinyl composition are placed in contact with a rubber compound under a pressure of 10 g/cm^2 for 20 h at 70 °C. At the end of this period all test pieces (including the control samples) are exposed to a specified light source (e.g. Xenon arc) for 4 h before assessing the degree of staining.

DIN 53476 1979[29] follows ISO 175 in laying down procedures for measuring the resistance of plastics to some fifty liquids.

In marine atmospheres, salt spray may be present in the atmosphere to a significant degree. It is not reckoned a serious hazard to plastics generally and there is no British Standard or ASTM test designed to assess degradation of plastics by salt spray. BS 4618, Section 4.4 (1973),[30] *Presentation of Plastics Design Data The Effect of Marine Exposure*, categorically states: 'Plastics do not suffer chemical corrosion in diluted (3 per cent) salt solution, the effects of which are certainly no greater than those of distilled water. There is no evidence of any effect on plastics of evaporated salt'. Despite this, provision for a salt spray test on plastics has been included in the international standard ISO 4611[1] (see §17.2).

17.3.3 Environmental Stress Cracking

So far, consideration of the effects of liquids on plastics has been confined to absorption, extraction and chemical attack (in the sense that they cause a change in physical properties), when 'ideal', unstressed test pieces are used. Unfortunately, in real life applications of plastics it is not uncommon for the material to be used under conditions of stress induced as a result either of the end-use of the moulding or fabrication technique. Plastics may be resistant to a chemical in the unstressed state but fail by cracking when exposed under stress to the same chemical. Before this phenomenon was recognised considerable problems were caused in the industry, particularly with injection-moulded polyethylene house-wares which cracked in contact with household detergents, and with the early vacuum-formed toughened polystyrene refrigerator liners which cracked when they came into contact with vegetable fat and fruit juices. Even today the exact mechanism of such 'environmental stress cracking', as it is called, is not fully understood but we now know that most, if not all, thermoplastics can show environmental stress cracking given the right conditions of stress and aggressive medium. The aggressive medium need not be a liquid; semi-liquids (e.g. greases) and gases are now also known to cause environmental stress cracking.

Clearly, the determination of resistance to environmental stress cracking is of great importance both for the designer and for quality control purposes. The polymer chemist has learned how to reduce the tendency to stress crack by various means and therefore the polymer manufacturer needs to know how to check his production to ensure that the procedure he is making is maintaining the designed level of environmental stress crack resistance.

A review of environmental stress cracking test methods is given by Brown.[31] Originally, the available test methods consisted of exposing a strained test piece to a liquid environment and measuring the time to failure as the environmental stress crack resistance. The failure might be defined as appearance of crazing or of physical cracks or, in some cases, complete rupture. Test methods have now been extended to include exposure at either constant stress or constant strain.

Simple Bent Strip Tests

ASTM D1693 (1980),[32] *Environmental Stress Cracking of Ethylene Plastics*, was developed from the well known Bell Telephone test and is generally accredited to the work of De Coste *et al.*[33] The test piece is a strip, nominally 38 mm long and 13 mm wide, in which a controlled cut (imperfection) is made with a blade held in a jig. The test piece is bent into a 'U' shape, placed in a simple retaining jig, exposed to the aggressive environment and examined at intervals for failure. For the purpose of the test, stress crack failure is defined as 'any crack visible to the observer with normal eyesight'; extension of the controlled imperfection is not to be construed as a failure. The preparation of the test pieces is carefully defined and the standard aggressive liquid is Igepal CO–630 (Antarox CO630) which is nonylphenoxy poly(ethylene-oxy) ethanol. Ten replicate test pieces are used and in the absence of other instructions, as for example in a product specification, the duration of the test is 48 h.

The standard clearly states that the method is intended for quality control purposes and the results should not be used for direct application to engineering problems. Roe and Gienewsky[34] have made a detailed and careful study of the method to identify sources of error and have suggested modification to improve variability. In particular they mention the need to control closely the moulding conditions, the pre-test conditioning and the length of time the test piece stands between moulding and the test. These authors also confirmed that variation in test results occurs when the test piece is bent without backing or guide so that its precise deformed shape and the strains realised are not closely defined or controlled.

Other variations of the simple bent strip have been proposed, all being characterised by the absence of any restraining member to define and control the degree of bending. Steinle and Pflasterer[35] used a channel to hold strips of material in an arc to give a much lower strain than the ASTM method and hence a more uniform and better defined strain. They also introduced a procedure of measuring the change of a physical property (elongation at break) as a function of exposure time to quantify the measure of degradation as well as observing cracking and fracture. Leghissa and Salvatore[36] used a simple bent strip and devised an optical system for measuring the degree of cracking or crazing, but this was only suitable for transparent material.

Defined Curvature Methods

The most obvious improvement to the ASTM test is to define precisely the imposed curvature by clamping the test piece to a rigid curved jig or, perhaps less satisfactorily by subjecting it in a fixed deflection jig to three-point loading. A logical extension of this is to bend the test piece over a jig having various radii over its length; both parabolic and elliptical jigs have been used in this way. Dempsey,[37] Ruhnke and Biritz,[38] and Bergen,[39] among others, have proposed tests and the value of such tests have been recognised by the publications of ISO 4599.[40] This calls for test pieces, in strip form, to be bent over a metal former in the shape of the arc of a circle and then conditioned in contact with the test medium (either solid, liquid or gas). The standard short-term test period is 22–24 h at 23 °C but the standard allows test temperatures up to 55 °C and longer periods of contact. The test pieces are then tested for mechanical properties and the results compared with the results on unconditioned test pieces. In the absence of any specified mechanical tests by the material specification, the determination of flexural stress at maximum bend (ISO 178) and tensile stress at rupture (ISO R527) is recommended.

The flexural strain on the extended surface of a flat test piece is calculated from:

$$\varepsilon = 100d/2r + d) \text{ per cent} \qquad \text{(see Fig. 17.1)}$$

DIN 53499 Part 3[41] follows ISO 4599.

Insert Methods

Instead of by bending the test piece, the strain can be introduced by impressing an oversize ball or pin into a hole drilled in the test piece. This method has been described by Pohrt[42,43] and others on several occasions. Carefully drilled and reamed holes are prepared, balls (or pins) of various known oversizes are inserted and the assemblies are exposed to the aggressive medium. Either the oversize equivalent to the critical strain can be noted from visual observation or changes in tensile or flexural properties can be measured. When measuring physical

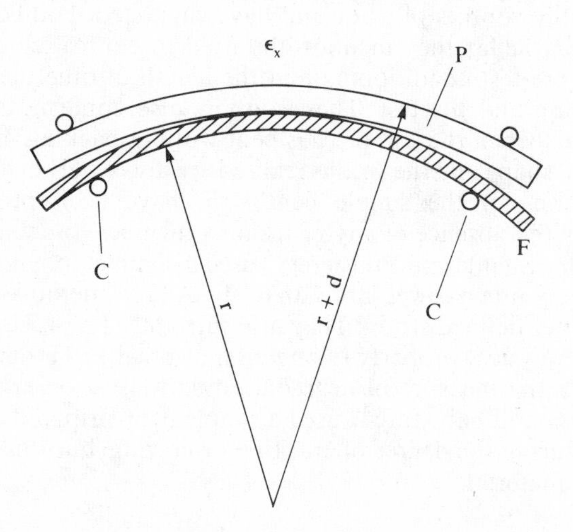

Fig. 17.1 Test specimen P with defined strain in the outer surface. P, test specimen; *d*, thickness of the test specimen; F, former; *r*, radius of former; C, clamps; ε_x, nominal strain in the extended surface

properties there are no problems because the test pieces are not bent after removal from exposure jigs.

This method has been standardised in ISO 4600[44] and this allows two methods, A and B. Method A requires the expression of results as the oversize of pin or ball at which the first crack or cracking at the 50 per cent level is observed or, for routine testing only, whether cracks are visible or not. For Method B the failure limit (as defined) is calculated from specified percentage reductions in maximum tensile force, maximum flexural force or tensile elongation at rupture caused by the exposure of the test piece to the test conditions.

DIN 53499[41] Part 1 follows the ISO method.

The actual performance of the test requires a high degree of skill as the diameters of the holes into which the pins or balls are impressed here to be measured to be within 0·005 mm. Balls or pins are also required which have a range of diameters from 3 to 6 mm with close tolerance (0·001 mm up to 4 mm diameter and 0·01 mm above 4 mm diameter).

Gnauck[45] evaluated the impressed ball method for use with PMMA and concluded that it was not satisfactory for materials at that level of modulus because the strains could not be maintained with enough precision. Both steel and rubber balls were embedded in polystyrene samples by Tsuey *et al.*[46] to examine the crazing around relatively soft and hard spherical inclusions. The test pieces were then strained in tension but the procedure was not intended to be used as a test method in the general sense.

Constant Stress Methods

The use of a dumb-bell stressed in direction tension by a constant load was proposed by Lander[47] as an alternative to the Bell Telephone test. He used an ASTM Type A tensile impact dumb-bell, cut from 1 mm sheet, loaded through a lever arm; the time was noted to complete failure of the test piece. Herman and Biesenberger[48] used effectively the same method. The essential difference between these tests and the bent strip test is that the latter operates at a constant strain but these were abandoned on the grounds that at constant strain the development of a crack changes the strain distribution near it and the stress relaxation continuously reduces the severity of the test.

There are many references in the literature to environmental stress cracking tests on test pieces subjected to both constant tensile stress and constant tensile strain. Some workers have used holes drilled in the test pieces to provide a preferential site for failure.[50,51] Haslett and Cohen[52] used a constant load applied to a three-point loading configuration (rather than in tension) and measured the time to complete failure of the test piece. They also considered the degradation which had occurred prior to complete failure by subjecting test pieces to static flexural loading, followed by measurement of cyclic fatigue life using a universal tensile machine. Apart from showing that fatigue life had been significantly decreased before visible signs of failure, they found evidence of a 'characteristic environmental stress', i.e. a stress, not necessarily the maximum stress, which induced maximum damage.

Judging by the interest being shown in the constant-load direct-tension methods by the standardisation bodies, this method is finding increased favour in the testing laboratory. ISO 6252[53] uses the usual tensile properties dumb-bell test piece (see

Ch. 8) but with all the dimensions halved, cut from 2 mm thick sheet. A constant load is applied, the test piece immersed in the aggressive medium and the method requires either the time to rupture for a given stress or the stress loading to rupture after 100 h to be reported. In the latter case the result is obtained by interpolation from a graph of time to rupture against stress.

BS 4618, Section 1.3.3 (1976),[54] *Recommendations for the Presentation of Plastic Design Data – Environmental Stress Cracking*, gives a constant tensile load method using two well tried test piece geometries which are not standardised elsewhere. It also recommends the use of both notched and unnotched test pieces.

ASTM D2552,[55] *Environmental Stress Rupture of Type III Polyethylenes under Constant Tensile Load*, is restricted to this particular type of material (as specified in ASTM D1248[56]). It uses a dumb-bell similar to that of Lander but with wider tab ends, presumably to improve gripping, and cut from 1 mm sheet.

It is clear that there are advantages and disadvantages in the types of test which have been described. Tests at constant strain (e.g. bent strip tests) are useful for quality control purposes and require simple apparatus, whereas tensile constant stress tests may give a better insight into the mechanism of environmental stress cracking by separating the effect of the environment on crack initiation and crack propagation. Nevertheless, it cannot be emphasised too strongly that if consistent results are to be obtained from any environmental stress cracking test great attention must be paid to all the experimental conditions. Figure 17.2 shows a typical apparatus for determining environmental stress cracking under constant tensile stress.

17.4 EFFECT OF GASES

In comparison with the effects of liquids on plastics, very little testing is carried out on the effect of gases (permeability is dealt with in Ch. 18), with the notable exception of air or oxygen which has been covered in §16.8 on heat ageing.

The effect of ozone on elastomers is well known and is important in rubber technology since many elastomers are adversely affected by exposure to ozone to a greater or lesser degree. Saturated polymers, which form the bulk of plastics materials, are quite resistant to ozone and there are no standard tests for plastics which specifically examine this property separately from its contribution to the general degradative influence of weathering (§17.5). Weiss[57] has reviewed ozone tests on plastics and Brown[58] has given a comprehensive account of ozone exposure apparatus for elastomers.

17.5 WEATHERING

17.5.1 General Consideration

Weathering effects are a combination, *inter alia*, of those due to heat (cf. Ch. 16) including infrared radiation, to damp (see §§17.2, 17.3), to ultraviolet radioactive radiation (u.v.), to oxidation, to ozone formed by the action of u.v. on atmospheric oxygen and perhaps to radioactive radiation from active fall-out or emitting sources.

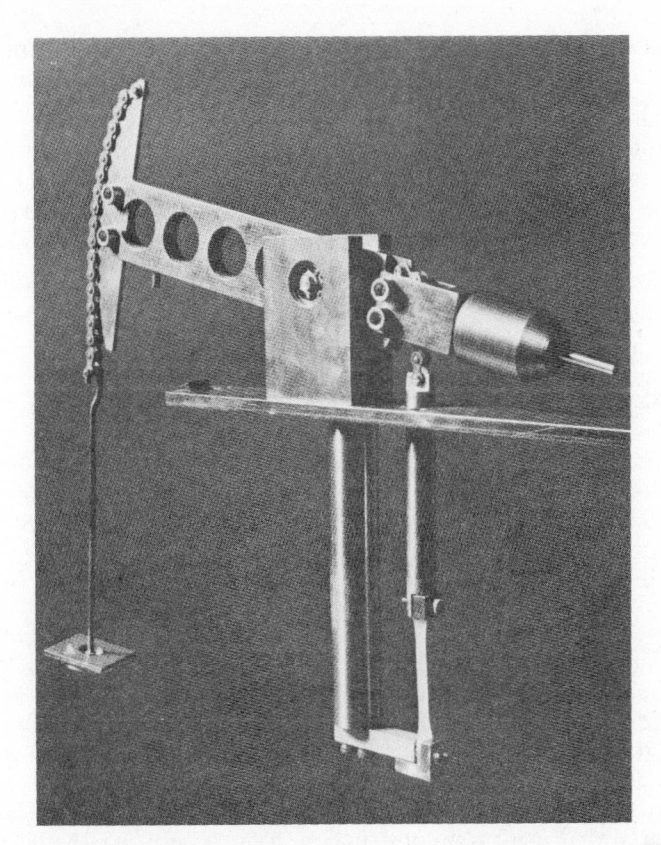

Fig. 17.2 Essential parts of constant stress environmental stress cracking apparatus

The last-mentioned is conveniently separated for consideration (§17.7) and those already discussed (i.e. heat and damp) may well be encountered as singular or combined degradative influences. However, u.v., oxidation and ozonolysis are unlikely to be hazards except as constituents of general weathering attack and are therefore dealt with together. The reason for this non-classical approach is that in general weathering the separate degrading influences affect the changes or rate of change brought about by each other. To this synergistic phenomenon must be added the varying intensity of the degradative influences, which variation may be more severe in action than a continuous high level of the influence. To take a simple example, repeated drying out caused by occasional saturation with water in a high-temperature environment may be far more damaging than continuous immersion at the same temperature.

Estevev[59] analyses the weathering of plastics into its separate components and lists the major objections to accelerated tests:

(i) the degree of acceleration in relation to 'natural' ageing is unknown;
(ii) the order of merit ranking of materials produced by an accelerated test may be quite different from that obtained in use;

(iii) phenomena may occur in accelerated tests which never occur naturally – and vice versa. (Point (ii) could easily result at least in part from point (iii).)

With regard to point (i), the very variability of the weather at any one geographical location, quite apart from the vast differences in average climate between various parts of the world, obviously militates against the chances of any direct correlation between accelerated weathering test results and the results of natural weathering.

The problem of predicting weathering of plastics, the methods and many test results are to be found in a set of symposium papers edited by Kamal.[60]

17.5.2 Natural Weathering and Ageing

If one can afford the time to wait, evaluation of resistance to weathering, and of associated effects such as fading, is most reliably undertaken using the natural elements. Even then the results obtained must be carefully related to the geographical location, time of the year and climatic conditions prevailing at the time. The degree of resistance to natural weathering can then be judged by the change in physical properties. This technique is standardised by ISO 877 (1976),[61] BS 2782 Method 540A (1977),[62] which are both entitled *Plastics – Determination of Resistance to Change upon Exposure under Glass to Daylight*. A similar method is given in ASTM G24,[63] *Practice for Conducting Natural Light Exposure under Glass.*

Test pieces, 20 mm wide and of convenient length, are used for determining colour change but other sizes of test piece are allowed for determining changes in other physical properties. The exposure case is closely specified and is set at an angle of 45° facing the equator (i.e. in the northern hemisphere it faces south) and is covered with 3 mm thick glass having specified light transmittance. The exposure case is so placed that there is no obstruction in an easterly, southerly or westerly direction subtending a vertical angle greater than 20° nor any northerly obstruction subtending an angle greater than 70°. The exposure case consists essentially of an open-bottomed shallow square box, mounted at 45° to the horizontal and supported on legs so that its lower front edge is 760 mm from the ground. A framed wire screen fits snugly into the 'case' box and on the upper surface of the wire mesh rest a rack which also fits snugly in the box. The rack comprises a frame with spaced horizontal battens across it, there being a 50 mm gap between each batten; on the battens 50 mm wide wooden flaps are hinged so that the lower 50 mm of the depth of each batten may be covered. Over this rack rests a lid, kept at a distance of 75 mm from the upper surface of the rack by a distance piece mounted in the centre of the rack. This lid contains the glass cover mentioned above.

Each test piece is mounted so that it is partly covered by the batten flaps and there is a space of not less than 6 mm between each test piece or between the test piece and the edge of the frame.

The standard allows the use of instrumental means for measuring the radiation dosage under the glass lid but recommends the use of the blue dyed wool standards to ISO 105/A (1984),[64] examined for colour change by the grey scale. No. 1 is of very low lightfastness and No. 7 of very high lightfastness, each standard being approximately twice as lightfast as the one below it. The grey scale ranges from No. 1 (greatest degree of contrast) to No. 5 (zero contrast), having two patterns of identical colour. In use, a set of the blue standards is exposed under the glass along with the best pieces. The blue standards are examined at intervals for change of colour between the exposed and unexposed portions, the change of colour to be defined as a difference in contrast equal to that shown by No. 4 on the grey scale. Each stage of the exposure is recorded as each wool standard fades to this degree of contrast. In the event of very long exposure which exceeds the time to fade No. 7 standard, then further No. 7 standards are successively exposed as required. A standard blue cloth No. 8 is now also available which has twice the lightfastness of No. 7 which may, in some cases, avoid the need to use successive exposures to the No. 7 cloth. Test pieces are examined at each stage of exposure as required and the results expressed as change in properties or colour at each stage. In British Standard practice it is common to examine the test pieces for colour change to grey scale No. 4 and then to report the blue standard which has faded to the same degree as being the colourfastness of the material under test. A similar procedure is followed by DIN 53388.[65]

In recent years there has been an increasing interest in determining the effect of total outdoor exposure ('natural weathering') on the properties of plastics and ISO, BS and ASTM have published standards covering this subject.

ISO 4607 (1978)[66] describes procedures for exposing plastics to natural weathering in order to assess changes produced after specific stages of such exposure. It lays down a series of preferred exposure times (ranging from 1 week to 6 years) and recommends methods for measurement of total radiation received by the exposed test pieces. The use of the dyed wool standards and grey scale is permitted for this purpose but instrumental means are recommended for finding the total energy and the energy in specified wavelength intervals.

BS 4618, Section 4.2 (1972)[67] lays down the procedures for presenting design data on the resistance of plastics to natural weathering. The standard provides for total exposure to the natural elements under five standard climatic conditions and gives procedures for reporting biological effects (§17.6) and changes in dimensions, appearance and mechanical properties.

Full exposure to the natural elements is allowed in ASTM D1435 (1985).[68] The general design of the weathering racks and their siting are covered and the method of storage of control samples is laid down (23 °C ± 1 °C and 50 ± 2 per cent r.h.) so that final control values for the material under test may be determined as well as initial values.

In many instances, change of physical appearance may be a sufficient criterion upon which to assess the effect of weathering (or other ageing); obviously this is particularly true if the intended application is basically decorative. When it comes to standardising techniques, however, degrees of colour change, darkening or fading, and changes in opacity need less subjective techniques. Alternatively, if it is a mechanical property (or several), or electrical performance which is of importance, changes in these as a result of weathering should be determined, following, for example, the guidelines laid down by the above standards. It should be noted that ISO 4607 refers to ISO 4582[69] for the specified procedures for the determination of changes in colour and appearance and variations in mechanical and other properties of plastics upon exposure to natural or artificial weathering. This standard details the methods of assessing changes in colour and appearance and gives the procedure for determining the changes in other properties. It also refers to three other international standards which give instrumental methods for evaluation of colour differences: ISO 2579,[70] ISO 3557[71] and ISO 3558.[72]

17.5.3 Accelerated Natural Ageing

The time factor involved in performing outdoor exposure tests on plastics (exposure periods of at least 2 years are not uncommon) and the difficulties of accurately predicting natural ageing from (accelerated) laboratory ageing tests (§17.5.4) has prompted several attempts to accelerate natural ageing. One way of doing this is to employ exposure sites which give severe weather conditions because of their geographical location. Ellinger[73] describes the use of the Florida climate for accelerating the weathering of paints in particular; South Florida has a consistent climate of a combination of warm temperatures, high daily relative humidity, heavy annual rainfall, heavy dews and high values of solar radiation. There are undoubtedly other locations enjoying similar weather conditions; however, apart from the problem of inaccessibility, the degree of acceleration is probably not very great.

Garner and Papillo[74] have described a device called 'EMMA' which is a combination of ten aluminium mirrors so arranged as to intensify natural sunlight tenfold, while the test pieces are prevented from becoming too hot by a stream of cooled air. Results obtained indicate acceleration of degradation ranging from threefold for unsaturated polyesters to ninefold for rigid PVC. Further description, including the addition of water spray (EMMAQUA), and results from such accelerated tests have been provided by Caryl.[75]

In 1978, Binder *et al.*[76] re-opened this subject and claimed a new procedure for accelerated weather testing by the use of stainless steel mirrors to magnify natural sunlight. Unfortunately, this paper is short, does not adequately describe the apparatus and, apparently, overlooks the earlier work of Garner *et al.*

Until recently such methods had not found favour with standardisation authorities but there is now a standard test method based on the ideas of Garner and Papillo published in ASTM D4363.[77] This specifies a Fresnel–Reflector test machine which is a follow-the-sun rack having ten flat mirrors so positioned that the sun's rays strike them at near-normal incident angles while in operation. The mirrors are arranged to simulate tangents to a parabolic trough, so that they reflect the sunlight uniformly onto test pieces in the target area.

Two methods are permitted: Method A specifies dry exposure and Method B allows the use of a water spray on the test pieces to simulate rain and exposure in subtropical, semi-humid and temperate regions.

17.5.4 Accelerated Weathering and Ageing

In general, natural weathering and ageing take too long to achieve results in an industry as fast developing as plastics. Furthermore, the results will predict service life only for a time period similar to that of the weathering trial as, in the present state of our knowledge, extrapolation must largely be based on guesswork and may bear no relation to actual long-term effects.

Many authors report work on the problem of correlating actual outdoor ageing with accelerated exposure tests in the laboratory using artificial light sources. A comprehensive survey is outside the scope of this book but a brief summary is given below.

In the past four types of lamp source have been used for artificial weathering in the laboratory. Low-pressure mercury vapour lamps give an emission which comprises a series of sharply defined bands including a very strong emission at 2537 Å. When this is compared with natural sunlight, of which all wavelengths below 2900 Å are filtered out by the earth's amosphere, it is obvious that the source of 2537 Å might easily cause a degradation or chemical change that would not be caused by sunlight. High-pressure mercury arc lamps give a better spectral distribution and are reasonably suitable for test purposes if the lower wavelengths are filtered out. Carbon arc lamps are still widely used and specified (particularly in the paint industry) but their spectrum bears no relation to that of sunlight. The fourth type of lamp source, the xenon arc, has become widely used in the plastics industry during the past 20 years. With the appropriate filter this source simulates sunlight fairly closely, though the degree of acceleration compared with natural sunlight is not great.

Kamal[78] provides a comprehensive survey of the complexity of weatherability studies using outdoor exposure, carbon arc 'Weather-Ometers', xenon arc 'Weather-Ometers' and fluorescent sunlamps. He points out that his own results, and those of other workers, show that the degree of correlation between the enclosed carbon arc and outside exposure was achieved when the carbon arc was modified by incorporating eight fluorescent sunlamps in the case of cellulose acetate butyrate, yet good correlation was obtained with daylight with an unmodified carbon arc when testing polypropylene filaments.

Despite the closeness of the xenon arc spectrum to that of natural sunlight there is no clear correlation between the two light sources which enables a positive statement to be made that x hours exposure to the xenon lamp is equivalent to kx hours outdoor exposure where k is a constant.

Kamal[78] quotes results by various workers indicating that 500 h in the Xenotest with high humidity and water spray is equivalent to 1 year of outdoor exposure as measured by colour change in a number of plastics. The manufacturers of the Xenotest apparatus suggest that exposure for 1 h to the lamp source is roughly equivalent to 10 h exposure to sunshine in temperate climates; however, Ruhnke and Biritz[79] claim that, for ABS plastics, exposure for 1 h to the xenon arc corresponds to 5 h sunshine.

An exhaustive bibliography was compiled by the Building Research Establishment[80] on weathering tests followed the RAPRA review[81] of the subject up to 1968. The references which may be found in these reviews cover detailed information on weathering tests both natural and artificial. Much work has also been done on the development of instrumental methods of monitoring ultraviolet radiation[82–84] and on the use of plastics films for this purpose.[85]

ISO Technical Committee TC61 has, in recent years, been giving consideration to the standardisation of methods to monitor the ultraviolet radiation during weathering tests other than by using the dyed wool blue standards. This committee appears to be favouring instrumental methods using commercially available apparatus such as solar radiometers. So far international agreement has not been reached on this topic and some countries prefer the method based on the use of polysulphone film.

BSI recommend the use of polysulphone film and a draft Method 540C of BS 2782[86] is already in existence. Davis *et al.*[87,88] have reported in depth on the use of this technique and have produced results which demonstrate the effectiveness of the method in both natural and accelerated light sources.

Although the problems of accelerated weathering, as outlined above, mean that such tests can only give arbitrary results and that great care is needed in their interpretation, there are a number of standardised test methods, some of which have been used for many years.

ISO 4892[89] gives methods for exposure of plastics to laboratory light sources. This standard replaces two earlier ISO Recommendations (R878 and R879) and describes the use of the xenon arc lamp, enclosed carbon arc lamp and the open flame carbon arc lamp. It does not include the use of fluorescent tube light sources because of insufficient information on the reliability and repeatability which may be obtained with this light source. Nevertheless the standard contains Annex D which gives useful information concerning exposure to the light from fluorescent tube lamps and it is hoped to include this source in future revisions of the standard.

Overall, ISO 4892 is a useful document and it gives advice (Annex A) on correlation of effects of exposure to artificial light sources and exposure to natural daylight. ISO 4892 has been included in BS 2782 as a dual numbered standard (Method 540B[90]). BS 4618 Section 4.3 (1974)[91] also refers to exposure of plastics to artificial light sources and this offers the choice of carbon arc, xenon and fluorescent lamps.

ASTM D1499 (1984)[92] and ASTM D2565[93] specify practices for operating carbon arc and xenon arc lamps, respectively, and makes reference to the general standards for recommended practices for exposure of non-metallic materials G23 and G26, respectively. Both require the change in the exposed test pieces to be evaluated by the applicable ASTM specification and neither allows the use of dyed wool standards for assessing the level of light dosage. Instead the standards require that the spectral irradiance at the sample location and the total irradiation be recorded during the test.

ASTM D4329 (1984)[94] makes specific reference to the use of fluorescent ultraviolet (u.v.) and condensation apparatus to simulate the deterioration caused by sunlight and water as rain or dew. This standard then lays down the construction of the apparatus which includes eight lamps in two banks of four and a condensation mechanism which ensures that the condensate runs off the test pieces by gravity and is replaced in a continuous process. Arrangements are provided to conduct the test at elevated temperatures in both non-condensing and condensing modes. It would appear that this test has been written around commercially available apparatus as it is required that the type and model of apparatus is cited in the report of any test carried out according to this standard. There is no requirement for the spectral radiation level to be measured during the test as there is in ASTM D1499 and ASTM 2565.

DIN 53389 (1983)[95] is similar to ISO 4892.

17.6 BIOLOGICAL ATTACK

Synethetic polymers are not usually attacked by micro-organisms (bacteria and fungi) but the picture is confused by the presence of additives, reinforcements, etc., which may themselves be attacked. The resistance of plastics to biological attack has been extensively reviewed by Heap[96] and Heap and Morrell.[97]

Wienert and Hillard[98] have discussed the errors in interpreting accelerated mould growth tests in terms of service life, using plastics films to demonstrate the limitations; this work has been substantiated by Dalton,[99] using PVC films.

Undoubtedly the subject of 'biological ageing' is best entrusted to the experts, although a number of published standards deal with the subject of microbiological attack on plastics.

ISO 846 (1978)[100] specifies the use of five strains of fungi over a period of at least 28 days at 30 °C ± 2 °C and 95–100 per cent r.h. and measures the results of the attack by visual examination for mould growth. Two test procedures are included: Method A for finding the ability of plastics to act as nutritive medium and Method B to determine the fungitoxic properties of plastics.

BS 4618, Section 4.5 (1974)[101] broadly outlines the variables in exposure conditions for soil burial and suggests ways of determining the results of exposure. This standard makes reference to ISO 846 and also gives a short bibliography, but it can only be regarded as a guide as it does not lay down precise procedures for exposure to biological attack.

At present there is no other British Standard that offers test methods for determining biological attack on plastics but BS 1204 (1979)[102] for synthetic resin adhesives for wood contains a test for measuring resistance to micro-organisms. BS 1982 (1968)[103] measures the effect of fungi, cellulose-attacking microfungi and moulds and mildew on manufactured building materials by visual examination and weight change.

ASTM presents two standard test methods. ASTM G21 (1985)[104] specifies the use of five fungi at from 28 °C to 30 °C and 85 per cent r.h. for a minimum period of 21 days and provides for measuring the effects on appearance and on optical, physical or electrical properties. ASTM G22 (1985)[105] specifies the use of the bacterium *Pseudononos aeruginosa*, incubated at 35 °C to 57 °C for a minimum of 21 days, the methods of assessing the result being similar those of G21.

17.7 RADIATION

Radiation is taken here to mean bombardment by atomic and nuclear particles, i.e. gamma rays, electrons, neutrons, etc. The intensity of such radiation at the earth's surface is not high enough significantly to affect plastics and tests are only required in connection with applications in atomic or nuclear plants or perhaps where radiation is used to induce crosslinking. Not surprisingly, such a specialised subject has not given rise to widespread standardisation of test methods. ASTM E1027,[106] *Practice for Exposure of Polymeric Materials to Ionising Radiation*, specifies a recommended practice for exposure of polymeric materials to various types of radiation and a detailed account of testing polymers for radiation resistance has been given by Metz.[107]

REFERENCES

1. ISO 4611 (1980) *Plastics – Determination of the Effect of Exposure to Damp Heat, Water Spray and Salt Mist.*

2. ASTM D756–78 (1983) *Practice for Determination of Weight and Shape Changes of Plastics under Accelerated Service Conditions.*
3. BS 2011 *Basic Methods for the Climatic and Durability Testing of Components for Telecommunications and Allied Electronic Equipment*; Part 2: 1Db (1981) *Damp Heat, Cyclic.*
4. BS 3900: Part F2 (1983) *Determination of Resistance to Humidity (cyclic condensation).*
5. Uijlenberg, H. (1960) *Plastics, London,* **25** (275), 359.
6. Braden, M. (1963) *Transactions of the Plastics Institute, London,* **31** (94), 83.
7. Blank, F. (1962) *Plaste und Kautschuk,* **9** (8), 391.
8. Blank, F. (1965) *Plaste und Kautschuk,* **12** (11), 657.
9. BS 771 (1980) *Phenolic Moulding Materials.*
10. Hauck, J. E. (1966) *Materials in Design Engineering,* **64** (5), 93.
11. ISO 62 (1980) *Plastics – Determination of Water Absorption.*
12. BS 2782 Methods 430A to 430D (1983) *Water Absorption of Plastics.*
13. BS 2571 (1963) *Flexible PVC Compounds.*
14. ASTM D570 (1981) *Water Absorption of Plastics.*
15. DIN 53495 (1984) *Testing of Plastics – Determination of Water absorption.*
16. BS 2782 Method 504A (1970) (1986) *Water Vapour Absorption of Cellulose Acetate Moulding Material.*
17. CEN 198 (1982) *Draft Specification for Baths for Domestic Purposes Made from Acrylic Materials.*
18. Jessup, J. N. (1967) *Modern Plastics,* **44** (7), 174.
19. ISO 175 (1981) *Plastics – Determining the Effect of Liquid Chemicals, including Water.*
20. BS 4618, Section 4.1 (1972) *Recommendations for the Presentation of Data on the Chemical Resistance of Plastics to Liquids.*
21. BS 3505 (1968) *Unplasticised PVC Pipe for Cold Water Services.*
22. BS 3506 (1969) *Unplasticised PVC Pipe for Industrial Uses.*
23. ASTM D543 (1982) *Resistance of Plastics to Chemical Reagents.*
24. ASTM D2299 (1982) *Determining the Relative Stain Resistance of Plastics.*
25. ASTM D1712 (1983) *Resistance of Plastics to Sulphide Staining.*
26. BS 1763 (1975) *Thin PVC Unsupported Calendered Sheeting.*
27. BS 2739 (1975) *Thick PVC Unsupported Calendered Sheeting.*
28. ASTM D2151 (1982) *Staining of Poly (Vinyl Chloride) Compositions by Rubber Compounding Ingredients.*
29. DIN 53476 (1979) *Testing of Plastics – Determination of the Behaviour against Liquids.*
30. BS 4618 Section 4.4 (1973) *Presentation of Plastics Design Data – The Effect of Marine Exposure.*
31. Brown, R. P. (1980) *Polymer Testing,* **1** (4), 267.
32. ASTM D1693 (1980) *Environmental Stress Cracking of Ethylene Plastics.*
33. De Coste, J. B., Malm, E. S. and Wallder, V. T. (1951) *Industrial and Engineering Chemistry,* **43**, 117.
34. Roe, R. J. and Gienewski, C. (1975) *Polymer Engineering Science,* **15** (6), 197.
35. Steinle, H. and Pflasterer, H. (1967) *Kautschuk und Gummi Kunststoffe,* **20** (9), 516.
36. Leghissa, S. and Salvatore, O. (1966) *Polymer Engineering Science,* **6** (2), 127.
37. Dempsey, L. T. (1967) *Polymer Engineering Science,* **7** (2), 86.
38. Ruhnke, G. M. and Biritz, L. F. (1970) *Plastics and Polymers,* August, 265.
39. Bergen, R. L. (1962) *SPE Journal,* Mune, 667.
40. ISO 4599 (1986) *Plastics – Determination of Resistance to Environmental Stress Cracking (ESC) – Bent Strip Method.*
41. DIN 53499 (1984) *Testing of Plastics – Environmental Stress Cracking.*
42. Pohrt, J. (1973) *Kunststoffe,* **63** (3), 163.
43. Pohrt, J. (1976) *Kunststoffe,* **66** (8), 481.
44. ISO 4600 (1981) *Determination of Environmental Stress Cracking (ESC) – Ball and Pin*

Method.
45. Gnauck, B. (1977) *Gummi und Asbest Kunststoffe*, **10**, 708.
46. Tsuey, T., Wang, M. and Kwei, T. K. (1971) *Journal of Applied Physics*, **42** (11), 4188.
47. Lander, L. L. (1960) *SPE Journal*, **16** (12), 1329.
48. Herman, J. N. and Biesenberger, J. A. (1966) *Polymer Engineering Science*, October, 341.
49. O'Connor, A. and Turner, S. (1962) *British Plastics*, September, 452, October, 526.
50. Hittmair, A., Mark, H. F. and Ullman, R. (1958) *Journal of Polymer Science*, **33** (126), 505.
51. Sternstein, S. S., Ongchin, L. and Silverman, A. (1968) *Applied Polymer Symposia*, **7**, 175.
52. Haslett, W. H. and Cohen, L. A. (1964) *SPE Journal*, **20** (3), 246.
53. ISO 6252 (1986) *Plastics – Determination of Resistance to Environmental Stress Cracking (ESC) – Constant Tensile Stress Method.*
54. BS 4618, Section 1.3.3 (1976) *Recommendations for the Presentation of Plastics Design Data – Environmental Stress Cracking.*
55. ASTM D2552 (1980) *Environmental Stress Rupture of Type III Polyethylenes under Constant Tensile Load.*
56. ASTM D1248 (1984) *Polyethylene Moulding and Extrusion Materials.*
57. Weiss, E. (1966) in *Testing of Polymers*, vol. 2 (Ed. J. V. Schmidtz), Interscience, Ch. 9.
58. Brown, R. P. (1986) *Physical Testing of Rubbers*, Elsevier Applied Science, p. 292.
59. Estevev, J. H. J. (1965) *Transaction of the Plastics Institute, London*, **33** (105), 89.
60. Kamal, H. R. (1967) Weatherability of plastics materials, in *Applied Polymer Symposium*, No. 4, Interscience.
61. ISO 877 (1976) *Plastics – Determination of Resistance to Change upon Exposure under Glass to Daylight.*
62. BS 2782 Method 540A (1977) *Plastics – Determination of Resistance to Change upon Exposure under Glass to Daylight.*
63. ASTM G24 (1980) *Practice for Conducting Natural Light Exposure under Glass.*
64. ISO 105/A (1984) *Textiles – Tests for Colour Fastness – General Principles.*
65. DIN 53388 (1984) *Testing of Plastics – Testing of Stability under Natural Light (Global Radiation Behind Glass).*
66. ISO 4607 (1978) *Plastics – Methods of Exposure to Natural Weathering.*
67. BS 4618, Section 4.2 (1972) *Recommendation for the Presentation of Plastics Design Data – Resistance to Natural Weathering.*
68. ASTM D1435 (1985) *Outdoor Weathering of Plastics.*
69. ISO 4582 (1980) *Plastics – Determination of Changes in Colour and Variations in Properties after Exposure to Daylight under Glass, Natural Weathering or Artificial Light.*
70. ISO 2579 (Draft) *Plastics – Instrumental Evaluation of Colour Differences.*
71. ISO 3557 (Draft) *Plastics – Recommended Practice for Spectrophotometry and Calculation of Colour in CIE Systems.*
72. ISO 3558 (Draft) *Plastics – Assessment of the Colour of Near-white or Near-colourless Materials.*
73. Ellinger, M. L. (1963) *Paint Technology*, **27** (12), 40.
74. Garner, H. I. and Papillo, P. J. (1962) *ACS Division of Organic Coatings and Plastics Chemistry, Atlantic City Meeting*, **22** (2), 110.
75. Caryl, C. R. (1976) *SPE Journal*, **23** (1), 49.
76. Binder, K., Vukovich, S. and Tschamler, H. (1978) *Chemie Kunststoffe Aktuell*, **32** (5), 209.
77. ASTM D4364 (1984) *Performing Accelerated Outdoor Weathering of Plastics using Concentrated Natural Sunlight.*
78. Kamal, H. R. (1972) *Colouring of Plastics* 6, *RETEC, Philadelphia, Pa.*, October, SPE Philadelphia Section, p. 4.
79. Ruhnke, G. M. and Biritz, L. F. (1972) *Plastics and Polymers*, **40** (147), 118.
80. BRE (1975) *Bre Library Bibliography*, no. 247, February.

81. Matthan, J., Scott, K. A. and Wiechers, M. (1970) *Ageing and Weathering of Plastics*, RAPRA.
82. Harris, P. B. (1973) *BRE Current Paper*, no. 17/73, June.
83. Capron, E. and Crowder, J. R. (1975) *BRE Current Paper*, no. 17/75, February.
84. Suga, S. and Katayanagi, S. (1981) *Polymer Testing*, **2** (3), 175.
85. Davis, A., Gordon, D. and Howell, G. V. (1973/74/75) *ERDE Technical Reports*, nos. 141, 186, 190.
86. BS 2782 Method 540C (Draft) *Determination of Ultraviolet Radiation Intensity using Polysulphone Film.*
87. Davis, A., Howes, B. V., Ledbury, K. J. and Pearce, P. J. (1979) *Polymer Testing*, **1** (2), 121.
88. Davis, A. and Gardiner, D. (1982) *Polymer Testing*, **4** (2), 145.
89. ISO 4892 (1981) *Plastics – Methods of Exposure to Laboratory Light Sources.*
90. BS 2782 Method 540B (1982) *Plastics – Method of Exposure to Laboratory Light Sources.*
91. BS 4618 Section 4.3 (1974) *Recommendations for the Presentation of Plastics Design Data – Resistance to Colour Change Produced by Exposure to Light.*
92. ASTM D1499 (1984) *Operating Light and Water Exposure Apparatus (Carbon Arc type) for Exposure of Plastics.*
93. ASTM D2565 (1985) *Operating Xenon Arc Type (Water Cooled) Light and Water Exposure Apparatus for Exposure of Plastics.*
94. ASTM D4329 (1984) *Operating Light and Water Exposure Apparatus (Fluorescent UV Condensation Type) for Exposure of Plastics.*
95. DIN 53389 (1983) *Testing of Plastics and Elastomers – Light Exposure Apparatus.*
96. Heap, W. H. (1965) *RAPRA Information Circular*, no. 476.
97. Heap, W. H. and Morrell, S. M. (1968) *Journal of Applied Chemistry*, **18** (7), 189.
98. Wienert, L. A. and Hillard, M. W. (1975) *Australian Plastics and Rubber Journal*, **26** (10), 15.
99. Dalton, D. L. (1977) *Polymer Paint and Colour Journal*, **167** (3964/5), 987/990.
100. ISO 846 (1978) *Determination of Behaviour under the Action of Fungi and Bacteria – Evaluation by Visual Examination or Measurement of Change in Mass or Physical Properties.*
101. BS 4618, Section 4.5 (1974) *Recommendations for the Presentation of Plastics Design Data – The Effect on Plastics of Soil Burial and Biological Attack.*
102. BS 1204 (1979) *Specification for Synthetic Resin Adhesives (Phenolic and Aminoplastic) for Wood.*
103. BS 1982 (1968) *Methods of Test for Fungal Resistance of Manufactured Building Materials made, of or Containing, Materials of Organic Origin.*
104. ASTM G21 (1985) *Determining Resistance of Synthetic Polymeric Materials to Fungi.*
105. ASTM G22 (1985) *Determining the Resistance of Plastics to Bacteria.*
106. ASTM E1027 (1984) *Practice for Exposure of Polymeric Materials to Ionising Radiation.*
107. Metz, D. J. (1966) in *Testing of Polymers*, vol. 2 (Ed. J. V. Schmidtz), Interscience, Ch. 5.

Chapter Eighteen

Fire Testing of Plastics

18.1 INTRODUCTION

Fire may be defined as 'the oxidative destruction of a material' and is characterised by the generation of heat and light.

Real life fires may pass through several stages starting with ignition, then growth, steady state burning and decay (Fig. 18.1). Growth is frequently described in terms of flame spread, heat release, and smoke and toxic gas generation rates. These parameters are frequently determined separately but it is important to realise that

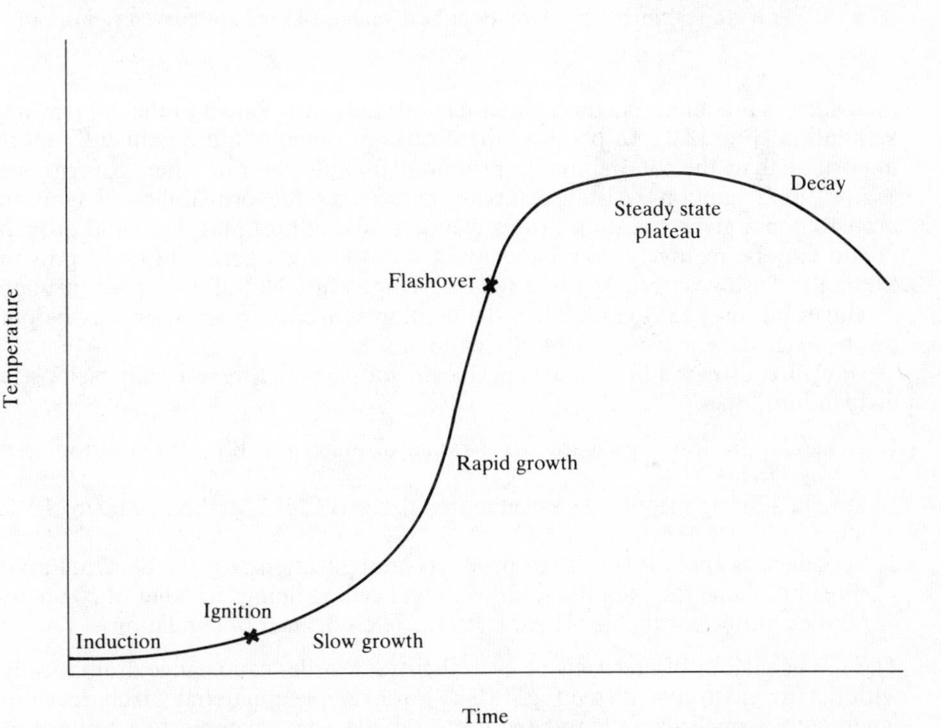

Fig. 18.1 Fire growth: many fire tests apply to ignition and pre-flashover period

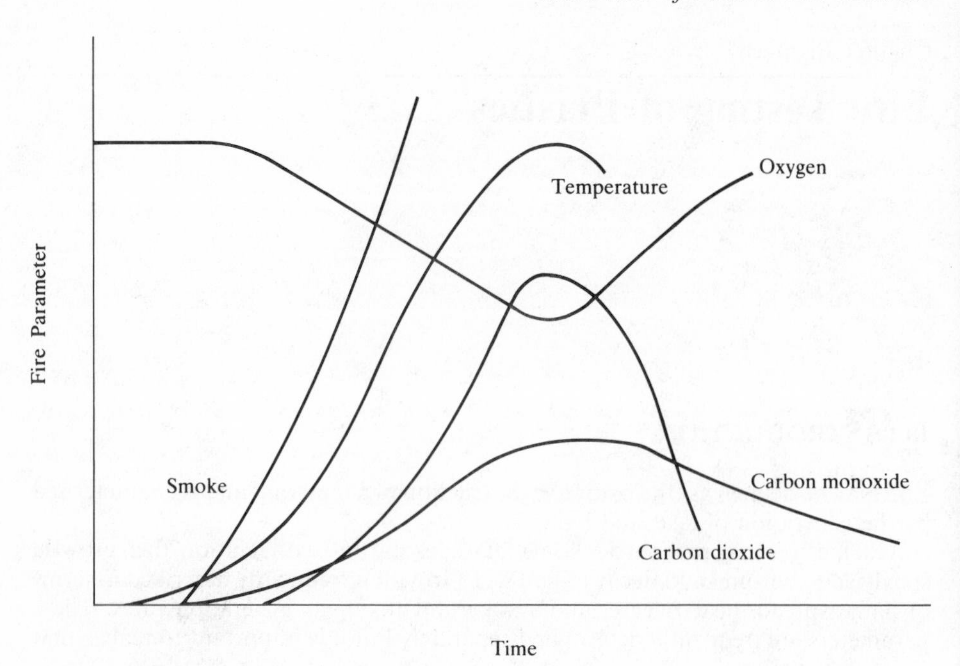

Fig. 18.2 Fire growth: interdependability of heat smoke and fire gas concentration rates

these factors are interrelated and interact with the environment of the fire and the ventilation (Fig. 18.2). In practice, these rates of generation are frequently more important than the total amounts produced because the rate often governs the escape time. Ignition resistance largely determines the probability of ignition occurring in a given situation. It is a characteristic of fires that the initial growth period may be relatively slow but may be followed by a period of rapid growth leading to flashover resulting in a fully developed fire. Not all fires pass through all stages but may extinguish when the ignition source is removed or may be put out by external actions, e.g. by a fire extinguisher.

Many fire tests exist but few are specifically for plastics. Fire tests may be loosely divided into types.

1. General fibre tests applicable to a range of materials. BS 2782 Method 141, *Oxygen Index.*[1]
2. Specific tests applicable to a given material, e.g. BS 2782 Method 140D for PVC film.[2]
3. Specific tests applicable to given products or applications, e.g. the BS 476: Part 7 spread of flame test[3] applies to *all* wall and ceiling linings whether of concrete, plastics, rubber or textile. These tests are often defined in regulations.

The fire behaviour of materials or even their rank order can change dramatically with the fire environment (see Fig. 18.3).[4] For example a material which does not ignite with a small flame may burn more rapidly when exposed to intense heat than a material which is ignited by the small flame. It follows that a fire test must

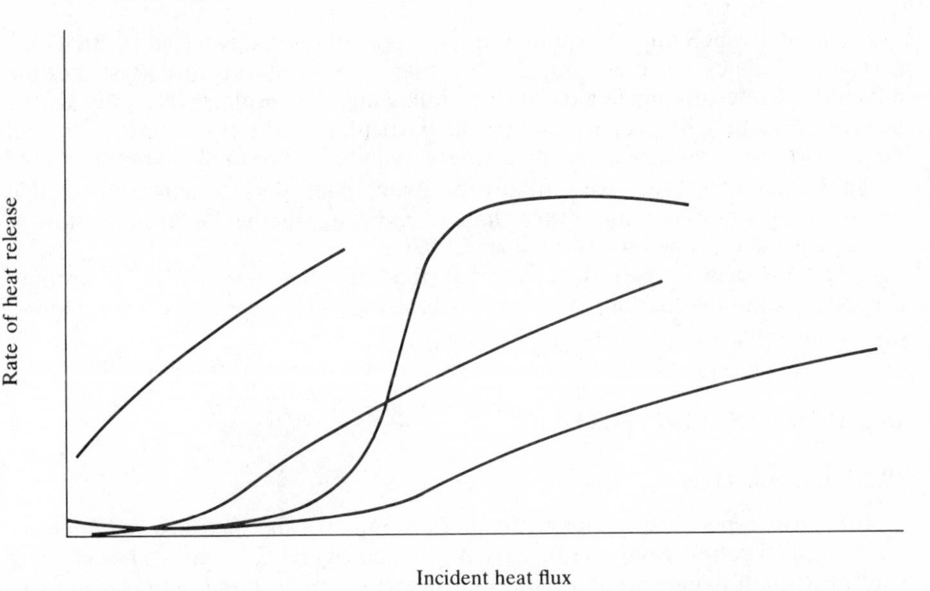

Fig. 18.3 Fire growth: change of rank order with fire environment (schematic representation of data[4])

be relevant to the anticipated end use hazard if misleading and potentially dangerous situations are to be avoided. It is important to understand the limitations of fire tests and not to extrapolate results to predict fire performance.

For these reasons, many fire tests use composite or model specimens and may involve complex fire models. These tests are rarely specific to plastics and are frequently product-based. Thus BS 476: Part 7, on surface spread of flame, is applicable to *all* materials composites and products used in the UK for wall and ceiling lining, applications defined in the DOE *Building Regulations.*[5] This has sometimes resulted in major difficulties for the plastics and rubber industries because a number of tests were developed at a time when few plastics were used for the relevant application or were not even available. The spread of flame test was in fact developed in 1933 and certain thermoplastics are virtually impossible to test because they melt and drip away from the fire test zone. The cry 'cannot test thermoplastics' is frequently heard and has resulted in the proliferation of test modifications, of alternative test methods and, especially in recent years, of the increased use of *ad hoc* or product-based, large-scale tests. Recent test developments have acknowledged this difficulty and, particularly within ISO, more scientific tests are being developed which permit a variety of orientations to be tested. Perhaps the ultimate solution to overcoming the practical difficulties of testing, and the complexities of interpreting results and their application to real life situations, lies in mathematical modelling. It will be some time before this becomes practical for all except the simpler situations. At present, reliance is based on combinations of simple tests, large-scale *ad hoc* tests and empirical observation from real fires the theoretical interpretations of which are embodied in codes of practice such as the *Building Regulations.*

The caveat which must be quoted in the report of materials tested to BS 4735[6] is typical of the caveat now quoted in the majority of fire tests and illustrates the difficulty of interpreting test data: 'The following test results relate only to the behaviour of the test specimens under the particular conditions of test; they shall not be used as a means of assessing the potential fire hazard of the material in use.'

Many small fires tests are essentially for quality control use and are not suitable for assessing fire behaviour other than to indicate whether a given section of material is likely to be ignited by a small flame.

It has frequently been demonstrated that the flammability of a product depends on the interactions between materials, design, orientation, size, fixings, decoration, etc.[7] (Table 18.1).

18.2 TYPES OF FIRE TEST

18.2.1 Ignition Tests

Ignition tests expose a specimen to an ignition source of specified size and intensity for a specified period. Sources range from a lighted cigarette, a spark, a hot glowing wire or a small flame through flames of different sizes to large intense flames to radiant heaters with or without pilot ignition. The results of such tests are usually expressed in terms of the duration and/or the distance of flaming or smouldering combustion after the source is removed.

In essence these tests answer the questions whether a material will ignite and burn with a given source and whether the material will sustain combustion after the source is removed. Materials which do ignite with a given source are likely to do so in real life situations but it must never be assumed that materials which do not ignite in a test cannot be ignited. Many examples exist where a material which resists a small source can burn disastrously when exposed to a larger source.

The most common ignition sources are flames which may be based on solvents (e.g. alcohol or hexane), or wood but more usually on gas. Methane, propane, butane, natural gas or coal gas may be used in different tests. Flames may be diffusion (yellow) or pre-mixed (blue) and may be as small as 10 mm long or as

Table 18.1 Factors affecting fire performance of products

Product	Factors affecting fire performance
Cellular polystyrene ceiling tiles	Surface decoration, type and area of adhesion to substrate[2] thickness
Composite wall panel	Surface layer, adhesive bond
Insulation panels	Thermal conductivity and capacity of material
Suspended ceiling panels Plastic stacking chairs	Softening point and melting range of polymer

large as 500 mm. Flame contact time may also range from a few seconds to several minutes. A very large number of small flames are used for different tests. The majority of them are superfluous since they all determine small flame ignitability but they will doubtless continue to be used because they exist for historical reasons or in product specifications.

18.2.2 Flame Spread

Flame spread is essentially a progressive series of ignitions from a continuous flame front moving over a material. Considerable overlap exists between ignition and flammability tests because of the methods of assessing results but the distance and/or duration of burning from the point of ignition are frequently measured. Materials which initially burn slowly may burn more quickly when exposed to radiant heat which may be generated by an external source or by the material itself (thermal feedback). Simple flame spread tests may involve flame ignition but more complex tests exist in which the specimen is exposed to radiant heat together with a pilot flame.

18.2.3 Rate of Heat Release Tests

Rate of heat release tests are still in their infancy and tend to be applied to products rather than to materials. Rate of heat release is arguably the most important fire parameter because it directly affects the rate of fire growth. In rate of heat release tests, materials are frequently exposed to radiant heat and burned at a pre-determined intensity. Results are assessed by direct temperature measurements, by comparing temperature measurements with a calibrated gas flame, or by 'topping-up' with a gas flame to a pre-calibrated temperature. More recently heat release rates have been determined by measuring the amount of oxygen consumed during a test. The heat of combustion determined in a bomb calorimeter is the maximum heat produced by complete combustion and does not necessarily relate to the rate of heat release in a fire.

18.2.4 Smoke Formation

Smoke formation is not a fundamental property of a material but depends on whether the material is pyrolised (i.e. non-flaming decomposition) or burned, the oxygen availability and the temperature. The way a material is burned, i.e. the fire model, is therefore of considerable importance.

Smoke tests usually contain the smoke within a cabinet and measure its density using a lamp and a photometric system with a sensitivity similar to that of the human eye, to determine the amount of light passing through the smoke. This is then expressed as 'optical density' or a 'specific optical density' which can be factored and is the smoke produced related to the area of specimen exposed. The smoke produced per mass of material burned may also be determined.

Optical density may be related to visibility but ignores irritant effects, coagulation of particles, deposition of soot, etc. A limited number of tests determine the smoke as the mass of smoke particles produced.

18.2.5 Toxicity

Toxicity tests may be used to compare particular combustion products of a series of materials but the tests alone do not assess the total fire hazard of a product, since this also depends on other factors including ignitability, rate of flame spread, burning, smoke generation, etc., as well as other environmental factors such as product and building design, ventilation, human factors, etc.

Toxicity is the poisonous quality of a substance which, through its chemical action kills, injures or impairs a living organism.

Although substances of known toxicity may be determined chemically and the results used to indicate possible toxicity, the resultant data only infers a toxic hazard but does not determine it.

Fires produce complex and variable mixtures of gases and vapours (more than 200 have been identified in some cases) and whilst it may be possible to identify and quantify each component, it is only possible to determine the toxicity of such mixtures by biological tests.

The fire model and especially the temperature and oxygen concentration can have a major effect on the composition of fire gases and research has established six fire models which may produce different fire gas from a single material. The fire model must therefore be relevant to the end use hazard (Table 18.2).

A number of different methods may be used to determine the chemical composition of smoke. It is extremely important to realise that before fire gases can be analysed they must be sampled and that the method of sampling can significantly alter the composition of the sample. It is often necessary to use several different sampling methods to ensure that the smoke is correctly sampled.

Methods of chemically determining fire gases may be general, e.g. total hydrocarbons, or may be specific.

In some tests a specified gas or vapour is determined using chemical reagent tubes. These are rarely specific and are frequently subject to interferences by other gases. The complex nature of fire gases means that some tubes may give results

Table 18.2 Summary of fire classifications

Classification	Oxygen (%)	CO_2	Temperature (°C)	Irradiance (kW/m^2)
(i) Smouldering (non-flaming)	< 21	N/A	400–1000	N/A
(ii) Oxidative decomposition (non-flaming)	5–21	N/A	< 500	< 25
(iii) Pyrolytic decomposition	< 5	N/A	< 1000	N/A
(iv) Developing fire (flaming)	10–15	100–200	400–600	20–40
(v) Fully developed (flaming, low ventilation)	1–5	< 10	600–900	> 40–70
(vi) Fully developed (flaming, well ventilated)	5–10	< 50	600–1200	50–150

which relate to a number of gases other than those specified and this fact, coupled with the limited number of tubes used, means that the toxicity index produced can only be used for a crude assessment of possible toxicity.

More precise methods may involve classical wet analysis methods such as titrametric, colourimetric or gravimetric methods. Instrumental methods may involve infrared, paramagnetic (for oxygen) and chemilumniscence techniques, while the detailed analysis and quantification of smoke samples uses gas chromatography and mass spectroscopy. A recent report[8] gives advice and details of preferred methods.

Animal protocols are frequently based on mice or rats although other animals such as monkeys may be used. Traditional methods of determining the time to death of the animals are being increasingly extended by determining the time to incapacitation with observations of effects such as irritancy.

18.3 FIRE TESTS

Included in this section are standard fire tests specifically intended for plastics and also standard tests which are frequently used with plastics and plastic products. Emphasis is placed in ISO, BS and ASTM tests. For convenience the tests are subdivided into ignition and flammability tests, rate of heat release, smoke and toxicity tests. A further section deals with large-scale or *ad hoc* tests. Fire tests for plastics materials are dealt with by ISO TC61 but other ISO committees deal with produces containing plastics, e.g.

TC20 Aircraft and Space Vehicles
TC22 Road Vehicles
TC38 Textiles
TC42 Building Materials Products and Structures
TC188 Small Craft Boats

18.3.1 Guidance Documents

ISO 3261[9] This standard is essentially a list of defined fire terms listed with English and French equivalents.

BS 4422[10] is a glossary of terms associated with the phenomenon of fire and provides simple basic definitions of acceptable terms to be used in fire tests and reports.

ASTM E176[11] also defines fire test terminology.

ISO TR6585[12] was published by TC92, *Fire Tests for Building Materials, Components and Structures* but is applicable to the wider use of fire tests. The report provides a general insight into the main points that should be considered when fire tests are being developed selectively and/or used.

BS 6336[13] is a most important standard and discusses the need carefully to assess the limitations of small-scale tests in order to avoid possible incorrect or misleading interpretation. Large-scale tests are frequently useful in establishing the performance of a product or material in a specific environment, suitable small-scale tests can then be selected to provide quality assurance and control.

BS 476: Part 10[14] is a guide to the principles and applications of fire testing as applied to materials, composites and products used in building construction. The BS 476 series of tests are not dealt with in detail here as they are not primarily used for plastics. The use of these tests and relevant performance requirements are specified in the *Building Regulations* and apply to all types of materials. Problems may arise with certain tests, e.g. BS 476: Part 7, because thermoplastics melt and drip and consequently for certain applications, for instance suspended ceilings, other tests of the BS 2782 series are specified.

18.3.2 Ignition and Flammability: Tests for Ignition Temperature

ISO 871 (1968)[15] sets out to find the temperature at which plastics begin to decompose to flammable gaseous products. One gram of plastics material is introduced into a chamber fitted with a nozzle set in a heated metal block. An igniter flame is played over the nozzle and, if after 5 min no ignitable product issues from the nozzle, the test is repeated with a fresh sample at a temperature 10 °C higher until ignitable gas is detected at the nozzle. The decomposition temperature is reported to the nearest 10 °C.

ASTM 1929 (1977),[16] (Fig. 18.4) *Test Method for Ignition Properties of Plastics,*

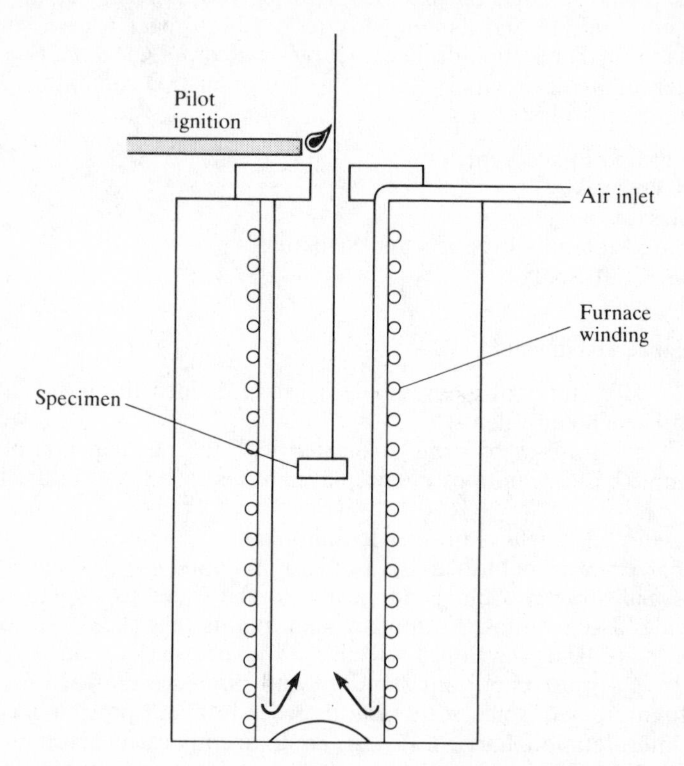

Fig. 18.4 Schematic diagram of equipment for determining flash and self-ignition temperatures (ASTM D1929)

uses a similar principle to that of ISO R871 except that a much more closely specified apparatus is used and a forced air flow is introduced into the heating chamber. Three criteria can be measured:

1. *Flash ignition temperature.* The lowest initial temperature of air passing around the specimen at which a sufficient amount of combustible gas is evolved to be ignited by a small external pilot flame.
2. *Self-ignition temperature.* The lowest initial temperature of air passing around the specimen at which, in the absence of an ignition source, the self-heating properties of the specimen lead to ignition or ignition occurs of itself, as indicated by an explosion, flame or sustained flow.
3. *Self-ignition by temporary glow.* A special case of self-ignition temperature where, in some cases, slow decomposition and carbonisation of the specimen result only in glow of short duration at various points therein without general ignition.

UL 94 Ignition temperature test[17]
Pieces of material approximately $\frac{1}{4}$ in (6·5 mm) square are dropped into a heated glass conical flask, 114 mm high, 60 mm diameter at the bottom and 28 mm diameter at the top. The pieces are dropped into the flask, at intervals of 5 °C and it is observed whether or not ignition takes place (without application of flame) within 2 min.

BS 476: Part 11[18] is essentially a simplified and developed version of ASTM D1929 in which the furnace is held at 750 °C. Specimens are classed as combustible if ignited. This is indicated by a 50 °C temperature rise or by flaming for more than 10 s. The test is applied to building materials and is unlikely to be passed by plastics. Similar tests are defined by ISO and ASTM standards.

Limiting Oxygen Index
This is a useful parameter which is widely used for quality assurance although its use to indicate flammability has been seriously doubted.

BS 2782 (1978) Methods 141A to 141D[1] cover the oxygen index of combustion determined on four different test pieces:

141A A bar of 10 × 4 mm nominal cross-section (rigid materials)
141B A bar of 6·5 × 3 mm nominal cross-section (rigid materials)
141C A bar 10 mm wide cut from rigid sheet 1·2–6·5 mm thick
141D A bar of 6·5 × 3 mm nominal cross-section of electric cable insulation or sheathing material

The oxygen index test has received much attention during the past few years and as it is claimed to give good repeatability and is useful for quality control purposes. It does not purport to predict performance of a material in real fire conditions and indeed must not be used for this purpose, nor does it necessarily indicate relative merits of, say, flame-retardant additives in a polymer under conditions other than those specified in the test method.

Basically, the test determines the minimum oxygen concentration necessary to support flaming combustion of the material under certain conditions. The test piece is clamped vertically at its base and supported in a glass chimney of specified dimensions. Arrangements are provided to ignite the top of the test piece with a propane flame and a mixture of nitrogen and oxygen (of known purity) is metered into the bottom of the chimney. Tests are then made to find the oxygen concentration in the gas mixture to cause the test piece, after ignition, to burn over a specified distance and time (at least 50 mm and at least 180 s). Fresh test pieces are used for each determination and the test is continued until the critical oxygen concentration for each burn is within 0·3 per cent on total volume of gas. Three such determinations are carried out and the mean value is reported as the oxygen index of the material under test.

ASTM D2863 (1977)[19] (Fig. 18.5) is similar to BS 2782 Method 141B.

ISO 4589[20] is also similar to BS 2782 Method 141.

NES 714[21] refers to BS 2782 Methods 141A to D but is frequently used in conjunction with NES 715[22] which determines the temperature at which materials will just burn in air, i.e. at 21 per cent oxygen. Two procedures are used, one in which the oxygen index is determined at various temperatures and the temperature at 21 per cent LOI is read off. Alternatively, the temperature of the furnace is adjusted until the specimen just burns in air.

Flame Tests

ISO 1210 (1982)[23] is a loosely defined test utilising a test piece in the form of a

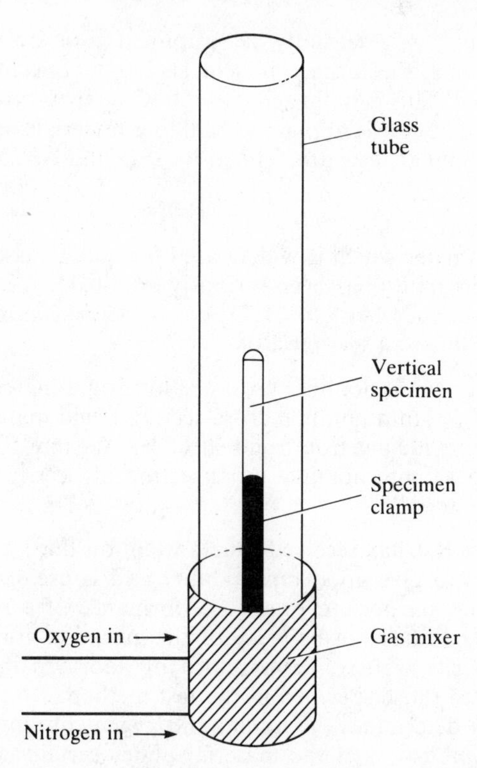

Fig. 18.5 Schematic diagram of oxygen index test ASTM D2863

bar 80 mm long, from 10 to 15 mm wide and from 3 to 5 mm thick. This is clamped horizontally and a 'standard' luminous bunsen burner flame, 100 mm long, is applied in a specified way to the test piece for 60 s. A stop clock is used to determine the time of burning after the flame is removed. 'Flammability' is then assessed to be in one of three categories: Category 1, if the test piece does not burn after removal of the flame; Category 2, if it burns for less than 15 s, some of it remaining unburnt; Category 3, if the test piece has burnt away or is still burning after 15 s. Five test pieces are subjected to test and the overall category of the sample is determined as the most flammable category found amongst the five test pieces.

BS 2782 1970 Method 508A[24] (Fig. 18.6) *Rate of Burning* (*Laboratory Method*), currently under review, is similar to ASTM D635 (1981).[25] A test piece in the form of a bar 150 mm long, 13 mm wide and 1·5 mm thick is scribed with reference lines across the test piece at 25 mm and 125 mm from one end. It is then clamped horizontally and with the plane of the width at an angle of 45°, and so mounted that the lower edge of the specimen is 6 mm above a sheet of metal gauze 130 mm square. A flame, from 13 to 19 mm in height, is applied to the free end of the test piece for 10 s and the time taken for the edge of the flame of the burning test piece to travel the distance of 100 mm between the two reference lines is measured with a stop-watch. At least three test pieces are used and the result (reported for each test piece) is expressed as: (a) The rate of burning (mm/min); or (b) that the flame does not reach the first mark; or (c) that the flame does not reach the second mark; or (d) for test pieces where the flame does not reach the first mark, the duration of flame or after flow after removal of the burner.

BS 2782 Method 140D[2] for polyvinyl chloride sheeting is a revision of the now discontinued BS 2782 Method 508C. A strip of material of the prescribed

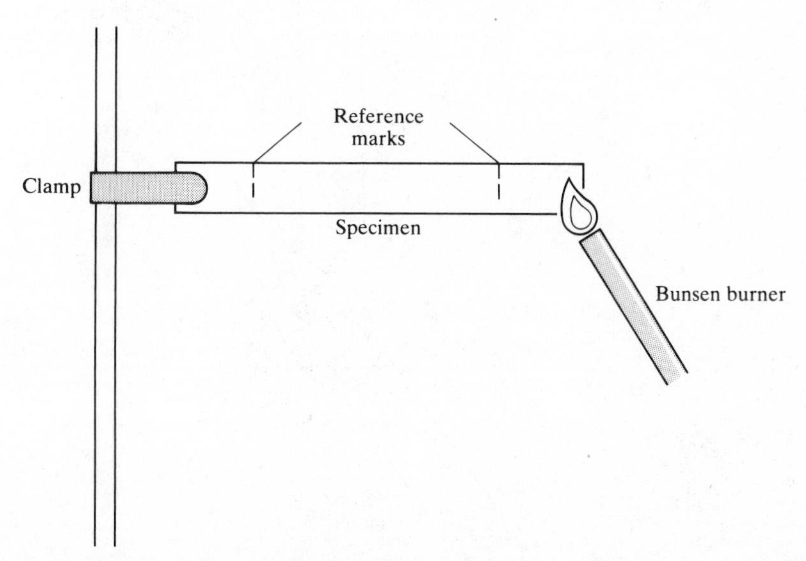

Fig. 18.6 Typical arrangement of small flame ignitability/flame spread test with horizontal specimen of rigid material, e.g. BS 2782 Method 508A

dimensions is clipped to a semicircular rack and one end of the strip is subjected to the flame from a specified volume of ethanol. The test result is reported as the distance over which the strip is burned or charred under these conditions.

BS 2782 Method 140E[26] is a revision of BS 2782 Method 508D (1970),[27] the alcohol cup test. The test piece of prescribed dimensions is supported on a simple brass and wire frame so that its major plane is at 45° to the horizontal. The flame from burning a specified amount of alcohol is allowed to impinge on the centre of the sheet and the result is expressed in terms of the amount of charring of the surface of the test piece and the duration of flaming or glowing.

UL94,[17] horizontal burning test for classifying materials as 94HB is a similar test to ASTM D635[25] and BS 2782 Method 508A.[24]

UL94[17] (Fig. 18.7), vertical burning test for classifying materials as 94V–0, 94V–1 or 94V–2, is somewhat similar to ASTM D568 but requires a pad of dry absorbent surgical cotton to be placed 305 mm below the test piece. Classifications 94V–0 and 94V–1 require that the ignited test pieces do not have any flaming particles which ignite the cotton. Classification 94V–2 is a lower one allowing some flaming particles to ignite the cotton. This classification test of UL94 has found favour in recent years in trade and insurance specifications in both the UK and USA.

UL94,[17] vertical burning test for classifying materials as 94–5V, is similar to the previous test above but uses a plaque 152 mm square. The test flame is applied

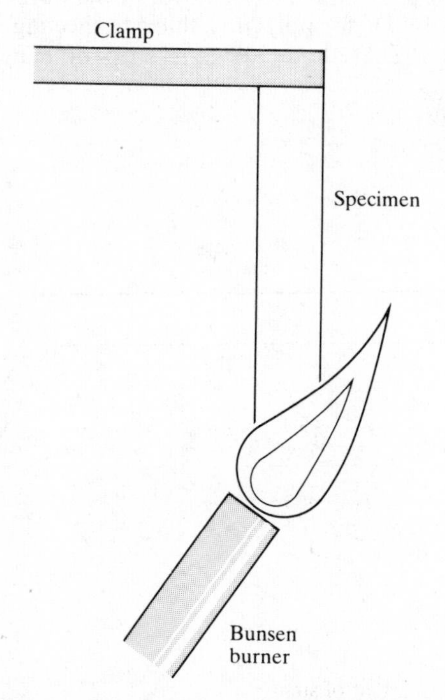

Fig. 18.7 Typical arrangement of small flame ignition test with vented specimen, e.g. UL94–5V

for 5 s and then removed for 5 s, this being repeated for five applications of the flame. To receive the 94–5V classification no test specimen must: (a) burn for more than 60 s; (b) drip, or (c) be destroyed to any significant degree in the area of the test flame.

ASTM D635 (1977)[25] determines the rate of burning in a horizontal position and is similar to Method 508A of BS 2782. The ASTM test more precisely specifies the method than the BS test and requires the result to be reported in terms of burning rate (cm/min), average time of burning (s), and average extent of burning (mm), together with any special effects noted during the test.

ASTM D568 (1977)[28] for rate of burning in a vertical position has a test piece 25 mm wide and 450 mm long which is marked by gauge marks to define a 380 mm length over which the burning rate is to be measured. The test piece is suspended vertically and ignited at the bottom with a specified Bunsen flame. The result is expressed in terms of burning rate (cm/min), average time of burning (s) and average extent of burning (mm).

ASTM D1433 (1977)[29] determines rate of burning of flexible thin plastics sheeting supported on a 45° incline. A test piece 76 mm wide and 228 mm long is supported on a metal frame inclined at 45° to the horizontal with 25·4 mm of the test piece hanging vertically over the lower end of the frame. A specified butane gas flame is applied to the centre of the hanging vertical portion and the results are expressed as the burning rate of the test piece (mm/min). In the test apparatus specified, the burn is timed automatically by microswitches activated by the burning through of nylon sewing threads set at a distance of 152 mm apart across the test piece.

ASTM D3801[30] measures the duration of flaming, after first and second, 10 s flame applications, the mean burning and flowing times. Standard deviations, are also determined for these results. The test is similar to UL94 but does not impose categories of burning behaviour.

BS 4735[6] (Fig. 18.8) is a small-scale laboratory test in which a strip of cellular plastics or cellular rubber material, no larger than 150 × 50 × 13 mm supported on horizontal wire gauze is exposed to a small flame. The distance and the duration of burning are reported. ISO 3582[31] is similar.

BS 2011: Part 2, p. 3, 1970[32] is a method of test for electrical components, equipment and electrotechnical products. Specimens are exposed to a propane burner, from 14 to 15 mm ID, generating a flame of 1100 °C for 15 s. The time of burning after the flame is removed and any violent or explosive type of burning is recorded.

BS 5946 (1980)[23] exposes a cube of phenolic foam to a Bunsen burner placed beneath the lower surface. The block is heated until the thermocouple placed in the block reaches 180 °C. The flame is removed and the foam is said to punk (smoulder) if a second thermocouple placed above the first exceeds 360 °C.

BS 4066: Part 1[34] exposes a vertical cable to a flame and the duration and distance of burning is limited. For specimens greater than 50 mm diameter, two burners are used. IEC 332–1 is similar.

Incandescent Bar Tests

ISO 181 (1981),[35] *Plastics – Determination of Flammability Characteristics of Rigid Plastics in the Form of Small Specimens in Contact with an Incandescent Rod*, is similar to BS 2782 Method 508E (1970).[36] A test piece, not less than 80 mm

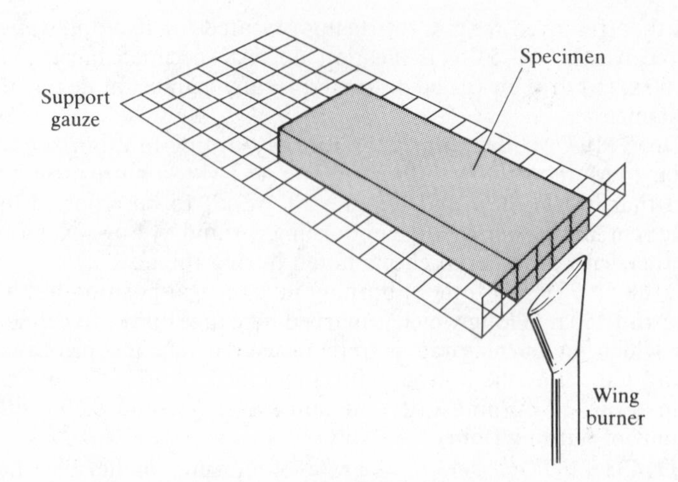

Fig. 18.8 Typical arrangement of small flame ignition test with flexible foam specimen supported on stainless steel gauze, e.g. BS 4735

long, 10 mm wide and 4 mm thick, is weighed and clamped horizontally (thickness in the vertical plane) so that its 10 mm wide edge can be brought into contact with an incandescent silicon carbide rod, 8 mm in diameter, maintained at a temperature of 950 °C. The incandescent bar is held in contact with the test piece with a force of 295 mN for a period of 3 min. The bar is removed, any flame is dry-extinguished and the test piece is re-weighed. It is then cleaned and polished so that the limit of the flame spread is apparent and the length of the test piece is measured to this point. The difference between the original and final lengths is termed the 'flame spread'.

BS 6334[37] which is a combination of the incandescent bar test and UL94 tests is applied to solid electrical insulating materials.

Radiant Heat Tests

ISO DP5657 (1982)[38] (Fig. 18.9), although primarily intended for building materials and products, is more generally useful in that horizontal specimens (165 mm square with a 140 mm exposed central disc) are exposed to a radiant heat flux of from 10 to 50 kW/m^2 with or without a pilot ignition flame. The time to ignition is recorded and results for specimens that do not ignite within 15 min are reported as 'no ignition'. The importance of this test is that specimens are exposed to high heat flux levels but thermoplastics materials do not melt away from the ignition source.

The test is defined in BS DD70[39] and the apparatus, without the pilot ignition flame, is used as the fire model in the ISO dual chamber smoke test.

Many tests are used for the testing of plastics, although originally designed for other materials. For example, BS 476: Part 7 (1987),[3] for surface spread of flame tests for building materials. A test piece 95 mm wide and 300 mm long and not more than 25 mm thick is mounted so that the width is vertical and the long axis is virtually at right angles to a gas-fired radiant panel so that the radiant intensity

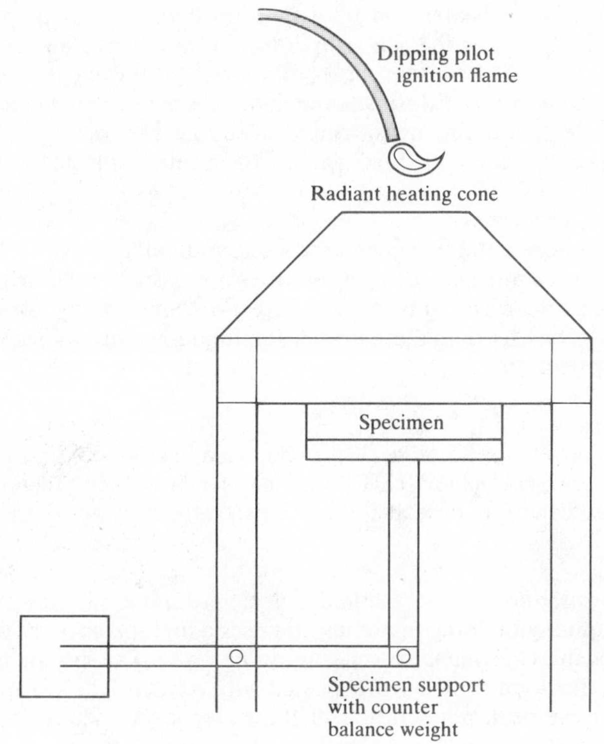

Fig. 18.9 Schematic diagram of ISO ignitability apparatus ISO DP5657

reaching the test piece is within the limits specified by the test method. As soon as the test piece is in position a gas flame is applied to the hotter end for 1 min. Measurements of the time of the flame spread along the test piece are made over a period of 10 min or until the flame has reached the far end of the test piece, whichever is the shorter. The flame spread so measured is used to classify the test material into one of four categories, depending on flame spread after $1\frac{1}{2}$ min and final flame spread.

A similar but smaller radiant panel test for building materials is being developed within ISO as ISO DP5658.[40]

18.3.3 Rate of Heat Release Tests

There are no widely accepted rate of heat release tests although these tests are arguably the most important for determining fire behaviour of products and materials. Although many rate of heat release tests have been suggested, few have been developed to full scientific or regulatory standards. However, numbers of heat release apparatus exist and are reviewed by M. Jannssens and R. Minne[41] in ISO TC92/SC1/104 and also by Brabrauskas.[43]

Ohio State University Apparatus

This apparatus consists of an insulated box containing a vertical specimen and a parallel electric radiant heater and pilot ignition flame. Air at a controlled rate flows through the box. The flue gas temperature is monitored and compared to a calibrated gas flame. Maximum heat flux on the specimen is 100 kW/m^2. Modifications to the OSU calorimeter include the measurement of heat release rate by oxygen depletion, the measurement of smoke, the use of different air flow rates and the use of a controlled oxygen/nitrogen mixture instead of air.

Factory Mutual Apparatus
This consists of a horizontal specimen exposed to radiant heat in a vertical chimney through which air or nitrogen/oxygen mixtures are passed at a constant rate. The rate of heat release is measured by oxygen depletion. Smoke may also be measured and Tewarson[42] has used this apparatus for fundamental studies to determine data for fire modelling.

The Cone Calorimeter[43]
This exposes a vertical or horizontal specimen to a heat flux of up to 100 kW/m^2, generated by a conical heater (cf. ISO cone). Air flows over the specimen at a constant rate and the rate of heat release is determined using oxygen depletion.

ISO 1716[44]
This standard specifies a test method for determining the calorific value of non-metal containing building materials and uses a high-pressure bomb in a water jacket immersed in a calorimeter vessel containing water. The specimen is contained in a crucible in the bomb which is pressurised with oxygen. The specimen is ignited electrically and the peak temperature of the water jacket determined. The gross or net calorific potential is then calculated using specified formulae. Composite specimens are tested after first separating into separate compartments.

18.3.4 Smoke Tests

Smoke tests may use photometric or gravimetric systems to determine optical density or mass density, respectively, and may be applied to static or dynamic smoke tests. The majority of smoke tests apply an optical measuring system to static smoke tests in which the smoke is generated, stored and measured within an enclosed cabinet.

BS 5111[45]
This is a relatively simple test in which a foamed specimen of 25 mm cube is burned in a gas flame in a cabinet measuring 308 × 308 × 924 mm high. The maximum smoke density (as per cent obscuration) is determined using a horizontal photocell/lamp system in the upper part of the cabinet. The maximum obscuration and the time at which it occurs are reported.

ASTM D2843[46]
This test uses the same XP2 cabinet and test method used in BS 5111 but is applied to a wider range of materials. A 25 mm square specimen is burned and the smoke obscuration/time curve is reported together with the maximum smoke obscuration amid the area under the smoke obscuration/time curve.

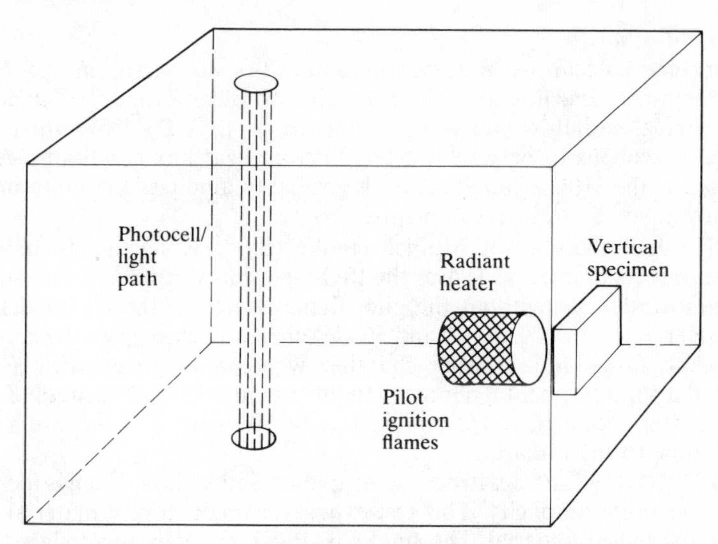

Fig. 18.10 Schematic diagram of NBS smoke test (ASTM E662)

BS6401[47] *ASTM E662*[48] (*Fig. 8.10*)

These are based on the NBS smoke test and are essentially the same test although some differences occur in various procedures. A 75 mm square specimen of up to 25 mm thick is combusted in a vertical orientation at 25 kW/m^2 incident heat flux. Tests are carried out with and without a series of pilot flames along the lower edge of the specimen. The smoke is contained in a cabinet of 0·51 m^3 and measured using a vertical photo multiplier/photocell system. The results are typically expressed as specific optical density which relates the optical density of the smoke to the volume of the cabinet, the length of the smoke-measuring path and the area of the specimen exposed in the test. Other methods have been used in which specific optical density is related to the mass of specimen combusted and/or to time.

NES 711[49]

This uses the NBS test cabinet but involves a number of fundamental modifications.

1. The cabinet includes a smoke stirring fan.
2. The pilot gas burner is different.
3. The first 4 min of the test are under non-flaming conditions and the remainder under flaming conditions.
4. The results are expressed as an index which is calculated from the time for the smoke transmittance to reduce to 70, 40, 10 per cent and the minimum value. In practice, this means that the smoke index is biased against materials which generate large amounts of smoke in the early part of the test.

ATS 1000.001[50]

This test is based on ASTM E662 and almost invariably involves the determination of toxic gases. It is dealt with in the next section.

Hovde Modification[51]
Major problems occur when some thermoplastics are tested in the NBS test because they melt and flow out of the specimen holder. Hovde[51] has developed a small conical radiator similar to that used in ISO DP5657 which permits horizontal specimens to be tested under different heat flux conditions. Although in its infancy, the Hovde modification is probably amongst the more important of the current smoke test developments.

The ISO dual chamber or Munich smoke tests[52] is starting to be used for building products in Europe. It uses the ISO ignitability apparatus (see section on flammability tests) but without the pilot flame ignitor as the fire model. Smoke may be generated at 10, 20, 30, 40 and 50 kW/m^2 and passes from the combustion chamber and flows into a larger chamber where it is stirred with a fan and determined using a horizontal photocell/light system with a path length of 360 mm. The horizontal specimen is 165 mm square but is masked to expose a 140 mm diameter zone to the radiator.

ASTM D4100 (1982)[53] describes the Arapahoe Smoke Box which is increasingly used because of its simplicity. This test exposes a small strip of material to a gas flame for a specified material. The smoke is drawn through a fine glass filter and is expressed as the percentage mass of soot to the mass of the material combusted.

Draft DIN 53436/DIN 53437[54] and NFT 51–073[55] are dynamic furnace tests in which smoke is generated in an air stream through a tubular furnace and measured using a photocell/light system. The DIN furnace is more frequently encountered as a toxicity test and is considered in the next section.

Modified Flammability Tests
A number of smoke tests exist in which the smoke produced by a flammability or heat release test is determined using photocell/light systems positioned in the smoke streams.

18.3.5 Toxicity Tests

ISO TC92/SC3 have set up a series of working groups to consider fire models (WG1), chemical analysis methods (WG2), biological assay (WG3), guidance documents (WG4) and bioanalytical methods (WG5).

BS PD 65031 (1982)[56] reports on the toxic hazard in fire and makes recommendations concerning problems related to the development of tests for the measurement of toxicity and combustion of products used in buildings, contents of buildings and transport. BS PD65031 contains background information useful to those concerned with fire fighting, the handling and use of flammable materials and carrying out the fire tests and interpreting their results.

ASTM E800[57] gives details of the methods of test and analytical procedures to be used in determining the main gas produced in fires.

ISO Technical Report 6543[8] describes the assessment of hazard caused by toxic fire gases and the development of suitable test methods. Tests currently under serious consideration include the NBS protocol (Potts Pot.) and DIN 53436.

NBS protocol.[58] The test material is decomposed in a small furnace heated to 25 °C above and to 25 °C below the material self ignition temperature. The furnace is mounted in the base of a cabinet of approximately 0·2 m^3 volume.

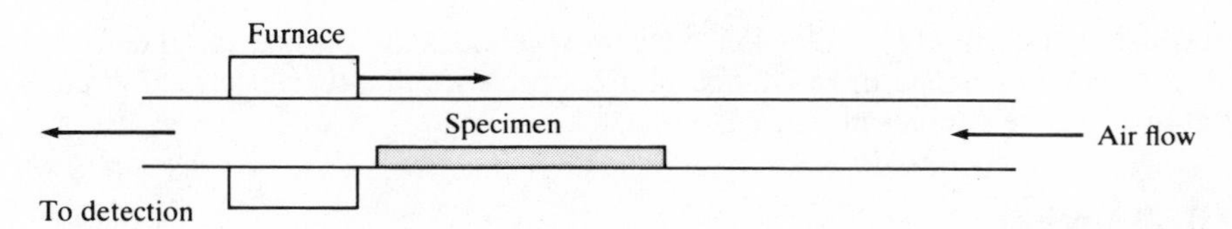

Fig. 18.11 Schematic diagram of tubular furnace for toxic gas studies DIN 53436

Fire gases circulate around the cabinet and are determined analytically or by animal protocols. Criticisms of the fire model have led to the development of a small radiant heater as an alternative to the original heated crucible.

DIN 53436[54] (Fig. 18.11) describes a 40 mm diameter tubular furnace in which the specimen is heated under pre-determined conditions of temperature, gas flow rate and oxygen concentration. The furnace moves over the specimen in the opposite direction to the air flow. Materials are typically tested at 300, 400, 500 and 600 °C in nitrogen and air. The fire gases are typically diluted with air prior to entering an animal exposure chamber and/or are determined analytically. Tests at temperatures between 200 °C and 1000 °C have also been reported.

Other fire models are used in various countries but the following methods are frequently used in the UK.

NES 713.[59] This is a relatively simple test in which, typically, 2 g of material are burned by a gas flame in a polypropylene-lined cabinet of about $1-1\frac{1}{2}\,m^3$. Simple fire gases (carbon monoxide, carbon dioxide, hydrogen cyanide, hydrogen fluoride, hydrogen chloride, hydrogen bromide, nitrogen oxides, sulphur dioxide, formaldehyde, acrylonitrile, phenol and phosgene) are determined using chemical reagent tubes. Test results are expressed as an index which is the sum of the ratios of each gas concentration (expressed as volume ppm per 1 m^3 per 100 g sample) to the concentration required to cause death in 30 min. This test is applied to materials for use in warships and fighting vehicles and occasionally for other applications.

ATS 1000.001[50] is based on the NBS Smoke test (ASTM E662) and determines simple fire gases using chemical reagent tubes or specific ion electrodes. Limits are specified for smoke and gas concentrations for the combustion products which are generated by exposing a 75 mm square sample to a radiant heat flux of 25 kW/m^2 with and without pilot flame ignition. Although intended for use with aircraft materials, this test is also applied to other products.

18.3.6 Corrosivity of Fire Gases

A number of tests have been developed to determine the corrosivity of fire gases and have resulted from damage caused to building materials and components, to machinery, metallic products and to electrical and electronic installations.

This damage is only important remote from the source of the fire because corrosive damage is negligible compared to the thermal damage caused in the fire zone.

Corrosivity tests frequently involve heating materials in a horizontal tubular furnace and determining the corrosivity by dissolving the gases in water and determining pH, or acids equivalent (e.g. HCl). Examples of this type of test which are frequently applied to cables include VDE 0472: Part 813.[60] An alternative approach is used in UTE C20–453[61] where specimens are burned in an enclosed cabinet and the change in resistance of a copper wire or other electrical circuits contained in the cabinet measured.

18.3.7 Large-scale Tests

As yet, there are no published national or international standards for large-scale fire tests for plastics. However, use is made of the large-scale standard tests which have been published for building materials. Typical examples of these are BS 476: Part 3 (1975),[62] *External Fire Exposure Roof Test*, and Part 8 (1972),[63] *Test Methods and Criteria for the Fire Resistance of Elements of Building Construction.* Part 3 requires test pieces not less than 1·5 × 1·2 mm and Part 8 requires furnaces which can give a temperature rise of up to 1193 °C and specimens of $2\frac{1}{2}$ m × $2\frac{1}{2}$ m for wall panels and $2\frac{1}{2}$ m × 4 m for floor and ceiling panels. Similarly, to test the burning characteristics of a complete product, e.g. furniture, requires a specially built room or a large ventilation duct, containing sensors to measure heat, smoke and toxic gases.[64]

ASTM E603[65] contains a *Guide for Room Fire Experiments* and gives advice on the methodology, instrumentation, test protocols, etc., for use in the full-scale testing of room linings and contents.

ASTM 3894,[66] *Evaluation of Fire Response of Rigid Cellular Plastics using a Small Corner Configuration*, is the standard which tests the flammability of the specimen arranged as two walls only, each of 1220 mm × 610 mm high or two walls plus a ceiling 1220 mm square which are exposed to a pre-mixed propane flame positioned in the lower corner. A specified temperature/time calibration is specified and results are assessed in terms of flame spread. Thermocouples are positioned at specified locations on the specimen surface and the times to maximum temperatures and the maximum temperatures are recorded.

Detailed discussion of large-scale fire tests is outside the scope of this book and those requiring any such fire tests to be carried out are advised to get in touch with a specialist organisation such as The Fire Research Station (Boreham Wood, England), Rapra Technology Limited, (Shawbury, England) or their equivalent organisations in other countries. A useful appraisal of large-scale tests and especially on the need carefully to design the tests to meet the end-use environment and also to establish the limitations of such tests is given in BS 6336.[13]

REFERENCES

1. BS 2782 Methods 141A to D (1978) *Plastics*: Part 1: *Thermal Properties*, Methods A to D *Oxygen Index of Combustion.*
2. BS 2782 Method 140D (1980) *Flammability of a Test Piece 550 mm × 35 mm. Method of Test of Thin Polyvinyl Chloride Sheeting.*
3. BS 476: Part 7 (1987) *Surface Spread of Flame for Materials.*
4. Rowan, J. W. and Lyons, J. W. (1978) *Journal of Cellular Plastics*, **14** (1), 25.
5. *The Building Regulations* (1976) HMSO.
6. BS 4735 (1974) *Assessment of the Horizontal Burning of Specimens of Cellular Plastics and Cellular Rubber Materials when Subjected to a Small Flame (Laboratory Method).*
7. Paul, K. T. (1984) *Plastics and Rubber Processing and Applications*, **4**, 221.

8. ISO TR6543 (1979) *The Development of Tests for Measuring Toxic Hazards in Fire.*
9. ISO 3261 (1975) *Fire Test Vocabulary.*
10. BS 4422 (1969) *Glossary of Terms Associated with Fire*: Part 1: *The Phenomenon of Fire.*
11. ASTM E176 (1981) *Terminology Relating to Fire Standards.*
12. ISO TR6585 (1979) *Fire Hazard and the Design and use of Fire Tests.*
13. BS 6336 (1982) *Guide to the Development and Presentation of Fire Tests and their use in Hazard Assessments.*
14. BS 476: Part 10 (1983) *Guide to the Principles and Applications of Fire Testing (for Building Materials and Structures).*
15. ISO 871 (1980) *Plastics – Determination of Temperature of Evolution of Flammable Gases (Decomposition Temperature) from a small sample of pulverised materials.*
16. ASTM D1929 (1977) *Test Method for Ignition Properties of Plastics.*
17. UL94 (1978) *Underwriters Laboratory Tests for Flammability of Plastics Materials in Devices and Appliances.*
18. BS 476 Part 11 (1982) *Method of Assessing the Heat Emission from Building Materials.*
19. ASTM D2863 (1977) *Measuring the Minimum Oxygen Concentration to Support Candle-like Combustion of Plastics (Oxygen Index).*
20. ISO 4589 (1984) *Laboratory Method of Test to Assess Oxygen Index Values for the Burning Characteristics of Small Vertical Specimens after Ignition by a Small Flame or in Candle Cut Surfaces.*
21. NES 714 (1981) *Determination of the Oxygen Index of Small Specimens of Materials.*
22. NES 715 (1981) *Determination of the Temperature Index of Small Specimens of Materials.*
23. ISO 1210 (1982) *Plastics – Determination of Flammability Characteristics in the Form of Small Flame.*
24. BS 2782 Method 508A (1970) *Rate of Burning (Laboratory Method).*
25. ASTM D635 (1981) *Test for Rate of Burning and/or Extent and Time of Burning of Self Supporting Plastics in a Horizontal Position.*
26. BS 2782 Method 140E (1982) *Flammability of a Test Piece 150 mm square not Exceeding 50 mm thick of a Plastics Material (Alcohol Cup Method) (Laboratory Method).*
27. BS 2782 Method 508D (1970) *Flammability Alcohol Cup Test.*
28. ASTM D568 (1977) *Test for Rate of Burning and/or Extent and Time of Burning of Flexible Plastics in a Vertical Position.*
29. ASTM D1433 (1977) *Test for Rate of Burning and/or Extent and Time of Burning of Flexible Items Plastic Sheeting Supported on a 45 Degree Incline.*
30. ASTM D3801 (1980) *Comparative Extinguishing Characteristics of Solid Plastics in a Vertical Position.*
31. ISO 3582 (1968) *Cellular Plastics and Cellular Rubber Materials Laboratory Assessment of Horizontal Burning Characteristics of Small Specimens Subjected to a Small Flame.*
32. BS 2011: Part 2 (1970) *Methods for Basic Environmental Testing Procedures*; Part 2.2.1 (PZ: 1970 *Flammability*).
33. BS 5946 (1980) *Determination of the Punking Behaviour of Phenol Formaldehyde Foam.*
34. BS 4066: Part 1 (1980) *Tests on Electric Cables Under Fire Conditions Method of Test on a Single Vertical Insulated Wire or Cable.*
35. ISO 181 (1981) *Plastics – Determination of Flammability Characteristics of Rigid Plastics in the Form of Small Specimens in Contact with an Incandescent Rod.*
36. BS 2782 Method 508E (1970) *Incandescence Resistance of a Specimen 10 mm × 4 mm × Approximately 100 mm of Rigid Thermosetting Plastics (Laboratory Method).*
37. BS 6334 (1983) *Method of Test for the Determination of the Flammability of Solid Electrical Insulating Intervals when Exposed to an Igniting Source.*
38. ISO DP5657 (1982) *Ignitability of Building Materials.*
39. BS DD70 (1981) *Method of Test for Ignitability of Building Products.*
40. ISO DP5658 (1977) *Spread of Flame of Building Materials.*
41. Janssens, M. and Minne, R. ISO TC92/SC1/104.

42. Tewarson, A. and Pion, R. F. (1976) *Combustion and Flame*, **26**, 85.
43. Brabrauskas, V. (1982) NBSIR 82-2611, US Bureau of Commerce, National Bureau of Standards, Washington.
44. ISO 1716 (1983) *Building Materials – Determination of Colorific Potential.*
45. BS 5111 (1974) *Laboratory Method for Testing a 25 mm Cube to Specimen of Low Density Material (up to 130 kg/m³) to Continuous Flaming Conditions.*
46. ASTM D2843 (1977) *Standard Test Method for Density of Smoke from Burning or Decomposition of Plastics.*
47. NS 6401 (1983) *Method of Measurement in the Laboratory of Specific Optical Density of Smoke Generated by Materials.*
48. ASTM E662 (1979) *Test for Specific Optical Density of Smoke Generated by Solid Materials.*
49. NES 711 (1981) *Determination of the Smoke Index of the Products of Combustion from Small Specimens of Materials.*
50. ATS 1000.001 (1984) *Airbus Industries Technical Specification, Fire, Smoke and Toxicity Test Specification*, Issue 3.
51. *J. Hovde Report*, no. STF25, A81002, Norges Branntekniske Lab. Trondheim, Norway.
52. ISO TR5924 (1980) *Fire Tests Reaction to Fire Smoke Generated by Building Materials.*
53. ASTM D4100 (1982) *The Arapahoe Smoke Test Apparatus.*
54. DIN 53436/DIN 53437 *Tubular Furnace for the Thermal Decomposition of Plastics.*
55. NFT 51–073 *Method of Measuring the Optical Density of Smoke.*
56. BS PD 65031 (1982) *Report and Recommendations on the Development of Tests for Measuring the Toxicity of Combustion Products.*
57. ASTM E800 (1981) *Guide for the Measurement of Gases Present or Generated During Fires.*
58. Levin, B. C. NBS IR82/2532 US Department of Commerce National Bureau of Standards, Washington.
59. NES 713 (1985) *Determination of the Toxicity Index of the Products of Combustion from Small Samples of Materials.*
60. VDE 0472: Part 813: *Test for Corrosivity of Fire Gases.*
61. UTE C20–453 (1976) *Determination of the Corrosivity of Fumes.*
62. BS 476: Part 3 (1975) *External Fire Exposure Roof Test* (Note UK *Building Regulations* are based on 1958 version).
63. BS 476: Part 8 (1972) *Test Methods and Criteria for the Fire Resistance of Elements of Building Construction.*
64. Sandström V. (1984) *Nordtest Project 143–78 Technical Project SP-RAPP-1984 16 ISSN 9280*, Boras, Sweden 1984.
65. ASTM E603 (1977) *Guide for Room Fire Experiments.*
66. ASTM 3894 (1981) *Evaluation of Fire Response of Rigid Cellular Plastics using a Small Corner Configuration.*

GENERAL REFERENCES

Mackower, A. D., *Flammability of Solid Materials, a Guide to the relevant British Standards*, BSI.

ASTM Fire Test Standards (1982) ASTM E5 Committee, Philadelphia.

Troitzsch, J. (1982) *Plastics Flammability Handbook, Principles, Regulations, Testing and Approval*, Hanser, MacMillan.

Hilando, C. J. *Flammability Handbook for Plastics*, Technomic Publishing Co. Inc.

Drysdale, D. (1985) *Fire Dynamics*, Wiley.

Chapter Nineteen

Permeability

19.1 INTRODUCTION

19.1.1 Basic Theory

No polymeric material forms an impervious barrier to gas or vapour molecules. The transmission of such molecules through a membrane is a result of intermolecular spaces, pin-hole defects, or porosity, or some combination of these three structural features of the material. In this chapter we are restricting the term 'permeation' to the movement of gas or vapour molecules through molecular scale voids. The quantity of permeant is usually measured as the volume at STP for a gas, or as the mass in the case of a vapour.

Permeation of gases and vapours takes place in two stages. In the first stage the permeant dissolves into the surface of the material. The solubility of the permeant in a polymer is defined as that volume (or mass) which dissolves in unit volume of the material under applied unit pressure. It is described by Henry's law:

$$c = SP \qquad [19.1]$$

where c is the volume at STP (or mass) per unit volume of polymer, S is the solubility coefficient, and P is the applied pressure.

The second stage is the diffusion of the dissolved molecules through the material under the action of a concentration gradient. The diffusion process is described by Fick's first and second laws. Fick's first law relates the flux to the concentration gradient

$$J_n = -D\frac{\partial c}{\partial x} \qquad [19.2]$$

where J_n is the permeant flux, i.e. the rate of flow per unit area; D is the diffusion constant or diffusivity; and $\partial c/\partial x$ is the concentration gradient.

Equations [19.1] and [19.2] can be combined to give

$$J_n = -Q\frac{\partial P}{\partial x} \qquad [19.3]$$

where $Q = DS$ is the permeability; and $\partial P/\partial x$ is the pressure gradient.

Fick's second law, from which the time-dependent concentration distribution may be calculated, is obtained from Eqn [19.2] and the equation of continuity

$$\frac{\partial J_{\mathrm{n}}}{\partial x} = -\frac{\partial C}{\partial t} \qquad [19.4]$$

The equation of continuity is an expression of the conservation of mass. The flux can be eliminated between Eqns [19.2] and [19.4] to give:

$$\frac{\partial c}{\partial t} = D\frac{\partial^2 c}{\partial x^2} + \frac{\partial D}{\partial c}\left(\frac{\partial c}{\partial x}\right)^2 \qquad [19.5]$$

If the diffusivity is independent of concentration this reduces to

$$\frac{\partial c}{\partial t} = D\frac{\partial^2 c}{\partial x^2} \qquad [19.6]$$

Equations [19.4] to [19.6] can also be written in terms of pressure instead of concentration.

Assuming that Q is independent of pressure and that the sample is homogeneous, when steady state conditions have been achieved Eqn [19.3] can be integrated to give

$$J_{\mathrm{n}} = Q\frac{\Delta P}{l} \qquad [19.7]$$

where ΔP is the pressure drop across the sample; and l is the sample thickness. If Q changes with pressure Eqn [19.7] would give an average value for the permeability. If there is any doubt as to the homogeneity of the material it is usual to calculate the gas transmission rate for the particular sample under test rather than to calculate a permeability. The gas transmission rate (GTR) is the flux per unit pressure drop:

$$\mathrm{GTR} = \frac{J_{\mathrm{n}}}{\Delta P} \qquad [19.8]$$

The use of polymer laminates in packaging is common and it is useful if a permeability for the laminate can be calculated from the permeabilities of the constituent layers. For a laminate of n separate layers the permeability is given by

$$\frac{1}{Q_{\mathrm{L}}} = \frac{1}{l}\sum_{i=1}^{n}\frac{x_i}{Q_i} \qquad [19.9]$$

where Q_{L} is the permeability of the laminate; l is the thickness of the laminate; Q_i is the permeability of the ith layer; and x_i is the thickness of the ith layer. This equation can be used if, and only if, the permeabilities are independent of pressure. If the permeabilities are pressure-dependent the final value would depend on the order of the layers and a more complicated calculation would be required.

The steady state equation for gas transmission only applies when the gases and vapours are sparingly soluble in the polymer and when there is no chemical association. This is true for the air gases, but often not for water and organic vapours which have pressure-dependent permeabilities. Vapours may also act as plasticisers and so increase diffusion.

For all polymer–permeant systems the permeability has a dependence on temperature which can be described by the Arrhenius equation:

$$Q = Q_0 \exp\left(-\frac{E_P}{RT}\right) \qquad [19.10]$$

where Q_0 is a constant for a particular system, E_P is the activation energy for the permeation process, R is the gas constant, and T is the absolute temperature. For both gases and vapours the diffusivity increases with increasing temperature. The solubility increases with temperature for gases and decreases for vapours.

The mechanism of diffusion in polymers and the effects of varying the environmental factors on the values of the material properties are discussed in, for example, Comyn[1] and McGregor.[2]

19.1.2 Units

A bewildering variety of units have been used in the literature and Huglin and Zakaria,[3] for example, reported twenty-nine different units for permeability. Yasuda and Stannett[4] give useful conversion factors as well as tables of selected values.

The SI unit for gas permeability is $m^3/m^2\,s\,(Pa/m)$. Rearranging gives: $m^2\,s^{-1}\,Pa^{-1}$ or $m^4\,s^{-1}\,N^{-1}$. The SI unit for vapour permeability is: $g\,m^{-1}\,s^{-1}\,Pa^{-1}$ or $g\,m\,s^{-1}\,N^{-1}$.

The SI unit for diffusivity is $m^2\,s^{-1}$.

19.2 GAS PERMEABILITY

In this section well established techniques used for the measurement of permeability and gas transmission rate are described. More detailes are given in, for example the review by Lomax.[5,6]

It is absolutely essential when measuring permeability that steady state gas (or vapour) transfer exists, i.e. when equal quantities of permeant enter and leave the polymer. Under these conditions the concentration gradient does not change with time, and Eqn [19.7] or Eqn [19.8] can be used. Before steady state conditions are achieved more permeant dissolves into the polymer than evaporates from it. The first stage of the permeation process (solubility) is dominant initially but, as the concentration gradient increases, the second stage (diffusion) becomes the major contributor. The attainment of steady state may take anything from a few hours to many days. If the measurement is made before steady state conditions have been achieved then the calculated permeability will be smaller than the correct value.

19.2.1 Manometric Method

The manometric method forms the basis of test standards ISO 2556 (1974),[7] BS 2782: Part 8 Method 821A (1979),[8] ASTM D1434 (1982),[9] and DIN 53380 (1969).[10] The quantity of gas which permeates through the test piece in a given time is measured as a change in pressure and volume. The test piece forms a barrier between two chambers in a gas transmission cell (Fig. 19.1). A constant high pressure (usually 1 atm) is maintained in one chamber and a low pressure (usually a vacuum) is initially established in the other chamber. The test piece is supported against the high pressure by a porous substrate. A manometer is coupled to the

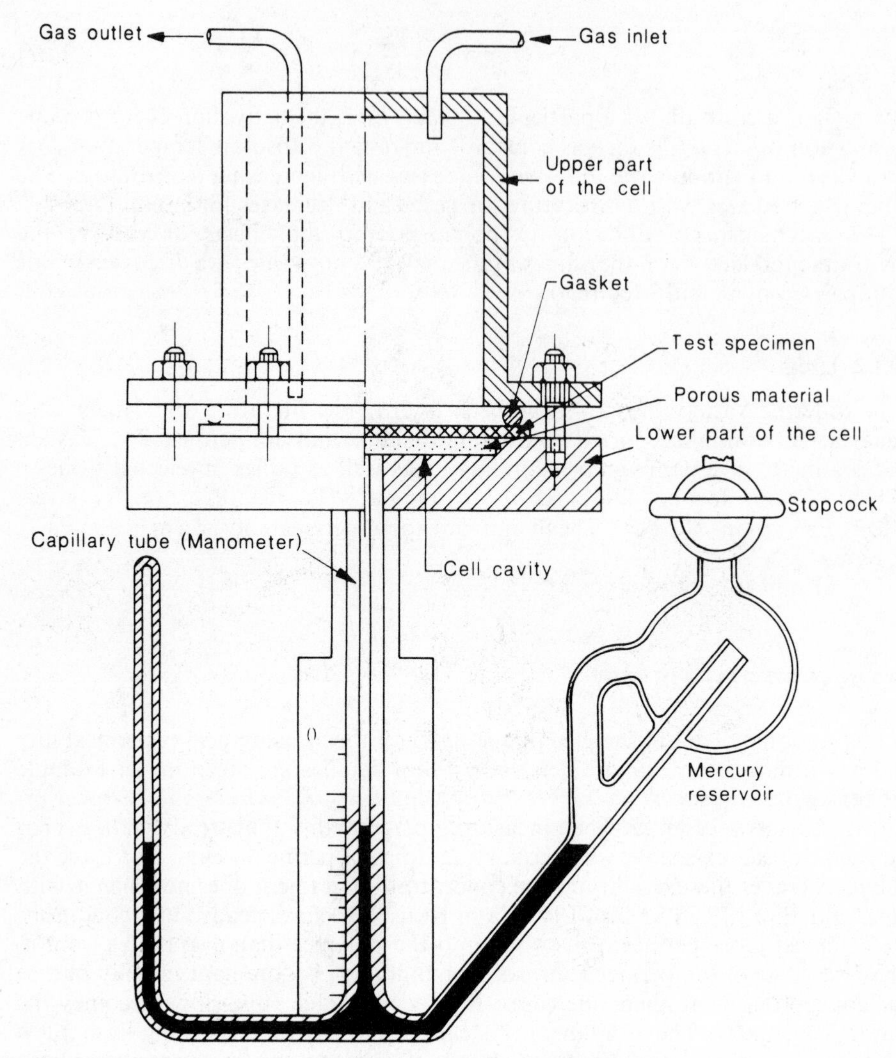

Fig. 19.1 Manometric apparatus

low-pressure chamber to measure the variation in pressure and volume over a specified test time. A plot of mercury displacement versus time gives the characteristic curve illustrated in Fig. 19.2. The initial build-up to steady state gas transfer is shown; the start of the steady state gas transfer is shown by the inflection in the curve.

From Eqn [19.8] it can be shown that the instantaneous gas transmission rate is given by

$$\mathrm{GTR} = \frac{273}{AT\Delta P}\left(P\frac{\partial V}{\partial t} + V\frac{\partial P}{\partial t}\right) \qquad [19.11]$$

where A is the permeated area, T is the absolute test temperature, ΔP is the pressure

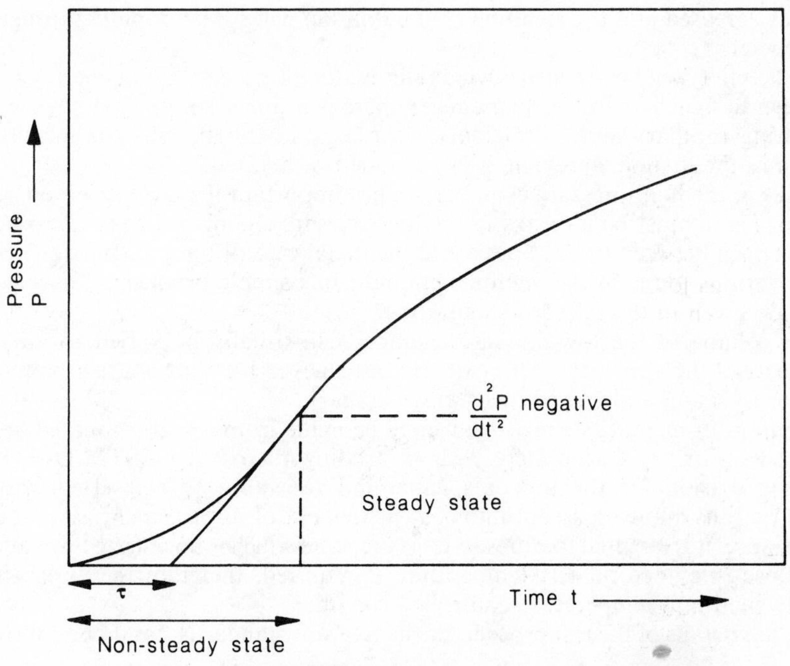

Fig. 19.2 Plot of pressure versus time

difference across the test piece, P is the instantaneous pressure in the low-pressure side of the apparatus; and V is the instantaneous volume. Equation [19.11] may be written in terms of mercury displacement:

$$\mathrm{GTR} = \frac{273}{AT(P_H - P_0 - h)}\left[(P_0 - h)a\frac{\mathrm{d}h}{t} + (V_0 + ah)\frac{\mathrm{d}h}{\mathrm{d}t}\right]$$

$$= \frac{273}{AT(P_H - P_0 - h)}[V_0 + aP_0 + 2ah]\frac{\mathrm{d}h}{\mathrm{d}t} \qquad [19.12]$$

where V_0 is the initial volume between the test piece and the datum line in the manometer, P_0 is the initial pressure in the manometer, P_H is the high pressure, h is the mercury displacement from the datum line, $\mathrm{d}h/\mathrm{d}t$ is the rate of change of mercury displacement, and a is the cross-section area of the measuring capillary.

ISO 2556 and BS 2782 use Eqn [19.12] to measure the gas transmission rate at a single point chosen on the steady state portion of the plot of mercury displacement versus time. ASTM D1434 uses an approximation to the integrated form of Eqn [19.12] for an extended time interval.

In practice it is essentially that the manometer and mercury are thoroughly clean to eliminate stick–slip effects in the movement of the mercury meniscus. A tipping manometer is frequently employed to transfer the mercury from the reservoir into the arms of the manometer after evacuation as in Fig. 19.1. During the tipping operation it is essential that the mercury does not reach the stopcock because this is always greased and the resulting contamination will spread rapidly through the manometer.[11]

A different type of manometer, which eliminates the need for tilting the apparatus, has been designed.[12] In this manometer there is a direct connection between the measuring capillary and the mercury reservoir and this permits the mercury to remain in the manometer while it is evacuated or aerated.

Leaks in the high-pressure chamber are not important if a large reservoir of gas is used. There must be no leaks in the low-pressure chamber. Leaks can occur at the interface between the test piece and the metal case of the transmission cell, or at the various joints in the manometer, and some simple producers for detecting leaks are given in the relevant standards.

The volume of the low-pressure chamber can usually be varied to alter the sensitivity of the apparatus. This can also be achieved by using a larger permeated area, or by using an inclined measuring capillary.

Readings of mercury displacement may be made from a scale mounted behind the capillary or, more accurately, with a travelling microscope. ASTM D1434 also makes provision for the use of a calibrated resistance wire inserted into the capillary, thus allowing a continuous measurement of displacement as a function of resistance; it is essential to choose a resistance wire which is not affected by mercury.

A close tolerance on test temperature is required, therefore the apparatus is usually sited in a temperature-controlled chamber.

For full details of the test procedures the relevant standards should be consulted.

19.2.2 Constant Volume Method

Although no current plastics test standard (ISO, BS, ASTM or DIN) is believed to use this method, its use in the field of research has been reported and it is also used in a test standard for rubbers.[13]

If the volume on the low-pressure side of the test piece is kept constant, Eqn [19.11] becomes

$$\mathrm{GTR} = \frac{273}{AT\Delta P}\left(V\frac{\mathrm{d}P}{\mathrm{d}t}\right) \qquad [19.13]$$

where V is the volume between the test piece and the measuring device. The steady state portion of the plot of pressure versus time may be near linear, hence the calculation of the gas transmission rate can be carried out either at a single point or over an extended time interval. The permeability coefficient may then be obtained from $\mathrm{GTR} = Q/L$ where L is the thickness. An estimate of the time required to reach steady state is obtained from the relationship $t = L^2/2D$ if the value of the diffusion constant, D, is approximately known.

The onset of steady state is indicated by an inflexion in the curve of pressure against time, similar to the results from the manometric method shown in Fig. 19.2. Strictly speaking, beyond the inflexion point conditions are again transient because the increasing pressure continuously reduces the flux from the sample. However, since the pressure change is very much smaller than the pressure drop across the sample this perturbation is usually negligible and after the inflexion point the curve of pressure against time is a straight line.

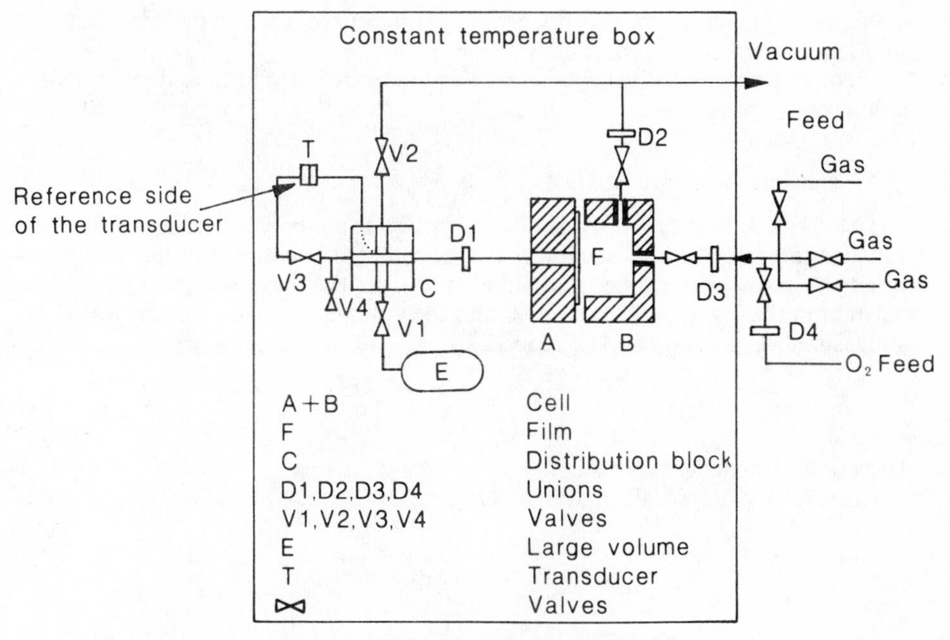

Fig. 19.3 Constant volume apparatus

The use of a differential pressure transducer instead of a manometer to measure pressure change makes the apparatus very much easier to use.[14,15] (Fig. 19.3). The reference side of the transducer is held at some arbitrary chosen pressure in the low-pressure chamber of the gas transmission cell at the start of a test. This low pressure can be in the high vacuum range, i.e. $<0{\cdot}07$ Pa ($<5 \times 10^{-4}$ mm Hg). The sensitivity range of the transducer depends on the choice of diaphragm thickness. The instrument is robust and permits operation of the apparatus at high pressures in safety. In this case high pressures from approximately 0·2–7 MPa have been employed. The volume change due to diaphragm deflection is considered insignificant in comparison with the volume of the low-pressure chamber, which is quoted at 12 cm^3. The volume of the latter can be increased by coupling with a larger vessel, to allow highly permeable materials to be tested over sufficiently long experimental times. Conversely, the high driving pressures reduce experimental times for materials of low permeability. However, it should be remembered that a material may behave differently under differing hydrostatic pressures.

An interesting practical detail of the transmission cell is the incorporation of a magnetic stirrer when the permeation rate of a gas mixture is investigated. This is used to prevent concentration polarisation effects at the surface of the test piece exposed to the high pressure.

The high-pressure chamber is flushed out with test gas prior to pressurisation. In addition, if low pressures other than vacuum are used then the low-pressure chamber and ancillary pipework must also be adequately purged with test gas to prevent bidirectional gas permeation. Leaks in this chamber manifest themselves as a falling off in the gradient of the pressure–time curve. The test piece is supported on a laminate of coarse and fine wire mesh.

A temperature-controlled cabinet houses the transmission cell and low-pressure circuitry.

19.2.3 Constant Pressure Method

ASTM D1434 (1982)[9] provides for a second method to measure the gas transmission rate of plastics, namely that of constant pressure (this technique also forms the basis of a rubber test standard[16]). In this case the pressure is kept constant in the low-pressure chamber and the volumetric change in permeated gas is measured. Equation [19.11] reduces to

$$\mathrm{GTR} = \frac{273}{AT\Delta P}\left(P\frac{\partial V}{\partial t}\right) \qquad [19.14]$$

where P = atmospheric pressure.

An example of a typical apparatus[17] is given in Fig. 19.4. The same basic design of

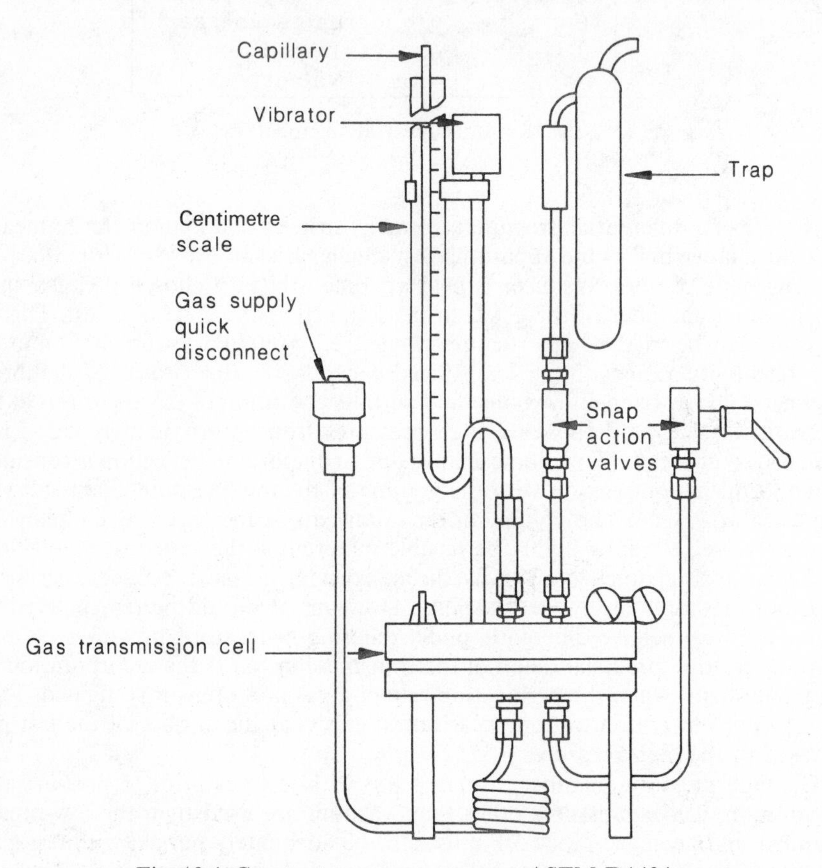

Fig. 19.4 Constant pressure apparatus ASTM D1434

gas transmission cell is used as in the manometric and constant-volume methods. The test piece forms a barrier between high- and low-pressure chambers. The test gas is maintained at atmospheric pressure on the low-pressure side of the test piece and at 0·2 MPa or greater on the high-pressure side. The volumetric change is registered by the displacement of a slug of manometer fluid, which does not dissolve the test gas, in a vertically or horizontally mounted measuring capillary. This displacement may be measured on a scale or by a travelling microscope. The bore of the capillary must be uniform to a close tolerance. Three diameters are suggested, 1 mm, 0·5 mm and 0·25 mm, for use with materials ranging from high to low permeability.

The volumetric displacement is plotted against time and the slope of the linear steady state portion is used in the calculation of gas transmission rate and permeability coefficient.

As in the previous method, it is essential that both test chambers and associated pipework be flushed out with test gas unless air is to be used. This method is very susceptible to temperature change, the capillary acting as an air thermometer. This effect is reduced by the maintenance of a close tolerance on the test temperature; a maximum variation of $\pm 0{\cdot}1\,^{\circ}C$ is required by the standard. An improvement is also obtained if the volume of the low-pressure chamber is kept to a minimum.

An alternative apparatus (Fig. 19.5) is also suitable. The main difference is that

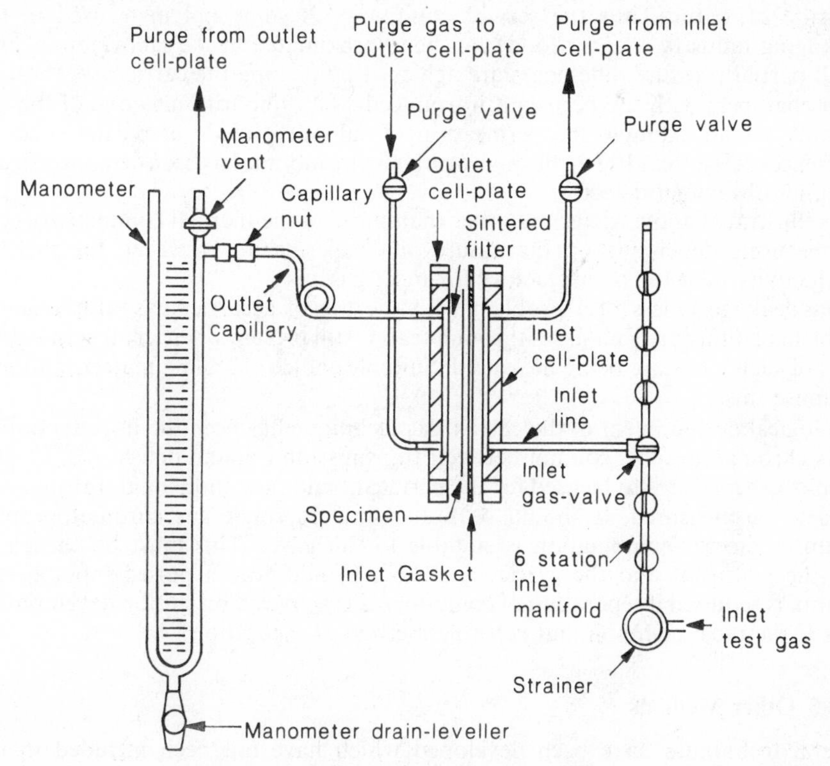

Fig. 19.5 Constant pressure apparatus using adjustable manometer: ASTM D1434

the measuring capillary is replaced by an adjustable manometer. The volumetric change is registered as a change in the height of the balanced liquid columns.

19.2.4 Carrier Gas Methods

Carrier gas techniques are commonly referred to as 'dynamic' because gas is permitted to flow across each side of the test specimen, at equal pressure. The gas transmission cell is similar to those of the previous methods, the test piece forming a barrier between two chambers. In the most basic of carrier gas methods the test gas flows at a constant rate through one chamber and a second gas, the carrier, flows through the other chamber at a constant rate. Test gas which permeates through the polymer is swept away to a detector which may be of the absorptiometric[18] or thermal conductivity type.[19,20]

An important feature is the lack of need for mechanical support of the test piece, as both gases are held at the same pressure, usually 1 atm. The polymer experiences the same hydrostatic pressure as in the vacuum manometric method. The difference in test gas partial pressure (the driving pressure) across the test piece is also the same, because the concentration of permeant in the carrier gas is kept small by suitable choice of flow rate.

Another feature of the technique is that small differences in partial pressure can be established across the polymer membrane while still maintaining a hydrostatic pressure of 1 atm. Thus the service conditions of some polymers used in the packaging industry (e.g. for food) can be approximated in the laboratory. These small partial pressure differences are achieved by passing the carrier gas through both chambers with the permeant introduced as an impurity into one of the gas streams. As an example the permeation of sulphur dioxide at partial pressure differences of less than 100 Pa through a polymer membrane has been demonstrated[21] using an ultraviolet detector.

As illustrated above, detectors other than those of the thermal conductivity type can be used, depending on the nature of the permeant. However, the thermal conductivity detector is satisfactory for most gases.

Gas leakage is less of a problem with the use of a carrier gas, although the problem of diffusion of air gases through seals is still present when small permeation rates of such gases are being measured. Suitable choice of sealing materials should minimise this.

A logical development of the carrier gas technique has been the introduction of a gas chromatographic column between transmission cell and detector.[22–25] This permits gas mixtures to be used for the permeant and their individual transmission rates to be measured. It should be noted that no single gas chromatographic column–detector combination is suitable for all gases. This must be chosen to suit the permeant and the carrier gas used. In addition, a pulsed input to the column is required if separation of components is achieved by elution development (this is the most common and versatile method of separation).

19.2.5 Other Methods

Several techniques have been developed which have not been included in the previous sections. Perhaps the most important of these from the research point of view is the use of a mass spectrometer,[26,27] as the detector. In this case a pressure differential may be maintained across the test piece by coupling the low-pressure chamber of the transmission cell directly to the inlet manifold of the instrument. The mass spectrometer offers high sensitivity and the facility to measure the permeation rates of several gases simultaneously. The cost of such an instrument naturally limits its use for such measurements, but where it serves several analytical functions such a use is a natural extension of its capabilities.

A colourimetric method[28] utilises the change in colour intensity of ammonia solution due to the formation of cupric ammonium ions by the permeation of oxygen. In this case the test piece is a produce (package) containing the solution and strands of copper. The intensity is measured with a colourimeter giving an electrical output which is a function of oxygen concentration. This method is suitable for quality control testing.

Two other techniques are based on the sorption of the permeant by the test material and hence operate under non-steady state conditions. It will be recalled from §19.1 that the non-steady state is described by Fick's second law. Of the two methods, one uses a highly sensitive torsion balance[29] to measure weight change in the polymer as gas is sorbed. The other is based on the phenomenon of radiothermoluminscence of polymers,[30] as a function of oxygen content. These last named methods determine the diffusion constant, but if the solubility constant is known or can be measured then the permeability constant may be obtained.

A radioactive isotopic method has also been used.[31] In this case the permeabilities of elastomeric membranes to carbon dioxide, labelled with carbon-14, were measured. The experimental set-up is basically the same as the constant-volume method but with a Geiger counter substituted for the pressure-measuring instrument.

19.3 VAPOUR PERMEABILITY

Methods used to measure the permeability of plastics to vapours fall broadly into two categories, namely those for water vapour and for volatile liquids. In both cases the most common measuring technique is the weight change or gravimetric technique on which several test standards are based. Other methods include techniques similar to those used for gases.

19.3.1 Water Vapour

Gravimetric Method

The transmission cell used in the gravimetric method is relatively simple, consisting of a light-weight metal dish with lid (Fig. 19.6). Depending on the procedure, a desiccant, such as calcium chloride, or water is introduced into the bottom of the dish and the test piece is sealed onto the lip of the dish using wax, which has a very low permeation rate to water vapour. The assembly is then placed in an atmosphere of controlled temperature and humidity. The weight gain or loss is measured at regular intervals of time and, when this becomes approximately constant, the rate of weight change is used to calculate the quantity of water vapour which has permeated through the unit area of the test piece in unit time.

The experimental techniques are described in ISO 2528 (1974),[32] BS 3177

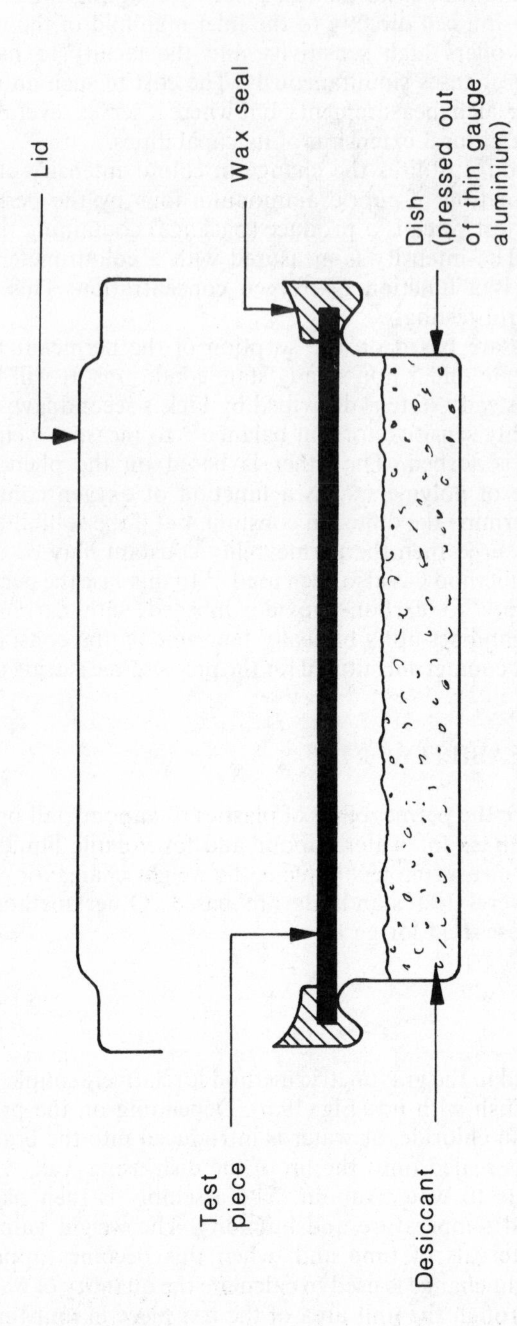

Fig. 19.6 Water vapour transmission apparatus – dish method: BS 2782 Methods 513A, B

(1959)[33] and ASTM E96–80.[34] The other British Standard BS 2782: Part 5, Methods 513A/B (1970) has been withdrawn and will shortly be replaced by BS 2782: Part 8 Method 820A and will be drafted in line with ISO 2528.

In all cases the desiccant in the dish effectively reduces the water vapour pressure to zero on one side of the test piece, while the water methods raise the vapour pressure to a maximum for a given test temperature. The water contact method specified in ASTM E96 also tests for capillary transfer as well as diffusion.

Thicker sheet materials and rigid cellular materials are accommodated in separate test standards: BS 4370: Part 2 Method 8 (1973),[35] and ISO 1663 (1981).[36]

In all the previously mentioned standards for water vapour transmission, the use of wax to seal the test piece into the dish requires care and dexterity if minute leaks are to be avoided. Such leaks are difficult to detect and it is useful to test the sealing technique using a metal template in place of the polymer.

An alternative to the dish method is that which uses the test material in the form of a sachet. This is suitable for thermoplastic materials in thin sheet form capable of being heat-sealed. The transmission of vapour through the seals must be considerably less than through the sheet. BS 2782: Part 5 Methods 513C/D (1970),[37] provided for such a test using the desiccant technique. This will shortly be replaced by BS 2782: Part 8 Method 822A.

Several definitions are applied to the transfer of water vapour through a polymer film. BS 3177 and BS 2782 define it as follows: 'permeability' is the mass transfer rate of water vapour per unit area (g/m^2 24 h) (see §19.1).

BS 3177 extends this definition, when the material is homogeneous and the transfer rate is inversely proportional to thickness, as an equivalent permeability defined as follows: 'equivalent permeability' is permeability × the thickness of the test piece in thousandths of an inch ($g\,mil/m^2$ 24 h).

The ASTM standards adopt the following definitions: *water vapour transmission rate* is the mass transfer rate of water vapour per unit area (g/m^2 24 h); *permeance* is the ratio of the water vapour transmission rate to the difference in vapour pressure between the surfaces of the test piece (g/m^2 24 h mm Hg = metric perms). This is equivalent to the gas transmission rate; *permeability* is the product of the permeance and the thickness of the test piece, assuming that the permeance is inversely proportional to thickness for homogeneous materials ($g\,cm/m^2$ 24 h mm Hg = perm-centimetre). Since the adoption of SI units the water vapour permeability may also be expressed in the units of microgram metre per newton hour ($\mu g\,m\,N^{-1}\,h^{-1}$ or $\mu g\,m\,m^{-2}\,Pa^{-1}\,h^{-1}$).

Other Methods

Refinements of the weighing technique have been made by enclosing the dish and a sensitive weighing balance in a chamber whose temperature and relative humidity are controlled.[38,39] This eliminates the need to remove the cell from its controlled environment for weighing. However, this is still a time-consuming test.

Electrical and chemical detection systems have also been employed.[38] In addition, radioactive labelling of water and its subsequent detection by liquid scintillation counter have been demonstrated.[40,41]

19.3.2 Volatile Liquids

The vapour transmission rates of other volatile liquids may also be measured from weight change data. This technique is also being considered for an international standard; a suitable weighing vessel is illustrated in Fig. 19.7. The test piece is sealed into the vessel by a screw-on lid having an aperture which defines the permeated area. To prevent distortion of the material, a metal ring separates the lid from the specimen. The ring and lid are free to rotate with respect to one

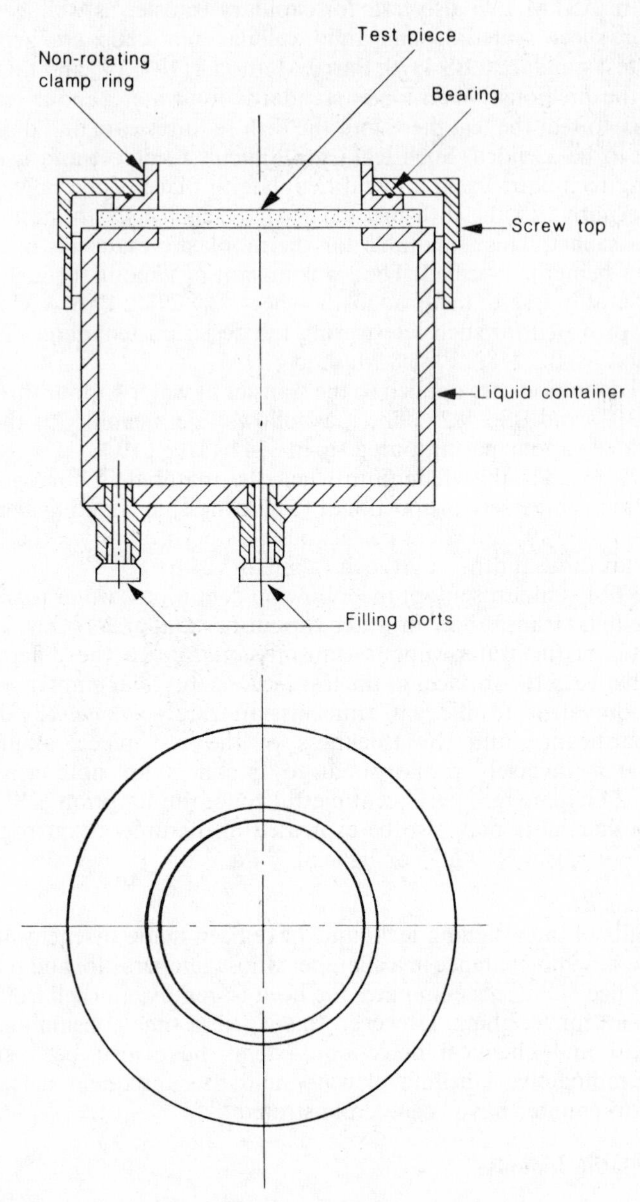

Fig. 19.7 Vapour transmission cell

another on ball bearings. The vessel is fitted with inlet and outlet ports to allow replenishment of the test liquid. This is particularly important if the liquid is made up of more than one volatile component having differing permeation rates. Thus the composition may be maintained constant throughout the test. When testing, the vessel is usually inverted so that polymer and liquid are in contact.

The result is quoted as a transmission rate with the test conditions stated. It should be remembered that some polymers swell when in contact with organic liquids, hence the thickness and the permeability are not constant throughout the test. Steady state conditions must also have been achieved before the transfer rate is calculated.

Other methods make use of the manometer[42] and gas chromatograph[43] (see also gas chromatography tests, §19.2.4[22–25]). The mass spectrometer perhaps offers the greatest potential in a single instrument for both water and other volatiles.

REFERENCES

1. Comyn, J. (Ed.) (1985) *Polymer Permeability*, Elsever Applied Science.
2. McGregor, R. (1974) *Diffusion and Sorption in Fibres and Films*, vol. 1, Academic Press.
3. Huglin, M. B. and Zakaria, M. B. (1983) *Die Angewandte Makromolekulare Chemie*, **117**.
4. Yasuda, H. and Stannett, V. (1975) *Polymer Handbook*, 2nd edn (Eds. J. Brandrup and E. H. Immergut), Wiley.
5. Lomax, M. (1980) *Polymer Testing*, **1** (2), 105.
6. Lomax, M. (1980) *Polymer Testing*, **1** (3), 211.
7. ISO 2557 (1974) *Plastics – Determination of the Gas Transmission Rate of Films and Thin Sheets under Atmospheric Pressure – Manometric Method.*
8. BS 2782: Part 8 Method 821A (1979) (1986) *Determination of the Gas Transmission Rate of Films and Thin Sheets under Atmospheric Pressure (Manometric Method).*
9. ASTM D1434 (1982) *Standard Test Methods for Gas Transmission Rate of Plastic Film and Sheeting.*
10. DIN 53380 (1969) *Testing of Plastics Film: Determination of the Gas Transmission Rate.*
11. Hems, G. (1987) Private communication.
12. Linovitski, V. (1977) *Kunststoffe*, **67** (8), 433.
13. BS 903: Part A17 (1973) *Determination of the Permeability of Rubber to Gases (Constant Volume Method).*
14. Pye, D. G., Hoehn, H. H. and Panar, M. (1976) *Journal of Applied Polymer Science*, **20**, 1921.
15. Lomax, M. L. (1987) *Polymer Testing*. To be published.
16. BS 903: Part A30 (1975) *Determination of the Permeability of Rubber to Gases (Constant Pressure Method).*
17. Sweeting, O. J. (Ed.) (1971) *The Science and Technology of Polymer Films*, vol. II, Wiley Interscience.
18. Landrock, A. H. and Proctor, B. E. (1952) *Modern Packaging*, **25**, 131.
19. Pasternak, R. A., Schimscheimer, J. F. and Heller, J. (1970) *Journal of Polymer Science*, Part A2, **8**, 467.
20. Ziegel, K. D., Frendsdorff, H. K. and Blair, D. E. (1969) *Journal of Polymer Science*, Part A-2, **7** (5), 809.
21. Davis, E. G. and Rooney, M. L. (1975) *Journal of Applied Polymer Science*, **19**, 1829.
22. Caskey, T. L. (1967) *Modern Plastics*, **45** (4), 148.
23. Kapanin, V. V., Lemanik, O. B. and Reithninger, S. A. (1974) *Vysokomolekulyarnye Soyedineniya*, **A16** (4), 911.
25. Senich, G. A. (1981) *Polymer Preprints*, **22** (2), 343.

26. Invashchenko, D. A., Krotov, V. A., Talakin, O. G. and Fuks, Ye. V. (1972) *Vysokomolekulyarnye Soyedineniya*, **A14**(9), 2109.
27. Eustache, H. and Jacquot, P. and (1968) *Modern Plastics*, **45**, June, 163.
28. Speas, C. A. (1972) *Packaging Engineering*, October, 78.
29. Volabuyev, P. V., Kupryazhkin, A. Ya. and Suetin, P. Ye. (1972) *Vysokomolekulyarnye Soyedineniya*, **A14** (2), 489.
30. Kiryushkin, S. G. and Gromov, B. A. (1972) *Vysokomolekulyarnye Soyedineniya*, **A14** (8), 1715.
31. Kirshenbaum, A. D., Streng, A. G. and Dunlap, W. B., Jr (1954) *Rubber Age*, March, 903.
32. ISO 2528 (1974) *Sheet Materials, Determination of Water Transmission Rate – Dish Method.*
33. BS 3177 (1959) *Method for Determining the Permeability to Water Vapour of Flexible Sheet Materials used for Packaging.*
34. ASTM E96–80 *Standard Test Methods for Water Vapour Transmission of Materials in Sheet form.*
35. BS 4370: Part 2 Method 8 (1973) *Methods of Test for Rigid Cellular Materials – Measurement of Water Vapour Transmission.*
36. ISO 1663 (1981) *Plastics – Determination of Water Vapour Transmission Rate of Rigid Cellular Plastics.*
37. BS 2782: Part 5 Method 513C/D (1970) *Water Vapour Permeability using Sachet of Material under Test.*
38. Leslie, H. J. (1973) *Plastica*, **26** (12), 544.
39. Comyn, J., Cope, B. C. and Werrett, M. R. (1985) *Polymer Communications*, **26**, 294.
40. Matsui, E. S. (1970) *US NECL Technical Report R674*, April, p. 33.
41. Mozisk, M. (1979) *Gummi und Asbest Kunststoffe*, **32** (11), 856.
42. Izydorezyk, J., Podkowka, J., Salivinski, J. and Grzyuna, 2. (1977) *Journal of Applied Polymer Science*, **21**, 1835.
43. Duskova, D. (1974) *Plasty a Kautcuk*, **11**(3), 72.

Chapter Twenty

Non-destructive Testing

20.1 INTRODUCTION

The essential attraction of non-destructive testing (NDT) is that, by definition, the part is still intact when the testing is complete. Hence its first application is where destructive testing of large expensive products is impracticaly, and this includes particularly the in-service inspection of critical components to detect the onset of failure.

NDT may also have the advantage of being able to detect faults which destructive tests may miss. This is the case when destructive methods would be used on discrete samples or positions whilst an NDT technique would examine the whole product. This indicates a further potential advantage of NDT in that by avoiding test piece preparation, etc., and by automating the process, testing may be very rapid and hence inexpensive thus allowing, for example, 100 per cent inspection for quality control purposes.

Against these advantages there is one big problem: the most powerful NDT techniques are either expensive or extremely expensive in terms of capital investment to the extent that their use for many products would simply not be cost effective. However, before rejecting NDT on cost grounds it is worth reflecting on the fact that many of the more traditional tests are essentially non-destructive. In fact we all use NDT in making dimensional measurements. Testing of electrical and optical properties, many dynamic and thermal tests are other examples. Any mechanical proof test which does not produce permanent change in the material is non-destructive and clearly there is still much scope for the imaginative application of relatively inexpensive tests indirectly to assess the integrity and structure of plastics.

Because of cost restrictions, it is not surprising that the more sophisticated NDT techniques have been primarily applied to plastics for critical applications, in particular for composites used in the aerospace industry. Indeed, the application of glass and carbon fibre laminates to critical structures has provided the impetus for the development of techniques particularly suited to these materials. Plastics welding and adhesive jointing have also provided areas of interest. The metals industry has long used NDT on a routine basis and it is frequently called for in product specifications. As the application of plastics expands to fill roles traditionally the preserve of metals it would not be unreasonable to assume that NDT requirements for plastics products will become more commonplace.

To a greater or lesser extent, all components have faults; it is only where they are of a size or number to detract significantly from the service performance of the component that it is necessary to detect them. This critical flaw size must be calculated considering the service conditions and the design safety factor, as it will set the sensitivity requirement for the NDT technique to be used.

When applying NDT it is also essential to know the nature and likely location of the fault to be found. Faults tend to fall into one of three classes; first, voids or holes of macroscopic dimensions; second, discontinuities where the separation of the surfaces is extremely small, and finally, areas where there is no discontinuity but where the mechanical properties of the material have been degraded.

Typical examples of the first class are voids in the thick sections of mouldings and gaps in thick adhesive layers. Examples of the second class are lack of 'wetting out' in GRP components and incomplete fusion of welds. In the third class, orientation of thermoplastics and lack of cure of thermosets are typical. A fault not included in the above classes is the misplacement of reinforcing members, e.g. resin-rich areas in GRP.

The NDT technique selected must suit the need, i.e. it must detect the type of fault expected to the required level of sensitivity and must be practically applicable to the component in question. Unfortunately for the non-specialist there are a number of NDT techniques and many variants. Their scope of application and relative efficiency is not always easily apparent and to many the situation is confusing.

Indeed, the problem with writing a chapter on NDT is that it cannot do the subject full justice – this would require a complete book. There are no international or British test methods for NDT of plastics to reference and although the range of industrial applications may be relatively narrow, a great many techniques and applications have been considered. Consequently, discussion here will be restricted to sources of published information and brief outlines of major techniques. Tests which are essentially non-destructive but which are covered elsewhere in the book will not be considered further.

20.2 OVERVIEW OF NDT METHODS

The literature on non-destructive testing is vast and even in the more restricted reference to polymers there are hundreds of published papers. Amongst these are a number of reviews of the techniques available and consideration of newer developments. Reviews of applications to composites[1–8] are most common, reflecting the fact that these materials are the polymers most frequently tested non-destructively. Reynolds[4] lists 155 references, Harris and Phillips,[5] 60, Scott and Scala,[3] 82 and Lemascon[6] (in French), 42. Application to polymers more generally are reviewed by Berger[9] and Geerans[10] (in Dutch) and to tyres by Trivisonno[11] (149 references). Consideration to recent developments for plastics is given by Eyerer[12,13] (in German), the increased range of applications of NDT is reviewed by McClung[14] and the future path of NDT developments is predicted by Posakony.[15]

The major, long-established NDT techniques are usually given as magnetic, dye penetrants, eddy current, radiography and ultrasonics. Magnetic methods are only applicable to ferromagnetic materials and hence not useful for plastics. Dye-penetrant testing, whilst very commonly used on metals, is not often applied to plastics. The technique detects cracks on the surface only and could be said to be an extension of visual inspection. Where the detection of fine surface cracking is important this simple but sensitive method could well be more widely used. Eddy current testing is basically a surface method although the penetration may be a matter of millimetres and is material-dependent. The material is required to be conducting and hence application to plastics is restricted. However, the technique has been applied to carbon fibre laminates to detect delamination[16] and to indicate variations in fibre lay up.[17] It is also used to measure the thickness of coatings on metals.

Ultrasonics and radiography are, by contrast, techniques which allow the detection of defects which occur within the bulk of the material. The sound or radiation passes through the material and defects are detected by differential absorption or reflection. Both have found considerable application for polymers and will be outlined in subsequent sections.

In terms of the extent of applications or the number of published papers, acoustic emission and optical methods are now also major techniques for polymers. Acoustic emission occurs when the material is stressed, defects giving rise to characteristic stress waves which can be detected and analysed. The method is primarily applied to composite materials. Optical methods using coherent light also rely on a degree of stressing of the material so that interference occurs between images of the article stressed and unstressed, variations in the pattern indicating possible defects or uneven construction. There is now a somewhat bewildering number of optical methods. Some of these, and acoustic emission are outlined in subsequent sections.

Thermography is now also a well established technique although it does not as yet appear to have realised its full potential. It has in a sense a similarity to coherent light techniques in that effects at the surface resulting from variations in structure internally are detected. In this case variations in surface temperature arising from thermal or mechanical induced heating are measured.

It has been pointed out that measurement of dielectric properties is non-destructive and as these properties will vary with, for example, state of cure, hence these electrical tests can be used to monitor the crosslinking process as, for example, described by Kranbuehl *et al.*[18] In principle a similar approach could be taken no monitoring the progress of any chemical or mechanical process which affected dielectric properties. Spark testing can be used to detect pinholes in film or even welds.[19] The mapping of electrostatic charge decay has also been suggested[20] to detect flaws.

Microwave radiation of a millimetre wavelength has been used to detect voids in GRP structures.[21] The sensitivity of this technique is limited by the wavelength used, although it could possibly be useful for large voids and thickness measurements.

Beta-radiation gauges are in common use as on-line thickness gauges for some continuous processes. Their ability to provide a measure of mass per unit area enables measurement of porosity, wall thickness, etc., to be made. Damage in a material is associated with a drop in stiffness. Adams[22] describes a vibrational technique applied to carbon fibre reinforced plastics which analyses the changes in natural frequencies to detect broken fibres and delamination.

The *Quality Technology Handbook*[23] edited by the National NDT Centre at Harwell gives information on organisations supplying NDT equipment and services, standards, certification authorities and training organisations.

20.3 ULTRASONIC TESTING

Ultrasonic testing comprises a range of methods which make use of mechanical oscillation at frequencies above 20 kHz. These waves behave in a similar manner to light: they obey the laws of reflection and refraction. The *Quality Technology Handbook*,[23] for example, gives an account of the physics and applications to ultrasonics when applied to materials testing.

Ultrasonics is used to find imperfections in a component, to measure mechanical moduli and to measure thickness. In flaw detection, pulses of ultrasound are deflected from boundaries to build up a 'picture' of the interior of a component. These pulses are produced and received by transducers which are normally piezoelectric crystals. With this type of transducer, it is possible to use the same transducer for both transmission and reception. Figures 20.1 and 20.2 show various configurations of transducers for flaw detection.

There are two basic types of sound wave: (a) longitudinal (compression), where the mechanical energy is transmitted through the material, exciting the molecules in the direction of the sound wave; and (b) transverse (shear), where the excitation is normal to the wave.

For a given material the two types of wave have different values of velocity and attentuation. The shear wave has a lower velocity and receives much higher attenuation. In addition there are Rayleigh waves which propagate on the surface of the material and Lamb (or plate) waves which occur in thin sheets.

Compared with metals, plastics have low sound velocities. Table 20.1 gives values for some common materials. Attenuation of the shear wave in plastics is very high. hence it is rarely used for flaw detection. Some of the more flexible materials, such as LDPE and plasticised PVC, will not support a shear wave.

An important property is acoustic impedance, which is the product of the sound velocity and the density of the material. Again, Table 20.1 gives some values.

When a sound wave hits the boundary of two materials, the proportion of its energy which is reflected is a function of the acoustic impedance of the two materials:

$$R = \frac{(W_1 - W_2)^2}{(W_1 + W_2)^2} \qquad [10.1]$$

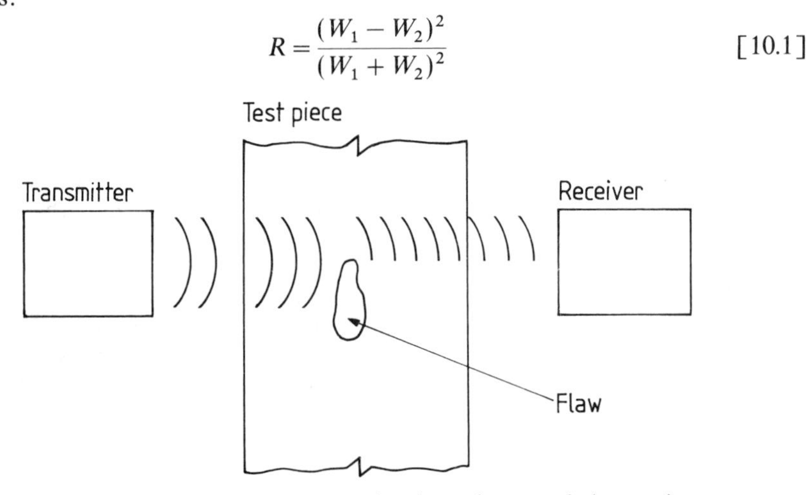

Fig. 20.1 Transducer arrangement for through transmission testing

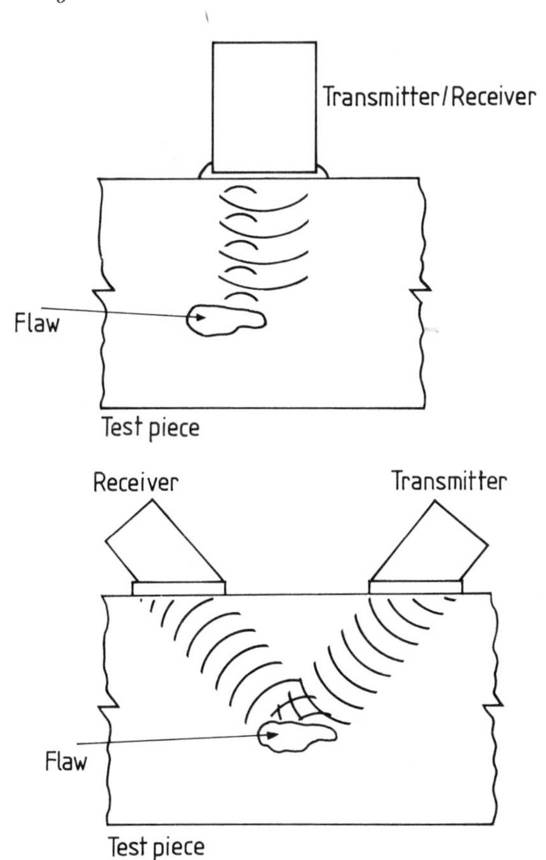

Fig. 20.2 Transducer arrangements for pulse echo testing – one and two transducers

where R is the reflection coefficient, and W_1 and W_2 are the acoustic impedances of the two materials.

When the boundary comprises a gas and a solid, reflection within the solid is virtually 100 per cent even if the thickness of the gas normal to the wave is microscopic. Generally a liquid-coupling material is required between the transducer and the test object.

Once the reflected pulse is received from the flaw, it must be displayed to the operator in a convenient form. The methods of achieving this are called 'scans' and the three most common are termed 'A', 'B' and 'C'.

'A' scan is the most simple and commonly used. The received pulse is displayed on a cathode-ray tube as an amplitude against time, the time sweep being started by the transmitted pulse. The displacement of the flaw echo from the start of the trace gives the depth of the flaw, and the height of the pulse gives some indication of the size of the flaw normal to the beam. A typical 'A' scan trace is shown in Fig. 20.3.

When using 'A' scan, the operator moves the transducer over the component's surface and notes its position when a flaw is found.

Table 20.1 Acoustic properties of some common materials

Material	**Acoustic velocity (longitudinal wave)** ($m/s \times 10^3$)	**Acoustic impedance** ($kg/m^2 s^{-1} \times 10^6$)
Aluminium	6·32	17
Brass	3·83	33
Glass (crown)	5·66	14
Steel	5·90	46·5
ABS	2·0	2·2
Acetal	2·4	3·6
Nylon 66	2·6	2·9
HDPE	2·4	2·3
Polycarbonate	2·0	2·4
PMMA	2·74	3·2
Polypropylene	2·4	2·1
Polystyrene	2·3	2·4
Rigid PVC	2·3	3·2
Poly(4-methyl pentene)	2·0	1·66
PTFE	1·5	3·2

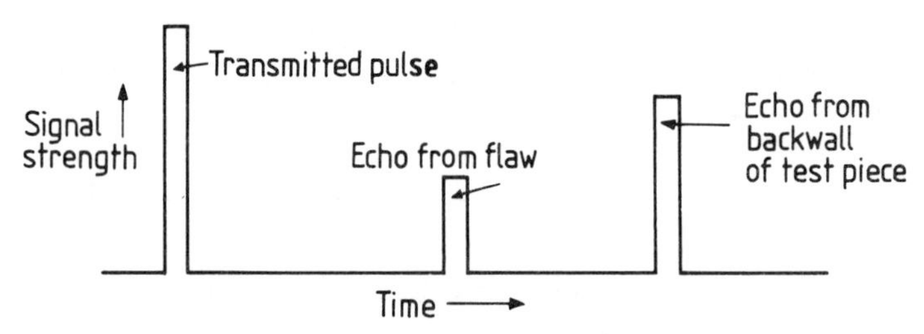

Fig. 20.3 'A' scan oscilloscope trace from a pulse echo test

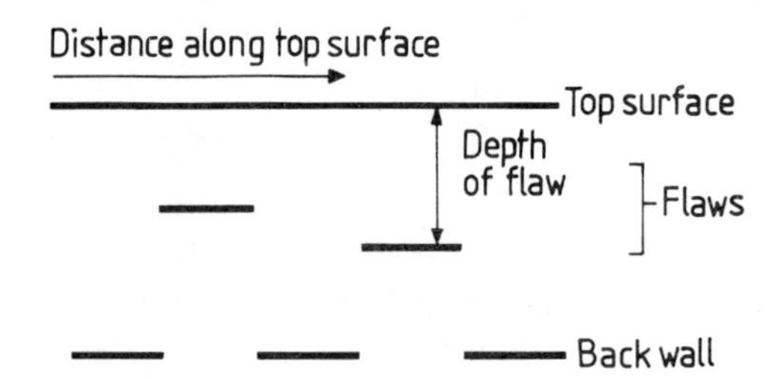

Fig. 20.4 'B' scan recorder output from a pulse echo test

'B' scans give a display representing a section through the object so that flaw depth and position along the line of scan is shown (Fig. 20.4). 'C' scans give a map view of the component showing flawed areas (Fig. 20.5). These methods require the transducer to be fitted to a mechanical scanning device and normally

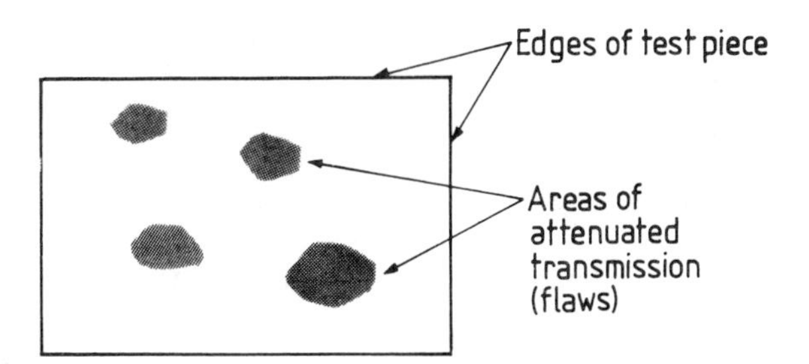

Fig. 20.5 'C' scan recorder output from a through, transmission test

present their display on paper. Both 'B' and 'C' scan systems are well suited to automatic inspection methods.

There are several variations on the way in which the signal can be displayed and the use of a microcomputer allows both increased sophistication and greater flexibility. It is also possible to produce ultrasonic holograms by optically reconstructing the interference of ultrasonic beams and to produce the ultrasonic equivalent of tomography (§20.5). Ultrasonic spectroscopy involves analysis of the frequency distribution of ultrasonic signals.

Ultrasonic testing has been applied to a variety of flaw detection and thickness measurement problems. The 'C' scan method has been used successfully to detect areas of voids in fibre reinforced plastics components.[24–27] For the inspection of welds in thermoplastics, the 'A' scan method has proved to be of value.[28–30] Recent papers include consideration of wall thickness measurement,[31] 'C' scan analysis of filament wound spheres,[32] disbonds in laminate/metal joints,[33,34] coating adhesion[35] and inspection of plastic gas pipes.[36] The ultrasonic impedance plane method[37] and the ultrasonic stress wave factor[38,39] have been applied to composites.

Ultrasonic spectroscopy, in which the frequency content of the received pulse is examined, has been found to be capable of distinguishing areas of delamination in fibre-reinforced plastics,[40,41] and has been used to characterise thermoplastics.[42]

The velocities at which the longitudinal and transverse waves travel through a material are determined by the elastic constants of that material. Measurement of the velocities allows the calculation of all the moduli and Poisson's ratio.[43] This is of particular importance when dealing with composite materials, as the ability to direct the ultrasonic beam through the material along different axes allows the calculation of the complete set of elastic constants.[44–46] This is also useful in assessing the degree of anisotropy in mouldings.[47] The results obtained from this method are not normally comparable to conventional testing because of the frequency-dependence of the moduli; however, agreement has been found in measurements on epoxy resin.[48] Even PMMA filled with iron spheres has had moduli measured ultrasonically.[49]

Applications of ultrasonic techniques go beyond conventional flaw detection and moduli determination. Measurement of ultrasonic velocity and/or attenuation has been used in degradation studies,[50,51] estimating mechanical damage,[52] monitoring polymerisation,[53,54] measuring dispersion in a polymer melt,[55] as an indication of melt fracture,[56] to characterise IPNs[57] and to measure density.[58]

Sonic, rather than ultrasonic, frequencies can also be used for non-destructive testing and have been applied to structural changes in PETP fibres,[59] and to the characterisation and fibrilation of polypropylene films.[60,61]

Ultrasonics clearly has a very wide range of applications to polymers although its use on a routine basis is still relatively restricted. In practice the techniques are not always easy to apply and may require considerable operator skill. The high attenuation of many plastics is a problem and the complex shapes of most plastics products adds to the difficulty of placing and coupling the transducer.

20.4 ACOUSTIC EMISSION

Certain materials, when subjected to stress, exhibit the phenomenon of acoustic or stress wave emission. This is the release of stress wave pulses, in a broad frequency range, due to deformation and fracture. The frequency range from 50 kHz to 1·5 MHz has been found to be most useful.

Because some deformation must occur to produce acoustic emission, the method is not wholly non-destructive; however, if this deformation and subsequent emission are produced by the first loading of the product, and loads are kept to those used in service, no extra damage will occur.

The stress waves emitted from the component are detected by one or more sensitive piezoelectric transducers. The electrical signal produced is amplified and passed to the signal analysis equipment. The analysis equipment characterises the emission in terms of pulse count, height, length, energy content, frequenty content, etc.

An important factor in the use of acoustic emission is the Kaiser effect, which is simply the observation that loadings of the component which cause no further damage are free from acoustic emission, i.e. when repeated loadings to a fixed stress level are performed, only the first loading will produce real emission. This provides a means of determining whether the emission is a genuine measure of the damage on the component or is a spurious emission due to frictional sources.

Figure 20.6 shows the total stress wave emission count and its associated load

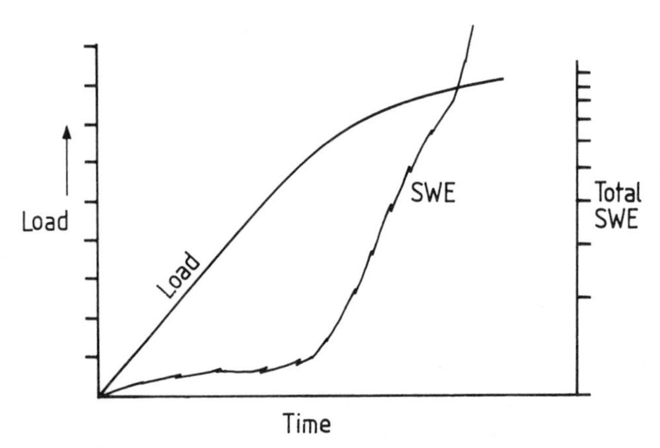

Fig. 20.6 Typical acoustic emission output showing load and total stress wave emission versus time

curve against time. The total emission begins to rise sharply as the component approaches its critical load point. Finding this point enables a proof load to be determined which, when applied to faulty components, will give rise to characteristic emission evidencing poor quality.

The great majority of acoustic emission studies have been performed on composite materials. The first practical application was perhaps pressure testing of GRP Polaris Missile chambers by Green *et al.*[62] Since then there have been a great many papers covering fibre reinforced composites of various types and forms. The majority of these are concerned with laboratory studies and there is no universal agreement as to the consistency of the technique in practical applications. Cole[63] gives the basics of the technique and considers established and potential applications. Theory is outlined and instrumentation described by Mitchell and Taggaet[64] who also discussed the test applied to fibre glass bucket truck booms. Application to automotive composites,[65] continuous filament composites,[66] proof testing of a graphite/epoxy dome,[67] testing large reinforced plastics structures,[68] rejection criteria for FRP vessels,[69] prediction of long-term performance of pipes,[70] and analysis of acoustic emissions[71] represent recent reviews and studies.

Relatively few studies have been reported on polymers without fibre reinforcement. Shirouzu *et al.*[72] give results of measurements on a range of materials and contrast amorphous and crystalline polymers. They also tested notched PMMA specimens during crack propagation.[73] Nishiura *et al.* investigated the fracture mechanism[74] and crack growth in PMMA.[75]

20.5 RADIOGRAPHY

The discovery by Röntgen of X-radiation early in this century has provided non-destructive testing with one of its most useful tools.

Electromagnetic waves in the visual part of the spectrum have little penetration power – hence the opacity of most of the objects about us. The wave energy is reflected from, or absorbed by, the object. However, as we move beyond the visible spectra through the ultraviolet, the energy of the emitted photon increases. After passing through the far ultraviolet we reach the X-ray spectra with wavelengths between 500 and 0.1 Å. These waves are highly penetrative, the shorter wavelength more so than the longer.

The amount of X-ray attenuation within a material is a characteristic of that material. Hence, just as it is possible to see voids within a translucent material, so, with X-rays, it is possible to study the interior structure of an object.

X-rays are produced electrically by means of a tube similar to a common cathode ray tube. Whereas the domestic television set uses accelerating potentials of around 9 kV, X-ray tubes use potentials between 10 and 800 kV to accelerate the electrons. In the television set, the accelerated electrons strike the phosphor of the screen and cause visible light to be emitted; the higher energy electrons in the X-ray tube gives rise to the higher energy X-ray when they hit a metal plate called the 'target'. The X-radiation from this target covers a broad band of frequencies. Different frequencies are attenuated differently by a material, and different materials have different attenuation characteristics. For this reason, the choice of accelerating potential is important. Compared with steel, plastics are quite transparent to X-rays.

In order to detect fine detail in the structure of plastics components, it is usually necessary to use low energy. To enable a tube to produce these low-energy X-rays, it must possess a beryllium 'window' through which the radiation can emerge, since the usual glass tube envelope absorbs these energies. The final selection of accelerating potential is made by trial on the component to be tested with a view to obtaining the most detailed picture possible.

As the human eye is not able to see X-radiation, a means must be provided to view the picture produced. Normally a photographic film is used which is exposed to the X-radiation emerging from the item under test. The disadvantages of film are its high cost and its lengthy processing time. Systems are produced which do not have these problems; the fluorescent screen, which emits visible light when struck by X-rays is the most popular. Such systems, especially when linked to closed-circuit television, provide an efficient and rapid solution, and can be combined with image enhancement techniques.

Unlike photography, no lenses are used in radiography hence the geometry of the test set-up is important. The general form of an X-ray system is shown in Fig. 20.7. The X-ray source may be regarded as a point. The X-rays emerge from the window and diverge. The film is placed immediately behind the object. The film-to-source distance (FTSD) is important in determining the resolution of the radiography. The higher the ratio of FTSD to object thickness, the higher will be the resolution of the radiography. Ratios of 5:1 are adequate for detecting large voids and for gross inspection, 10:1 is suitable for detecting small voids in mouldings. Ratios of the order of 100:1 are required for optimum resolution of very small flaws in critical components.

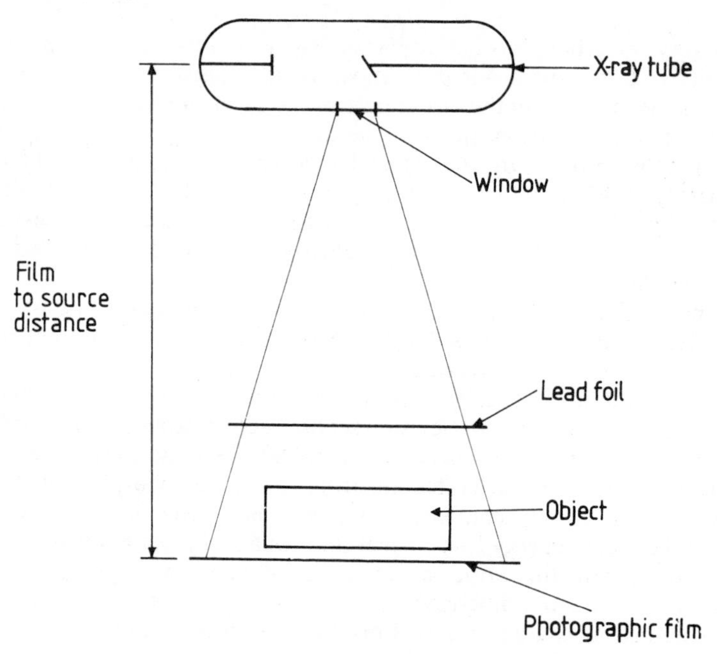

Fig. 20.7 General form of an X-ray system

A further criterion which determines the X-ray system required is the maximum size of object to be inspected. For large objects, e.g. pressure vessels, chemical plant, aircraft structures, a mobile X-ray source in a shielded room with remote controls is required. For smaller items (up to about 760 mm square) a fixed source in a shielded cabinet will usually suffice. This cabinet-type system has the advantage that it requires little installation, operators require the minimum of training and the units are inherently safe.

Gamma-radiation is essentially the same as X-rays but originates from radioactive isotopes. Gamma sources are often used in the metals industry as an alternative to X-radiation especially where portability is a requirement but are less suitable for plastics because of the higher radiation energies produced.

The contrast in a radiograph is formed by differential absorption of the X-radiation by materials of different atomic number. Hence, voids and metal inserts can be readily seen but there is likely to be difficulty in obtaining sufficient contrast to distinguish different organic materials or discontinuity. Perhaps because radiography is quite straightforward for simple problems such as voids but has definite restrictions for more difficult flaws, there are relatively very few papers on application to plastics (although rather more for rubber tyres). X-ray testing of carbon fibre laminates has been discussed by Altmann and Lembke,[76] determination of glass content by Schmeling and Sandell[77] and fibre orientation and content by Cherek.[78]

Computed tomography is a technique whereby a picture of one plane in the object is obtained, the information from all other planes being eliminated. The picture is constructed using a narrow pencil of radiation which is moved relative to the object to produce a number of images which are then processed by computer. Originally developed for medical needs, the technique has been described[79] and applications to polymers discussed[80] by Persson and Ostman. They indicate that much better resolution can be obtained than with conventional radiography.

Neutron radiography is the same in principle as X-radiography but the attenuation mechanisms are very different. Neutrons are heavily attenuated by hydrogen-containing materials but not by much heavier materials such as steel. Hence it is possible to investigate plastics components even when they are surrounded by metal. Spowart[81] has reviewed the technique, its application to adhesive bonds has been considered by Sancaktor[82] and its application to composite structures discussed by Dance and Middlebrook.[83]

20.6 OPTICAL METHODS

20.6.1 Visual Inspection

The non-destructive testing technique most ignored or taken for granted is visual inspection – the oldest technique and the one that utilises the finest item of non-destructive apparatus available, the human eye. Visual inspection provides an assessment of shape, surface finish, surface flaws and colour and especially when augmented by instrumental optical techniques (see Ch. 14) or dimensional measurement (see Ch. 7) can give a comprehensive picture of the quality of a component.

The least that is required is that the inspection area is clean, uncluttered and well lit and that the operator is comfortably situated. Lighting should be strong but not throw flare at the inspector. Failure to provide these basic requirements will inevitably result in substandard production being passed. For small components an illuminated magnifier is recommended and for some purposes a low-powered microscope or projection microscope may be necessary or an advantage. For inspection of cavities and bores, etc., various introscopes, fibre optical devices and even miniature television cameras are available. Dye penetrants enable very fine cracks or porosity to be identified.

Unfortunately, visual inspection is labour-intensive and can easily suffer from poor efficiency due to tired or casual operators. There are now automated systems capable of monitoring size, displacement, etc., and may incorporate computerised image processing and pattern recognition, but such instrumentation is expensive. The other, obvious limitation to visual inspection is that it does not directly detect any internal defect (expect in transparent materials). However, in a number of cases even internal defects in plastics mouldings can be deduced from the surface condition.

20.6.2 Photoelasticity

As discussed in Ch. 14, many plastics are optically active when stress is applied. When they are examined with polarised light, the direction and magnitude of the stress within the material may be deduced. Apart from the straightforward case of examination of a transparent moulding to assess the residual stresses, a technique has been developed to examine the strain distribution, in use, of opaque and optically inactive materials. A thin coating of optically active material is applied to the component in the unstressed condition, the test loading is applied and the surface coating is examined using reflected light. Instruments called 'photoelastic reflection polariscopes' are available to measure the strains in the structure under test.

20.6.3 Coherent Optical Techniques

The development of the laser allowed the practical use of a number of well known optical phenomena. The ability of the laser to produce an intense, parallel and coherent beam of monochromatic light made feasible the application of interferometry techniques to component testing. The measurement of small displacements using light wave interference, e.g. by the Michelson interferometer, has been known for many years. The technique of holography, the storage of three-dimensional information on photographic medium, was also well known but the lack of a suitable light source prohibited its application. The laser enables these two techniques to be applied together to provide the method of holographic interferometry. Using the optical arrangement shown in Fig. 20.8, a photographic plate is partially exposed with the object in the unstressed condition, the component is stressed and the plate further exposed. The plate now contains three dimensional information on the component in both the stressed and unstressed states. When the hologram is reconstructed using the same optical arrangement but with the component removed, two superimposed images of the component are formed. Where the images are different, i.e. where the stress has caused movement of

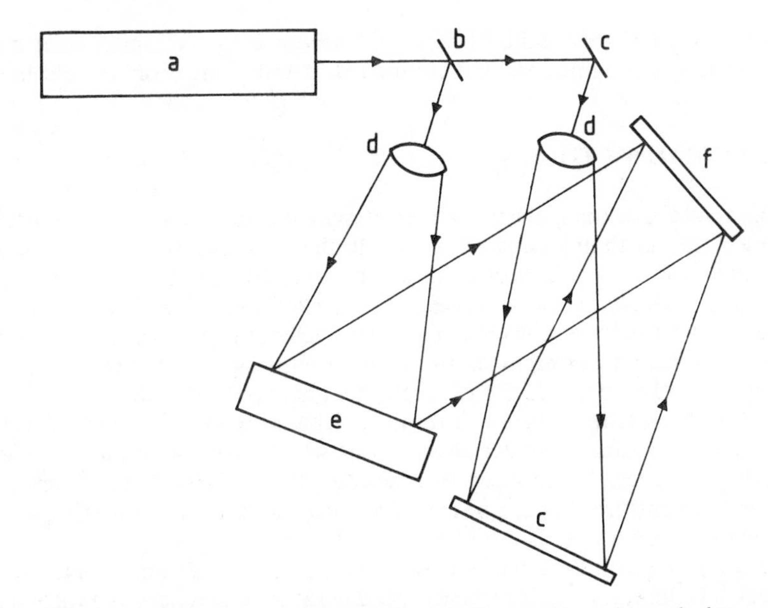

Fig. 20.8 Optical arrangement for holographic interferometry: a, laser; b, beam splitter; c, mirror; d, lens; e, object; f, photographic plate

the surface, an interference fringe pattern is formed, the density of the fringes indicating the amount of displacement. This is known as 'lapsed time holographic interferometry'.

The technique of real time holographic interferometry uses the same basic optical arrangement but the photographic plate is developed after the first exposure and the resulting image is projected onto the component as the stresses are applied, hence the interference pattern is formed in real time.

Both techniques require very stable optical and mechanical arrangements, real time holography being the more critical as gross movement between the components of the system must be limited to one-eighth of the wavelength of the light. The test stress may be applied mechanically or thermally.

Where there is an interest in the effect of resonant vibration within the component, the technique of time-average holographic interferometry, as described by Stetson and Powell[84] and Stetson and Taylor,[85] can be utilised.

When an object is imaged with coherent light a speckle pattern results caused by random interference from scattered light. The interference of speckle patterns produced before and after movement of the object produces a series of fringes. These fringes represent lines of constant displacement in the plane of the surface and hence allow a measurement of surface strains.[86,87]

Applications of these techniques to plastics are wide ranging and there is a considerable volume of literature. However, it is not too clear which applications have been found to be practically viable in industry. Examples of reported studies are detecting faults in laminated fuel tanks,[88] investigation of crazing,[89,90] disbonding in sandwich plates,[91] flexural stiffness of GRF,[92] evaluations of structure solids (PMMA and PU foam)[93] adhesive joints[94] interfacial gaps in GRP,[95] adhesion of laminates and chipboard,[96] and elongation at weld lines.[97]

20.7 THERMAL METHODS

Thermography is essentially the measurement of temperatures to reveal characteristics of an object, rather than having an interest in the temperature itself. Normally the surface temperature is scanned using an infrared non-contact detector and the temperature profiles viewed via closed-circuit television.

There are essentially two ways of producing temperature rises in the object: by applying heat from an external source or by generating heat by cyclic mechanical deformation of the object. If, for example, heat is supplied to the back of a sheet sample an uneven temperature will result, at least initially, on the front surface if there are any structural or material variations which give rise to different thermal transport properties. When heat is generated through mechanical action, the maximum temperature rises can be expected where stresses are highest, for example in the vicinity of flaws.

The exact procedure used can be chosen to optimise the detection of the features of interest. The limitations are clearly the temperature differences and their spacing that can be detected as well as the magnitude of the effect on the surface temperature of any flaw. Sensitivities of 0·1 °C are claimed and resolutions of 1 mm.

As observed for other NDT techniques, the vast majority of published papers on thermography applied to plastics refer to composites, although Florin[98] discusses application to the plastics industry generally. Studies on composites[99–105] include both methods with external application of heat and by mechanical vibration. Detection of polymer/metal adhesion failures was studied by Dom *et al.*[106] and void detection in polyurethane foam by Clarke and Mack.[107]

Thermal waves induced by optical radiation have been used to study composites[108,109] and a relatively simple and inexpensive way to monitor temperature distribution is by the use of liquid crystals.[110,111]

REFERENCES

1. McGowen, P. (1986) *Standard News*, **14** (7).
2. Puttick, K. E., *Short Fibre Reinforced Thermoplastics Conference, Uxbridge, 17th–18th June 1985*, paper 7.
3. Scott, I. G. and Scala, C. M. (1982) *NDT International*, **15** (2).
4. Reynolds, W. N. (1985) *SAMPE Quarterly*, **16** (4).
5. Harris, B. and Phillips, M. G. (1983) *Developments in GRP Technology*, vol 1, Applied Science Publishers.
6. Lemascon, A. (1984) *CETIM Informations*, no. 84.
7. Vipond, R. and Daniels, C. J. (1985) *Composites*, **16** (1).
8. Kulkani, S. B. (1983) *Machine Design*, **55** (9).
9. Berger, H., *ACS Rubber Div. 119th Meeting, Minneapolis, June 2nd–5th, 1981*, paper 44.
10. Geerans, H. A. (1984) *Kunstchaff En Rubber*, **37** (10).
11. Trivisonno, N. M. (1985) *Rubber Chemistry and Technology*, **58** (3).
12. Eyerer, P. (1985) *Kunststoffe*, **75** (10).
13. Eyerer, P. (1985) *Kunststoffe*, **75** (12).

14. McClung, R. W. (1986) *Standard News*, **14** (7).
15. Posakony, G. J. (1986) *Standard News*, **14** (7).
16. Altmann, O., Winter, L., Rucker, D. and Schroeder, G. (1985) *Kunststoffe*, **75** (6).
17. Prakash, R. and Owston, C. N. (1976) *Composites*, **7** (2).
18. Kranbuehl, D., Delos, S., Yi, E., Mayer, J., Jarvie, T., Winfree, W. and Hon, T. (1986) *Polymer Engineering and Science*, **26**, no. 5.
19. Forbes, K. (1978) *Anti-corrosion Methods and Materials*, **25**, 5.
20. Ming-Kie, Tse and Suh, N. P. (1981) *SPE ANTEC Conference, Boston, May 4th–7th*, paper 012.
21. Botsco, R. J. (1963) *Plastics Design and Processing*, November 12.
22. Adams, R. D. (1984) *Plastics and Rubber Processing Applications*, **4** (2).
23. *Quality Technology Handbook* (1984), 4th edn, Butterworth Scientific.
24. Stone, D. E. W. and Clarke, B. (1975) *Non-destructive Testing*, June, 137.
25. Jones, B. R. and Stone, D. E. W. (1976) *Non-destructive Testing*, April, 71.
26. Martin, B. G. (1976) *NDT International*, October, 242.
27. Van Dreumel, W. H. M. (1978) *NDT International*, October, 233.
28. Herrmann, H. (1971) *Kunststoffe*, 61, November, 839.
29. Herrmann, H. (1973) *Kunststoffe*, 63, August, 535.
30. Kolb, K. (1976) *Kunststoffe*, **66** (6), 357.
31. Rosenberg, R. (1986) *Plastverarbeiter*, **37** (3).
32. Brosey, W. D. (1985) *Composites Science and Technology*, **24** (3).
33. Scramm, S. W., Daniel, I. M. and Hamilton, W. G. (1981) *Conference Working Together for Strength, Washington DC, Feb. 16–20,* Session 23-D.
34. Claus, R. and Rogers, R. T. (1981) *Conference, Physicochemical Aspects of Polymer Surfaces, New York, August 23–28, Proceedings*, vol. 1, p. 1101.
35. Good, M. S., Nestleroth, J. B. and Rose, J. L. (1981) *Conference Adhesion Aspects of Polymeric Coatings, Minneapolis, May 10–15, Proceedings*, p. 623.
36. Anon (1985) *Plastics in Building Construction*, **9**, no. 2.
37. Botsco, R. J. and Anderson, R. T. (1984) *Adhesives Age*, **27** (7).
38. Rebello, C. J. and Duke, J. C. (1986) *Journal of Composites Technology and Research*, **8** (1).
39. Duke, J. C., Henneke, E. G., Stinchcombe, W. W. and Reifsnider, K. L. (1983) *Second International Conference on Composite Structure, Paisley, Sept. 14–16, Proceedings*, p. 53.
40. Change, F. H., Yee, W. G. W. and Couchman, J. C. (1974) *Non-destructive Testing*, August, 194.
41. Cousins, R. R. and Markham, M. F. (1977) *Composites*, July, 145.
42. Matsushige, K., Shiroruzu, S., Taki, S. and Takemura, T. (1984) *Reports of Progress in Polymer Physics in Japan*, vol. 27.
43. Markham, M. F. (1969) in *Conference, Ultrasonics for Industry*, Iliffe, p. 1.
44. Markham, M. F. (1970) *Composites*, **1** (3), 145.
45. Dean, G. D. and Turner, P. (1973) *Composites*, **4**, 174.
46. Dean, G. D. (1974) in *Conference, Composites – Standards, Testing and Design*, National Physical Laboratory, April, p. 126.
47. Thomas, K. and Meyer, D. E. (1976) *Plastics and Rubber; Materials and Applications*, **1** (3), 136.
48. Smith, A., Wilkinson, S. J. and Reynolds, W. N. (1974) *Journal of Materials Science*, **9** (4), 547.
49. Piche, L. and Hamel, A. (1986) *Polymer Composites*, **7** (5).
50. Ishai, O. and Bar-Cohen, Y. (1980) *Composites*, **11** (4).
51. Opperman, W., Crostack, H. A. and Engelhardt, A. H. (1983) *International Conference Evaluation and QA of Composites, Guildford, Sept. 13–14, Proceedings*, p. 277.
52. Knollman, G. C., Martinson, R. H. and Bellin, S. L. (1980) *Journal of Applied Physics*, **51** (6).

53. Sladky, P., Parmah, and Zdrazil, J. (1982) *Polymer Bulletin*, **7** (8).
54. Hauptmann, P., Dinger, F. and Sauberlich, R. (1985) *Polymer*, **26** (11).
55. Erwin, L. and Dohner, J. (1984) *Polymer Engineering and Science*, **24** (16).
56. Herranen, M. and Savolainen, A. (1984) *Rheologica Acta*, **23** (4).
57. Haeusler, K. G., Hauptmann, P., Klemm, E., Haase, L. and Schubert, R. (1985) *Plaste under Kautschuk*, **32** (4).
58. Piche, L., Hamel, A. and Kelly, P. Y. (1984) *Symposium Quantitative Characterisation of Plastics and Rubber, Hamilton, June 21–22*, p. 134.
59. Hinrichsen, G., Sadat-Darbandi, S. M. and Irobaidi, A. (1985) *Polymer Bulletin*, **13** (1).
60. Ibrahim, A. M., Wedgewood, A. R. and Seferis, J. C. (1986) *Polymer Engineering and Science*, **26** (9).
61. Raab, M., Hnat, V. and Kudrna, M. (1986) *Polymer Testing*, **6** (6).
62. Green, A. T., Lockman, C. S. and Steele, R. K. (1974) *Modern Plastics*, **41**, July, 137.
63. Cole, P. T. (1983) *International Conference on Composite Structures, Paisley, Sept. 14–16, Proceedings*, p. 61.
64. Mitchel, J. R. and Taggart, D. G. (1984) *Composites go to Market Conference, New York, Jan. 16–19,* Session 16B.
65. Brown, T. S. and Mitchel, J. R. (1980) *SPI Conference, New Orleans, Feb.*, Section 26B.
66. Brown, T. S. (1982) *Polymer News*, **8** (9).
67. Hamstad, M. (1982) *NDT International*, **15** (6).
68. Mitchel, J. R. (1984) *Plastics Engineering*, **40** (1).
69. Fowler, T. J. and Scarpellini, R. S. (1980) *Chemical Engineering*, **87** (23).
70. Gillette, J. M. (1984) *Composites go to Market Conference, New York, Jan. 16–19,* Session 10D.
71. Betteridge, D., Connors, P. A., Lilley, T., Shoko, N. R., Cudby, M. E. A. and Wood, D. G. M. (1983) *Polymer*, **24** (9).
72. Shirouzu, S., Shichijyo, S., Taki, S., Matsushige, K. and Takahashi, K. (1984) *Reports on Progress in Polymer Physics in Japan*, **27**, 349.
73. Shirouzu, S., Sakurada, Y., Matsushige, K., Takemura, T. and Takahashi, K. (1984) *Reports on Progress in Polymer Physics in Japan*, **27**, 353.
74. Nishiura, T., Joh, T., Okuda, S. and Miki, M. (1981) *Polymer Journal* (Jap.), **13** (1).
75. Nishiura, T., Joh, T., Okuda, S. and Miki, M. (1983) *Polymer Journal* (Jap.), **15** (11).
76. Altmann, O. and Lembke, B. (1980) *Kunststoffe*, **70** (6).
77. Schmeling, P. and Sandell, S. (1983) *International Conference, Testing, Evaluation and Control of Composites, Guildford, Sept. 13–14, Proceedings*, p. 193.
78. Cherek, H. (1985) *Advances in Polymer Technology*, **5** (1).
79. Persson, S. and Ostman, E. (1986) *Polymer Testing*, **6** (6).
80. Persson, S. and Ostman, E. (1986) *Polymer Testing*, **6** (6).
81. Spowart, A. R. (1972) *Journal of Physics, E*, **5**, 497.
82. Sancaktor, E. (1981) *International Journal of Adhesion and Adhesives*, **1** (6).
83. Dance, W. E. and Middlebrook, J. B. (1978) *ASTM Special Technical Publication*, 696.
84. Stetson, K. A. and Powell, R. L. (1965) *Journal of the Optical Society of America*, **55**, 1694.
85. Stetson, K. A. and Taylor, P. A. (1972) *Journal of Physics, E*, **5** (9), 923.
86. Denby, D. and Leendertz, J. A. (1974) *Journal of Strain Analysis*, **9** (1).
87. Ennos, A. E. and Achbold, E. (1976) *Plastics and Rubber Materials and Applications*, September.
88. Grunwald, K. and Fritzsch, W. (1973) *Kunststoff-Rundschau*, **20**(12), 593.
89. Peterson, T. L., Ast, D. G. and Kramer, E. J. (1974) *Journal of Applied Physics*, **45** (19), 4220.
90. Krenz, H. G., Kramer, E. J. and Ast, D. G. (1975) *Polymer Letters*, **13**, 583.
91. Erdmann-Jesnzitzer, F. and Winkler, T. (1981) *International Journal of Adhesion and Adhesives*, **1** (4).

92. Snell, M. B. and Marchant, M. J. (1984) *Journal of Strain Analysis and Engineering Design*, **19** (4).
93. Laes, R. S., Gorman, D. and Bonfield, W. (1985) *Journal of Materials Science*, **20** (8).
94. Vallat, M. F., Smigielski, P., Martz, P. and Schultz, J. (1985) *Journal of Applied Polymer Science*, **30** (10).
95. Sargent, J. P. and Ashbee, K. H. G. (1985) *Composites Science and Technology*, **22** (2).
96. Newmann, W. and Breuer, K. (1979) *Kunststoffe*, **69** (3).
97. Michel, P. and Potente, H. (1986) *Plastics and Rubber Processing Applications*, **6** (3).
98. Florin, C. (1983) *Kunststaffe Plastics*, **30**(6).
99. McLaughlin, P. V., McAssey, E. V. and Dietrich, R. C. (1980) *NDT International*, **13** (2).
100. McAssey, E. V., Koert, D. N. and McLaughlin, P. V. (1981) *SPI Conference, Working Together for Strength, Washington DC, Feb. 16–20*, Session 10D.
101. Russell, S. S. and Henneke, E. G. (1983) *International Conference, Testing, Evaluation and Quality Control of Composites. Guildford, Sept. 13–14, Proceedings*, p. 282.
102. Pye, C. J. and Adams, R. D. (1981) *NDT International*, **14** (3).
103. Reynolds, W. N. (1985) *Quality Control of Composite Materials by Thermography*, UKAEA, Harwell.
104. Reifsnider, K. L. and Henneke, E. G. (1984) *Developments in Reinforced Plastics*, vol. 4, Elsevier Applied Science Publishers, p. 89.
105. Milne, J. M. and Reynolds, W. N. (1985) *Application of Thermal Pulses and Infrared Thermal Imagers for Observing Sub-Surface Structures in Metals and Composites*, UKAEA, Harwell.
106. Dom, B. E., Evans, H. E. and Torres, D. M. (1981) *Conference Adhesion Aspects of Polymeric Coatings, Minneapolis, May 10–15, Proceedings*, p. 597.
107. Clarke, W. D. and Mack, R. T. (1985) *SPI Conference Magic of Polyurethane, Reno, Oct. 23–25, Proceedings*, p. 200.
108. Eyerer, P. and Busse, G. (1983) *Kunststoffe*, **73** (9).
109. Busse, G., Rief, B. and Everer, P. (1986) *Antec 86, Boston, April 28, May 1, Proceedings*, p. 386.
110. Charles, J. A. (1978) *ASTM Special Publication*, 696.
111. Altmann, O. and Winter, L. (1983) *Kunststoffe*, **73** (3).

Chapter Twenty-one

Testing Products

21.1 INTRODUCTION

Moulding powders are, of course, products, as are moulded sheet and laminates, but in the context of the bulk of testing subjects considered in this book they are products for which the normal, and the easiest, approach will be to produce standard test pieces. For 'finished' products, it is often desirable to cut or machine standard test pieces by the methods discussed in Ch. 3, rather than specially to mould test pieces or sheet. The reasons are self-evident: there is no chance of the test material being different from that of the product and the properties measured will not be changed by a different processing procedure being used for the test pieces. There are, however, the disadvantages of the extra effort required to produce test pieces and the fact that machining may affect the property in question.

These considerations apart, it is the product which has to perform and the surest way to test for product performance is to forget standard test pieces and test the whole product. This is not always easy or convenient and often judgements on performance will still have to be made on the basis of standard tests. (These tests will of course also continue to be valuable in quality assurance, material evaluation, etc.).

Tests on products are standardised in product specifications, although perhaps not as often as would be desired, but a great many more product tests are invented and used on an *ad hoc* basis. It is not possible in this book to consider every product test known to the authors but it is appropriate that attention is drawn to the types of test which can be carried out, with advantage, on the product itself. Anyone contemplating tests on products is advised to check the ISO and the appropriate national standards yearbooks for tests on the same or similar products which have already been standardised.

21.2 SEMI-FINISHED PRODUCTS

Semi-finished products can be taken to include sheet, film, laminates and coated fabrics which will later be used to fabricate other articles. In general, all the usual laboratory tests from tensile strength to gas permeability can be made on test pieces cut from these products, the only deviation from standard procedures perhaps being the need to use test pieces which are the thickness of the product in question rather than that given in the standard method. As this may affect the results, care should be taken when comparing data.

In the case of films, test methods for mechanical properties such as tensile strength and impact have been specially formulated and standardised and have been mentioned in earlier chapters. Coated fabrics are treated as being a little apart from either rubbers or plastics and a large number of test methods have evolved specifically for them. In Britain the tests which have been standardised are published as parts of BS 3424.[1-14] The equivalent ISO standards are noted in these references.

Decorative and industrial laminates can be tested for many properties in the same manner as sheets by the process of cutting standard test pieces, but they will have particular requirements for properties such as abrasion resistance and the effect of hot objects, for which tests need to be specially formulated. Tests standardised to date can be found in specifications for laminates, for example those of ISO and British Standards.[15-22] For some properties, a variation on the standard methods used for plastics in general is specified for laminates. This may be included in the standard method of test for that property but may be published under a separate number; for example, in the BS 2782[23] series there are separate methods for temperature of deflection under load (121C), crushing strength after heating (1310), and dimensional change (106H).

21.3 FIBRE REINFORCED PLASTICS

The fibre reinforced plastics are really a type of plastics rather than a particular product but it is emphasised that this class of material does sometimes require a different approach to testing. This important class of materials is considered quite briefly here because most of the tests carried out on fibre reinforced materials are in principle the same as those carried out on other plastics. The need for special test piece geometries for such tests as tensile stress–strain properties must be noted; where these requirements have been incorporated into standard methods they have been discussed in Ch. 8.

There are a number of standard test methods specific to fibre reinforced materials. A list of properties and corresponding test methods is given in ISO 4899[24] for thermosetting plastics reinforced with textile glass. For several properties, methods are stated to be 'under study'. Although it is intended ultimately to include all tests for tensile properties in ISO 527,[25] currently glass reinforced materials are covered by ISO 3268[26] which is discussed in Ch. 8. There is a very similar European standard EN 61[27] which is reproduced as BS 2782 Method 1003 (1977).[28] This proliferation of similar standards (GRP is also covered in BS 2782 Method 220) is unfortunate and it is to be hoped that they will be rationalised in the future.

An unusual technique has been developed particularly for the evaluation of the raw materials of filament-wound structures. It involves the making of NOL (Naval Ordnance Laboratory) rings, which is described in ASTM D2291.[29] The rings can be made from dry reinforcement or pre-impregnated materials. The winding system for dry reinforcement is shown schematically in Fig. 12.1. After having been wound onto a mould or mandrel, the material is oven cured. Three sizes of ring are specified, each 146 mm internal diameter by 6·35 mm wide, the wall thicknesses

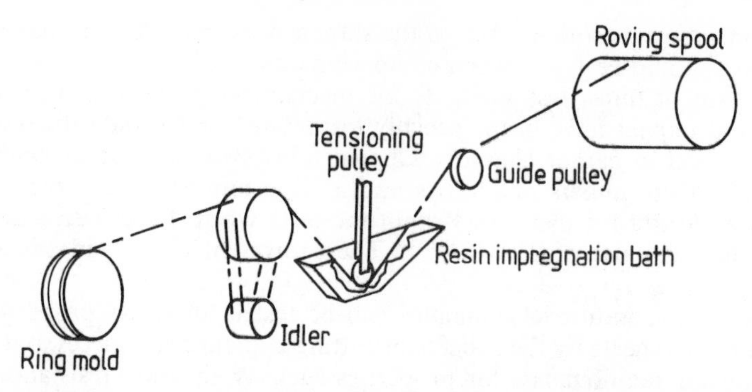

Fig. 21.1 Schematic diagram of wet roving ring winding apparatus: ASTM D2291

being in two cases 1·52 mm and the other 3·18 mm. The difference between the two apparently identical rings is that one is made over-thick and machined down, whereas the other is moulded to the correct wall thickness directly. The 3·18 mm ring is also moulded over-thick and machined down. The procedure for machining is not given.

Tests may be applied to these rings include split disc ring tensile strength, ring compression strength, flexural strength and interlaminar shear strength. The interlaminar shear test is given in ASTM D2344,[30] and is essentially the method given in BS 2782 Method 341A[31] (see Ch. 8), using test pieces cut from the rings. Flexural strength can be measured using segments of the ring and compression strength measured on the complete ring. The split ring tensile test is given in ASTM D2290[32] and uses the complete ring. A test fixture to be used for straining the ring is shown in Fig. 21.2.

There are two ISO standards for tests on rods made from textile glass rovings ISO 3597[33] for the determination of flexural strength and ISO 3605[34] for compressive strength, both of which give details of test piece preparation. ISO 3597 uses a modification of the general flexural strength method, ISO 178, to suit rods, while ISO 3605 simply specifies compression to failure at 1 mm/min using jigs to support the ends of the test piece. There is also a number of standards for testing the textile glass reinforcement but consideration of these is outside the scope of this book. Methods for the preparation of panels of glass reinforced plastics for test purposes have been considered in Ch. 3.

European standard test methods exist for glass reinforced plastics and these have also been reproduced in the BS 2892 series. EN 59[35] covers measurement of hardness using the Barcol impressor (Ch. 8). EN 60[36] is substantially in agreement with ISO 1172[37] and is intended to give a measure of glass content, although the results will be influenced by the presence of ingredients other than glass and resin. The method for tensile properties, EN 61[27] has been mentioned above and is largely in agreement with ISO 3268.[26] All these European standards refer to EN 62,[38] *Standard Atmospheres for Conditioning and Testing*, which is effectively a copy of those internationally standardised atmospheres which are relevant to European climatic conditions. EN 63[39] for flexural properties is equivalent to the content of ISO 178[40] (see Ch. 8) which is applicable to glass reinforced plastics.

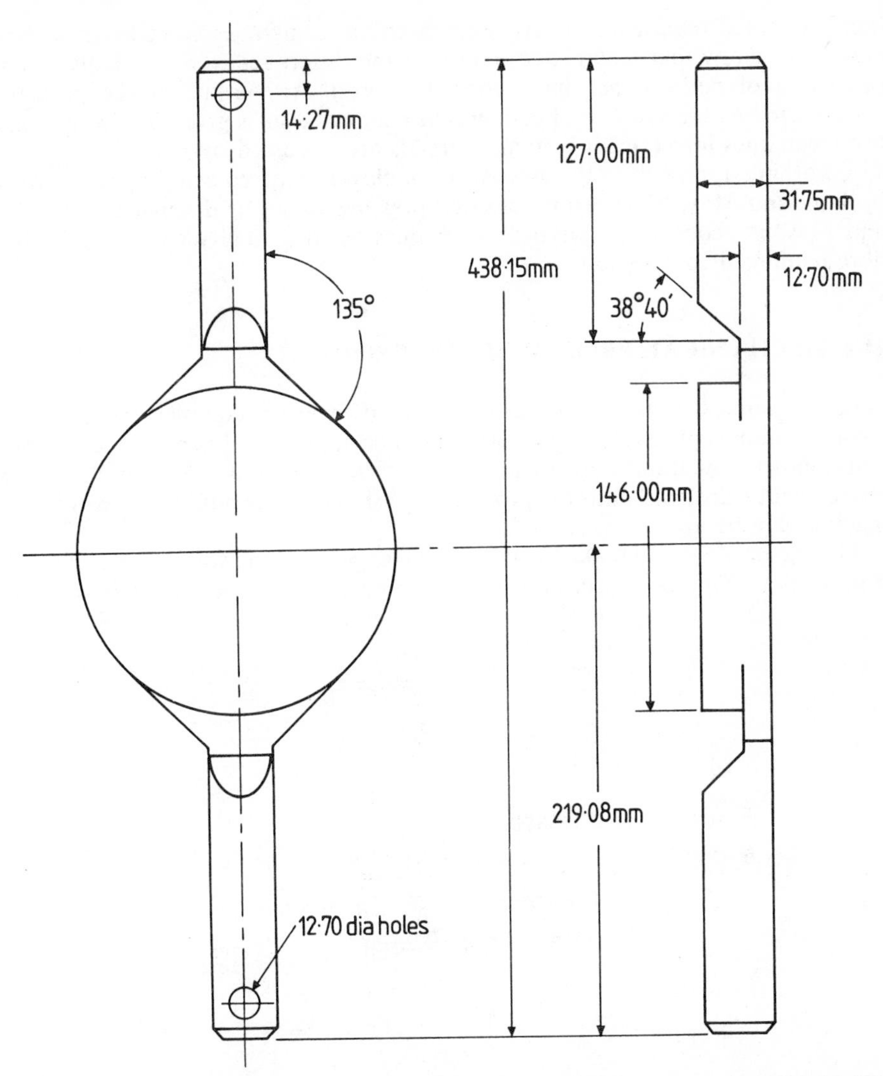

Fig. 21.2 Split ring test fixture for parallel fibre reinforced ring specimens: ASTM D2291

The BS 2782 series also includes Method 1006[41] for determination of volatile matter and resin content, which has greater scope than, and differs from, EN 60.

From this short consideration of tests for fibre reinforced plastics it is clear that standards committees have neither treated such materials as a separate entity nor wholly included them with plastics in general. The result is that a little care is needed to search out the most appropriate standard test method for a given property of the fibre reinforced material. The fact is that, although in most cases the basic type of test will be the same, the conditions needed for fibre reinforced materials will differ from those for other plastics and this will be especially so for highly oriented reinforcement. In research and in companies specialising in these materials, many particular test routines and approaches to evaluation and prediction of performance have been developed. An outline of the problems encountered with a number of common mechanical tests is given by Ewing[42] and test techniques for filament-wound materials are discussed by Bert.[43]

A working approach to the assessment of glass reinforced cladding is presented by Lant and May,[44] the testing of GRP pressure vessels is discussed by Nava[45] and Mason[46] considers analytical techniques for the quality control of carbon fibre reinforced composites.

21.4 MECHANICAL PROPERTIES OF PRODUCTS

Some mechanical tests can be readily adapted to test a complete product. The most common is probably impact, which can be applied to dustbins, buckets, pipe and even window frames. Either the product itself can be dropped or a weighted striker can be dropped onto the product; occasionally, a pendulum type of impact machine can be used.

The choice of how to make the impact will depend on the convenience of testing a particular shape or object and on how impacts in service are likely to occur.

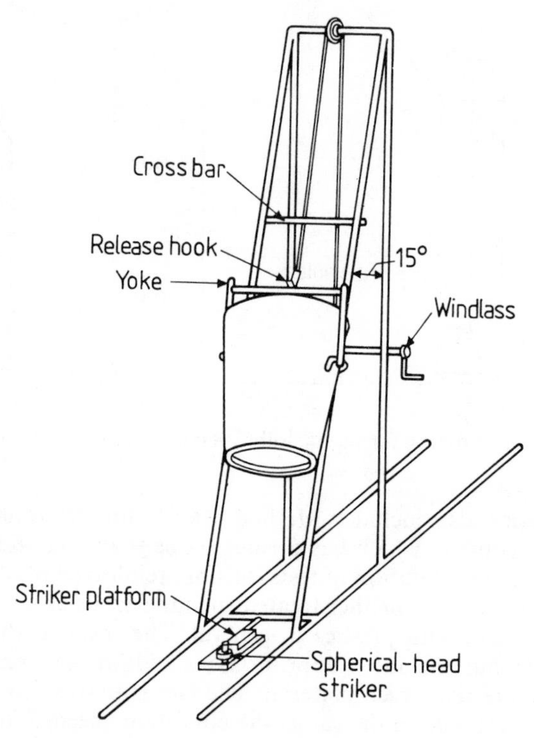

Fig. 21.3 General layout of dustbin test rig: BS 4998

For a dustbin or a bucket it is sensible both for convenience and for simulating service to drop the product. An example of where such a test has been standardised is BS 4998 (1974)[47] for dustbins. A rig is specified (Fig. 21.3) to enable the drop to be made smoothly and to ensure that the bin strikes at the right attitude. Similar tests are used for many containers but often no advice is given on how to align and release the test product. This can be a little difficult when impacts onto an edge are required.

For pipes it is usual to drop a weight onto the product and the apparatus is exactly the same in principle as that used for laboratory-prepared test pieces (see Ch. 8). An example can be found in BS 4991[48] where the test is required to be made at 0 °C. With a little ingenuity, rigs can be devised to hold a variety of products to be impacted in this manner. Falling weight impact apparatus can be instrumented to measure force and deceleration and this much more complicated and expensive apparatus is used, for example, for testing crash helmets.[49]

If a large enough stressing frame is available it is quite feasible to monitor the strains induced in a product as large as a fibre reinforced plastics car body (Fig. 2.14) by the use of strain gauges attached to the product. Many similar products can likewise be tested using a more humble universal tensile testing machine. A very

Fig. 21.4 Large stressing frame

simple example is a bending test suggested for the handles of domestic hollow-ware[50] and many extrusions can readily be subjected to a similar test without test piece preparation.

Pressure testing must be carried out on the product and is often required for pipes, drums, bottles, etc. In the simple forms of test, a proof pressure is applied internally for a given time and there may be both long-term and short-term tests as in BS 4991 (1974),[48] for example. The pressurising medium is usually water. By testing at a series of pressures the time to failure can be determined as a function of pressure. For some purposes the pressure may be applied externally, as in a test for leak-proofness of pipe joints.[51]

In service, products may be subjected to rapidly varying pressures which will reduce their fatigue life and more advanced pressure testing systems can be programmed to give cyclic pressurisation. The dynamic fatigue of PVC pipe formulations has been considered by Gotham and Hitch,[52] while Stapel[53] has compared the pressure variations occurring in water distribution systems with the fatigue properties of PVC.

21.5 ENVIRONMENTAL RESISTANCE

Testing the environmental resistance of plastics products presents the same difficulties as similar testing on laboratory samples – the long exposure times needed or, alternatively, the uncertainties of using accelerated conditions. The very size and value of some products will add to the difficulty or the cost of testing, and simulation of the stress realised in service is more complicated than when stressing test pieces.

The environmental stress cracking resistance of products exposed to liquids can be estimated by simply immersing the products, but in most cases it is desirable to apply stress externally. The impressed ball method discussed in Ch. 17 can be applied to almost any object or in many cases some other form of mechanical restraint can be devised. Bottles can very conveniently be tested by applying an internal pressure and such a method has been standardised in ASTM D2561.[54] Of three methods given in D2561 for polyethylene blow-moulded containers, procedure C involves pressurisation. Test methods for environmental stress cracking, including those on products, have been critically reviewed by Brown.[55]

As the size of the product increases so the difficulty or cost of environmental testing escalates. For exposure to heat, cold or high humidity, conditioned rooms become necessary and very large facilities for artificial weathering and exposure to corrosive atmospheres are not often available. Natural weathering is very time-consuming but product specifications do on occasions require exposure at sites having extreme climatic conditions. One advantage of embarking on such exposure trials before the product is marketed is that any degradation which may occur can be noted before products in service reach the same state.

21.6 SIMULATED SERVICE TESTS

Perhaps the ultimate proof in the laboratory of product performance is a simulated service test, which usually involves some type of test rig designed specifically for that product. The stressing frames and impact tests mentioned in §21.4 (mechanical tests) could be said to be in this category and there is no sharp division between tests covered in that and the environmental section and those considered here.

It may not be necessary to use any apparatus in the usual sense – for example to measure the stackability of drums requires only a flat surface, and material to fill the drums. Generally, however, some sort of device is needed to subject the product to the types of stress and strain it will receive in service. Often it is fatigue which worries the designer, in which case product testing rigs need to have a dynamic action. A simple example is the fatigue test specified in BS 4083[56] for plastics handles for hollow-ware (Fig. 21.5). Packaging materials can be subjected to vibration by using an electromagnetic shaker.

When cyclic action has been built into the test rig there arises the question of the effect of the environment on fatigue life and the apparatus if further complicated by housing it in conditioned cabinets or rooms. In fact, the mark of the more sophisticated simulated service test is that it seeks to combine the static and dynamic stresses on the product with all the environmental factors which themselves may be cyclic in nature. Ultimately the rig may become excessively costly as attempts are made to simulate all the possible combinations of conditions. Consequently, when designing product test rigs, a compromise is always sought between uneconomic complication and failure to cover factors which may prove of overriding importance in service. Generally, the more critical the application the more viable an extremely costly test rig becomes so that for products such as large pressure pipes for the petrochemical industry very sophisticated test facilities have been developed.[57]

In some cases it may be environmental effects rather than direct mechanical actions that are of greatest interest. In pipe systems for water distribution, for example, temperature cycling of the water being conveyed may be more likely to cause failures than any direct mechanical action on the pipe. A BPF publication[58] gives guidance on a test procedure which subjects a pipe system to cyclic and fluctuating temperatures.

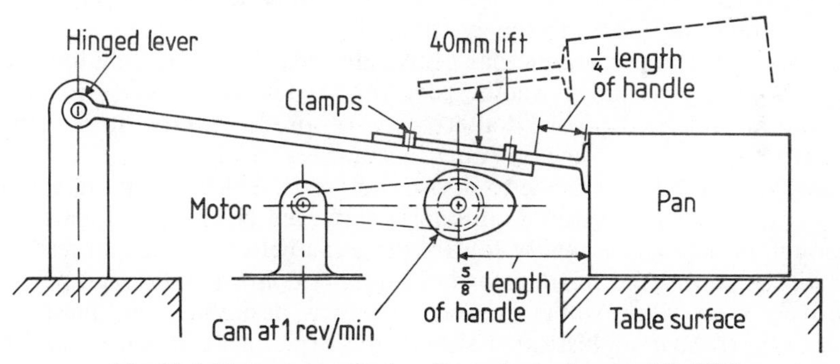

Fig. 21.5 Fatigue test for handle and attachment: BS 4083

21.7 WHEN TO TEST PRODUCTS

If our knowledge of the properties and behaviour of plastics and hence our design rules were such that we could predict the performance of the product accurately from tests on laboratory test pieces then product testing would rarely be needed. We can overcome the changes which the manufacturing process introduces by obtaining test pieces from the product rather than from specially prepared sheets and this practice should be followed much more frequently than it is because the manufacturing effects can be very large. However, the fact is that our understanding of the properties of plastics is simply not good enough to make performance predictions reliably in a great many cases even if the test piece comes from the product. Hence there often will be need to test the whole product since this is the only way to be sure that it will perform satisfactorily.

In the case of a new design it can be more expedient, and certainly effective, to subject prototypes to real service rather than to develop simulation tests. However, there are many cases when this is simply not sensible for time, cost or safety reasons. So, when real service trials have to be ruled out and prediction from laboratory tests cannot be relied upon there must be whole product testing.

As we have seen, it can be extremely difficult and/or expensive to devise tests to simulate service adequately and the justification for investment will be in proportion to the importance of the product in risk and/or sales terms. There is clearly much skill involved in designing rigs and test schedules which give maximum information at minimum cost. In practice there is a danger of spending very large amounts and still not getting the simulation accurate enough but most commonly the pressure is to underdesign the apparatus and curtail the program to cut costs. By far the most difficult factor is any form of acceleration to reduce the time scale, and the problems associated with the validity of accelerating processes are the same as when conducting fatigue or ageing studies on test pieces.

The same principle applies to quality-control testing, but here there is a much greater probability that the experience gained from proving the product initially will allow the quality of subsequent production to be reliably judged on the basis of tests on test pieces. Sometimes a product test will give more valuable assessment of quality for the same testing cost as needed for test pieces. This would be true, for example, for impact resistance of a bucket because the cost of moulding test pieces would be little different from the value of the bucket and the testing costs would be equal. Impact testing the bucket would actually be cheaper than machining standard impact test pieces from it.

When the value of the product far outweights the cost of making test pieces it is again necessary to judge whether control on test pieces gives us sufficient confidence to reject the costly alternative of product tests. It is here that non-destructive tests on the product become especially attractive and, not surprisingly, great effort is made to devise such tests, which give more confidence than the use of test pieces and additionally may even be cheaper to carry out.

Although in the case of quality control there is a better chance that tests on test pieces will be considered reliable, for both quality control and design purposes it is relatively clear when whole product testing is desirable. The question then becomes one of how sophisticated should the experiment design be and this is generally answered (with great difficulty) by weighing cost against the risks and values involved.

It is again emphasised that tests on products are specified in many national and international standards and these, or an appropriate adaptation, should be applied or at least carefully considered whenever relevant. Additionally, valuable guidance may be obtained from published studies. Papers since 1980 include consideration of bottles,[59,60] containers,[61,62] pipes,[63–70] artificial sports surfaces,[71] heat-shrinkable products,[72] windows,[73–77] surface coatings,[78] gears,[79] dentures,[80] skin patches,[81] moulded parts,[82–83] and packaging regulations.[84–86]

REFERENCES

1. BS 3424: Part 0 Method 1 (1982) *Foreword and General Introduction.*
2. BS 3424: Part 1 Method 4 (1982) *Conditioning and Selection of Test Specimens* (ISO 223).
3. BS 3424: Part 3 Methods 5A-C (1982) *Determination of Mass per Unit Area* (ISO 2286).
4. BS 3424: Part 4 Method 6 (1987) *Determination of Breaking Strength and Elongation at Break* (ISO 1421).
5. BS 3424: Part 5 Methods 7A-C (1982) *Determination of Tear Strength.*
6. BS 3424: Part 6 Methods 8A, B (1982) *Determination of Bursting Strength* (ISO 3303).
7. BS 3424: Part 7 Method 9 (1982) *Determination of Coating Adhesion Strength* (ISO 2411).
8. BS 3424: Part 8 Methods 10A–C (1983) *Determination of Low Temperature Performance* (ISO 4646, ISO 4675).
9. BS 3424: Part 9 Methods 11A–D (1984) *Determination of Resistance to Damage by Flexing* (ISO 7854).
10. BS 3424: Part 11 Method 13 (1982) *Determination of Resistance to Blocking* (ISO 5978).
11. BS 3424: Part 14 Method 16 (1985) *Determination of Colour Fastness to Wet and Dry Rubbing and Determination of Resistance to Printwear.*
12. BS 3424: Part 18 Methods 21A, B (1986) *Determination of Resistance to Wicking and Lateral Leakage.*
13. BS 3424: Part 22 Method 25 (1983) *Determination of Fusion of PVC Coatings and the State of Cure of Vulcanised Rubber Coatings* (ISO 6451).
14. BS 3424: Part 26 Methods 29A–C (1986) *Determination of Resistance to Penetration by Water* (ISO 1420).
15. ISO 1642 (1979) *Industrial Laminated Sheets Based on Thermosetting Resins – Basis for Specifications.*
16. BS 4965 (1983) *Decorative Laminated Plastics Sheet. Veneered Boards and Panels.*
17. BS 3953 (1976) *Glass Fabric Laminated Sheet.*
18. BS 2572 (1976) *Specification for Phenolic Laminated Sheet and Epoxide Cotton Fabric Laminated Sheet.*
19. BS 5102 (1974) *Phenolic Resin Bonded Paper Laminated Sheets for Electrical Applications.*
20. BS 3794: Parts 1 and 2 (1986) *Decorative High Pressure Laminates (HPL) based on Thermosetting resins.*
21. BS 6128 *Industrial Laminated Rods and Tubes based on Thermosetting Resins.* In several parts.
22. ISO 4568: Parts 1 and 2 (1981) *Decorative Laminated Sheets based on Thermosetting resins.*
23. BS 2782 *Methods of Test for Plastics.*
24. ISO 4899 (1982) *Textile Glass Reinforced Thermosetting Plastics – Properties and Test Methods.*
25. ISO 527 *Determination of Tensile Properties.* In course of publication.
26. ISO 3268 (1978) *Glass Reinforced Materials – Determination of Tensile Properties.*
27. EN 61 (1977) *Glass Reinforced Plastics. Determination of Tensile Properties.*
28. BS 2782 Method 1003 (1977) *Glass Reinforced Plastics. Determination of Tensile Properties.*

29. ASTM D2291 (1976) *Fabrication of Ring Test Specimens for Reinforced Plastics.*
30. ASTM D2344 (1976) *Apparent Horizontal Shear Strength of Reinforced Plastics by Short Beam Method.*
31. BS 2782 Method 341A (1977) *Determination of Apparent Interlaminar Shear Strength of Reinforced Plastics.*
32. ASTM D2290 (1976) *Apparent Tensile Strength of Ring or Tubular Plastics by Split Disc Method.*
33. ISO 3597 (1977) *Textile Glass Reinforced Plastics – Composites in the Form of Rods made from Textile Glass Rovings – Determination of Flexural (Cross-breaking) Strength.*
34. ISO 3605 (1978) *Textile Glass Reinforced Plastics – Composites in the Form of Rods made from Textile Glass Rovings – Determination of Compressive Strength.*
35. EN 59 and BS 2782 Method 1001 (1977) *Glass Reinforced Plastics. Measurement of Hardness by means of a Barcol Impressor.*
36. EN 60 and BS 2782 Method 1002 (1977) *Glass Reinforced Plastics. Determination of Loss on Ignition.*
37. ISO 1172 (1975) *Textile Glass Reinforced Plastics – Determination of Loss on Ignition.*
38. EN 62 and BS 2782 Method 1004 (1977) *Glass Reinforced Plastics. Standard Atmospheres for Conditioning and Testing.*
39. EN 63 and BS 2782 Method 1005 (1977) *Glass Reinforced Plastics. Determination of Flexural Properties. Three Point Method.*
40. ISO 178 (1975) *Determination of Flexural Properties of Rigid Plastics Materials.*
41. BS 2782 Method 1006 (1978) *Determination of Volatile Matter and Resin Content of Synthetic Resin-impregnated Textile Glass Fabric.*
42. Ewing, P. D. (1974) in *Conference, Composites, Standards Testing and Design, National Physical Laboratory, April*, p. 144.
43. Bert, C. W. (1974) *Composites*, **5** (1), 20.
44. Lant, T. P. R. and May, J. O. *PRI Design and Specification of GRP Cladding Conference, London, October 1978*, p. 31.
45. Nava, H. R., *SPI, Reinforced Plastics/Composites Institute, Thirty-seventh Annual Conference, Washington DC, January 1982*, Session 14-D, p. 1.
46. Mason, A. J., *International Conference on Testing, Evaluation and Quality Control of Composites, Guildford, September 1983*, p. 102.
47. BS 4998 (1985) *Moulded Thermoplastic Dustbins.*
48. BS 4991 (1974) *Propylene Copolymer Pressure Pipe.*
49. BS 6658 (1985) *Specification for Protective Helmets for Vehicle Users.*
50. BS 6743 (1987) *Performance of Handles and Handle Assemblies Attached to Cookware.*
51. ISO 3459 (1976) Polyethylene (PE) *Pressure Pipes – Joints Assembled with Mechanical Fittings – Internal Under-pressure Test Method and Requirement.*
52. Gotham, K. V. and Hitch, M. J. (1978) *British Polymer Journal*, **10**, 38.
53. Stapel, J. J. (1977) *Pipes and Pipelines International*, February, 11; April, 33.
54. ASTM D2561 (1970) *Environmental Stress–Crack Resistance of Blow-moulded Polyethylene Containers.*
55. Brown, R. P. (1980) *Polymer Testing*, **1** (4), 267.
56. BS 4083 (1981) *Uncoated Aluminium Hollow-ware Cooking Utensils.*
57. Lawrence, C. C. and Choo, V. R. S. (1986) *Polymer Testing*, **6** (4).
58. British Plastics Federation (1976) *Guide for the Determination of Performance for Thermoplastics Pipes and Fittings for use in Hot and Cold Water Applications*, BPF Publication no. 132/1.
59. Troy, E. J. and Shortridge, T. J. (1985) *Plastics Engineering*, **41** (11).
60. Anon. (1985) *Plastics South Africa*, **15** (1).
61. Thomas, R. H. (1986) *Polymer News*, **11** (1).
62. Gaynes, C. (1984) *ASTM Standard News*, **12** (8).
63. Greig, J. M. (1981) *Plastics and Rubber Processing Applications*, **1** (1).

64. Gedde, U. W., Tersellius, B. and Jansson, J. F. (1981) *Polymer Testing*, **2** (2).
65. Gross, R. E. and Kyle, P. D. *SPI Conference, Washington DC, February 1981*, Sessions 3-A, pp. 1–9
66. Skarelius, J. *Quality Assurance of Polymeric Materials and Products Symposium, Nashville, Tenn., March 1983*, p. 118.
67. Ifwarson, M. and Eriksson, P. (1986) *Kunststoffe*, **76** (3).
68. Schwencke, H. F. (1984) *Plastics and Rubber Processing Applications*, **4** (1).
69. Marshall, G. P. *PRI Conference, Use of Plastics and Rubber in Water and Effluents, London, February 1982*, paper 24.
70. Chosh, S. and Tully, D. K. (1982) *Plastics Industry*, **9** (1).
71. Brown, R. P. (1982) *Polymer Testing*, **3** (2).
72. Kleinheins, G., Stark, W. and Nuffen, K. (1984) *Kunststoffe*, **74** (8).
73. Poschet, G. (1985) *Kunststoffe Im Bau.*, **20** (2).
74. Provan, T. and Younger, J. (1986) *Building*, **4** (7429).
75. Schwabe, A. (1985) *Kunst. Bau.*, **20** (2).
76. Dalev, R. and Schmid, J. (1985) *Kunst. Bau.*, **20** (2).
77. Boysen, M. (1985) *Kunst. Bau.*, **20** (2).
78. Sellars, I. C. and Mcloskey, M. (1982) *Polymer Paint and Colour J* **172** (4083).
79. Crawford, R. J. and Brown, D. (1981) *Journal of Testing and Evaluation*, **9** (5).
80. Ahmed, R., Bates, J. F. and Lewis, T. T. (1982) *Biomaterials*, **3** (2).
81. Leeper, H. M. and Enscore, D. (1982) *Adhesives Age*, **25** (2).
82. Anon. (1981) *Plastics and Rubber News*, July.
83. Foy, H. (1981) *Plast. Res Process.*, **21** (6).
84. Anon. (1985) *Packaging (USA)*, **30** (4).
85. Sheldon, L. (1985) *European Chemical News*, July.
86. Gut, M. (1984) *Swiss Chem.*, **6** (112).

APPENDIX B: National Standards Bodies – ISO Members

ALBANIA/ALBANIE (KÇSA)
Komiteti i Çmimeve dhe i Standardeve
Tirana

ALGERIA/ALGÉRIE (INAPI)
Institut algérien de normalisation
et de propriété industrielle
5, rue Abou Hamou Moussa
B.P. 1021 – Centre de tri
Alger

ARGENTINA/ARGENTINE (IRAM)
Instituto Agentino de
Racionalización de Materiales
Chile 1192
1098 Buenos Aires

AUSTRALIA/AUSTRALIE (SAA)
Standards Association of Australia
Standards House
80–86 Arthur Street
North Sydney – N.S.W. 2060

AUSTRIA/AUTRICHE (ON)
Österreichisches Normungsinstitut
Heinestrasse 38
Postfach 130
A-1021 Wien

BANGLADESH (BSTI)
Bangladesh Standards and Testing
Institution
116-A, Tejgaon Industrial Area
Dhaka-1208

BELGIUM/BELGIQUE (IBN)
Institut belge de normalisation
Av. de la Brabançonne 29
B-1040 Bruxelles

BRAZIL/BRÉSIL (ABNT)
Associação Brasileira de Normas Técnicas
Av. 13 de Maio, nº 13–28º andar
Caixa Postal 1680
CEP: 20. 003 – Rio de Janeiro-RJ

BULGARIA/BULGARIE (BDS)
Comité de la qualité
auprès du Conseil des Ministres
21, rue du 6 Septembre
1000 Sofia

CANADA (SCC)
Standards Council of Canada
350 Sparks Street, Suite 1200
Ottawa, Ontario
K1P 6N7

CHILE/CHILI (INN)
Instituto Nacional de Normalización
Matias Cousiño 64 – 6º piso
Casilla 995 – Correo 1
Santiago

CHINA/CHINE (CSBS)
China State Bureau of Standards
P.O. Box 820
Beijing

COLOMBIA/COLOMBIE (ICONTEC)
Instituto Colombiano de Normas Técnicas
Carrera 37 No. 52–95
P.O. Box 14237
Bogotá

CUBA (NC)
Comité Estatal de Normalización
Egido 602 entre Gloria y Apodaca
Zona postal 2
La Habana

CYPRUS/CHYPRE (CYS)
Cyprus Organization for Standards and Control of Quality
Ministry of Commerce and Industry
Nicosia

CZECHOSLOVAKIA/TCHÉCOSLOVAQUIE (CSN)
Uřad pro normalizaci a měření
Václavské náměsti 19
113 47 Praha 1

DENMARK/DANEMARK (DS)
Dansk Standardiseringsraad
Aurehøjvej 12
Postbox 77
DK-2900 Hellerup

EGYPT, Arab Rep. of/ÉGYPTE, Rép. arabe d' (EOS)
Egyptian Organization for Standardisation and Quality Control
2 Latin America Street
Garden City
Cairo-Egypt

ETHIOPIA/ÉTHIOPIE (ESI)
Ethiopian Standards Institution
P.O. Box 2310
Addis Ababa

FINLAND/FINLANDE (SFS)
Suomen Standardisoimisliitto SFS
P.O. Box 205
SF-00121 Helsinki

FRANCE (AFNOR)
Association française de normalisation
Tour Europe
Cedex 7
92080 PARIS LA DÉFENSE

GERMAN DEMOCRATIC REPUBLIC/RÉPUBLIQUE DÉMOCRATIQUE ALLEMANDE (ASMW)
Amt für Standardisierung, Messwesen und Warenprüfung

Fürstenwalder Damm 388
DDR-1162 Berlin

GERMANY, F.R./ALLEMAGNE, R.F. (DIN)
DIN Deutsches Institut für Normung
Burggrafenstrasse 6
Postfach 1107
D-1000 Berlin 30

GHANA (GSB)
Ghana Standards Board
P.O. Box M-245
Accra

GREECE/GRÈCE (ELOT)
Hellenic Organization for Standardization
Didotou 15
106 80 Athens

HUNGARY/HONGRIE (MSZH)
Magyar Szabványügyi Hivatal
1450 Budapest 9
Pf. 24

INDIA/INDE (BIS)
Bureau of Indian Standards
Manak Bhavan
9 Bahadur Shah Zafar Marg
New Delhi 110002

INDONESIA/INDONÉSIE (DSN)
Dewan Standardisasi Nasional – DSN
(Standardization Council of Indonesia)
Gedung PDII-LIPI
Jalan Gatot Subroto
P.O. Box 3123
Jakarta 12190

IRAN Islamic Rep. of/Rép. islamique d' (ISIRI)
Institute of Standards and Industrial Research of Iran Ministry of Industries
P.O. Box 15875–4618
Tehran

IRAQ (COSQC)
Central Organization for Standardization and Quality Control
Ministry of Planning
P.O. Box 13032
Aljadiria
Baghdad

IRELAND/IRLANDE (NSAI)
National Standards Authority of Ireland
Glasnevin
Dublin-9

ISRAEL/ISRAËL (SII)
Standards Institution of Israel
42 University Street
Tel Aviv 69977

ITALY/ITALIE (UNI)
Ente Nazionale Italiano di Unificazione
Piazza Armando Diaz 2
I-20123 Milano

IVORY COAST/CÔTE D'IVOIRE (DENT)
Direction de l'environnement, de la normalisation et de la technologie
Ministère de l'industrie
B.P. V65
Abidjan

JAPAN/JAPON (JISC)
Japanese Industrial Standards Committee
c/o Standards Department
Agency of Industrial Science and Technology
Ministry of International Trade and Industry
1-3-1, Kasumigaseki, Chiyoda-ku
Tokyo 100

KENYA (KEBS)
Kenya Bureau of Standards
Off Mombasa Road
Behind Belle Vue Cinema
P.O. Box 54974
Nairobi

KOREA, Dem. P. Rep. of/
CORÉE, Rép. dém. p. de (CSK)
Committee for Standardization of the Democratic People's Republic of Korea
Zung Gu Yok Seungli-Street
Pyongyang

KOREA, Rep. of/CORÉE, Rép. de (KBS)
Bureau of Standards
Industrial Advancement Administration
2, Chungang-dong, Kwachon-city
Kyonggi-do 171-11

MALAYSIA/MALAISIE (SIRIM)
Standards and Industrial Research Institute of Malaysia
P.O. Box 35, Shah Alam
Selangor

MEXICO/MEXIQUE (DGN)
Dirección General de Normas
Calle Puente de Tecamachalco N.º 6
Lomas de Tecamachalco
Sección Fuentes
Naucalpan de Juárez
53 950 Mexico

MONGOLIA/MONGOLIE (MSC)
State Committee for Prices and Standards of the Mongolian People's Republic
Ulan Bator II

NETHERLANDS/PAYS-BAS (NNI)
Nederlands Normalisatie-instituut
Kalfjeslaan 2
P.O. Box 5059
2600 GB Delft

NEW ZEALAND/
NOUVELLE-ZÉLANDE (SANZ)
Standards Association of New Zealand
Private Bag
Wellington

NIGERIA (SON)
Standards Organisation of Nigeria
Federal Ministry of Industries
4 Club Road
P.M.B. 01323
Enugu

NORWAY/NORVÈGE (NSF)
Norges Standardiseringsforbund
Postboks 7020 Homansbyen
N-0306 Oslo 3

PAKISTAN (PSI)
Pakistan Standards Institution
39 Garden Road, Saddar
Karachi-3

PAPUA NEW GUINEA/
PAPOUASIE-NOUVELLE-GUINÉE (PNGS)
National Standards Council
P.O. Box 3042
Boroko

PERU/PÉROU (ITINTEC)
Instituto de Investigación Technológica Industrial y de Normas Técnicas
Prolongación de la Av. Guardia Civil
esquina Av. Canadá, s/n
Lima 41

PHILIPPINES (BPS)
Bureau of Product Standards
Department of Trade and Industry
361 Sen. Gil J. Puyat Avenue
Makati
Metro Manila 3117

POLAND/POLOGNE (PKNMiJ)
Polish Committee for Standardization, Measures and Quality Control
Ul. Elektoralna 2
00-139 Warszawa

PORTUGAL (IPQ)
Instituto Português da Qualidade
Rua José Estêvão, 83-A
1199 Lisboa Codex

SAUDI ARABIA/ARABIE SAOUDITE (SASO)
Saudi Arabian Standards Organization
P.O. Box 3437
Riyadh – 11471

SINGAPORE/SINGAPOUR (SISIR)
Singapore Institute of Standards and Industrial Research
Kent Ridge
P.O. Box 1128
Singapore 9111

SOUTH AFRICA, Rep. of/ AFRIQUE DU SUD, Rép. d' (SABS)
South African Burea of Standards
Private Bag X191
Pretoria 0001

SPAIN/ESPAGNE (AENOR)
Asociación Española de Normalización y Certificación
Calle Fernandez de la Hoz, 52
28010 Madrid

SRI LANKA (SLSI)
Sri Lanka Standards Institution
53 Dharmapala Mawatha
P.O. Box 17
Colombo 3

SUDAN/SOUDAN (SSD)
Sudanese Standards Department
Ministry of Industry
P.O. Box 2184
Khartoum

SWEDEN/SUÈDE (SIS)
SIS – Standardiseringskommissionen i Sverige
Box 3 295
S-103 66 Stockholm

SWITZERLAND/SUISSE (SNV)
Swiss Association for Standardization
Kirchenweg 4
Postfach
8032 Zurich

SYRIA/SYRIE (SASMO)
Syrian Arab Organization for Standardization and Metrology
P.O. Box 11836
Damascus

TANZANIA/TANZANIE (TBS)
Tanzania Bureau of Standards
P.O. Box 9524
Dar es Salaam

THAILAND/THAÏLANDE (TISI)
Thai Industrial Standards Institute
Ministry of Industry
Rama VI Street
Bangkok 10400

TRINIDAD AND TOBAGO/ TRINITÉ-ET-TOBAGO (TTBS)
Trinidad and Tobago Bureau of Standards
Century Drive
Trincity Industrial Estate
Tunapuna
P.O. Box 467
Port of Spain

TUNISIA/TUNISIE (INNORPI)
Institut national de la normalisation et de la propriété industrielle
B.P. 23
1012 Tunis-Belvédère

TURKEY/TURQUIE (TSE)
Türk Standardlari Enstitüsü
Necatibey Cad. 112
Bakanliklar
Ankara

UNITED KINGDOM/ ROYAUME-UNI (BSI)
British Standards Institution
2 Park Street
London W1A 2BS

USA (ANSI)
American National Standards Institute
1430 Broadway
New York, N.Y. 10018

USSR/URSS (GOST)
USSR State Committee for Standards
Leninsky Prospekt 9
Moskva 117049

VENEZUELA (COVENIN)
Comisión Venezolana de Normas
Industriales Avda. Andrés Bello-Edf.
Torre Fondo Común
Piso 11
Caracas 1050

VIET NAM, Socialist Republic of/
République socialiste du (TCVN)
Direction générale de standardisation,
de métrologie et de contrôle de la
qualité
70, rue Tràn Hung Dao
Box 81
Hanoi

YUGOSLAVIA/YOUGOSLAVIE
(SZS)
Savezni zavod za standardizaciju
Slobodana Penezića Krcuna br. 35
Post. Pregr. 933
11000 Beograd

ZAMBIA/ZAMBIE (ZABS)
Zambia Bureau of Standards
National Housing Authority Building
P.O. Box 50259
Lusaka

ISO Correspondent Members

BAHRAIN
Ministry of Commerce and Agriculture
P.O. Box 5479
Bahrain

BARBADOS/BARBADE
Barbados National Standards
Institution (BNSI)
"Flodden" Culloden Road
St. Michael

CAMEROON/CAMEROUN
Service de la normalisation
Direction de l'industrie
Ministère de l'Économie et du plan
B.P. 1604
Yaoundé

ECUADOR/ÉQUATEUR
Instituto Ecuatoriano de
Normalización
Casilla 3999
A. Baquerizo Moreno 454
Quito

HONG KONG
Industry Department
Hong Kong Government
14/F, Ocean Centre
5 Canton Road
Kowloon
Hong Kong

ICELAND/ISLANDE
Technological Institute of Iceland
Standards Division
Keldnaholt
IS-112 Reykjavik

JORDAN/JORDANIE
Directorate of Standards and Measures
Ministry of Industry and Trade
P.O. Box 2019
Amman

KUWAIT/KOWEÏT
Standards and Metrology Department
Ministry of Commerce and Industry
Post Box No. 2944
13030 Kuwait

MADAGASCAR
Service du contrôle des qualités
et du conditionnement
Ministère du commerce
B.P. 1316
Antananarivo (101)

MALAWI
Malawi Bureau of Standards
P.O. Box 946
Blantyre

MAURITIUS/ÎLE MAURICE
Mauritius Standards Bureau
Ministry of Industry
Reduit

OMAN
Directorate General for Specifications and Measurements
Ministry of Commerce and Industry
P.O. Box 550
Muscat

UNITED ARAB EMIRATES/ ÉMIRATS ARABES UNIS
Directorate of Standardization and Metrology
P.O. Box 433
Abu Dhabi

URUGUAY
Instituto Uruguayo de Normas Técnicas
San José 1031 P. 7
Galeria Elysée
Montevideo

Index